Contributions from Science Education Research

Volume 1

For further volumes:
http://www.springer.com/series/11180

Catherine Bruguière • Andrée Tiberghien
Pierre Clément
Editors

Topics and Trends in Current Science Education

9th ESERA Conference Selected Contributions

Springer

Editors
Catherine Bruguière
EA 4148 S2HEP
University of Lyon 1, ENS-Lyon
University of Lyon
Villeurbanne Cedex, France

Andrée Tiberghien
UMR ICAR, CNRS
University of Lyon 2, ENS-Lyon
Lyon Cedex, France

Pierre Clément
EA 4148 S2HEP
University of Lyon 1, ENS-Lyon
University of Lyon
Villeurbanne Cedex, France

ISSN 2213-3623 ISSN 2213-3631 (electronic)
ISBN 978-94-007-7280-9 ISBN 978-94-007-7281-6 (eBook)
DOI 10.1007/978-94-007-7281-6
Springer Dordrecht Heidelberg New York London

Library of Congress Control Number: 2013947739

Printed on acid-free paper

Springer is part of Springer Science+Business Media (www.springer.com)

Foreword

I am delighted to have been asked to write the preface for this volume which brings together 35 selected and peer-reviewed contributions to the 9th European Science Education Research Association (ESERA) Conference which was held in Lyon, France, in 2011. This book is the first in a series of ESERA publications with Springer. The plan is to publish one book every two years in a series entitled 'Contributions from Science Education Research'.

The biennial ESERA conference has become an unmissable feature of the science education calendar, and Lyon 2011 was no exception. Unlike other science education conferences, no single culture or country dominates, and the collegiality of ESERA, which has been fostered since its inception in 1995 through its alternating conferences and summer schools, has become its trademark.

Lyon was a fitting host – a UNESCO World Heritage site, a city redolent with culture and famed for its gastronomic virtues. It is also a city which benefits from its science and technology industries which provide employment for many of its citizens. The history of science education owes much to work done in France over many years. The conference provided an opportunity to reflect on France's contributions in the past and on what it has to offer for the future.

I congratulate Catherine Bruguière, Andrée Tiberghien and Pierre Clément and their team of strand chairs for putting this collection together. The papers reflect high-quality science education research and offer an insight into the field at the end of the first decade of the twenty-first century.

King's College London

Justin Dillon
President of ESERA (2007–2011)

Contents

Part I
Overview of the Book

Chapter 1
Introduction

Catherine Bruguière, Andrée Tiberghien, and Pierre Clément

This book is the first volume in a new series *Contributions from Science Education Research*, created by Springer in partnership with ESERA in July 2012. Our goal is to publish at least one volume per year in this series:

- For each biannual ESERA conference, the conference chairs will be editors of a volume based on the event. This volume will be composed of representative high-quality contributions chosen from the hundreds given at the conference. The chairs for the conference will be the book's editors.
- In the years with no conference volume, there will be a publication coordinated by the ESERA editorial board.

These edited books aim to present the best research studies representing the current orientations in science education research. The books will include studies drawn from different educational traditions from all over the world, which will considerably enrich them and broaden their appeal. At the same time these books will present this wide scope in a structured way, making them easily accessible to specialist and generalist readers alike.

This first volume offers an overview of the 9th ESERA Conference held in Lyon, France. Out of the 1,200 proposals for this conference, less than 1,000 were selected as oral communications (sometimes grouped together in symposia) or as posters.

C. Bruguière (✉) • P. Clément
EA 4148 S2HEP, University of Lyon 1, ENS-Lyon, University of Lyon, Villeurbanne Cedex, France
e-mail: catherine.bruguiere@univ-lyon1.fr; clement.grave@free.fr

A. Tiberghien
UMR ICAR, CNRS, University of Lyon 2, ENS-Lyon, Lyon Cedex, France
e-mail: andree.tiberghien@univ-lyon2.fr

C. Bruguière et al. (eds.), *Topics and Trends in Current Science Education: 9th ESERA Conference Selected Contributions*, Contributions from Science Education Research 1,
DOI 10.1007/978-94-007-7281-6_1, © Springer Science+Business Media Dordrecht 2014

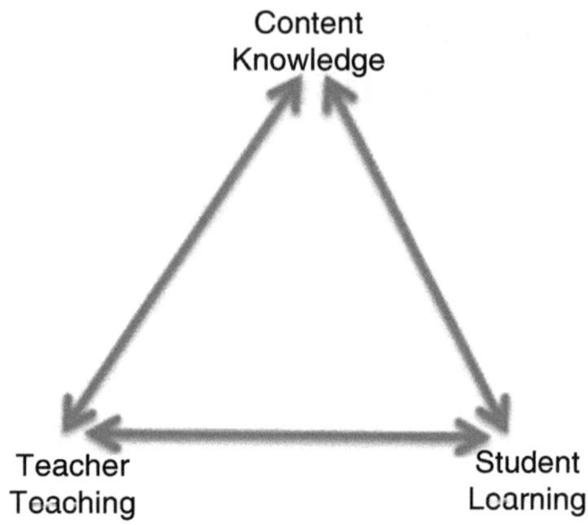

Fig. 1.1 The didactic triangle

They were organised into 14 strands depending on their focus. Moreover, the conference featured four plenary talks by well-known invited guest lecturers. After the conference, each strand chair was asked to select a limited number of presentations among the best in her or his strand. From this pool of presentations, the editors of the book selected 35 chapters on the basis of the originality of their contribution to science education research and their scientific quality.

The structure of this book is not directly based on the organisation of the conference, which consisted of 14 strands that can be grouped into four categories: teaching science, learning science, citizenship, and theoretical perspectives on teaching and learning. We have instead chosen a more generic structure, based on the classical didactic triangle of student–teacher content introduced at the end of the nineteenth century[1] which formed the basis of the French definition of the field of "Didactics of Sciences" in the 1970s also known as "Science Education". This triangular structure emphasises that the relation between the teacher and the students is not direct, but is always linked to scientific knowledge or other content (such as citizenship values). This triangle is a schematic representation of the didactic system of interrelations between the three poles, and it is generic to the extent that it covers a variety of theories (Fig. 1.1).

This didactic triangle helps us to categorise the research activity in science education and also to identify the evolution of the research within this field over the past 40 years. The reader should bear in mind that the poles are never treated in isolation, but the analysis that follows is made in terms of the main focus of the various research studies considered.

[1] The works of Johann Friedrich Herbart, quoted in Peterssen, W. H. (1989). *Lehrbuch Allgemeine Didaktik*. Munchen: Ehrenwirth as well as in Kinnunen, P., Meisalo, V., Malmi, L. (2010). Have we missed something? Identifying missing types of research in computing education. *ICER* 10, 8 pp., Aarhus.

In the second half of the twentieth century, *the "Student" pole*, which more generally corresponds to learning, was the most widely investigated with many analyses of the students' conceptions (also called representations) that emphasised the relationships between the students and the content rather than between the student and the teacher. Today, while still present, this trend is less dominant and has diversified into innovative perspectives where the relationships between the students and the teachers are much more developed. Part IV of this book focuses on multiple perspectives with respect to student learning, such as assessments of student competences (e.g. evaluations like Pisa), students' attitudes, students' understanding and the role of environment in learning, in school as well as out of school. The chapters in this section illustrate the variety of points of view centred on the students, with their studies of the relationships between the students' learning or the students' outcomes and the content in relation with the teaching that can be enlarged to include the educational system.

During the same 40 years, *the "Content" pole*, initially limited to "Scientific Knowledge", also enlarged its orientations of research. One of them is the growing link between science education, history and philosophy of science, with researchers investigating the nature of science (NOS): science published by researchers as well as science taught after the selection and transposition of the initial scientific content. The general theme of this ESERA Conference "Science Learning and Citizenship" illustrates this opening up of the knowledge involved in science education to social life and citizenship values. Part II presents different perspectives on this movement, some more related to teachers, others to teaching resources, while most of them are wider innovative reflexions on the links between science and society, based on the analysis of socio-scientific issues (SSIs), including Socially Acute Questions (SAQs).

The "Teacher" pole or more generally "Teaching" is a growing trend in science education research, with two complementary types of works, "Teachers' practices" and "teachers' professional development" (Part III). In "Teachers' practices", the studies identify the difficulties of teachers placed in innovative situations, such as science inquiries, use of ICT and debates related to SSIs, and develop propositions for their professional development. These studies emphasise the relationships between teachers and content. The teachers' conceptions and the development of their practices are also increasingly analysed and compared across different countries. In this type of research, the relationships are mainly between teachers' learning, where the teacher is considered as a student/learner, and teachers' practices.

The *relationships between teaching and learning* have their own specific part of this book given over to them (Part V). The studies of these relationships have considerably evolved. It is interesting to note that on one hand there was the development of projects of science education, led by high-level researchers recognised in their discipline as early as the 1950s, and on the other hand, in the 1970s, as already discussed above, there was a development of studies on students' conceptions. At the beginning, these two developments were not really connected, whereas nowadays numerous studies, both quantitative and qualitative, relate teaching and learning. In particular, many assessment projects of educational systems, of teaching sequences or resources establish links between them. Other important and more recent approaches are also represented in this part of the book such as those involving the investigation of what is going on in teaching situations in particular in the classroom.

Finally, the last part (Part VI) is about *teaching resources and curriculum*. The corresponding studies involve a design activity and therefore are based on the three poles of the didactic triangle. The numerous choices necessary for designing activities involve teaching, learning and content. It is interesting to note that the development of the activity of design has led to arguments for it to be recognised as a subject of study in its own right, and nowadays it constitutes a significant part of science education research. This activity is much more culturally dependent than the research on students' conceptions, for example. The European projects involving the transfer of resources developed in one country to another one are particularly innovative, and reports on these initiatives are included in this part. Chapters in this part also illustrate the different types of resources that can be used; while most of the studies treat teaching sequences, progressions or curricula, others concern more specific practices, like examples of laboratory work or the use of storybooks.

We now introduce each part of the book individually.

1 Socio-scientific Issues (SSIs) and the Nature of Science (NOS)

This part mainly deals with issues of content, with a broadening of the scope from scientific knowledge to citizenship values which are also at stake in science education.

The first chapter is the plenary talk given by the philosopher of science Isabelle Stengers, which opened the ESERA Conference. Here, she develops the idea that citizenship is a matter of concern related to scientific education and that science is not simply a matter of facts. For her, science education must be developed not in terms of a required level of scientific literacy, but with reference to the need for a transformation of the relationship between scientists and a democratic society. Learners must know science in the making, in particular because "many of the didactically purified examples of scientific reasoning may serve as a template for pseudo-scientific claims", for "the industry of doubt", such as the "anti-Darwinian industry". She promotes the confrontation of science students with situations of socio-technical controversies. "Do the carefully purified, reproducible facts which one can derive from accepted, stabilized science help to understand the facts with which we, as citizens, are confronted?"

The second chapter, the plenary talk by Laurence Simonneaux, offers a convergent message, although this time rooted in science education research rather than philosophy of science. She situates "Socially Acute Questions" (SAQs or, in French: "Questions Socialement Vives") "at the heart of the problem of teaching and learning in an uncertain world influenced by the development of techno-science, and by environmental and health crises". SAQs are always controversial as they challenge social practices and reflect social representations and value systems. Rather than relying on the STS (science–technology–society) and SSI education promoted by Sadler, Zeidler and others, Simmonneaux suggests that "SAQs can develop high-level thinking, decision-making and critical thinking with a focus on promoting an engaged citizenship". SAQs can be considered the *hot pole* of SSI-based education. Among other points, Laurence Simonneaux proposes several didactic strategies to develop students' critical rationality and socio-scientific reasoning (SSR, a notion introduced by Sadler et al. 2007[2]).

The other five chapters of Part II analyse different cases of SSI-based education. These cases are located at a "cooler" pole than SAQs in the continuum of different SSI educational stakes, as defined by L. Simonneaux. The two last chapters investigate the interactions between SSIs, STS and NOS.

For instance, Domenech and Marquez, from Spain, analysed students' arguments related to the reintroduction of bears in the Pyrenees. The same reintroduction was one of the three French cases analysed by Simonneaux and Simonneaux (2009)[3] rooted in their theoretical framework of SAQs, which is not explicitly taken into account by Domenech and Marquez. These authors use a qualitative and quantitative approach to categorise the students' arguments "as social-oriented, ecological-oriented and moral-oriented" and observed that students use mainly arguments from only one of these categories and have difficulties in considering the different perspectives related to an SSI.

Mercan, Yakmacı-Güzel and Akarsu also explored students' arguments in the context of SSIs in Turkey. 36 students of one of the best Turkish school were interviewed in writing on three dilemmas: the energy crisis, global warming and water taxation. From a sophisticated quantitative and qualitative analysis of data, the themes that emerged were less linked to scientific or technical knowledge than to action and ethical or political dimensions, with most of the students considering human nature as individualistic and expressing the need for government regulation.

The chapter by Albe, Barrué, Bencze, Byhring, Carter, Grace, Knain, Kolstø, Reis and Sperling summarises a symposium, juxtaposing the presentations of different research on science teachers' viewpoints and teaching practices on SSIs in six countries: Australia, Canada, France, Norway, Portugal and the UK. Three empirical research projects analyse how science teachers negotiate their contributions to citizenship education and SSI classroom discussions: a study of teachers' views in France (using a questionnaire), a "meta-analysis of five case studies on Portuguese biology teachers" and an action research on teachers' professional development in

[2] Sadler, T. D., Barab, S. A., & Scott, B. (2007). What do students gain by engaging in socio-scientific inquiry? *Research in Science Education*, *37*, 371–391.

[3] Simonneaux, L., & Simonneaux, J. (2009). Students' socio-scientific reasoning on controversies from the viewpoint of education for sustainable development. *Cultural Studies of Science Education*, 4(3), 657–687.

Ontario. A second axis of this chapter analyses how SSIs and IBST (inquiry-based science teaching) converge to develop complex competences, using more or less complex socio-scientific environmental issues. In spite of the diversity of cases, theoretical backgrounds, methodologies and results, one thing that emerges is the importance of the involvement of teachers for a positive implementation of SSIs in their classes.

The last two chapters of Part II are more in continuity with the plenary talks of I. Stengers and L. Simonneaux (although they do not make any explicit reference to them), exploring the interactions between STS, NOS and SSIs.

Eilks, Nielsen and Hofstein discuss learning about the role and function of science in public debate. SSIs are rooted in an STS framework. Taking examples of debates over the chemical industry in Germany and in Israel, they argue that one should deal with societal debate in science education in more detail, including the way in which debates take place and decisions are made in society, for the purpose of democratic decision making. Their chapter presents several interesting theoretical models, particularly in relation to scientific information transfer.

Vázquez–Alonso, Manassero–Mas, García–Carmona and Bennàssar-Roig assess teachers' understanding of STS and NOS issues in a large sample of secondary and high school science teachers, encompassing pre-service and in-service teachers, who were asked to rate their level agreement with a long list of items reproduced from classical questionnaires used for research on STS or NOS. From a sophisticated analysis of the data, it appears that teachers' STS–NOS beliefs are more complex than usually described, being sometimes *appropriate* or *inappropriate*. In conclusion they suggest that science teachers' competences to teach STS–NOS issues need to be improved.

Finally, Part II illustrates the growing importance of SSIs and NOS issues in science education, to cope with less dogmatic science, as well as preparing responsible citizens in a democratic society, and also in order to motivate students for science learning, as is confirmed by other chapters dealing with SSIs in other parts this book.

2 Teachers' Practices and Teachers' Professional Development

This part focuses on teachers' practices, conceptions and professional learning, constituted around two main orientations. The first one provides models and tools to study teachers' professional development both in terms of the processes involved in professional learning (Van Driel) and in terms of teachers' CK (content knowledge) and PCK (pedagogical content knowledge) (Tepner et al.). The second orientation aims to provide different aspects identified as promising for professional development: Some point out the impact of experimenting in classroom practices (Tepner and Dollny; Crawford et al.; El Hage and Buty), while others demonstrate the need for appropriate materials or information (Engeln et al.; Blonder et al.) or taking into account the teachers' conceptions (Clément and Caravita).

In the first chapter, based on his plenary talk, Van Driel (the Netherlands) makes us aware of the need for science teacher education and professional development programme to make connections to teachers' professional and personal contexts and to focus on student learning of the subject matter. First, the author discusses models of professional learning from the literature and then selects an interactive model the "Interconnected Model of Teacher Professional Growth" (IMTPG; Clarke and Hollingsworth 2002) as a framework for examining the professional learning of pre-service and in-service science teachers through two case studies. According to this model, teachers' professional learning is represented by changes in four domains – the personal domain, the external domain, the domain of practice and the domain of consequence – through the mediating processes of "reflection" and "enactment". The results show that working with the IMTPG as an analytical tool proved to be helpful, giving useful insight into the processes involved in professional learning. The IMTPG appears to help to make the often tacit and implicit change pathways explicit, and, furthermore, it makes it possible to identify important elements within professional learning programmes.

Tepner and Dollny (Germany) present a large-scale test instrument based on direct measurement for quantifying chemistry teachers' content knowledge (CK) and pedagogical content knowledge (PCK). The authors study what differences in chemistry content knowledge are observed in teachers teaching at intensified- and non-intensified-level schools as well as the correlation between these teachers' chemistry CK and PCK. The results show that teachers' chemistry CK varies according to the type of school in which they work. Teachers teaching at non-intensified-level schools show limited chemistry CK. Furthermore, the results show a higher correlation between CK and PCK at the non-intensified level while a high CK does not affect a high PCK to the same extent in the intensified ones.

The chapter by Crawford, Van Driel, N. Lederman, J. Lederman, Luft, Wong, Tan, Lim, Loughran and Smith (the USA, the Netherlands and Singapore) presents different professional development projects. They develop empirical evidence from studies focused on supporting teachers in engaging children in inquiry science learning. The results show that science teachers' professional learning is supported by providing opportunities to experiment with new teaching approaches in their classroom, sometimes in combination with authentic experiences to learn science, as well as opportunities to reflect on these experiences, both individually and collectively. The goal of the authors is to work towards constructing an evidence-based framework for effective professional development.

El Hage and Buty (France) focus on the teacher's discourse during a teaching sequence where the teacher is using a computer-based simulation. The authors analyse this discourse from a threefold point of view: the science modelling processes, the moves between several different semiotic registers and the affordances or learning opportunities offered to the students. The results show that the meaningful relationship between the semiotic registers, the modelling levels and the activities of the sequence enables the teacher to offer new learning opportunities to the students or strengthens learning opportunities already offered. The use of ICT assists the teacher in building bridges and creating links between different sessions and activities in the sequence.

In their chapter, Blonder, Parchmann, Akaygun and Albe (Israel, Germany, Turkey and France) present four different professional development programmes which aim at enhancing in- and pre-service teachers' backgrounds in nanoscience and nanotechnology. The teachers' training programmes focus specifically on the use of models to teach and learn important nano-techniques. The results show the aspects identified as promising for such programmes: connecting nanoscience to socio-scientific and ethical issues, using simple teaching models in the professional development that can be transferred to school teaching and introducing leading instrumentation in the field as part of the teachers' course.

Engeln, Mikelskis-Seifert and Euler (Germany) study the potentials and challenges of implementing inquiry-based learning (IBL) in mathematics and science education. The authors present results from a baseline survey and another on the effects of using material being developed by a team of teachers and researchers. The results show the variety of IBL approaches in mathematics and science teaching across Europe. The implementation of inquiry-based education strongly depends on the subject, the culture and the individual teacher. The evaluation demonstrates the potential for improvement in mathematics, whereas in science classes the implementation of IBL-oriented material does not show the expected effects. The positive effects of moving to more IBL-oriented teaching practices are less pronounced when elements of IBL practices are already at a higher initial level. The authors conclude that teaching practice can be oriented towards more inquiry-based education by making appropriate material available.

Clément and Caravita analyse teachers' conceptions of environment and human rights as informed by possible interactions between their scientific knowledge, their values and their social practices. In the first section, they list the values linked to education for a sustainable development (ESD) and describe the degree of implementation of ESD in several countries all around the world. In a second section, they compare teachers' conceptions from important and similar samples in 24 countries. The differences between countries are very important, with teachers of the less developed countries displaying more anthropocentric concerns and exhibiting relatively little awareness of the problems of limited resources. These anthropocentric conceptions are significantly correlated with several intolerant conceptions related to human rights and also with a high degree of belief in God and religious practice.

3 The Students: Multiple Perspectives

Breaking with the large amount of prior research on students' conceptions, often describing their "misconceptions", these six chapters present new trends of research.

Some still take a cognitive perspective but nevertheless represent new approaches. Two studies are focused on students' competences and on the different ways to assess them, by a critical analysis of assessments proposed by PISA (Le Hebel et al.) or by the proposition of different designs for the assessment of experimental competences in physics (Theissen et al.). One study adopts a longitudinal approach focused on learning progression in chemistry (Sevian et al.).

The other chapters are rooted in theoretical backgrounds other than the classical cognitive perspective. One is based on a semiotic approach used to analyse students' conceptions of electromagnetic induction (Yeo). Another mobilises interactional ethnography to analyse the influence of certain configurations of masculinity on the development of learning opportunities (Julio and Vaz). The last one presents, develops and uses the concept of learning affordance (Shin et al.).

In a cognitive perspective, Le Hebel, Montpied and Tiberghien (France) link the theoretical background to the mental and behavioural processes of students when solving problems or answering questions. Using audiotaped and/or videotaped data from nine pairs of students answering PISA items, and also using interviews and written questionnaires, they analyse the representations and strategies that students develop while solving PISA items, showing a difference between high and low achievers. They also show that the competences which students use when they construct their answer may differ from the competences that PISA claims to evaluate. This last result puts into question the consistency between the competences that teachers a priori believe that they are assessing in an evaluation and the competences effectively involved when students answer questions.

Theyßen, Schecker, Gut, Hopf, Kuhn, Labudde, Müller, Schreiber and Vogt (Germany, Switzerland and Austria) present their symposium focused on the modelling and assessment of experimental competences in physics. After a presentation of modelling experimental competences, this chapter mainly juxtaposes the four communications of the symposium, each one with its own design, method of research, results and discussion. Each of these contributions briefly illustrates a sample of assessment methods for experimental competences of physics students, ranging from written tests to product-based analysis of hands-on experiments, to process-based analysis of hands-on activities, and computer simulations of experiments.

Sevian, Talanquer, Bulte, Stacy and Claesgens (the USA and the Netherlands) analyse how students' understandings of core ideas in chemistry develop over time, from a learning progression perspective. Three different approaches to learning progression are described in this chapter, respectively focused either on students' reasoning in terms of cognitive constraints or on mapping the students' understanding and assessment framework, or on promoting sequences of types of activities. In spite of the great diversity of the objects and methods found in these approaches, their results converge to suggest a similar progression of core ideas about the nature of matter, and the authors argue for the need for coordinated curricula and assessments that structure the development of student learning in chemistry.

The chapter of Yeo (Singapore) uses the social semiotic multimodal framework as an alternative lens to analyse student' conception of electromagnetic induction. After a fine presentation explaining the interest of this theoretical background, Yeo illustrates it with a case study based on two students having contrasting explanations of the same phenomenon. This case study helps to illuminate the differences in meaning-making between the two students, albeit both predicting the outcome of the phenomena correctly.

The chapter where Julio and Vaz (Brazil) focus on masculinity in an analysis of learning opportunities is particularly original. Their concepts and methodologies are inspired by interactional ethnography. In the first part, they give interesting definitions of masculinity and of learning opportunities that are central concepts in their work. They observed all the physics lessons of two grade 10 classes, occasionally making audio and video recordings. They present a microanalysis of a group of three boys over a six-lesson activity, as a typical example of their observations during the year. Analysing the influence of certain configurations of masculinity on the development of learning opportunities, they conclude that resistance patterns, power relations and collaboration are factors that intervene in students' group learning and that these patterns are modified when the teacher alternates between small- and large-group dynamics.

In their chapter, Shin, Park and Kim (South Korea) present and develop the concept of "learning affordance", based on the concept of affordance initially defined by Gibson (1979) as all action possibilities latent in the environment but dependant on the actor's capabilities. They analyse the case of a science museum and define learning affordance as the intersection of the affordances of exhibits and visitors' goals and abilities. Their data collected come from observations, video recordings, interviews and participants' written accounts. They finally differentiate four groups of affordance of exhibits – learning affordance, hidden affordance, false affordance and correct rejection – and also discuss a creative zone for some visitors' interactions with exhibits which have hidden or false affordance. Their work shows the pertinence of these concepts for understanding the visitors' learning in a science museum. It might be interesting to use the same theoretical approach in a school environment.

In summary, this part presents multiple and innovative perspectives on research concerning students' learning.

4 Relationship Between Teaching and Learning

This part, focused on the relationships between teaching and learning, illustrates some of the many possible approaches for studying this question. Two main orientations can be distinguished, first the studies of the relationships between a curriculum, a teaching learning sequence, or specific teaching activities and the students' learning outcomes (Banner and Ryder; Boer, Lee-Cheng; Olympiou-Zacharias, Steffen-Schaal), and second the studies focused on the classroom, the students', teachers' or classroom practices and the methodologies to study the classroom according to different perspectives (Kelly, Venturini et al., Souto Silva-Munford,).

In a chapter based on his plenary talk, G. Kelly (the USA) focuses on studies of science classrooms presenting theoretical arguments for the study of spoken and written discourse associated with methodological developments illustrated by two cases. These examples are situated at different levels of teaching, one at the university level with the teaching goals oriented towards disciplinary inquiry and learning to use evidence in science writing, and the other at teacher education level with the goal of developing processes of reflection through viewing and commenting on video of the student's own teaching. Five methodological themes are identified: (1) discourse processes situated in social practice, (2) exploration of different ways of representing social action and practices by "zooming in" to understand instances of action and "zooming out" to view patterns of activity, (3) how to make analyses systematic, (4) how pedagogical change in specific educational settings can be informed through research and (5) the importance for research of developing reflexivity and recognising the contingency of our own language.

The chapter of Venturini, Tiberghien, von Aufschnaiter, Kelly and Mortimer (France, Germany, the USA, Brazil) is also methodologically oriented. The studies focus on classroom activities with analyses of teachers' and students' discourses and actions recorded on video. They show that different domains of research can be involved. Three are presented: conceptual change, meaning-making and inquiry-based science education (IBSE). This chapter reveals the common methodological aspect of these studies that is the necessity of examining *both* the moment-to-moment interactions where meanings are constructed *and* the longer-term practices that stabilise meaning through conversation within a collective. In other terms, the methodological approaches that move across timescales and interactional spaces to examine how what counts as science in a milieu is constructed over time appear to be crucial to better understanding scientific and teaching practices.

Souto Silva and Munford (Brazil) examine how argumentation occurs in science classroom as part of "ordinary" events in science lessons in order to learn about a novice teacher's practices when she or he interacts with students in argumentative contexts. The authors focus on the teacher's argumentation and the procedures involved in resolving differences of opinion. They select three lessons conducted by one teacher involving argumentation during whole-class discussions, in which the resolution of the difference of opinion took a significant amount of time. They develop different types of representation to better visualise how arguments were constructed through interaction and how argumentation differs in the various contexts. The chapter shows differences in argumentation among events from various instructional contexts in relation to (a) relationships among the differences of opinion, (b) the nature of the differences of opinion, (c) the structure of the argumentation, (d) the roles of the participants and (e) whether the components of argumentation were made explicit or were implicit during discursive interactions. This research makes it evident that the field of the theory of argumentation can contribute to the development of new methods for investigating argumentation and science learning.

Concerning the studies of the relationships between a curriculum, a teaching–learning sequence, or specific teaching activities and the students' learning outcomes, we can distinguish between the studies involving a curriculum designed by others, like the official curriculum, and those where the teaching sequence or the teaching activities are developed by the researchers.

The study by Banner and Ryder (UK) belongs to the first category. Here they investigate students' experiences of a science curriculum with a greater emphasis on context-based science. The authors distinguish between two types of students (14–16 years old), students taking the more academic pathway and students taking the less academic pathway. The chapter shows that the new context-based curriculum appears to influence the way some students talk about their experiences in science lessons. The students taking the more academic pathway tend to want to learn more canonical science since this interests them and is seen to be useful for the future. Students taking the less academic pathway tend to want to learn more "real-life" science since this would be useful in their future lives. However, the content of "real-life" science is hard for many students to define, but is generally not the broadly canonical science they typically experienced in the classroom. It clearly appears that a context-led curriculum written by adults does not necessarily seem "real life" to students. The authors conclude that there is a need for both guidance and flexibility in the provision and teaching of science courses so that the student's voice is heard and acted upon and at the same time students are provided with a broad and balanced curriculum to maximise opportunities for students both in everyday life and for their careers.

Boer, Prins, Goedhart and Boersma (the Netherlands) develop a coherent chemical and biological curriculum unit based on an authentic practice in order to test whether students (16–17 years old) can experience coherence between chemistry and biology. The unit, which used a design research approach by regarding learning and teaching strategy, operationalises this coherence in two ways, both by connecting the macro–micro thinking in chemistry with the thinking in terms of levels of organisation in biology and by emphasising relationships between chemical and biological concepts. The findings show that students do experience coherence between chemistry and biology in terms of propositions linking chemical and biological concepts organised in a conceptual structure.

Lee and Cheng (Hong Kong) evaluate students' learning concerning the main conceptual aspects of a teaching plan on chemical bonding. This teaching plan has been designed on the basis of a previous curricular model and is well documented by research on students' alternative conceptions. The researchers used interviews at several steps of the teaching sequence to evaluate the students' learning. The questions asked were based on a deep analysis of the knowledge to be taught and the students' alternative conceptions. The results show that some aspects of the teaching content were not acquired (the application of Coulombic electrostatics to the learning of covalent bonding). Their analysis also shows that one reason was the need to fulfil the requirement of the official curriculum and public examination in which students would have to draw electron diagrams based on the octet rule. This illustrates that a larger view than that of the classroom is needed to interpret some results on students' outcomes.

Olympiou and Zacharias (Cyprus) analyse the relationships between teaching situations and learning in terms of affordances, a concept mobilised in other studies in other parts of this volume. The authors compare physical manipulations, virtual manipulations and a blended combination of physical and virtual ones by analysing their respective affordances. When the manipulations are blended, the physical and virtual characteristics are both available, and so the students can switch mode of experimentation (from physical to virtual and vice versa) and can acquire the associated knowledge and skills. The relationships between these three teaching conditions and the learning outcomes were studied with tests given before, during and after the teaching sequence. These manipulations were involved in a teaching sequence at undergraduate level previously designed by other researchers. It is interesting to note that like Lee and Cheng, this study uses teaching resources previously designed by another research group. This analysis was associated with a profound analysis of the learning objectives of each experiment and of the affordances provided by each manipulation, particularly those unique to only physical or virtual manipulations. The results confirm that blended manipulations are more conducive to student learning than the use of physical or virtual ones alone.

Schaal (Germany) analyses the effects of a health education programme focused on pre-service teacher training at the levels of pupils (10–17 years old) and of general school development. The 3-year longitudinal study aims to understand to what extent external support could initiate and sustain the adoption of health promotion approaches (HPS) in school under the direction of biology and science education. The HPS development process is analysed using structured interviews with the different groups of those in charge of the school community and questionnaires completed by more than 2,000 pupils. The results show that pupils from the schools with the intervention report a more favourable state of health-related well-being, less unfavourable nutritional behaviour, higher nutritional self-efficacy and fewer symptoms of physical stress. It appears that the intervention was not standardised across every participating school. Nevertheless, the outcomes of the programmes can be attributed globally to the specific health promotion approach supporting its generalisation, although without necessarily any standardised programme. This diversity of programmes may be one reason why knowledge construction in the domains of nutrition and physical activity did not succeed. These results show how the various components of the educative system can influence student outcomes very differently.

This part illustrates that studying the relationships between teaching and learning can have very different orientations: the evaluation of a curriculum, a programme, or a teaching sequence; the planning of the learning role of specific teaching resources; or the methodology of studying the "site" – the classroom – where this relationship takes place.

5 Part VI Teaching Resources, Curriculum

This part involves two main sets of themes. The first set, under which we can group three chapters, deals with teaching resources used by teachers and/or students in the classroom. The second set, which also includes three chapters, is more original and deals with the processes of adapting teaching materials from one country to another.

Acher and Arca (Italy) designed a learning progression concerning matter aimed at developing scientific knowledge construction across different years starting with early years science education. The theoretical basis of their learning progression was based on three main choices. The first two choices, based on analyses of students' prior knowledge, allowed the selection of two ideas on which the students could construct their scientific knowledge and the material contexts in which these ideas could be articulated and developed. The third choice was related to teachers' practices but still in relation with the first two choices. The progression was successively refined and the results obtained in classrooms led the researchers to improve the components of the learning progression related to the three main choices.

Bruguière and Triquet (France) propose fictional stories as teaching resources based on a theoretical position stressing the importance of narratives as a powerful and persuasive way of conveying ideas. They tested the role of fictional stories in elementary school and show how a story relevant to the theme of the teaching sequence – for example, the concept of a species – can trigger students' research for rational arguments and thus favour the passage from belief to knowledge.

Cheng and Wong (Hong Kong) present a comparative analysis of two physics curricula and textbooks in mainland China (Guangzhou) and in Hong Kong concerning the implication of the nature of science (NOS). The analyses are based on categories constructed from previous research studies of NOS; in other words, the theoretical framework is based on an analysis of previous studies and not on a theory. In this case, unlike the two previous chapters, there is no design activity, but an analysis of existing teaching resources.

The three other chapters deal with very different aspects of teaching resources, the design and adaptation from one country to another of teaching materials in middle or high schools. This type of research is developed in the framework of projects funded by the European Union that emphasises the sharing of teaching practices between member countries.

Pintó, Hernández and Constantinou present a collaboration between Spain and Cyprus. Their theoretical framework is composite; first, they use the literature on reform to argue that organisational culture, capability, management and policy can influence reform. They also consider knowledge transfer as a process of sharing individual and collective experience that can lead to a reconstruction of the transferred resources. The transfer constrained participants, researchers and teachers alike to define and document the processes involved. Thus, all the stages were documented: how the local group, where the teaching–learning material was designed, made the core features of the teaching–learning sequences explicit; how these features were then debated with the host group and refined; and also how the two groups worked together in the adaptation processes. The results present the main changes that occurred during the adaptation processes between the two groups concerning both the scientific knowledge to be taught and the pedagogical content knowledge.

The two next chapters of this second set are related to the same European project but at two different stages. The first one concerns the design of learning environments (LEs) that are all related to socio-scientific issues. In this project, the designed LEs are interactive web-based environments for data-driven inquiry and use the same platform STOCHASMOS. These chapters show that the framework of the project deals with the design and adaptation processes of the LEs, the perspective of adaptation being included at the design stage. These processes involved close collaboration between researchers, practicing teachers, scientists and educational technologists.

In the chapter by Redfors, Hansson, Kyza, Nicolaidou, Asher, Tabak, Papadouris and Avraam (Sweden, Cyprus, Israel), the results are at several levels. A first level presents the variety of LEs. Even if the four cases presented are related to socio-scientific issues and developed in the same framework with the same web-based platform, they present different emphases on the main objectives like conceptual understanding, argumentation, skills in assessing the credibility of evidence and skills in using graph synthesis. This variety mirrors the different interests of the researchers and teachers involved in the project and different views on the reasons for using SSI in the teaching of science.

The chapter by Kyza, Herodotou, Nicolaidou, Redfors, Hansson, Schanze, Saballus, Papadouris and Michael (Cyprus, Sweden, Germany) focuses on the adaptation from the country where the LE was initially designed and enacted to another one. The four same LE cases were presented as in the previous chapter. They compare the LEs designed and enacted in one country to their enactment in the host country. This comparison focuses on three points: (1) science content and skills to be taught, (2) pedagogy and (3) fit with the local curricular framework. First, the results show that the adaptations were done positively. In the adaptation process, they also demonstrate not only the importance of the country's school culture and teacher culture but also the role of the teacher's own practice. This type of result indicates the possibility for the transfer of best practices from one cultural context to another but not without serious investment as this transfer is resource- and cost intensive. Thus, the authors advocate using the knowledge embodied in the curricular designs and accompanying materials as well as that acquired in such multinational efforts as the basis for other curricular innovations in Europe. In conclusion, this part is characterised by the variety of the contributions, variety in terms not only of resources, curricula and textbooks but also teaching–learning sequences also known as learning progressions or learning environments, typical teaching–learning situations that, here, involve the use of a narrative.

Variety in the theoretical frameworks is not only in terms of their content but also in terms of their nature: some are based on the results of previous research, some on learning hypotheses, content analysis and teacher practices, while some also include hypotheses on the transfer of resources from one context to another. This variety, particularly for the five chapters involving the design of teaching–learning resources, is probably due to the complexity of the design activity that necessitates choices involving all three poles of the didactic triangle, teaching, learning and knowledge. However these five chapters all suggest means for refining the designed materials based on classroom observations (audio or video recordings) made when the resource was put into action, even if some of these studies also included pre/post-tests and interviews. This part illustrates the importance of the design activity, broadened to include the study of the adaptation of a resource created in one culture context to another one. As emphasised in the last chapter, this activity is costly in several respects, but the exchange and use of resources designed by other researchers represents interesting new research opportunities.

Please note that each “Part” is organised in the following manner. The plenary talks are placed at the beginning, and the chapters follow in alphabetical order, based on the family name of the first author.

Part II
Socio-scientific Issues

Chapter 2
The Need for a Public Understanding of Sciences

Isabelle Stengers

I am a philosopher, not a specialist in science education. But I will not intervene as a 'philosopher of science' as this specialization is usually understood. I will address neither science in the terms of a 'theory of knowledge' nor in more inclusive terms, that is, I will not try to define science in a way that could and should, as such, be part of a scientific education. As a philosopher, I address science not as something to be defined, what we may call a matter of fact, but as a matter of public concern.[1] Correlatively, I will address the question of the relations between citizenship and science education not as a matter of pedagogy, rather as related to this very concern. To paraphrase Karl Marx, the point may well be not to understand science and its role in our society, but to contribute to the possibility of a change in the relation between scientific and social, public concerns.

More than 40 years ago, the philosopher Paul Feyerabend famously wrote: 'While an American can now choose the religion he likes, he is still not permitted to demand that his children learn magic rather than science at school. There is a separation between state and church; there is no separation between state and science' (Feyerabend 1975, p. 299).

I propose that we take seriously Feyerabend's challenge and wonder how to answer it. Can we argue that the authority of the church should be restricted to its members, while sciences produce public knowledge, the legitimacy of which may be explained, even verified, by anyone competent enough to do so? Certainly, at the time when religion education was mandatory, there was no didactics of religion, even if catholic theological controversies were quite beyond the understanding of laypeople. Can we argue that this difference is sufficient to claim that science

[1] The distinction between matters of concern and matters of fact has been forcefully and fruitfully introduced by Bruno Latour. See, for instance, Latour 2004.

I. Stengers (✉)
Université Libre de Bruxelles, CP 175, 50 avenue Franklin
Roosevelt, Brussels 1050, Belgium
e-mail: istenger@ulb.ac.be

C. Bruguière et al. (eds.), *Topics and Trends in Current Science Education: 9th ESERA Conference Selected Contributions*, Contributions from Science Education Research 1,
DOI 10.1007/978-94-007-7281-6_2, © Springer Science+Business Media Dordrecht 2014

demands to be understood, while religion defines believers as a flock, having to faithfully follow the shepherd?

The argument does not work with Protestants, however, who address humans as free minds, having to take a personal position. And Feyerabend's challenge insists again. Science didactics certainly does not address a flock, but does it address free minds? Is not its aim to help teachers to present a knowledge which does not demand that a personal position be taken? Didacticians, teachers and students all understand that they deal with a knowledge that has been produced elsewhere, by people whom they know they must trust because they are not in a position to contest scientific knowledge.

When science education is related to the question of citizenship, the question becomes still harder. Indeed, what matters then is not the kind of success teachers may be rightly proud of – that some of their students get a taste for science and specialize in a scientific field. What matters is rather the majority of adolescents, girls and boys, who will never learn from working scientists how to participate in the production of scientific knowledge: they are the future citizens who, in our democratic societies, should be able to participate in choices about the techno-scientific options that will determine their future.

This question is both simple and hard. It is simple because, as both a matter of fact and a matter of concern, such choices are not really democratic ones. Citizens are rather asked to trust that the future is taken care of by those who know and that techno-scientific choice will bring progress. The hard question of the relation between science education and citizenship may then be: can science education at school make a difference in this situation? Can it contribute to empowering citizens to refuse that scientists take their concerns with the tolerant patience of those who know better? And here indeed a public understanding of science, that is, an understanding of the way scientists work, is important.

I am quite conscious that I am going here against the classical argument about the relation between citizenship and the need for a public understanding of science, which is an aim of scientific education.

According to this argument, citizens should have some scientific literacy in order to understand the world we live in. This argument is often mobilized when the public resists a transformation which scientists define as positive. The usual diagnosis then points to a lack of such understanding. For instance, the public would not understand that the genetic modification of plants is not very different from what was previously done by farmers for many thousands of years, the only point being that it is much more efficient.

Certainly, those who define the problem in terms of a deficit of understanding by the public do not deny the citizens' legitimate right to accept or refuse an innovation, but they insist that citizens should do so on sound grounds, that is, without confusing scientific facts and the values they believe in.

A correlative usual argument about the importance of science education is that citizens themselves need to practise careful observation, hypothesis, verification, objections and the like. Learning in such terms how scientific knowledge is produced would thus achieve a double goal: that citizens understand the difference between objective knowledge and subjective judgement and that they become able to apply themselves to the only kind of reasoning that may lead to valid, objective knowledge.

Both arguments validate what has become today the motto of public authorities when they confront what is a new but growing problem: citizens' relative mistrust or scepticism about the role of scientists in society. We need to reconcile the public with its science. The 'its', a possessive, is precisely what school should transmit, that the reasoning of scientists is something anyone can make their own – if I was observing the same facts as Galileo, Maxwell or Darwin, I would derive the same conclusions.

However, all specialists in didactics know that in order to produce such conviction, they have to radically reconstruct the situations of Galileo, Maxwell or Darwin, and more generally of scientists who participate in sciences in the making, with their controversies and conflicts. They have to address stabilized, made sciences from which they extract the elements for a rational, clean definition of the facts and the problem, leading to the correct scientific conclusion.

The question is then:

> Do the carefully purified, reproducible facts which one can derive from accepted, stabilized science help to understand the facts with which we, as citizens, are confronted?
>
> Do rationally reconstructed situations help us to address situations that demand that we hesitate and think, not derive and conclude, because they communicate with what Bruno Latour calls 'matters of concern'?

What I would call a 'matter of concern' would be for instance GMOs as quite distinct from the ones which are defined in experimental laboratory, because the situation in the fields entangles biological processes such as gene transfer and the appearance of resistant pests, which can be detected only at real field scale and over a long period. Other concerns may be added such as the genetic modification allowing the patenting of seeds, the loss of crop diversity or the heavy use of fertilizers and pesticides.

More generally, a matter of concern implies that there is not *one* good solution but choices to be made, sometimes hard and difficult choices, which should be made after a process of careful, well-documented deliberation.

Should not science education in such cases mean learning to situate the scientific contribution to this process – a contribution which, however important, may never be equated with the definition of the good solution?

But situating a possible scientific contribution to an issue which is a matter of collective concern often means also resisting claims by scientists that what they propose be recognized as the right solution to the issue when this issue is defined in rational, objective terms. In the case of GMOs, for instance, biologists loudly claimed they were providing the means to feed the growing human population. As a consequence, a last question is: should not future citizens also discover and discuss at school the manner in which scientific researchers think about their own role and the role of citizens?

Addressing this question, I will first describe my experience with students for which science education at school has been a success, since they have chosen to specialize in some scientific field – more precisely in some so-called hard science.

As you know well, when it is a question of successfully attracting adolescents to science, or of science didactics, or again of the social responsibility of scientists, not all sciences are equally concerned. History, social sciences, psychology and communication sciences are quite usually forgotten. To me this rather dual judgement is part of the problem, but the problem is a real one. The experience I will describe concerns the so-called hard science students and the specificity of their training.

In my country at least, students who have chosen a hard science specialization will be taught by working scientists, who are usually rather indifferent to didactics. As for citizenship, there will be some classes devoted to general approaches of science, maybe history of science, or science and society, or even philosophy of science, usually at the very beginning of their curriculum.

I have taught such classes. My experience was with first-year students (mandatory class) but also with students in their final year (optional class). In both cases the 'and' that appears in 'science education and citizenship' was rather felt as an 'either … or'.

In the first year, the situation was a bit schizophrenic. Either the students situated themselves as 'mere citizens', that is, as ignorant, seemingly forgetting that they were in the process of becoming one of those scientists they described as separated from citizens by an insuperable distance, or they were already on the other side, talking about how citizens misunderstand science. But in both cases, the overwhelming feeling was that they were acting out a role, something they had to do for my class but which would be of no consequence for their future course of study. They knew that while other classes were crucial for what would follow, mine was not. They even knew, I believe, that if they ever raised, in one of their 'true' science classes, a question I asked them to pay attention to, they would get some dirty looks from their teacher and would even hear explicit doubts about their future as scientists.

As for those final-year students who chose my master's seminar as an option, many chose it because they were hesitating in front of what they saw as an insuperable 'either… or' alternative: either giving up science or committing themselves and forgetting about any larger question about the world they live in and the role of science in that world.

I am not criticizing, rather diagnosing the situation which makes the relation between science education and citizenship such an ambitious and crucial one. My strongly felt conviction is that the point is not that specialist scientific education lacks some general culture, which would be needed in order to address the concerns of citizens. I rather believe that scientific education, since the nineteenth century at least, is effectively incorporating tacit normative judgements about what matters and what does not.

And I believe that as long as such normative judgements rule, that is, as long as they do not become counterproductive for their career and ambitions, scientists are bound to feel that they know better than citizens what really matters for those very citizens.

Let us take as an example of such an attitude a short text which could well be deciphered at school. It is an excerpt of the synthesis report of the *Etats généraux de la Recherche* held in France in 2004, quoted in the daily newspaper *Le Monde*. The author of the paper, P. Le Hir, quite judiciously takes it as an example of the researchers' anxious cautiousness. Indeed, one may read that 'citizens are expecting from science the solution of many social problems: unemployment, oil depletion, pollution, cancer… The path that leads to the answer to these questions is not as direct as the goal-directed vision of research would have us believe … Science can progress only if it works out by itself its own questions, protected from the urgency and the distortion inherent to economic and social contingencies' (Le Hir 2004, p. 18).

Now it is important to remember that this report was meant to convey the accepted, collectively discussed position of the assembled researchers. It is clear that the assembly wanted to communicate their anxiety as they felt the burden of all urgent questions they were supposed to answer. But the way they express this anxiety is very interesting. To begin with, they attribute to citizens the belief that science can solve nearly all questions, including unemployment, which may seem rather strange. But what is far stranger is that they seem to endorse such a belief. It seems that science will indeed find a path leading to the answer to such questions but if, and only if, it is not asked to answer what we may already consider to be their distorted formulations. True scientific solutions will transcend what is characterized as distorting social and economic contingencies.

In other words, matters of collective concerns are defined here in terms of distortions. When science will be able to give answers, it will not be in terms of a contribution to such concerns but as the answer to an at last well-formulated question. As a consequence, citizens are right to trust an all-powerful science, but they should understand that scientists have to remain deaf to their anxious cries and demands.

I would add that in this text researchers are in fact talking over the head of citizens: anxiously indeed, they are addressing the public authorities in charge of scientific policy about the goal-directed reorganization of research which is implemented under the name of the knowledge economy. And they try to resist this reorganization using the traditional 'goose that lays the golden egg' argument – you will kill the goose that lays the golden egg if you do not keep your distances. You must feed the goose without conditions and wait for the eggs.

I would argue that the goose that lays the golden egg already inhabits the imagination of science students together with the normative judgement that scientists should safely ignore matters of common concern because they can only answer them by following their own creative path. The normative judgement contrasting the rational, at last scientific answer, that is, the golden egg which should eventually be produced, with social and economic distorting concerns thus takes for granted the beneficent character of scientific progress. As is always the case when the goose that lays the golden egg cries, some questions are ignored. For instance, for whom will the eggs be golden? Who will enact scientific propositions? Were not, in the past, those for whom the eggs were golden also responsible for what is now recognized as an unsustainable development, the consequences of which are threatening our future?

Now you may well wonder:

> Should science education really address such questions, which are not about sciences, only about scientists, who are as imperfect as anybody else? Should it not rather concentrate on selected, well-founded, scientific achievements, which may be purified, separated from regrettable socio-psychological bias or ideology that in no way may question these achievements?

This is a hard question indeed, and again I would never dare to discuss it in front of an assembly of specialists of didactics if our encounter did not include citizenship as a matter of concern related to scientific education.

Let me make it clear, however, that I am NOT proposing to think in terms of 'either… or': either science as an edifying model of rationality or scientists as guided like everyone else by interest and ambition.

In fact, it is my conviction that the very aim of scientific learning and education in their relation with citizenship is to escape this 'either… or', which indeed is haunting the whole scene.

Scientists themselves are well aware that science as it is made, as it is their job to make it, has not a lot to do with the purified fiction they themselves learned at school. And they know that is not a question of mere approximation, as when Galileo, for instance, considers that his measurements are only approximate due to the friction he can diminish but never entirely eliminate. The situation would rather be analogous to a description of the flight of birds in a frictionless ideal world, a world where the air would be eliminated. As we know flight is possible *only* because in the case in point the air is involved in a new and very complicated hydrodynamic effect called 'lift'. Scientists know that, just as birds in an environment devoid of air would fall to the ground, their own practice would be impossible without what is simplified away when we represent facts as reproducible from the start, scientific reasoning in terms of general rationality or scientific research as indifferent to its social valorization. They know that, but they also know that nevertheless science is not a mere social business as any other.

They know it, but they do not trust citizens to understand it. Rather, they firmly believe that if non-scientists knew how science is actually made, their confidence would be destroyed and they would reduce science to mere arbitrary opinion. In other words, for scientists, non-scientists should be kept at a distance from science as it is made, like children who must believe that the world is simple and that adults know what to do.

To me, this is the true challenge at the heart of our theme, 'Science Education and Citizenship'. Citizenship has meaning only if citizens are not treated as children but are trusted to be able to understand the world we live in. Citizens should thus be trusted and not considered as children!

But, it will be objected, can they indeed be trusted? Is not democracy just like the flight of birds? The flight depends on the air, which is a defect from the ideal point of view authorizing the general definition of the way heavy bodies fall. Would it not be the case that democracy also works because of a defect from the ideal, because some questions remain outside public debate, left to the responsibility of scientific experts? That is, because citizens are led to admit that they should not meddle with what they cannot understand anyway. If this is the case, no doubt should indeed disturb their indifferent confidence.

In this perspective, the purified version of science learned at school would be quite functional. It would select those who love well-formulated problems with beautiful, intellectually satisfying solutions. And it would convince the others that science is really not their cup of tea but also that those who like science can be trusted to reliably solve the problems they address. Scientific education would thus play its rightful role in the formation of the citizen through the division between those who will come to learn about science in the making and those who will trust it and not interfere.

But is this solution really functional? Or rather, what is the cost of this functionality?

Years ago, the French physicist Jean-Marc Lévy-Leblond characterized such costs in these terms: 'Uncultivated science easily turns into occult science or into the cult of science' (Lévy-Leblond 1984, p. 97). For Lévy-Leblond, both the public and scientists are vulnerable to the temptation of confusing science with an access to the mysterious occult truth of things or of lending it some kind of religious authority. And he claimed that protection against such temptations requires an environment in which scientific productions are an object of living and critical interest – an environment in which an active scientific culture is developed (Lévy-Leblond 1984, p. 94).

What Lévy-Leblond calls a scientific culture is not to be confused with some general scientific literacy, knowing 'something' about physical laws, electrons, atoms, molecules, DNA, biological evolution and all that. A cultivated science should produce not only specialists but connoisseurship as is the case, for instance, in sport or music, or software production, that is, in domains where producers know that they have to take into account the existence of people who are able to evaluate the products, to assess the kind of information they get, to discuss its relevance and to differentiate between mere propaganda and daring option. As a consequence, producers in these domains know the danger of just skipping the weak points, as connoisseurs know how to pay attention both to what is claimed and to what is neglected or omitted.

It is important to disentangle Lévy-Leblond's proposition from the false ideal of some generalized connoisseurship, everybody being able to entertain a connoisseurship relation with everything. The ideal would rather be that of a dense multiplicity of connoisseurs, what may be called a distributed connoisseurship, such that anybody who is not a connoisseur in one domain may trust that if she/he ever comes to be concerned by it, she/he will be able to connect with it through the connoisseurship that has developed about it.

It is also important to disentangle Lévy-Leblond's connoisseurship from the question of autodidacts who desperately look for professional recognition of – or, at least, objections against – their theory offering a solution to some great problem. Connoisseurs are not defending alternative knowledge; rather, their interest for a produced knowledge is distinct from the interests of its producers. They may pay attention to dimensions which played no significant role in the production, but that may come to matter in other contexts. Their eventual objections will concern the way scientists may be tempted to claim that what they do not have to take into account should be recognized as generally insignificant. In so doing, they have a role the crucial character of which should be recognized by anyone who cares about rationality – the role of assessing what is produced for its originality and relevance and resisting any claim of general authority. They are agents in the production of what Donna Haraway calls 'situated knowledge' (Haraway 1991).

Connoisseurship is thus the reverse of the kind of indifferent confidence that must be protected against doubts. Connoisseurs constitute a demanding environment which, because it is interested in science, may also produce an effective critical voice. Their critique may constrain scientists to give up their normative judgements about what matters and what does not. It may oblige them to present their claims in a lucid way, to actively situate them in relation to the questions they were able to ask, and not as bringing rational answers to matters of common concern.

For instance, a connoisseur environment would have constrained French researchers to think twice before presenting themselves as they did in 2004.

Now, it could be asked how scientific education can foster such connoisseurship.

Many settings are possible, the common point being that it should be around a real issue, or a true matter of concern, with no pedagogical trick, no 'right answer' known by the teacher, which students should eventually reach.

I can give you an example: a 3-year experiment we have been able to conduct in my university. We have confronted science students with situations of socio-technical controversies, leaving to them the task of exploring by themselves the resources available on the web, in order to discover the conflicting arguments and the wide variety of facts which are mobilized by the parties. While, as I have already remarked, science students are generally indifferent to the offer to think about science, those students who explored by themselves such controversies appear to have been really interested by their exploration. They produced reports emphasizing how surprised and happy they were to have learned, in particular, that as soon as one leaves experimental laboratories, the divide between facts and values does not apply any longer. They did not draw sceptical conclusions about science, but were rather amazed by the off-hand way scientists could select convenient facts and authoritatively define as 'non-scientific' possibilities and perspectives that escaped their disciplinary frameworks but were a matter of serious concern from different, well-grounded, perspectives.

I would not say that those students were vaccinated once and for all against the 'science versus irrational opinion' opposition which is so common among scientists that they would never endorse in the future the goose's cry that the formulation of 'the' true scientific answer needs to ignore 'distorting' concerns. But what matters is that, for the first time in their curriculum, they did encounter real-world situations, situations marked by uncertainty and the entanglements of dimensions they were used to separating and that they were not dismayed or confused, rather positively interested.

I would even say that they did seem to discover with some relief the possibility of understanding how concrete situations determined the relevance of scientific knowledge. They had until then felt bound by an 'either… or' constraint – either behaving as scientists should behave or taking seriously so-called non-scientific concerns. They had been used to opposing scientific facts and values, which are a matter of ethics. They discovered that scientific facts were liable to enter into conflict with each other, because any consequential fact, able to sustain an argument, is always related to a particular definition of what matters in a situation, the best example being experimental facts, able to produce agreement between scientists. Such facts are not to be characterized as just reproducible observable facts. They are rather achievements fulfilling the ambition which matters above all for experimenters – that the facts they obtain be recognized by their colleagues as able to enter into a proof, that is, as gifted with a power which belongs to no usual fact, the power to define the way they should be interpreted.

It may well be that science education in general would benefit from such direct, non-mediated, confrontation with socio-technical issues, exploring controversies about problems that can neither receive a scientific solution nor ignore scientific contributions, situations where sciences are part of the construction of the problem but do not command the answer. Rather than 'systemic' descriptions of a complex reality, with its many correlated dimensions, which often produce a reassuring feeling that 'we know', but lead to no 'connoisseur' appetite, it may well be that conflicting real-world accounts, when diverging partial knowledge claims pretend to be crucial for a given issue, would have future working scientists and future citizens thinking together.

Using Internet and other sources of debatable information and argument is thus an empowering way to discover what Pierre Clément calls the KVP interactions, the interactions between scientific knowledge, values and practices (Clément 2004, 2006). And it is also an example of a possible answer to the challenge schools have to face at a time when they are no longer the sole source of information for citizens and future citizens. Such an answer would mean that the initiation to science would not be, or not only be, through well-controlled staging of settled knowledge.

However, the crucial point is that the attention paid to controversial or dubious claims should not have for its essential aim to recognize that scientific knowledge is not immune at all to ideology, that scientists entertain values just like anybody else, which may lead to a sceptical view not to a critical interest. What is usually called the need for a public understanding of science cannot be identified with some general scepticism, all the more so as it will verify scientists' belief that the public is unable to differentiate between serious science and dubious claims, that it takes advantage, for instance, of the bias frequent in weak sciences (such as the brain sciences, which are empirical rather than experimental in spite of their sophisticated instrumentation) in order to disallow serious science. Then will resound the challenge – show me the 'values' – when the identification of the DNA's triplets coding for an amino acid is concerned or when neutrinos were eventually attributed a mass.

I would claim that if scientific education is to lead to a public understanding of science, it should rather activate the awareness that there is no miracle. A scientific demonstration is always selective. The condition of its authority is the many questions its selected object allows to be ignored. In some cases assessing what the given answer asks us to ignore may indeed result in the conclusion that we deal with weak, even fake, science. In other cases, such as GMOs, it demands that biologists' capacity to produce genetic modifications be recognized, that the power and limitation of this capacity with regard to multigenic traits, epigenetics, the role of the environment and so on be assessed, and the question as to how specialists felt free to claim that genetically modified crops is the rational solution to the problem of hunger in the world be discussed.

In other words, the fact that any scientific knowledge is situated both by the questions it answers and the questions it ignores should not be considered as a general epistemological principle – the kind of thing students learn and then forget. The importance of this fact depends on the active interest and on the perplexing questions it raises and which may be the initiation of connoisseurship, activating empowering habits of thought, the awareness of the need to critically assess the way scientists present their knowledge and the relevance of that knowledge for common matters of concern.

I will now address another aspect of the concern for a public understanding of science and scientists. I will address three new questions which make this concern more urgent now than ever.

First, we now face the advent of what is called the 'knowledge economy', that is, a redefinition of public research which makes partnership with industry a requirement for public financing and which considers the patenting of the products of this research as a most desirable outcome.

Second, we also face the advent of the Internet, which is, as we have seen, a formidable possibility for the flourishing of connoisseurship *but is also* a powerful vehicle for the dissemination of scepticism, doubts, rumours and crank theories.

Finally, we face what is called the industry of doubt, which turns the demand for scientific proof into a means of refusing the conclusions of important scientific research.

I will show that in all three cases, Jean-Marc Lévy-Leblond's proposition has taken on a new importance. This proposition was originally meant to situate science, to give it a both interested and critical social environment, which would activate a more lucid and demanding relation between scientists and the knowledge they produce. Today, it may well be that it is also a matter of helping scientists to better resist what threatens them. In other words, a public understanding of science as associated with connoisseurship could become crucial for science itself.

Turning to the knowledge economy, I would first remark that the citizens whom French researchers invoked in 2004, those who trust scientists to solve the problems of common concern, certainly still exist. However this trust can less and less be taken for granted. When it will be well understood that scientists' work is from the start conditioned by some industrial partnership, this trust might quite easily swing over to mistrust and hostility, to the general denunciation of scientists as liars and betrayers, sold out to private interests.

On the other hand, it is urgent now, in the context of a knowledge economy, that scientists admit that citizens may help them not to become the mere providers of opportunities for industrial strategies. They have to stop demanding general support to 'save research' as if it was only a question of money and jobs. They have to work for an empowering public understanding of the situation of their sciences and what they must be saved from.

Some 10 years ago, at the time of the GMO affair, some interesting experiments indicated that when so-called normal citizens are asked to discuss a problematic situation without being addressed as ignorant and incompetent, they produce interesting and relevant objections (Marris 2001). The relevance of such objections indicates the kind of support citizens could give to scientists who are afraid that their science may be enslaved to industrial interests and lose the freedom without which there is no scientific reliability. But they will not get this support if citizens do not trust scientists to effectively take their objections seriously. The blind confidence that still prevails today will turn much more easily into blind mistrust and general denunciation than into a lucid assessment of the alliances that may be possible with anxious scientists.

Let us turn now to the second point.

I have already emphasized how the resources made available by Internet can be put to good use in science education. But Internet is also a privileged vehicle for propagating rumours, tales of dreadful conspiracies and extravagant claims. Here again, we face a rather new and testing situation. Certainly scientific education is meant to be linked with the learning of critical, discerning mind, but is that education able to really address what we freely and widely meet on the web?

This is a general problem for schools, but it is especially challenging in science education. Indeed, I would go so far as to claim that many of the didactically purified examples of scientific reasoning may serve as a template for pseudoscientific claims. Both those examples and those claims look like science, presenting facts that should authorize general agreement about the conclusions. What they both ignore is that facts which authorize scientific reasoning are themselves the product of hard work. Reliable facts and reliable conclusions are discussed, tested and determined together – an expensive process in terms of time and resources, in which scientists usually engage only if they feel it to be worthwhile.

However, this raises another thorny issue. A selection of the questions which scientists assess as worth investing time, workforce and money may indeed be needed. But what are the criteria employed by scientists?

Scientists are not used to explaining themselves. What they do not consider worthwhile they claim the right to ignore or dogmatically exclude. They consider, as the French researchers pleaded in 2004, that only scientists are able to select the path of fruitful research. But the web is now transforming the situation, giving a wide audience to those who claim that they have facts but that their colleagues ignore or even suppress what these facts demonstrate.

Further, the web is also full of stories about conflicts of interest, when scientists are accused of serving the interests of those who pay for their research, or worse, pay them directly. And this is not surprising. It may well be that usually, scientists' reasons for not considering a proposition as worth their attention and work are good scientists, but sometimes they may be less so. And they may become ever less so in a knowledge economy where scientists depend more and more upon private interests.

In short, a new public image is in the process of gaining credibility, which assimilates science with a dogmatic, dishonest enterprise and cranks with courageous fighters for freedom. There is no easy solution here, but I would claim that a public understanding of science in the making is a common requisite for all non-catastrophic outcomes. By this I mean both a general understanding that choices are needed and the existence of a demanding connoisseurship able to hear about and to discuss the reasons for the choices that are made.

If scientists had to account for what they prioritize and what they neglect in terms that do not insult demanding connoisseurs, the situation on the web would be different. Their arguments would be less vulnerable to the accusation of dogmatism and the defence of settled interests. They would produce new food for thought and induce new, much more interesting debates, in which connoisseurs would be able to relay, mediate or discuss their position, no longer leaving the whole room to the blind game of defence of, or attack on, scientific authority. It is even possible, because accountability demands imagination about alternatives, that scientists' criteria about what deserves to be taken seriously would become a bit wider, less determined by settled interests, fashionable prestigious priorities or conformist habits.

I am quite conscious that the idea of addressing such questions may hurt teachers. Not only are they not prepared to deal with them, but they may fear that the very point of science, the reliable knowledge it produces about the world, will be lost, leading to general confusion, opening the door to an irresponsible politicization of what deserves general agreement.

My last point will thus be to affirm that the door is already wide open to politicization, and deliberately so, by what can be called an 'industry of doubt' heavily benefitting public ignorance about the functioning of science.

I will first allude to the anti-Darwinian industry, the refrain of which is that the evolution of species is not proved and thus must be recognized as an opinion only, together with other opinions about what explains life on Earth.

This refrain takes advantage of an idea of proof which may be a good approximation when we deal with experimental sciences, but not at all where field sciences and historical sciences are concerned. It is in experimental sciences only that one can demand that a theory explain all facts that fall into its dominion. Field facts do not authorize such theories. What scientists call proofs when Darwinian evolution is concerned would in no way be called proofs in experimental science. What matters is rather the growing number and variety of new and old facts, observations, aspects and data which become intelligible in the Darwinian perspective. And this is quite sufficient to make the difference with creationism and intelligent design approaches, which have no impetus of their own, as the designer or creator is able to explain anything and everything. In other terms, the anti-Darwinian industry of doubt benefits from an idea of facts and proofs that fits experimental sciences only.

Here we meet a point which should be central in science education, that is, the plurality of sciences. It is true that plurality is usually downplayed by the scientists concerned, for whom the model of experimental science is a source of authority and prestige. But the model holds only as long as there are no enemies turning it against its users. In short, the anti-Darwinian industry needs the science of evolution to present itself with the authority of facts, in order to be able to oppose other facts that seem to challenge an evolutionary explanation. It would lose its grasp if biologists learned to share the growing interest and fecundity of the evolutionary perspective, the way it creates new questions and new insights and the way also it discovers that evolution is not a matter of one theory explaining every fact but of many entangled contributions to the understanding of the history of life on Earth.

In this case again, connoisseurship is needed in order to cultivate an idea of science which does not rest upon the model of a science deducing its theories from indisputable facts. It is this connoisseurship which Stephen J. Gould developed against the idea that biology is replaying the Copernican story, overturning the traditional beliefs by the power of scientific reason and hard facts. He demonstrated how much biologists would benefit from the help of historians and anthropologists who know about the intrinsic entanglement of causations and concerns which makes historical and cultural facts both intelligible and unique.

But the industry of doubt is fast expanding. It has already been active against scientific work showing the danger of tobacco smoking and the effects of acid rain. Over many years it succeeded in presenting the debate as still not settled and the danger as widely exaggerated. And we may measure its power when we see it at work in the recent attack on climate change science, which has succeeded in creating the impression that scientists are effectively divided about the subject and has led many – even the majority of US citizens – to conclude that, as a consequence, we should wait for scientific agreement before taking the problem into serious account. Till then we should accept it as a hypothesis only, all the more suspect as it pretends to be demonstrated (Oreskes and Conway 2010). In the USA today, doubters even systematically demand that any public discussion about climate change be balanced, with the presence in the panel of a voice doubting climate change counterbalancing a voice affirming climate change.

In all these cases, the attack against scientists again exploits the idea that the only sound science is the science that obeys the model of experimental sciences. Merchants of doubt turn this model against researchers who do their very best but must deal with the intrinsic messiness of field data, the unavoidable unknowns and the awful complexity of intercorrelated processes and not with the clean, well-controlled data which follow from an experimental design made to answer a precise question.

Here again a public understanding of the plurality of sciences, and of what can be legitimately demanded of each science, is a matter of crucial importance. All the more so as, in these cases, there is a crucial need not to wait for uncontroversial proofs to impose agreement. Indeed, such proof would not be an achievement, an event opening new possibilities of questions and knowledge production, as is the case in experimental sciences. It would probably mean that it is very late, even too late, to avoid the damage. Doubters themselves will then come to a late retrospective wisdom, as did the research director of the British American Tobacco Company who finally recognized that 'Scientific proof, of course, is not, should not be and never has been the proper basis for legal and political action on social issues. A demand for scientific proof is always a formula for inaction and delay and usually the first reaction of the guilty. The proper basis for such decisions is, of course, quite simply that which is reasonable in the circumstances' (Oreskes and Conway 2010, p. 274).

But there is another aspect of the situation that must be emphasized.

The title of Al Gore's book, *An Inconvenient Truth*, is relevant to designate the field where the industry of doubt very effectively works. Obviously, the merchants of doubt were paid by industries defending their interests. But the scientific claims they attacked are inconvenient for everybody because they make our world appear more dangerous than we would like to think it is. All of us would like the prospect of climate change to vanish, from the humble citizen to governments, and also the tenants of a boundless human development or of a free market economy. We are thus all vulnerable to the temptation to accept that there is room for doubt, that scientists exaggerate and that more research is needed.

In this case, I would claim that citizens must not only resist the temptation but act against it. They must understand that climate scientists today and other proponents of 'inconvenient truths' tomorrow need help and public support. Indeed, we cannot demand from these scientists to bravely face the attacks, to fight back and to publicly denounce the dishonest behaviour of other scientists. They are no heroes. They were not selected for their ability to stand up at the price of facing harassment, malevolent accusations and even vicious personal attacks. Worse, as scientists they are especially badly prepared to publicly defend themselves, as they have been educated to consider that their only true job is knowledge production and that everything else, including protesting against misrepresentation of their work, is a waste of time. Here again a public understanding of the way scientists work is required, together with the connoisseurship needed to identify merchants of doubt and to disqualify them as harshly as we do with racist propagandists or warmongers.

It is time to conclude, and I would conclude my talk with a dream about science learning and science education.

We do not know what science and scientists may be capable of, because what we know is the result of the way scientists are educated, with the message that somehow to take seriously public matters of concern is to demonstrate that one does not have the right stuff for science. The process of transforming this message must already begin at school level. The way science is learned at school matters because it gives an image of science that will attract some and put others off. I am afraid that today those who are attracted are the ones who love well-formulated problems with clean solutions.

I dream of a way to learn science at school that would also attract to scientific education students who take a lively interest in the many entangled, messy situations which scientists encounter, students who would demand that their instruction include what would make them reliable partners in questions of public interest. I also dream that those who do not choose further scientific education understand science in such a way that they feel empowered to assess the way sciences and scientists claim to answer common concerns. In short, I dream of a time when we may be able to trust scientists when they present sciences as serving the best interests of citizens.

References

Clément, P. (2004). Science et idéologie: Exemples en didactique et épistémologie de la biologie. *Actes du Colloque Sciences, médias et société*. ENS-LSH: 53–69 (http:--sciences-medias.ens-lsh.fr).

Clément, P. (2006). *Didactic transposition and KVP model: Conceptions as interactions between scientific knowledge, values and social practices* (pp. 9–18). Braga: ESERA Summer School, IEC, Univ Minho.

Feyerabend, P. (1975). *Against method. Outline of an anarchistic theory of knowledge*. London: NLB.
Haraway, D. (1991). *Simians, cyborgs, and women. The reinvention of nature*. London: Free Association Books.
Latour, B. (2004). Why has critique run out of steam? From matters of fact to matters of concern. *Critical Inquiry, 30*, 225–248.
Le Hir, P. (2004). La société en mal de sciences. 22 Dec 2004 issue of Le Monde.
Lévy-Leblond, J.-M. (1984). *L'Esprit de sel*. Paris: Le Seuil.
Marris, C. (2001). Public views on GMOs: Deconstructing the myths. *EMBO Reports, 2*(7), 545–548.
Oreskes, N., & Conway, E. M. (2010). *Merchants of doubt*. New York: Bloomsbury Press.

Chapter 3
Questions Socialement Vives and Socio-scientific Issues: New Trends of Research to Meet the Training Needs of Postmodern Society

Laurence Simonneaux

1 Introduction

In this chapter, we first present the French framework of *Questions Socialement Vives* in education and show its underpinning links to the risk society and in the field of post-normal science. We develop two essential approaches to address education issues through Socially Acute Questions: a socio-epistemological approach and a psychosocial approach. Secondly, we present the diversity of educational stakes and pedagogies that can cover the teaching of socio-scientific issues as they are 'heated' or 'chilled' in the classroom and then we locate within this panorama the French field of Socially Acute Questions. We present the different epistemological positions that can influence the construction of didactic strategies in the teaching about SAQs. We describe four types of didactic strategies: doctrinal, problematising, critical and pragmatic. We propose that a critical pedagogy can be used to develop students' socio-scientific reasoning in the perspective of sustainability. We describe an analytical grid for this. Finally, we discuss the challenges raised by the implementation of a post-normal education.

L. Simonneaux (✉)
Ecole Nationale de Formation Agronomique – Université de Toulouse,
UMR EFTS, Toulouse, France
e-mail: laurence.simonneaux@educagri.fr

C. Bruguière et al. (eds.), *Topics and Trends in Current Science Education: 9th ESERA Conference Selected Contributions*, Contributions from Science Education Research 1,
DOI 10.1007/978-94-007-7281-6_3, © Springer Science+Business Media Dordrecht 2014

2 Socially Acute Questions and Socio-scientific Issues

2.1 Definition of Socially Acute Questions

Many science educators believe that one of the goals of science education is to help students develop their understanding of how society and science are mutually dependent. The notion of a socio-scientific issue (SSI) has been introduced as a way of describing how social dilemmas impinge on scientific fields (Gayford 2002; Kolsto 2001; Sadler 2004; Sadler and Zeidler 2004; Sadler et al. 2004; Zeidler et al. 2002). These issues are very often controversial and they have implications in one or more of the following fields: biology, sociology, ethics, politics, economics and/or the environment. Sadler (2009) posits that SSIs include two necessary elements: conceptual and/or procedural connections to science and the element of social significance.

Legardez and Simonneaux (2006) coined the term *Questions Socialement Vives* – in English, 'Socially Acute Questions' (SAQs) – to describe complex open-ended questions that bring out the uncertainties embedded in ill-structured problems relating to SSIs. These questions are at the heart of the problem of teaching and learning in an uncertain world influenced by the development of techno-science and by environmental and health crises. These questions situate social and scientific controversy, complexity, the building of expertise, assessment of evidence, uncertainty and risk at the very heart of the teaching-learning process. Because real-world contexts (global or local) are perceived through individuals' identities (linked to gender, culture, political position, profession), SAQs are positioned within the framework of situated learning. In fact SAQs are always controversial as they challenge social practices and reflect social representations and value systems that many in society believe are important to discuss. Consequently, because of their controversial nature, they have the potential for generating debate in classrooms.

SAQs are perceived as 'acute' when they are located in the following three areas:

- In society: They have the potential to stimulate debate – there is often media coverage of such issues and consequently students may have some superficial knowledge.
- In research and professional fields: This is when competing points of view and controversies can lead to debates on the production of reference knowledge within academia – for example, in human sciences, various paradigms can be in conflict and in sciences, this type of question forms part of 'frontier' science where the results are discussed in the scientific community.
- In the classroom: Often, they are perceived to be 'acute' because they are encountered during discussions about society and research. In this situation, teachers often feel unable to deal with them in class discussion as they cannot solely rely on the use of scientific facts and may be afraid that they lack the ability to manage students' reactions. Consequently some teachers choose not to teach them or will neutralise them ('to cool them down'), while others will decide to activate them ('to warm them up'). It seems that teachers will position themselves according to the 'degree of acuteness' they perceive and to the 'teaching risk' they can tolerate.

Because SAQs integrate learning of content knowledge both in science *and* in the humanities, and learning about the construction of knowledge (about humanities *and* science) they provide opportunities for an interdisciplinary approach. In the French context, SAQs have a wider educational span as they integrate Socially Acute Questions in humanities (e.g. globalisation or financial crises) and Socially Acute Questions in science. In this chapter, the latter group is the focus of discussion.

It is apparent that such questions are found more and more frequently in science curricula and are often referred to in the guiding principles of 'education for' programmes, e.g. education for the environment, for sustainability, for health, for (eco-)citizenship, for choice and for consumption. Because of this relevance, Tytler (2012) argues that science education might be, and should be, framed in order to engage students in a science that is relevant and powerful for them as future citizens.

2.2 The Underpinning Links of Socially Acute Questions

The French approach to SAQs emphasises the degree of acuteness of the question in the world of research and/or society. This SAQ approach has a common aim with the science, technology, society (environment) (STS(E)) model (Hodson 2003); it aims for students to be committed to make responsible decisions about SAQs. Although SAQs may contribute to scientific literacy, they also have the potential to develop students' political literacy by including such topics as risk analysis, analysis of patterns of political and economic governance as well as decision-making and action. Even though Zeidler et al. (2005) have provided evidence that SSI education is a better way than the STS movement to integrate the nature of science, arguments, values and moral judgements, Hodson has recently (2011) critiqued both of these approaches and asserts that STS and SSI education have given too low a priority to the promotion of critical thinking. He asserts that neither STSE- nor SSI-oriented teaching goes far enough.

The emphasis on how we try to minimise risk in our society means that SAQs provide an indication of the problems of our risk-aversion society. In his work on the risk society, Beck (1986, 2001) suggests that these days we are emotionally aware of man-made hazards. This does not mean that life is more dangerous to postmodern humanity but that the postmodern society is concerned about the risks posed by techno-responses to past problems. He says that the production of new scientific knowledge has to resolve the multiple impacts (waste, pollution, new diseases) that have been generated by techno-science. By placing uncertainties and risks in the centre of this discussion of philosophy of science, Beck supports a critical approach to scientific rationality. He postulates that institutions, including science, are struggling with the effects of what they have created, and even though they have begun to change, the scientific enterprise cannot be locked into a single theoretical production of scientific knowledge. Instead, he suggests that it is necessary that research anticipate the consequences, uncertainties and risks of scientific advances through what he calls 'reflexive scientification'. Beck comments that science answers questions that are not really raised by society so it ignores the real issues that are the source of its anguish. Furthermore, he claims there is a difference between scientific rationality and social rationality, although he does not deny that there are overlaps and dependencies when scientists are required to comply with social expectations and values when exploring the impact of industrial hazards and their evolution. From his perspective, it appears that the risks associated with SAQs are not generally understood and that unknown risks could generate outcomes that are irreversible, with consequences that are difficult to repair, perhaps affecting the whole world and future generations. Consequently, using Beck's analysis of our postmodern society, it seems that scientific rationality would not be sufficient to justify any techno-science and accompanying such analysis there would lead to a reflexive criticism of its impact.

In our view, SAQs lie within the field of post-normal science (PNS) as defined by Funtowicz and Ravetz (1993), because such a positioning acknowledges its strong links to human needs that involve large uncertainties, major issues and values and require urgent decisions. In fact we assert that the social dimension of this positioning of science is emphasised within PNS. Funtowitcz and Ravetz emphasise that within PNS, the decision process should involve open dialogue with everyone concerned and propose the concept of an 'extended peer community'. According to Ravetz (1997), the question 'what if?' justifies a strong consideration of 'extended facts', that is to say, data from sources outside the dimensions of orthodox research. We believe that it is important to train students to participate within the 'peer extended community' so that different perspectives can be taken into account.

2.3 The Socio-epistemological Approach

It is also important to acknowledge a socio-epistemological approach when examining SAQs. We assert that scientific research, cultural norms, sociopolitical contexts and their applications influence each other. This viewpoint has been backed up by researchers in science studies (Duschl and Grandy 2008; Latour 1999; Nersessian 2008) who have identified the complex, contextual, contingent and cultural representational practices that are used to establish and validate science knowledge.

SAQs also question the foundations of science and the rationalist utopia according to which reason and truth emerge from the confrontation of ideas. Therefore, if we are consistent with Beck's argument (1986/2001), we must go beyond the 'successive attempts to rescue the "underlying rationality" of scientific knowledge' (p. 360) which occurs whenever science is confronted with failure or adverse effects.

We assert that the traditional image of academic science has changed and the dividing line between the sciences and their application is becoming blurred with the acknowledgement of the increasing importance of techno-sciences. In fact trends in science are now criticised as being more and more determined by economic interests, and numerous studies have shown the links that exist between science, politics and business. For example, Salter (1988) uses the term 'mandated science', Ziman (1996) acknowledges these wider dimensions with the term 'post-academic science', and Slaughter and Leslie (1997) call such influences 'academic capitalism'. However, Beck denounces the view that research should be increasingly at the service of an economic project, noting that the sciences have 'entered into a polygamous union with the economy, politics and ethics – or, more precisely, they live in a sort of "sustainable cohabitation" with all of these areas' (Beck 2001, p. 53). Ziman recognises these tensions when he asserts that (1998, p. 1813): 'Universities and research institutes are no longer deemed to be devoted entirely to the pursuit of knowledge for its own sake. They are encouraged to seek industrial funding for commissioned research, and to exploit to the full any patentable discoveries made by their academic staffs, especially when there is a smell of commercial profit in the air'. Hodson (2011, p. 126) is also aware of such influences when he notes that 'The vested interests of the military and commercial sponsors of research (...) can often be detected not just in research priorities but also in research design, especially in terms of what and how data are collected, manipulated and presented. More subtly, in what data are not collected, what findings are omitted from reports and whose voices are silenced'. Acknowledging the difficulty in trying to distinguish science from its technological applications, Ravetz (1997, p. 7) sees science as 'a total system including research, R&D and innovation'. He asserts that it 'is primarily valued for its contribution to industry and economic growth; and it is from just those forces that our major environmental problems arise. To enforce an assumption that the scientific tools should be studied in isolation from their practical uses, is to create an innocence that is artificial, temporary and ultimately false'. These writers seem to agree that the domination of capital over science does not weaken as knowledge, and nature itself, are seen as 'goods' and are turned into saleable and purchasable things. This raises the question of the moral responsibility for the use of scientific applications. Who is responsible? Society? Scientists? Technologists? The state? Ravetz (1975, p. 46) raises this issue in his own way: 'Scientists take credit for penicillin, but Society takes the blame for the Bomb'.

We consider that many different actors take part in knowledge production. These include scientists, citizens, philosophers and even whistle-blowers. Consequently, we assert that the knowledge involved in SAQs can be conceived as plural (polyparadigmatic) and/or engaged (analysing the controversies, uncertainties and risks) and/or contextualised (observing empirical data within a given context) and/or distributed (constructed by different knowledge producers).

The socio-epistemological inquiry into the SAQs is not easy to carry out; Latour talks about constructing a cartography of controversy. This process is always subjective and involves continuous updating. However, such a construction is not simple. How does one conduct a socio-epistemological inquiry? How does one know when and where to temporarily close it off? How does one assess the expertise, the 'evidence', the risks? How does one identify the uncertainties?

Agnotology (the science of ignorance) has emerged (Proctor and Schiebinger 2008); this discipline studies the techniques used by certain economic interests to cast doubt on well-established knowledge. This area is connected with the industry of doubt that was presented at this conference by Isabelle Stengers, when discussing the issue of creationism.

Then a question may be raised about SSIs/SAQs: do they develop the critical study of controversy or do they contribute to the spread of ignorance? And how can one contribute to the former instead of to the latter?

2.4 The Psychosocial Approach

Given the nature of SAQs, it is also necessary to analyse the psychosociological factors that determine the positions and behaviour of the actors involved in the activity. We assert that learning through SAQs has affective and social dimensions. For example, one learns from the points of view of one's social groups and from one's identities. Consequently we assert that one's perspectives are influenced by value systems, cultural and socio-professional identities, perceptions of norms, cultural bias in particular concerning risk perceptions (Douglas 1992) and future projections.

We note that exploring futures with students about SAQs can sometimes be counterproductive and can lead to feelings of despair. On the other hand, exploring alternative futures also helps students 'to clarify their hopes and fears about the future in order to move beyond passive forecasting about "how it is likely to be" to the generation of ideas about the sort of future they want in the basis for planning and action' (Hodson 2011, p. 159).

Also it is important to be able to analyse the impact of emotional points of view. Many studies have been made from this perspective, in particular Sadler and Zeidler (2008), who emphasised the important role of moral reasoning. They note that decisions related to SSIs can reflect the moral principles and qualities of virtue in learners. They also emphasise the role of education in the formation of conscience and the development of character. They assert that emotions and moral-ethical reasoning play an important role when taking stances and action on SSIs.

As already noted, context and identity are of importance when studying SAQs and various contextualisation impacts linked to the imprinting of values on learning have been identified.

For example, if the context presented to the students contradicts their system of values, it can hinder knowledge learning and critical reasoning, blind the participants to the issues as well as build a resistance to changing their minds. However, if the context allows them to defend their sociocultural positions, it stimulates their critical analysis (Simonneaux and Simonneaux 2009).

3 Curriculum Orientations: To 'Cool Down' or to 'Heat Up' the Questions

3.1 Diversity of Educational Stakes and Pedagogies

In the literature on the teaching of SSIs, we can observe very different objectives. There are many different dimensions to the concept of an SSI. Similarly there is variation in the extent to which teachers 'heat up' or 'cool down' these issues.

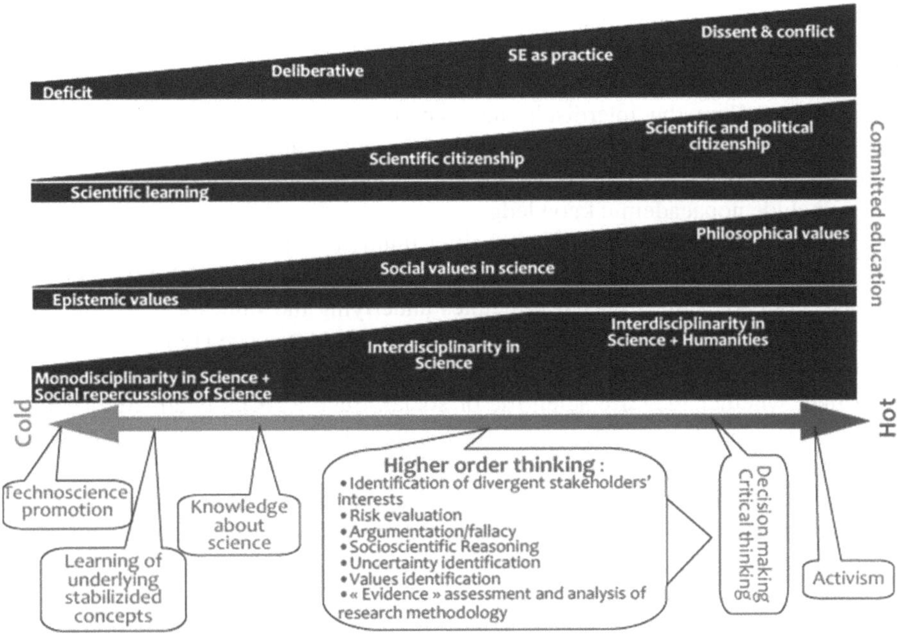

Fig. 3.1 Educational stakes and pedagogies beyond SSI education

There are different ways of teaching SSIs according to the teacher's view of the main educational stakes. In an attempt to map out the landscape, these different dimensions are represented as continuums in Fig. 3.1.

At the 'cold end', an integration of SSIs into a teaching programme is used to motivate students learning science or even to convince them of the merits of techno-sciences. At the 'hot end' of the continuum, the teaching focus goes beyond the purpose of developing conceptual and procedural knowledge of science to the nurturing of activist commitments amongst learners. Between these two ends of the continuum, a mix of educational issues is involved in the teaching and learning of scientific concepts that can contribute towards the development of critical thinking. When critical thinking occurs, the focus moves towards the 'hot end'. In the current field of French education, the educational stakes are high and it is asserted that SAQs can develop high-level thinking, decision-making and critical thinking with a focus on promoting an engaged citizenship.

At the 'cold end', knowledge mobilised in the classroom is single-disciplinary science. At the 'hot end', it is discussed on an interdisciplinary basis in science and humanities. Between the two ends, it is interdisciplinary science.

We assert that the study of SAQs forces education to transcend disciplinary divisions, particularly between 'hard' science and human science. When examining (techno-)sciences, we have realised that many characteristics go beyond the boundaries of the disciplinary divisions and these divisions are as much the result of a social construction as of epistemological specificities. With a French SAQ approach, we argue for real interdisciplinarity where science and humanities are integrated in order to account for the complexity of the reality linked to SAQs. This interdisciplinary approach is also advocated by Hodson (2011). Recently Eastwood et al. (2011) have described and analysed an ambitious 4-year interdisciplinary university programme. Here the interdisciplinary approach goes beyond the single social impact, or the impact of values, or even the impact of culture. The approach requires hybridisation of knowledge between the humanities and natural sciences and may often include nonacademic knowledge.

We assert that values may be explicit or implicit in the teaching of SSIs. At the 'cold end', only epistemic values may be mobilised (validity, reliability, accuracy). At the 'hot end', philosophical principles underlying the values are explained and discussed. Between the two ends, the social values in science (Longino 2002) are identified and acknowledged. In fact in the French field of SAQs, values must be clarified, whether they are scientific or social. Such a focus is aimed at helping students to identify the values of different stakeholders, as well as their own, in their decision-making.

Beyond science learning, the challenge may be to develop scientific citizenship and even political citizenship so that teaching about SAQs may lead to the combination of both science education and political education – and thus the development of scientific, and even political, literacy.

Levinson (2010) has identified a number of democratic participation frameworks that can be used in the teaching of SSIs. For example, within the *deficit framework*, he observes that:

- Students construct their socially relevant scientific knowledge with the help of the teacher – the hierarchy is scientist–teacher–student.
- Science is the corpus of knowledge; where there are uncertainties and tentative knowledge, this resides in expert authority – 'hard' science diffuses out into applied science.
- Students and laypeople are unlikely to have the requisite knowledge and understanding to engage in controversial issues.
- In addition to science content, they can be taught about the methods of science and controversies both within the scientific and socio-scientific domains – consequently, authority of knowledge resides within science and the teacher as science's representative.
- Once citizens become aware of the technical complexity, there will be an enhancement of trust and confidence in expert decision-making.
- Ethical and political aspects and value differences have only a marginal role.

Whereas in the *deliberative framework:*

- 'Jurors' are informed of technical information by scientists and hear evidence from a range of non-professional experts from interest groups.
- Rhetoric, argument and testimony are the main forms of presentation.
- Techno-science is uncertain and constrained by value positions – understanding of knowledge claims, critical thinking skills and underpinning empathy are likely to be prerequisites.
- Information about science is likely to come from expert scientist(s) rather than from the teacher.

While in the *science education as praxis framework:*

- Knowledge is distributed and emerges through praxis; knowledge is both situated and emergent – through legitimate participation, participants become inducted into more sophisticated and shared techniques of problem solving.
- Scientific knowledge is contestable and open to participant reflexivity;
- All participants, scientists and nonscientists, subject their views to communal questioning and reflection.

In the *science education for democracy through conflict and dissent framework:*

- Knowledge is distributed and emerges through praxis.
- The emphasis is on political literacy, identifying and analysing the sources of social injustice and both using and producing knowledge to address techno-scientific issues related to injustice.
- The emphasis is on campaigning and activism through scientific citizenship for social and political change.

Taking all of these views into account, we should not consider the continuums presented in Fig. 3.1 as if they coexisted; instead, these provide a visual summary of trends towards a more engaged education. It is interesting to put them into perspective with the four levels of sophistication for an issue-based education as described by Hodson (2010):

- Level 1: appreciating the societal impact of scientific and technological change and recognising that science and technology are, to a significant extent, culturally determined
- Level 2: recognising that decisions about scientific and technological development are taken in pursuit of particular interests and that benefits accruing to some may be at the expense of others; recognising that scientific and technological development are inextricably linked with the distribution of wealth and power
- Level 3: developing one's own views and establishing one's underlying value positions
- Level 4: preparing for and taking action on socio-scientific and environmental issues

3.2 Epistemological Stances

The diversity of functions that the actors in the education system attribute to science reveals their epistemological stances, and Simonneaux (2011) has established four categories, based on previous work in epistemology and in the sociology of science, which can be used to describe them. These are:

- The *scientistic* stance, which is inspired by Ernest Renan (1890). Here, science is considered to be essential to progress, and disciplinary and academic construction is the basis of this posture. The confidence placed in the scientific approach contributes to the sacralisation of science, where the researcher is the essential actor. The disciplinary content, and the way science is divided up, constitutes the basis of all learning from a hierarchical viewpoint. This view is widespread in schools and academic institutions, where learning is delivered by the teacher and disciplinary expert downwards to the student. The teaching of the agronomic and economic principles of the Green Revolution took on this stance. The Green Revolution refers to a strategy designed to transform agriculture in developing countries, which is based mainly on the principle of developing intensive farming methods and the use of high-yielding varieties of cereal grains. In the scientistic posture, new scientific knowledge in the field of plant breeding has led to technological innovation and progress.

- *Utilitarianism* constitutes a second stance. It can be defined as referring to the utilitarianism of John Stuart Mill. In this case, knowledge takes on meaning through the actions that it helps to produce. Here the operational dimension is paramount and the value of knowledge lies in the power to act on reality. Knowledge is considered from a production angle and is viewed as a resource. The institutions in which this knowledge is transmitted (a business, the market, a vocational training centre) have a function connected to production. The expert, engineer or administrator, who makes the right decisions, is the emblematic figure of this stance which fosters innovation. The teaching of precision agriculture is an illustration of this stance. Precision farming aims to optimise the management of a plot of land from an agronomic point of view (e.g. by adapting cultivation methods as closely as possible to the nitrogen requirements of the plant), from an environmental point of view (e.g. by limiting the leaching of excess nitrogen) and from an economic point of view (e.g. by increasing competitiveness through better management of nitrogen fertilisers).
- *Scepticism* constitutes a third stance, which can be linked to the works of Habermas (1987), Beck (1986) and Bourdieu et al. (1968). Events, with a more or less catastrophic impact in relation to the techno-sciences, have shaken our confidence in scientific progress and the gap between scientists and society is widening. The sciences produce breakthroughs but also controversies and risks. The scientists' questions and doubts are no longer confined to research but fuel public debate and are relayed by the media and citizens' associations. A person assuming this stance believes that scientific research is guided by political and economic choices. The educational intentions underlying this stance mean that the teacher will aim to promote citizenship training and critical thinking. Teaching on the potential environmental risks of producing insect-resistant GMOs illustrates this epistemological stance.
- *Relativism*, the fourth stance, finds its reference in the work of Feyerabend (1979), who considers that science cannot proclaim itself to be a superior form of knowledge because no universally validated method can be attributed to the sciences. It thus becomes difficult, or even impossible, to distinguish a scientific approach from any other belief or myth. Teaching the anthroposophist principles of biodynamics (a method used in organic farming) is a relativist stance. Anthroposophy is a school of thought and a form of spirituality founded in the early twentieth century by Rudolf Steiner. According to him, it is a spiritual science, an attempt at studying, experiencing and describing spiritual phenomena

with the same precision and clarity with which science both studies and describes the physical world. The use of the term 'science' in connection with this approach has been disputed by proponents of the scientific method.

The characteristics of these four stances are summarised in Table 3.1.

We argue that these epistemological stances impact the construction of didactic strategies in education about SAQs and will be explained in the next section.

3.3 Didactic Strategies

Didactic strategies reflect the stakes chosen by the school or by the teachers. Simonneaux (2011) identified four possible didactic strategies:

- A *doctrinal* strategy that aims to develop the acceptance of the ideas presented in the high authority of the teacher, who leaves little room for interaction with students.
- A *problematising* strategy that focuses on students' cognitive activity – here, students take an active part in the construction of an *issue* and develop a line of reasoning rather than finding *the* solution.
- A *critical* strategy aims to develop a critical sense – here, the educational purpose is to teach students how to argue and to assess expertise and different stances on complex issues which carry both uncertainties and risks.
- A *pragmatic* strategy is based on involving the students in an activity – here, the challenge is to stimulate student *action.*

Of course, several strategies can be used within the same teaching situation. Generally, there will be a dominant strategy, for instance critical and problematising strategies, which can bolster a dominant pragmatic strategy.

We argue for precognition, that is to say, developing a pedagogy which engages students into a knowledge inquiry, instead of retrocognition, where the traditional pedagogy is based on stabilised scientific 'works' and conceptual development that uses conventional teaching methods (Ladage and Chevallard 2010). As a consequence, we argue for a strategy that develops a critical rationality beyond a techno-scientist rationality and we believe that there are at least two key issues in the didactics of SAQs that should encourage critical analysis from the perspective of scientific citizenship: they are ethical reflection and epistemological questioning.

Several pedagogical strategies may be used to develop critical rationality. They include debate, role play, simulation of a citizens' conference, epistemological disturbance, an identification of a contextualised problem situation, analysis of media coverage, analysis of local projects, online cross-cultural cooperative work on local and global issues and the development of serious games. We have used all these types of pedagogical strategy and analysed them using various theoretical frameworks (Habermas 1987; Boltanski and Thévenot 1991; Moscovici 1976; Beck 1986; Douglas 1992; Simonneaux et al. 2012; Simonneaux 2001; Simonneaux and Chouchane 2011; Morin et al. 2011; Vidal and Simonneaux 2010).

Table 3.1 Characteristics of epistemological stances

	Scientism	Utilitarianism	Scepticism	Relativism
Relationship to the sciences	Sacralisation	Relevance of science is linked to its effects on the world	Science produces knowledge, controversies and risks	All types of thought deserve attention and can be uttered
Goals attributed to sciences	Progress, rationality	Advice, help with decision-making, innovation, development	Understanding the world but scientific reflexivity necessary	An understanding of the world amongst others
Authors	Renan	John Stuart Mill	Habermas, Beck, Bourdieu	Feyerabend
Institutions	Schools, education authorities	Business, market, vocational training centres	Citizens' associations, media	
Chosen communication model	Hierarchical and academic teaching	Innovation model, expertise	Civic debate, scientific café	All types
Examples of application in the field of agriculture	Green Revolution	Precision agriculture	Environmental and sanitary repercussions of GMOs	Biodynamics

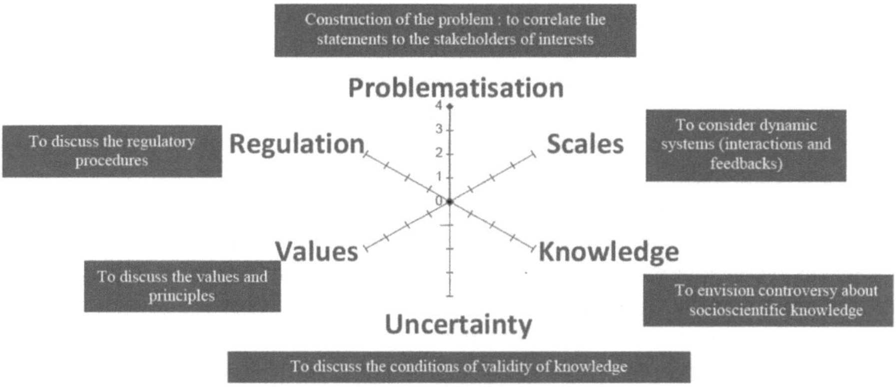

Fig. 3.2 Six dimensions of the SSR set within the perspective of sustainability

Within this perspective of procognition, we have tried to develop students' socio-scientific reasoning (SSR). We assert that context and identity have a great impact on SSR. Various studies have contributed to a characterisation of SSR, including *The Quality Gradient of Argument* (Grace 2009), *Informal Reasoning* (Sadler and Zeidler 2005), *The Development of Reflective Judgement* (Zeidler et al. 2009), *Socio-Scientific Reasoning* (Sadler et al. 2007) and *SSR and Sustainability* (Saoudi 2009; Morin and Simonneaux 2011).

At this point, the question we must ask is: are there patterns of reasoning broadly applicable to all or to most SSIs/SAQs? Sadler et al. (2007) developed a model of socio-scientific reasoning as a means of understanding student practices relative to the invariant features of SSIs that could assess these practices. They hypothesised that after negotiating specific SSIs, students would have the knowledge and skills to respond to the implications of the invariant features of diverse SSIs. But their results (Sadler et al. 2011) and ours (Simonneaux and Simonneaux 2009) do not confirm this hypothesis. We think that context and identity dimensions impact on SSR differently, depending on the issues under debate.

To support our viewpoint, it is important to note that our grid for the analysis of SSR within the perspective of sustainability (see Fig. 3.2) has been elaborated progressively and can be used to compare the SSR of different types of public, in different contexts and in different cultures. It can even be used to assess the evolution of SSR after an activity. In its latest (but still provisional) version, the grid is constructed in six dimensions (problematisation, scale, knowledge, uncertainty, value, regulation) and each one is divided into four levels and is explained as follows:

Within the problematisation dimension, points of view are identified as ranging from a person seeing only one point of view or everyone having the same point of view to persons who are capable of seeing that there are different points of view and are able to present these viewpoints from differing perspectives. The highest level of the problematisation dimension is the perception of a controversy about the stakes and being able of put forward perspectives through the various assertions that are reflected by each stakeholder's interest.

The levels of scale range from those people that see a limited effect – both spatial (local or perhaps generally or globally) and temporal (over a short period) – to those who see a complex effect of time and environment interacting in ways that involve a feedback system. The highest level of the scales' dimension is the conception of dynamic systems (integrating spatial interactions with diverse scales as well as temporal feedback).

In the knowledge dimension, viewpoints range from a point of view that is naïve, where knowledge is seen as unproblematic (it may be academic or not), to points of view which consider that multiple knowledge systems can contribute to the analysis of a socio-scientific problem. The highest level of the knowledge dimension is the integration of different types of knowledge (academic or not) in science and humanities, and the perception of controversy about socio-scientific knowledge.

In the uncertainty dimension, the viewpoints range from those who have no problem with the information and consider it to be the truth to those who hold an epistemic view of knowledge building and apply this to knowledge used in the resolution of socio-scientific issues. At the highest level is discussion of the conditions for validity of the reference knowledge that is an awareness of epistemological doubt.

In the values dimension, the range includes those who are unaware that arguments can be underpinned by different values to those who are aware of different value systems and the potential conflict between them. At the highest level is the perception of possible conflicts within these values.

The perspective of regulation ranges from a view that there is already a solution (based on law or ethics or accepted techno-scientifically) or one which does not consider the possibility of any other viewpoints and interactions to those who are aware that regulation must take into account all the stakeholders' positions. At the highest level is discussion of the regulation procedures between the categories of actors or the governance (the modalities of decision-making).

The impact of various didactic strategies on the SSR has been analysed in various studies. In our different studies, we noticed that students' commitment to reasoning was linked:

- With their rationality (scientific, social or techno-scientific) – the more they expressed techno-scientific rationality, the less they developed sophisticated reasoning
- With their personal conviction (environmental in the field of education for sustainability, ethical in the field of health)
- With their epistemological position (expressing doubt or 'blind' confidence in science)

4 Challenges for Future Post-normal Education

What knowledge will be relevant to future generations?

Is there a process of desacralisation of a more humanistic science?

Our research shows that there is a need to highlight successful practices, to provide more teacher professional development and to encourage more cross-curricular and school-based prioritisation of the value of SSIs as central to a contemporary curriculum (Prain 2012).

Whether coming from epistemological, sociological or anthropological areas, more and more voices are being raised to highlight the way in which social and economic factors interweave with scientific activity (Beck, Callon, Lakatos, Latour, Stengers). We acknowledge that this shift in perspective initially came from post-normal science (Funtowicz and Ravetz 1993) which legitimises post-normal education. It is significant to re-emphasise that this is where we situate SAQs.

We continue to advocate that critical reflexivity is essential to citizenship education. Similarly to Levinson (2010), we believe that it is the educational centrepiece of scientific literacy which can lead to political literacy (Levinson 2010). Such a purpose emphasises the need to move towards a 'critical-constructivist' pedagogical paradigm (Tutiaux-Guillon 2008). Here, the question of knowledge should be the central question for society and we cannot ignore its social and political dimensions. Once we can recognise how science works, we can consider other purposes and other forms of teaching science. Because we believe that teaching SAQs requires going beyond disciplinary divisions, we argue that the questioning of traditional didactic methods could provide a way of showing how to make the transition towards post-normal education.

In short, when implementing post-normal education, it is not sufficient just to learn and to understand. Instead, the central purpose is to encourage participation and action in the scientific activity. We realise that science has a limited and temporary validity which is marked by social interactions; therefore, we need to consider using a different model to develop understanding rather than the downward transmission that is normally practised. Instead we need to develop a model that gives priority to more active and participatory methods.

Overall the goal of a didactic for SAQ is to get the students to take an active part in the scientific process. This is a process which cannot be separated from the social process.

Acknowledgements The author would like to thank Bev France (University of Auckland) and Jenny Lewis (University of Leeds) for their help in the translation.

References

Beck, U. (1986). *La société du risque, sur la voie d'une autre modernité*. (Flammarion, Paris, 2001 French translation).

Boltanski, L., & Thévenot, L. (1991). *De la justification – Les économies de la grandeur*. Paris: Gallimard.

Bourdieu, P., Chamboredon, J.-Cl., & Passeron, J.-Cl. (1968): *Le métier de sociologue* (5th ed., 2005). Paris: Mouton de Gruyter.

Douglas, M. (1992). *Risk and blame: Essays in cultural theory*. London: Routledge.

Duschl, R., & Grandy, R. (Eds.). (2008). *Teaching scientific inquiry: Recommendations for research and implementation*. Rotterdam: Sense Publishers.

Eastwood, J. L., Schlegel, W. M., & Cook, K. L. (2011). Effects of an interdisciplinary program on students' reasoning with socio-scientific issues and perceptions of their learning experiences. In T. D. Sadler (Ed.), *Socio-scientific issues in the classroom – teaching, learning and research* (pp. 89–126). Dordrecht/Heidelberg/London/New York: Springer.

Feyerabend, P. (1979). *Contre la méthode*. (Translation: *Against Method* (1975)). Paris: Éditions du Seuil.

Funtowicz, S. O., & Ravetz, J. R. (1993). Science for the post-normal age. *Futures, 25*(7), 739–755.

Gayford, C. (2002). Controversial environmental issues: A case study for the professional development of science teachers. *International Journal of Science Education, 24*(11), 1191–1200.

Grace, M. (2009). Developing high quality decision-making discussions about biological conservation in a normal classroom setting. *International Journal of Science Education, 31*(4), 551–570.

Habermas, J. (1987). *Théorie de l'agir communicationnel, Tome 1: Rationalité de l'agir et rationalisation de la société*. Paris: Fayard.

Hodson, D. (2003). Time for action: Science education for an alternative future. *International Journal of Science Education, 42*(6), 645–670.

Hodson, D. (2010). Science education as a call for action. *Canadian Journal of Science, Mathematics, and Technology Education, 10*(3), 197–206.

Hodson, D. (2011). *Looking to the future – building a curriculum for social activism*. Rotterdam/ Boston/Taipei: Sense Publishers.

Kolsto, S. D. (2001). To trust or not to trust… pupils' ways of judging information encountered in a socio-scientific issue. *International Journal of Science Education, 23*, 877–901.

Ladage, C., & Chevallard, Y. (2010). *La pédagogie de l'enquête dans l'éducation au développement durable*. (Paper presented « *Éducation au développement durable et à la biodiversité* », IUT de Provence, Digne les Bains, 20–22 October 2010).

Latour, B. (1999). *Pandora's hope: Essays on the reality of science studies*. Cambridge: Harvard University Press.

Legardez, A., & Simonneaux, L. (2006). *L'école à l'épreuve de l'actualité – Enseigner les questions vives*. ESF, Issy-les-Moulineaux.

Levinson, R. (2010). Science education and democratic participation: An uneasy congruence? *Studies in Science Education, 46*(1), 69–119.

Longino, H. (2002). *The fate of knowledge*. Princeton: Princeton University Press.

Morin, O., & Simonneaux, L. (2011). Aborder des Questions Environnementales Socialement Vives pour une Éducation au Développement Durable: La grille PREVIS pour suivre les acquisitions. In L. Simonneaux & A. Legardez (Eds.), *Développement durable et autres questions d'actualité. Les questions socialement vives dans l'enseignement et la formation* (pp. 129–144). Dijon: Educagri Editions.

Morin, O., Simonneaux, L., Tytler, R., & Simonneaux, J. (2011). *A Framework for Considering Cross-Cultural Exchanges as a Way to Develop Reasoning about SSIs*. Lyon: ESERA.

Moscovici, S. (1976). Psychologie des représentations sociales. *Cahiers Vilfredo Pareto, 14*, 409–416.

Nersessian, N. (2008). Model-based reasoning in scientific practice. In R. Duschl & R. Grandy (Eds.), *Teaching scientific inquiry: Recommendations for research and implementation* (pp. 57–79). Rotterdam: Sense Publishers.

Prain, V. (2012). Acting on sustainability. *Research in Science Education, 42*(1), 149–154.

Proctor, R. N., & Schiebinger, L. (Eds.). (2008). *Agnotology. The making and unmaking of ignorance*. Stanford: Stanford University Press.

Ravetz, J. R. (1975). …et augebitur scientia. In R. Harré (Ed.), *Problems of scientific revolution. Progress and obstacles to progress in the sciences* (pp. 42–57). Oxford: Clarendon.

Ravetz, J. R. (1997). Simple scientific truths and uncertain policy realities. *Studies in Science Education, 30*(1), 5–18.

Renan, E. (1890). *L'Avenir de la science*. Paris: Garnier-Flammarion (1995 edition).

Sadler, T. D. (2004). Informal reasoning regarding socio-scientific issues: A critical review of research. *Journal of Research in Science Teaching, 41*(5), 513–536.
Sadler, T. D. (2009). Situated learning in science education: Socio-scientific issues as contexts for practice. *Studies in Science Education, 45*(1), 1–42.
Sadler, T. D., & Zeidler, D. L. (2004). The morality of socio-scientific issues: Construal and resolution of genetic engineering dilemmas. *Science Education, 88*, 4–27.
Sadler, T. D., & Zeidler, D. L. (2005). Patterns of informal reasoning in the context of socio-scientific decision-making. *Journal of Research in Science Teaching, 42*(1), 112–138.
Sadler, T. D., Chambers, F. W., & Zeidler, D. L. (2004). Student conceptualisations of the nature of science in response to a socio-scientific issue. *International Journal of Science Education, 26*(4), 387–410.
Sadler, T. D., Barab, S. A., & Scott, B. (2007). What do students gain by engaging in socio-scientific inquiry? *Research in Science Education, 37*, 371–391.
Sadler, T. D., Klosterman, M. L., & Topcu, M. S. (2011). Learning science content and socio-scientific reasoning through classroom explorations of global climate change. In T. D. Sadler (Ed.), *Socio-scientific issues in the classroom – teaching, learning and research* (pp. 45–78). Dordrecht: Springer.
Salter, L. (1988). *Mandated science: Science and scientists in the making of standards*. Boston: Kluwer.
Saoudi, H. (2009). *Etude de la relation biodiversité-transgénèse végétale à l'université selon une double perspective: socioscientifique et durable. Contribution à l'éducation à l'environnement pour le développement durable. Thèse en Cotutelle avec l'Université de Toulouse-Le Mirail et l'Université de Tunis.*
Simonneaux, L. (2001). Role play or debate to promote students' argumentation and justification on an issue in animal transgenesis. *International Journal of Science Education, 23*(9), 903–928.
Simonneaux, J. (2011). *Les configurations didactiques des questions socialement vives économiques et sociales, Habilitation à diriger des recherches.* Aix en Provence: Université de Provence.
Simonneaux, L., & Chouchane, H. (2011). The reasoned arguments of a group of future biotechnology technicians on a controversial socio-scientific issue: Human gene therapy. *Journal of Biological Education, 45*(3), 150–157.
Simonneaux, L., & Simonneaux, J. (2009). Students' socio-scientific reasoning on controversies from the viewpoint of education for sustainable development. *Cultural Studies of Science Education, 4*(3), 657–687.
Simonneaux, L., Panissal, N., & Brossais, E. (2012). *Students' perception of risk about nanotechnology after an SAQ teaching strategy*. International Journal of Science Education. doi:10.1080/09500693.2011.635164.
Slaughter, S., & Leslie, L. L. (1997). *Academic capitalism: Politics, policies, and the entrepreneurial university*. Baltimore: The Johns Hopkins University Press.
Tutiaux-Guillon, N. (2008). Interpréter la stabilité d'une discipline scolaire: l'histoire-Géographie dans le secondaire français. In F. Audigier & N. Tutiaux-Guillon (Eds.), *Compétences et contenus* (pp. 117–146). Bruxelles: De Boeck Université.
Tytler, R. (2012). Socio-scientific issues, sustainability and science education. *Research in Science Education, 42*(1), 155–163.
Vidal, M., & Simonneaux, L. (2010). *Using companion modelling on authentic territories in the teaching of biodiversity*. Braga: ERIDOB.
Zeidler, D. L., & Sadler, T. D. (2008). The role of moral reasoning in argumentation: Conscience, character, and care. In S. Erduran & M. P. Jimenez-Aleixandre (Eds.), *Argumentation in science education – perspectives from classroom-based research* (pp. 201–216). Dordrecht: Springer.
Zeidler, D. L., Walker, K., Ackett, W., & Simmons, M. (2002). Tangled up in views: Beliefs in the nature of science and responses to socio-scientific dilemmas. *Science Education, 27*, 771–783.

Zeidler, D. L., Sadler, T. D., Simmons, M. L., & Howes, E. V. (2005). Beyond STS: A research-based framework for socio-scientific issues education. *Science Education, 89*, 357–377.

Zeidler, D. L., Sadler, T. D., Applebaum, S., & Callahan, B. E. (2009). Advancing reflective judgement through socio-scientific issues. *Journal of Research in Science Teaching, 46*(1), 74–101.

Ziman, J. (1996). Post-academic science: Constructing knowledge with networks and norms. *Science Studies, 9*(1), 67–80.

Ziman, J. (1998). Essays on science and society: Why must scientists become more ethically sensitive than they used to be? *Science, 282*(5395), 1813–1814.

Chapter 4
Teachers' Beliefs, Classroom Practices and Professional Development Towards Socio-scientific Issues

Virginie Albe, Catherine Barrué, Larry Bencze, Anne Kristine Byhring, Lyn Carter, Marcus Grace, Erik Knain, Dankert Kolstø, Pedro Reis, and Erin Sperling

1 Introduction

During the recent past, efforts have been devoted in several countries to introduce socio-scientific issues (SSIs) in science curricula in an attempt to democratise science in society and promote scientific literacy for all: global climate change and energy systems, genetically modified food, nanotechnologies, gene therapy, pharmaceuticals, etc. Associated with such concerns are debates, for example, about

V. Albe (✉) • C. Barrué
STEF, Ecole Normale Supérieure de Cachan, Cachan, France
e-mail: virginie.albe@stef.ens-cachan.fr; catherine.barrue@ac-rennes.fr

L. Bencze • E. Sperling
Department of Curriculum, Teaching and Learning, Ontario Institute for Studies in Education (OISE), University of Toronto, Toronto, Canada
e-mail: larry.bencze@utoronto.ca; sperlimail@gmail.com

A.K. Byhring • E. Knain
Norwegian University of Life Sciences, Ås, Norway
e-mail: Anne.Kristine.Byhring@umb.no; erik.knain@umb.no

L. Carter
Trescowthick School of Education, Australian Catholic University, Banyo, Australia
e-mail: l.carter@patrick.acu.edu.au

M. Grace
Southampton Education School, University of Southampton, Southampton, UK
e-mail: M.M.Grace@soton.ac.uk

D. Kolstø
Department of Physics and Technology, University of Bergen, Bergen, Norway
e-mail: stein.dankert.kolstoe@ift.uib.no

P. Reis
Instituto de Educação, Universidade de Lisboa, Lisbon, Portugal
e-mail: preis@ie.ul.pt

C. Bruguière et al. (eds.), *Topics and Trends in Current Science Education: 9th ESERA Conference Selected Contributions*, Contributions from Science Education Research 1,
DOI 10.1007/978-94-007-7281-6_4,

regulation of corporate actions. The educational aim is to prepare students for active participation in society. Empirical research has been performed on pupils' and students' reasoning, concept learning, argumentation processes and decision-making (for literature review, see Sadler 2004, 2009). The design and analysis of teaching sequences and to a lesser extent the intentions of science teachers for teaching such socio-scientific controversies have also been under investigation (Lee et al. 2006; Sadler et al. 2006). As they have an essential role in supporting the development of young people's ability to engage with such issues, science teachers' viewpoints and teaching practices on SSIs also need to be investigated. A key agent of any reform is likely to be the teacher, whose instruction may be influenced by her/his general perspectives and awareness of fruitful teaching strategies. Helms (1998) has noted, for instance, that teachers tend to emphasise teaching and learning of 'content' (e.g. laws and theories) because science 'subject matter' tends to be highly integrated into their professional identities. Consequently, they may not have strong identities with SSI activities to support citizenship education and research-informed activism.

2 Rationale: Teachers' Commitments to SSI Activities

Researchers and national curricula call for the inclusion of SSI teaching in science education because of its potential for creating a more real, humane image of scientific activity and for promoting competencies essential to active and responsible citizenship (Kolstø 2001; Millar and Hunt 2002). They consider that the understanding of what science is and how it is produced is crucial for citizens' participation and involvement in the evaluation of science and technology: a key element of democratic societies. Several studies have shown the usefulness of classroom discussion of SSIs both in terms of science learning (its contents, processes and nature) and in terms of the students' cognitive, social, political, moral and ethical development (Sadler 2004). Despite mandates for it and corresponding research and development efforts, SSI education in school science remains relatively marginalised. Only some science teachers implement these activities, even when the SSIs comprise the curricular content. The discussion of SSIs in schools has been found to depend on several elements: (a) teachers' management skills related to classroom discussions; (b) teachers' knowledge about the nature of science and the sociological, political, ethical and economic aspects of those issues; (c) teachers' conceptions regarding science and science education; and (d) evaluation systems that value the discussion of SSIs (Levinson and Turner 2001; Newton 1999; Reis and Galvão 2009).

School science systems tend to didactically emphasise instruction in products of science (e.g. laws and theories). Such approaches can limit students' exposure to contentious issues (e.g. *corporatism*; Carter 2005) that might discredit fields of professional science (e.g. Hodson 2008). Where there *is* attention in school science systems to SSIs, it often appears to be limited to asking students to negotiate contentious issues and to develop arguments to defend their positions on them. It seems much rarer to ask students to take *action(s)* – such as petitions to power brokers – to address issues (e.g. Lester et al. 2006). There are numerous reasons to support development of a more activist citizenry through school science. Many suggest, for example, that potential serious harm to individuals, societies and environments is sufficient warrant (e.g. Hodson 2003). It also may be that students' understanding of and argumentation surrounding SSIs may be deepest when they enact plans of action to address them, because deep/committed learning appears to require close, reciprocal relationships between *phenomena* (e.g. toxins in foods) and *representations* (e.g. lobbying letters problematising toxins in foods) of them (Wenger 1998). There are, undoubtedly, a myriad of factors to consider when imagining school reform like greater student-led, research-based, sociopolitical activism. SSIs involve concepts and arguments from diverse fields of knowledge, several stakeholders with different perspectives and normally also uncertain knowledge and value judgments. Participating in discussions on SSIs involves process or methodological competencies: claims, arguments and values need to be articulated, discussed, critically examined, assessed and applied in arguments and decision-making. Moreover, ideally participants should to be able to develop increased insight through reading, knowledge building and participation in community efforts to contribute to knowledge development (Roth and Lee 2004). This combination of conceptual knowledge and application of these in diverse processes and diverse arenas of debate implies that fruitful dealing with SSIs is a complex competence. A key concern for SSI teaching is to develop students' ability to cope with this complexity. Interestingly, the different processes and competencies involved overlap significantly with the practices and competencies sought, developed and advanced in inquiry-based science teaching (McVaugh 2010). As inquiry-based methods are not restricted only to using first-hand empirical evidence as data, it sometimes involves the collection and analysis of evidence, claims and arguments put forward by others (e.g. by different stakeholders in debates on SSIs). Moreover, IBST also seeks to develop a willingness to question and modify ideas, ask questions, seek evidence and use evidence in argumentation and varied communication for different purposes and audiences (Carlson et al. 2003). Such habits of mind are also desirable for fruitful participation in debates on SSI, thus strengthening the relevance of inquiry as an approach for SSI teaching.

When SSIs and IBST meet, there are two key concerns that need to be addressed. One concern is that, in order to prepare for SSIs, the challenge that students encounter in IBST needs a certain level of complexity in terms of open-endedness, methods, resources and outcome (product). If we take seriously insights from research on situated cognition, competencies sought and developed are structured according to types of situations. A similar conclusion follows from literacy studies (Barton 2007). Thus, a competence must be trained and assessed in situations with some relevant similarity to the arenas where they are to be used later. The second concern opposes the first. Participating in SSIs demands a range of more specific competencies that are orchestrated according to the demands of the situation, the resources available and the interests of the speaker and the audience, as more general literacy skills (Kress 2003). Successful teaching and assessment (both formative and summative) calls for explicit instruction, modelling and students' reflections on specific competencies if they are going to be developed. Moreover, if all skills and competencies needed are to be developed simultaneously and when working with complex issues, many students will experience cognitive overload. These concerns mean that complexity needs to be reduced. This causes a dilemma when teaching SSIs by IBST. On the one hand, the more specific competencies become relevant and useful in the light of the purposes of the authentic, full complexity. This calls for a top-down approach (from full problem to specific competencies to handle it). But for the competencies to be teachable, it would be better to teach one at a time (a bottom-up approach). A solution to this challenge is to show that it is possible to develop a range of complexities through IBST.

In this chapter, based on five contributions to an ESERA 2011 symposium named 'SSI Active Participation: Challenges and Possibilities', two research questions have been investigated:

1. How do science teachers negotiate their contribution to a citizenship education and SSI classroom discussions and activism?
2. How do SSIs and inquiry-based science teaching (IBST) meet to develop complex competencies?

To address the first question, three empirical research projects are convoked: a qualitative study on teachers' views on citizenship and SSI teaching in France (Barrué and Albe 2011), a meta-analysis of five in-depth case studies centred on Portuguese biology teachers that integrate SSI discussions in their classes (Reis and Galvão 2009; Galvão et al. 2010) and an action-research project in Ontario focused on teachers' professional development towards promotion of student-led, research-informed activism to address SSIs (Bencze et al. 2011; Sperling and Bencze 2010).

To address the second question, an action-research project that sought to develop a group of teachers' abilities to teach SSIs by means of IBST and to develop students' scientific literacy in handling complex socio-scientific environmental issues has also been developed and analysed (Kolstø and Knain 2011; Knain and Byhring 2011). With a theoretical framework built on Wallace (2004), which introduces three foundational concepts for scientific literacy – 'authenticity', 'multiple discourse' and 'third space' – it is here considered that inquiry approaches in complex environmental issues offer rich opportunities for transformations across different authenticities, varied use of language and third space and thus literacy practices relevant in out-of-school discourses on environmental issues.

3 Methodology

3.1 *Documenting Teachers' Contribution to a Citizenship Education and SSI Classroom Discussions and Activism*

With the aim of identifying trends in French middle school teachers' contributions to citizenship education through SSIs, a questionnaire was built around seven items addressing pedagogical activities (1, 2), practices for teaching (3, 4, 5, 6) and teachers' contributions, consistent with the different academic subjects through 'convergence themes' (7a) and with citizenship education (7b). This seventh item was attached to a short extract of official texts. These questionnaires were submitted to 69 teachers.

Moreover, five case studies were selected for a meta-analysis from a larger group published during the last decade (Reis and Galvão 2009; Galvão et al. 2010) and under a line of research and intervention aimed at supporting the implementation, in Portugal, of new science curricula that call for the discussion of SSIs as a way of preparing students for an active, informed participation in society. The selection of the five case studies was based on the fact that the involved teachers highlighted these activities in their plans as a context and a pretext to develop students' competencies. The Portuguese biology teachers who integrate the discussion of SSIs in their classes had a wide range of teaching experience: 33, 33, 8, 5 and 3 years respectively.

All the selected in-depth case studies follow the same methodology, involving the triangulation of information gathered through semi-structured interviews and classroom observation. The interviews were carried out in school by the researchers before and after the observation of a sequence of successive classes. The observed sequence of classes (planned and implemented by the participating teacher) focused on topics considered by each teacher (during a previous interview) suitable to address SSIs. The observation was designed to identify activities used by the teacher in addressing these topics and to find out whether (and how) he or she made use of the discussion of SSIs. The researchers always adopted the role of direct, nonparticipant observers. Both classroom observation notes and full transcriptions of the interviews were subjected to content analysis through a model of analytical induction (Bogdan and Biklen 1992). The meta-analysis of the five case studies was centred on identifying factors that influence positively the classroom discussion of SSIs.

Towards helping teachers to develop deep commitments to promotion of student activism for addressing SSIs, a class of 16 student teachers was engaged in a 9-week elective course in Ontario (Bencze et al. 2011; Sperling and Bencze 2010). The student teachers were required to conduct primary (e.g. correlational studies of students' iPod™ uses) and secondary (e.g. Internet searches about toxins in electronics) research as possible influences on plans of action to address SSIs of their choosing. Concurrent assignments helped them to consider the nature of research and activism and to consider ways in which these activities may align with their instructional identities. Data collected for understanding effects of student teachers' research-informed activism projects on their orientation towards promotion of such projects in their future teaching balanced naturalistic and rationalist research perspectives (Guba and Lincoln 1988).

Semi-structured interviews were conducted with student teachers (for 60 min) prior to, during and after the course. Questions focused on their views about science, teaching and actions in relation to (i) their ongoing primary and secondary research and (ii) a stereotypical depiction of science in relation to technology and society. During interviews, we explored student teachers' trust in their primary and secondary research.

Moreover, samples of products generated by student teachers were collected, including issue descriptions, research plans, data collected, written reports, project reflections and action plans (including justification from primary and secondary research and other sources).

As student teachers worked on in-class and between-class activities relating to the project focus, field notes were also written and later transcribed. Regarding analyses, each researcher independently and repeatedly coded data for relevant categories and then developed encompassing themes – using constant comparative methods based on constructivist grounded theory (Charmaz 2000). Codes, categories and themes were then negotiated until consensus was reached. Member checks with participants were then conducted to help ensure the *trustworthiness* of claims, each of which was based on a minimum of three corroborating data sources.

3.2 An Action-Research Project Based on IBST as the Way and as the Goal to Deal with the Complexity of SSIs

A first round in an action-research cycle was conducted in a first-year class at an upper secondary school north of Oslo. The school is a combined vocational and general track school located in a rural area, but still within short driving distance from Oslo. The teachers wanted to provide some thematic direction on students' work, but also to provide freedom for the students to go in a direction of their own choosing and to start an inquiry into issues that were important and interesting to them. These concerns were addressed by creating a starting page in a wiki with paragraphs, each describing a broad issue primarily consistent with the natural science curriculum (6 issues), but also with issues from the social science curriculum (2 issues). Each paragraph contained a link to an empty page – the start of students' further inquiries. In the first cycle, the wiki introduced had a potential for inquiry processes: students could engage in each other's texts and they could create hyperlinks to each other's pages. This happened to some degree. However, tensions in the tacit understandings of the goals and methods of the project among the teachers as well as weak scaffolding of the process made the wiki's potential for passive reproduction become significant. Its similarities in appearance to Wikipedia may have made students associate it with the genre of the lexicon article, and it was possible to write the texts that would become the final product right away. The students did, as we all do, transform the task given into something that looked like what they had previously experienced. The teacher-controlled dialogue, expecting short answers that are quickly evaluated, seemed to be a resource also when students discussed in pairs in front of the computer, resulting in short exchanges that rapidly changed focus. Students' wiki pages tended to be slightly modified cut-and-paste texts from a single source, Wikipedia. Thus, there were few opportunities for transformations across authenticities and little variations in language use and third space was not important.

To solve the problems mentioned with the wiki tool that became a kind of all-in-one tool and the process dimension that became invisible in the overall frame of the project, the following steps were taken for the second cycle of the action research: each project was requested to have log page, a statement of problem to be enquired into, a planning page and a results page. Each project was requested to include an empirical investigation.

4 Results

4.1 Teachers' Contribution to Citizenship Education

The distribution of the different teachers' answers to the questionnaire on the choice of quoted convergence themes assumes a logic of academic subjects. For example, technology and physics teachers declared that they teach energy and sustainable development, and biology teachers said they teach sustainable development and health. Several teachers said they build specific activities for this teaching: in conjunction with relevant professionals for some of them, in multidisciplinary teaching for others. The responses show that teachers are aware of the need to involve different practices, according to the activities that they design with other teachers of the same academic subject or with other persons. Either they are alone, or they delegate to a professional participant working outside the school, whom they consider as most qualified to provide knowledge on selected topics. For item 7a, three categories have emerged from data analysis: teachers who say that they do not contribute to consistency between different subject matters, those who say that they contribute to consistency by collaborative work and those who stick to the official curriculum. With regard to their contribution to citizenship education (item 7b), a majority of teachers declared that they contributed to citizenship education by developing pupils' behaviours, others by developing new practices and a minority considered that they contributed by building pupils' skills. The majority of teachers expressed promoting pupils' attitudes such as civility, respect for democratic rules and awareness of society's challenges. Several teachers underlined the need to change school practices to contribute to citizenship education. Their common thinking is to build a bridge between school knowledge and what they called 'real life'. They considered that it is necessary to link teaching to pupils' everyday lives to give meaning to knowledge and to develop debate about 'convergence themes' which are identified by teachers as involving nonestablished knowledge. Finally, a minority of teachers states that they contribute to citizenship education by trying to develop in pupils skills such as searching for and evaluating information, argumentation and 'a critical mind'. Their aim is to develop pupils who are able to build their own argued opinion. This view of citizenship education may be understood as critical emancipatory citizenship education.

4.2 Factors Influencing Implementation of Classroom Discussions About SSIs

The meta-analysis of the five case studies shows that the implementation of classroom discussions of SSIs is related to (a) the teachers' conceptions about science, science education, curriculum and the educational relevance of these activities and (b) the knowledge needed for their design, management and assessment.

All teachers reveal a conception of the curriculum that stresses the possibility for teachers to manage content and to choose the educational experiences according to the needs of society, students' specific characteristics and the unique contexts in which they live. In line with the latter, the teachers assume the role of curriculum constructors (and not just consumers/executors) and are more concerned with how to develop specific competencies that they consider relevant, than with the lengthy curricular contents themselves. So, conceptions about the curriculum (and not the curriculum itself) emerge as an important inhibitor of the attention that teachers pay to the discussion of SSIs.

In all cases, teachers' strong personal beliefs about the importance of promoting discussion of SSIs and explicitly addressing aspects of the nature of science, together with their in-depth knowledge of the subject matter, the aims of science education and the strategies to carry it out, allow them to overcome any obstacles to the implementation of discussion activities about SSIs. A strong knowledge about science (its contents, nature and processes) associated with a pedagogical knowledge about how to manage and assess classroom discussions are decisive factors in succeeding with those activities. Teachers' beliefs and professional knowledge grant them a remarkable capacity to interpret the curriculum so as to address the topics and carry out the activities that they consider important and relevant.

Another interesting finding is that the ability to implement classroom discussions about SSIs is not exclusive to those who have most teaching experience. Several situations had a positive impact on teachers' personal and professional development regarding the discussion of SSIs. These situations provided the opportunity to experience, implement and evaluate completely new approaches under the supervision of science education experts. In two cases, previous experience as scientific researchers and in-depth knowledge about the nature of science and its interactions with technology and society seemed to be important factors in the teachers' ability to promote the discussion of SSIs with students.

4.3 Complex Student Teachers' Research and Activism Choices

Data suggest that student teachers' commitments to promotion of student-led, research-informed activism on SSIs resembled a 'normal' curve – with most indicating 'moderate' commitments. Most chose educational actions (e.g. posters, digital slides) and some lobbying (e.g. letters to power brokers). Indicators of research-informed activism commitments included passionate statements of concern about SSIs, clear statements of intention to implement researched activism and statements indicating self-efficacy beliefs. Myriad factors associated with the teacher, students, curriculum and learning contexts seemed to influence student teachers' activist choices. A major finding of this research, however, was that student teachers' commitments to activism depended less on findings of their primary research than was anticipated – with findings from their secondary research also significantly influencing them. As the course proceeded, student teachers' trust in primary research findings seemed to decline somewhat – particularly as surprising results occurred and as they realised the complexity of primary research. Their apparent confidence in findings from secondary research increased slightly, although they also felt some distrust for many of these findings – particularly for those from websites. Relationships among student teachers' primary and secondary research and activism choices were more complex and unpredictable than imagined. Their activism choices seemed to depend on the extent and nature of 'positive' and 'negative' relationships among their primary and secondary research and their actions. Findings from this study suggest that although student-led research can significantly influence student teachers' choices about and commitments to actions to address SSIs, there appear to be complex and unpredictable relationships among their primary and secondary research and their activist choices. Given the complexity of decision-making in general and in fields of science and technology more particularly (e.g. Sismondo 2008), these results are, perhaps, unsurprising. Arguably the most significant finding from this research pertains to student teachers' potential attachments to and identities with research-informed actions to address SSIs. Prior to the course studied here, none of the student teachers included research-based actions to address SSIs as part of their normal pedagogical repertoires. Data collected here, however, suggest that they emerged from the course with visions of and attachments to research-based actions on SSIs. For example, when discussing her future plans as a teacher in light of her primary research findings, 'Christie', a member of a group promoting reductions in trans-fat usage, said this about the importance of school-level activism: 'I think it is good to instil [a sense of activism] in kids when they are young because it is very easy to be apathetic – a slactivist. … [That is a person who says] "Oh, yeah, I am an activist. I joined this Facebook™ group." But, they don't do anything about it' (May 26, 2010). Earlier, she had indicated that her vision of such a project drew from the structure of the course's research-based activism mini-project. She said that members of her major project group kept asking:

'Is this like what we did with the [travel] mugs [study and action]?' (May 26, 2010). Given, as described above, that the dominant school science paradigm appears to be antithetical to this vision of science education, some of these student teachers may have developed, through the course described here, visions of the possible, which they might enact in the event that contextual variables align in appropriate ways.

Table 4.1 How issues with different levels of complexity in general put constraints on the openness in planned learning outcomes and the adequate level of teacher guiding

Complexity of issue	Typical issues dealt with	Typical types of learning outcomes	Characterisation
Low	Scientific concepts (e.g. laws of elmag radiation and its effect on cells)	Scientific concepts and scientific reasoning	Teacher-guided inquiry towards correct explanations
Intermediate low	Scientific laws (e.g. how to calculate and measure elmag radiation)	Scientific methodology (e.g. control of variables), practical skills, scientific concepts, scientific reasoning	Half-open inquiry towards well-known empirical relations
Intermediate high	Technological quality (e.g. comparing air and dug-down powerlines)	Scientific methodology (e.g. identification of variables), practical skills, scientific concepts	Open testing towards loosely defined learning outcomes
High	Socio-scientific issues (e.g. what to do with powerlines through residential areas)	Handle disputed claims, collect, examine and integrate information in co-operation, relevant scientific concepts	Open inquiry towards personal judgments

4.4 *Several Types of IBST and Possibilities for SSI Teaching*

IBST topics and issues used for inquiry have different levels of complexity. On the one hand, some IBST projects are rather narrow, using short demonstrations and experiments to drive a teacher-guided search for answers to a scientific question posed, e.g. 'Is there an upward force on things at rest on a table?' (NRC 1996). The number of variables in these issues is small, the learning outcome strictly and narrowly defined and the teacher might structure and even lead the inquiry. On the other hand, some issues for IBST projects are complex and without any known or definite answer. They might typically involve application of scientific knowledge in a complex situation or a real-world SSI. Such projects are open ended and ill defined, and students might go for different approaches. Thus, it is not possible to predefine a specific set of conceptual learning outcomes. Consequently it is not adequate for the teacher to try to synchronise the thinking of all students, and thus, the teachers guiding and structuring have to be at a general level. Moreover, the students have to work more autonomously and to practise a greater variety of competencies. As indicated in Table 4.1, it is also possible to identify intermediate types of projects. In short, in inquiry projects, the level of complexity of issues dealt with tends to covary with required openness in the planned learning outcomes and adequate level of teacher guiding.

A main idea in Table 4.1 is that issues with different levels of complexity involve and develop different process competencies. Specifically, issues with low and intermediate levels of complexity are open for teaching specific and singular competencies, while issues with high complexity demand an ability to orchestrate specific competencies according to the demands of an authentic, out-of-school situation. The wise combination of different types of IBST therefore holds the possibility of solving the dilemma discussed above.

4.5 *Inquiry-Based Teaching to Handle Complex Environmental Issues*

A strong idea in this research is that inquiry-based approaches to complex environmental issues offer rich opportunities for transformations across different authenticities, varied use of language and third space and thus literacy practices (Barton 2007), which are relevant in out-of-school discourses on environmental issues.

4.5.1 The First Cycle

The teachers wanted to provide some thematic direction on students' work, but also to provide freedom for the students to go in a direction of their own choosing and to start an inquiry into issues that were important and interesting to them. These concerns were addressed by creating a starting page in the wiki with paragraphs, each describing a broad issue primarily consistent with the natural science curriculum (6 issues), but also with issues from the social science curriculum (2 issues). Each paragraph contained a link to an empty page – the start of students' further inquiries.

In the first cycle, the wiki introduced had a potential for inquiry processes: students could engage in each other's texts and they could create hyperlinks to each other's pages. This happened to some degree. However, tensions in the tacit understandings of the goals and methods of the project among the teachers as well as weak scaffolding of the process made the wiki's potential for passive reproduction become significant: its similarities in appearance to Wikipedia may have made students associate it with the genre of the lexicon article, and it was possible to write the texts that would become the final product right away. The students did, as we all do, transform the task given into something that looked like what they had previously experienced. The teacher-controlled dialogue, expecting short answers that are quickly evaluated, seemed to be a resource also when students discussed in pairs in front of the computer, resulting in short exchanges that rapidly changed focus. Students' wiki pages tended to be slightly modified cut-and-paste texts from a single source, Wikipedia. Thus, there were few opportunities for transformations across authenticities and little variations in language use and third space was not important.

4.5.2 The Second Cycle

To solve the problems mentioned with the wiki tool that became a kind of all-in-one tool and the process dimension that became invisible in the overall frame of the project, the following steps were taken:

- Each project was requested to have log page, a statement of problem to be enquired into, a planning page and a 'results' page.
- Each project was requested to include an empirical investigation.

In the second cycle there was a marked shift in students' oral and written texts. Students worked with authentic data in a more complex problem formulation that could not easily be answered by science alone. Students needed to consider several semiotic and contextual resources, which added complexity to the problem they worked on, but also scope and force to their dialogues. Students' use of varied resources enabled a richer field context in terms of personal beliefs, authentic data and content knowledge. The following contextual resources are identified in one of the transcripts:

- The demands of cohesion in written text
- The genres at hand
- Personal beliefs
- Authentic data
- Content knowledge

In the first example from cycle 1, simple cut-and-paste reproduction of factual genres was prevalent. In cycle 2, a more complex problem area and explicit instruction on language use for both process and product resulted in more complex genres such as evaluating, arguing, collecting evidence and concluding from a problem formulation. In the example discussed (in this chapter), the demands of bringing different knowledge domains and sources into a coherent text scaffolded according to principles of inquiry-based teaching opened for literacy practices important to participating in socio-scientific issues and dealing with complexity: students' sustained engagement in conversations on complex problems, judging evidence and different value positions.

5 Conclusions and Implications

Our first question aims to clarify how science teachers negotiate their contribution to a citizenship education and SSI classroom discussions and activism. Questionnaires submitted to teachers provided significant views of their envisaged contribution to citizenship education, in connection with civility and rules to respect or to develop pupils' skills such as searching for and evaluating information, performing argumentation and critical thinking, and participation in debates and arguing their own opinions. Results also showed that the teachers' answers assume a logic of academic subject matters. A minority of teachers aimed for critical citizen education through interdisciplinary teaching. This suggests that teachers may not feel competent for this teaching or, alternatively, that they show a willingness to forge links between school context and society. The meta-analysis of biology teachers' classroom activities reported in this chapter shows that the implementation of SSI discussion depends decisively on (a) the teachers' conceptions about science, science teaching, curriculum and the educational relevance of the activities and (b) the knowledge needed for their design, management and assessment. In all cases, teachers' classroom practice is influenced by (a) a conception of science education focused both on knowledge construction and on the development of skills and attitudes required for citizens' intellectual autonomy and for exercising their citizenship and (b) an understanding of the curriculum allowing for levels of decision-making suited to the needs of society. All teachers assume the role of curriculum creators (and not just consumers/executors) of competencies which stresses the possibility for teachers to manage content and choose the educational experiences according to the needs of society, students' specific characteristics and the unique contexts in which they live. These teachers are more concerned with how to develop specific competencies that they consider relevant than with the lengthy curricular contents themselves. The development of these conceptions and competencies was triggered by the teachers' professional development, i.e. previous experience as scientific researchers and/or in-service training opportunities in which the teachers experienced, implemented and evaluated new approaches under science educators' supervision.

Studying the factors of success regarding classroom discussions of SSIs provides crucial information for the design of intervention processes capable of supporting teachers in the planning and implementation of such activities and consequently in attaining the goals of science curricula.

Towards helping teachers to develop deep commitments in SSI students' activism, a class of 16 student teachers was engaged in a 9-week elective course. Constant comparative analysis of qualitative data showed that, after the course focusing on SSI teaching, some student teachers seemed to develop moderate to strong orientations towards promotion of sociopolitical activism. Most chose educational actions (e.g. posters, digital slides) and some lobbying (e.g. letters to power brokers). Findings from this study suggest that although student-led research can significantly influence student teachers' choices about and commitments to actions to address SSIs, there appear to be complex and unpredictable relationships among the course activities and their activist choices. Unexpectedly, secondary and primary research findings appeared to equally influence their activist choices and attachments.

Another key concern for SSI teaching is to develop students' ability to cope with complexity. It is possible to identify several types of IBST in terms of complexity. In order to develop different competencies in dealing with complex SSIs, it seems necessary to involve students in IBST with different levels of complexity and SSI authenticity. In an action-research cycle on complex environmental issues, complexity was a resource in students' meaning-making by opening for students' use varied resources that enabled a richer field context in terms of personal beliefs, authentic data and content knowledge. However, developments between two cycles show that students needed explicit scaffolding on process to prevent complexity from fading away into factual reproduction by cut-and-paste strategies instead of active transformation of varied textual and contextual resources.

The involvement of teachers in experiencing the educational potential for the implementation of SSIs in their own classes can be a positive step forward in changing their conceptions about the relevance of this methodology and in developing the confidence and practices required for their use in the classroom.

References

Barrué, C., & Albe, V. (2011). New approach in french middle schools: A possible congruence between SSIs and citizenship education. Paper presented at the ESERA Conference, Lyon, 5th–9th September 2011.

Barton, D. (2007). *Literacy. An introduction to the ecology of written language*. Malden/Oxford: Blackwell.

Bencze, L., Sperling, E., & Carter, L. (2011). Students' research-informed socio-scientific activism: Re/visions for a sustainable future. *Research in Science Education, 48*(6), 648–669.

Bogdan, R., & Biklen, S. (1992). *Qualitative research for education*. Boston: Allyn & Bacon.

Carlson, M. O. B., Humphrey, G. E., & Reinhardt, K. S. (2003). *Weaving science inquiry and continuous assessment. Using formative assessment to improve learning*. Thousand Oakes: Corwin Press.

Carter, L. (2005). Globalisation and science education: Rethinking science education reforms. *Journal of Research in Science Teaching, 42*(5), 561–580.

Charmaz, K. (2000). Grounded theory: Objectivist and constructivist methods. In N. K. Denzin & Y. S. Lincoln (Eds.), *Handbook of qualitative research* (pp. 509–535). Thousand Oaks: Sage.

Galvão, C., Reis, P., & Freire, S. (2010). Enhancing the popularity and the relevance of science teaching in Portuguese science classes. *Research in Science Education*. doi: 10.1007/s11165-010-9184-3

Guba, E. G., & Lincoln, Y. S. (1988). Naturalistic and rationalistic enquiry. In J. P. Keeves (Ed.), *Educational research, methodology and measurement: An international handbook* (pp. 81–85). London: Pergamon Press.

Helms, J. V. (1998). Science – and me: Subject matter and identity in secondary school science teachers. *Journal of Research in Science Teaching, 35*(7), 811–834.

Hodson, D. (2003). Time for action: Science education for an alternative future. *International Journal of Science Education, 25*(6), 645–670.

Hodson, D. (2008). *Towards scientific literacy: A teachers' guide to the history, philosophy and sociology of science*. Rotterdam: Sense.

Knain, E., & Byhring, A. K. (2011). Coping with complex environmental issues by Inquiry-based teaching. Paper presented at the ESERA Conference, Lyon, 5th–9th September 2011.

Kolstø, S. (2001). Scientific literacy for citizenship: Tools for dealing with the science dimension of controversial socio-scientific issues. *Science Education, 85*(3), 291–310.

Kolstø, S. D., & Knain, E. (2011). Methodological competencies needed to deal with the complexity of SSIs: Inquiry-based methods as the way and as the goal. Paper presented at the ESERA Conference, Lyon, 5th–9th September 2011.

Kress, G. (2003). *Literacy in the new media age*. London/New York: Routledge.

Lee, H., Abd-el-Khalick, F., & Choi, K. (2006). Korean science teachers' perceptions of the introduction of socio-scientific issues into the science curriculum. *Canadian Journal of Science, Mathematics, and Technology Education, 6*, 97–117.

Lester, B. T., Ma, L., Lee, O., & Lambert, J. (2006). Social activism in elementary science education: A science, technology and society approach to teach global warming. *International Journal of Science Education, 28*(4), 315–339.

Levinson, R., & Turner, S. (2001). *The teaching of social and ethical issues in the school curriculum, arising from developments in biomedical research: A research study of teachers.* Final Report to The Wellome Trust by The Science and Technology Group, Institute of Education, University of London.

McVaugh, J. (2010). Problem-based learning. In A. Stibbe (Ed.), *The handbook of sustainability literacy*. Skills for a Changing World.

Millar, R., & Hunt, A. (2002). Science for public understanding: A different way to teach and learn science. *School Science Review, 83*(304), 35–42.

Newton, P. (1999). The place of argumentation in the pedagogy of school science. *International Journal of Science Education, 21*(5), 553–576.

NRC. (1996). *Inquiry and the national science education standards. A guide for teaching and learning*. Washington, DC: National Research Council/National Academy Press.

Reis, P., & Galvão, C. (2009). Teaching controversial socio-scientific issues in biology and geology classes: A case study. *Electronic Journal of Science Education, 13*(1), 162–185. Available at http://ejse.southwestern.edu/volumes/v13n1/v13n1.pdf

Roth, W.-M., & Lee, S. (2004). Science education as/for participation in the community. *Science Education, 88*, 263–291.

Sadler, T. D. (2004). Informal reasoning regarding socio-scientific issues: A critical review of research. *Journal of Research in Science Teaching, 41*, 513–536.

Sadler, T. D. (2009). Situated learning in science education: Socio-scientific issues as contexts for practice. *Studies in Science Education, 45*(1), 1–42.

Sadler, T. D., Amirshokoohi, A., Kazempour, M., & Allspaw, K. M. (2006). Socio-science and ethics in science classrooms: Teacher perspectives and strategies. *Journal of Research in Science Teaching, 43*, 353–376.

Sismondo, S. (2008). Science and technology studies and an engaged program. In E. J. Hackett, E. Sperling, & J. L. Bencze (Eds.), (2010). *More than particle theory: Citizenship through school science. Canadian Journal of Science, Mathematics and Technology Education. 10*(3), 255–26.

Sperling, E., & Bencze, J. L. (2010). More than particle theory: Citizenship through school science. *Canadian Journal of Science, Mathematics and Technology Education, 10*(3), 255–266.

Wallace, C. S. (2004). Framing new research in science literacy and language use: Authenticity, multiple discourses and the 'Third Space". *Science Education, 88*(6), 901–914.

Wenger, E. (1998). *Communities of practice.* Cambridge: Cambridge University Press.

Chapter 5
Which Perspectives Are Referred in Students' Arguments About a Socio-scientific Issue? The Case of Bears' Reintroduction in the Pyrenees

Ana M. Domènech and Conxita Márquez

1 Introduction

1.1 Background and Rationale

Scientific literacy, considered by several researchers as what the general public ought to know about science including scientific knowledge and knowledge about the nature of science (Kolsto 2001), has become an internationally well-recognized contemporary educational goal (Nuangchalerm 2010). Although its definition is still controversial, students' ability to deal with Socio-scientific Issues thoughtfully has been recognized as one of the important components of scientific literacy (Sadler 2004).

For several years now, researchers have been proposing the debating of Socio-scientific Issues in the classroom in an effort to democratize science in society and provide citizens with the education they need to think critically about interactions between science, technology, and society, as well as to make informed decisions with respect to questions raised by the techno-sciences (Albe 2008). It is important to notice that, with the rapid development of science and technology, students, as the citizens in democratic society, may have lots of opportunities to encounter a variety of socio-scientific issues, and they and their parents may need to make some decisions or positions toward these issues (Wu and Tsai 2007). This type of decision-making process needs to continue being studied in science education, and it becomes important to know how students take decisions in these contexts and

A.M. Domènech (✉) • C. Márquez
Departament de Didàctica de la Matemàtica i les Ciències Experimentals,
Universitat Autònoma de Barcelona, Bellaterra, Barcelona, Spain
e-mail: AnaMaria.Domenech@uab.cat; Conxita.Marquez@uab.cat

C. Bruguière et al. (eds.), *Topics and Trends in Current Science Education: 9th ESERA Conference Selected Contributions*, Contributions from Science Education Research 1,
DOI 10.1007/978-94-007-7281-6_5,

which teaching strategies could be developed in order to promote students' analysis of the controversy from different viewpoints and to facilitate the scientific knowledge transfer to SSI decisions.

The focus of this chapter is to explain our research related to the analysis of the perspectives referred by students when they make up their opinions about bears' reintroduction. Before that, in the following introduction, to address a review of the framework related, Socio-scientific issues features and decision making in these contexts are delineated.

1.1.1 Socio-scientific Issues (SSI) in Science Education

During the recent past, Socio-scientific Issues (SSI) have been introduced in science classrooms and have been investigated by science education researchers (Albe 2008). In fact, the phrase Socio-scientific Issues was used in science education literature as early as 1986 (Fleming 1986a), but it did not come to represent a recognizable framework for research and practice until the late 1990.

SSI relative to health, environment, and techno-scientific innovations are defined as social dilemmas linked to science about which citizens have to make decisions (Molinatti et al. 2010). They are important for society and have a basis in science; involve forming opinions; are frequently media-reported; address local, national, and global dimensions with attendant political and societal frameworks; involve values and ethical reasoning; may involve consideration of sustainable development; and may require some understandings of probability of risks, and there are no "right answers" (Ekborg et al. 2009). Moreover, Zeidler et al. (2005) emphasize that in SSI, it is crucial to deal with moral and ethics. It is important to notice that, although issues such as those related to biotechnology and environmental challenges can be classified together as SSI, it is not meant to imply that science and society represent independent entities. On the contrary, all aspects of science are inseparable from the society from which they arise (Sadler 2004).

As the media confronts citizens almost daily with these kinds of issues, citizens are introduced to a different type of science from the one that is usually presented in science classes (Reis and Galvao 2009). Most formal science education focuses on a conventional, non-controversial, established, and reliable science and does not discuss its tentative nature while the media's news highlights a "borderline science" that is controversial, preliminary, and under debate (Zimmerman et al. 1999).

Taking into account all these features related to SSI, several educators in science have called for the inclusion of these issues' discussion in science curricula because of its potential for creating a more real, humanistic image of scientific activity and promoting scientific literacy (Reis and Galvao 2009) as it has been aforementioned.

Sadler (2004) pointed out that SSI research focuses on three main directions: relationships between the nature of science conceptualizations and socio-scientific decision making, ways of evaluating information regarding SSI, and socio-scientific argumentation (Jimenez-Aleixandre et al. 2000; Osborne et al. 2004; Simmoneaux 2001; Zohar and Nemet 2002). In several research studies about SSI, students have usually been working with an issue which typically includes a dilemma. Empirical data have been gathered through observations and/or interviews and students' written reports have been analyzed (Aikenhead 2005).

In the published *Second International Handbook of Science Education*, Sadler and Dawson (2012, Chap. 53, pp. 799–809) provide compelling evidence supporting the efficacy of SSI as contexts for science education by describing and synthesizing a focused sample of research related to student learning associated with several widely assumed goals for science education: science content knowledge, nature of science, interest and motivation, and argumentation.

As the present study analyzes students' opinions during a classroom activity designed, relevant literature related to making decisions on an SSI will be described in the following section.

1.1.2 Making Decisions on an SSI

The argumentation and decision making on socio-scientific issues can provide students with meaningful contexts for practicing their informal reasoning or thinking skill and applying what they have learned in science classrooms to solve real-world problems they have encountered in daily life. In general, dealing with SSI often involves argumentation and decisions making on these issues as these are typically open-ended and ill-structured problems which citizens will find in their lives. Students need to argue for or against different positions and to be able to voice opinions based on scientific understandings, demonstrate logical patterns of reasoning, and support their arguments with sound scientific evidence (Sadler and Zeidler 2005a, b).

This type of decision-making process needs to continue being studied in science education, and it becomes important to know how students take decisions in these contexts and how do they evaluate contradictory scientific information. As it is shown in Albe's (2008) literature review, several studies have shown the importance of students' personal experiences to interpret socio-scientific issues, to justify positions, and to frame argumentation (Fensham 2002; Fleming 1986b; Grace and Ratcliffe 2002; Patronis et al. 1999; Sadler et al. 2004; Zeidler et al. 2002), while some authors underlined that the overriding considerations are social or epistemological (Aikenhead 1985; Ryder 2002) and others have questioned the importance of using scientific knowledge when SSI have to be settled (Irwin and Wynne 1996; Kortland 2001; Lewis et al. 1999; Ratcliffe 1997). Acar et al. (2010) propose that explicit teaching of argumentation research should provide students a decision-making framework in which students can consider their values about a socio-scientific issue and assess different alternatives. During the past two decades, researchers have conducted several qualitative studies to identify the perspectives from which learners make their arguments on an SSI. In the context of science, reasoning historically referred to formal reasoning characterized by rules of logic and mathematics. Although the results of science may be presented in a language of formal reasoning and logic, the results themselves originate through informal reasoning (Sadler 2004). Informal reasoning involves the generation and evaluation of positions reasoning as they ponder causes and consequences, pros and cons, and positions and alternatives (Zohar and Nemet 2002). SSI are good contexts for the application of informal reasoning as, by definition, they are typically complex, contentious, open-ended, and ill-structured problems, and arguments can be constructed from multiple perspectives. Sadler and Zeidler (2005b) have recognized that college students demonstrated three patterns of informal reasoning on six genetic engineering scenarios, including rationalistic (describe reason-based considerations), emotive (describe care-based considerations), and intuitive forms (describe considerations based on immediate reactions to the context of a scenario). According to the information used during the problem task involving nuclear energy, Yang and Anderson (2003) also identified two reasoning modes – scientific-oriented reasoning (i.e., reason with scientific-oriented information) and social-oriented reasoning (i.e., reason with social-oriented information). In their system network for students' arguments on a socio-scientific issue, Patronis et al. (1999) also summarized students' qualitative arguments into four aspects: social, ecological, economic, and practical. That is, learners may generate arguments on a socio-scientific issue from social, ecological, economic, or practical considerations. Taking all these information into account, Wu and Tsai (2007) developed an integrated analytic framework in order to analyze informal reasoning on SSI considering that this type of reasoning is often used in situations where reasons exist both supporting and against the conclusion, such as making decisions about what to believe or what actions to take.

1.2 Objective of the Research

Giving the theoretical framework just presented, the specific objective this research addressed was to analyze from which perspectives do students make their arguments to base their decisions about the bear reintroduction and explore if students give more than one argument and analyze the SSI from multiple perspectives. In this study, in using the term "perspectives," we are referring to the type of information (e.g., as preference for either scientific or social information) participants used in the arguments generated in a decision-making process.

2 Methodology

2.1 Data Collection

2.1.1 Research Population

In order to achieve the objective just presented, an activity related to an SSI was designed and carried out in two secondary schools in Catalonia (Spain) in a science class in the context of formal education. Both schools are situated in towns near

Barcelona but students from school 1 come from a low social and economic bracket, with 65 % of them being immigrants mainly from South America and the Maghreb, while students from school 2 come from a medium to high and cultural bracket, with 10 % of them being immigrants. A total of 125 students (66 male and 59 female, aged 13-14) took part in this study.

2.1.2 SSI Classroom Activity Design

The first phase of the research consisted of designing the classroom activity and the open-ended questionnaire we would develop and use to collect research data. After knowing which students would participate, we selected with their science teachers which SSI would be the most suitable, positing the premise that incorporating these issues in science classrooms has to be related to the scientific knowledge studied (Barab et al. 2006). As a consequence, it has to be an issue related to what students had studied in previous lessons and to scientific contents and competences defined by Spanish curricula. Considering these requirements, ecology bloc was selected as one of the most suitable subject for carrying out our activity.

With serious threats to global and local biodiversity, education about animal and plant conservation is an important SSI, and the ability to make decisions about conservation issues is a prerequisite to making informed decisions about wider issues of sustainable development (Grace and Ratcliffe 2002). Considering this need, the fact that the United Nations declared 2010 to be the International Year of Biodiversity and students awareness about the issue (as it appears frequently in the media), bears' reintroduction in the Pyrenees was considered the most adequate subject for carrying out our activity.

In the early 1990s, the last bear disappears from the central Pyrenees. Only seven or eight individuals remain in the western core and the extinction of the bear in the Pyrenees thus seemed inevitable. In 1996 bear reintroduction in French Pyrenees was carried out within the framework of a European Union LIFE program that finances, among other things, the capture, transportation, and release of the bears. In 1996, three bears were captured in Slovenia to be released in the Central Pyrenees and four new bears were reintroduced in 2006 although there were different groups against this initiative. Since then, it became a local controversy as local and national pro-bear and anti-bear movements have expressed different opinions about the issue. According to opponents of the bear, the authorities did not take the opinions of the population into consideration as they are adamant that the bear is a danger to man and the pastoralism-predator coexistence is not possible as methods for protecting flocks (animal enclosures, guard docs, electric fencing) seem to be ineffective. On contrary, for others, not enough bears are being reintroduced to maintain the population. Some argue that bears are primarily vegetarian (eating berries, nuts, and roots) or that they eat mostly ants and that is not bears but dogs that are attacking flocks. When it comes to socioeconomics, some argue that the image of the bear will favor the development of the tourist economy, while others complain that its presence will drive tourists again.

As with all SSI, politics, economics, and cultural aspects play an important role in decision making about bear's reintroduction in the Pyrenees. This issue is a social dilemma impinging scientific fields because people have to decide if this initiative continues, and more bears are reintroduced in this territory where bear disappeared as a result of hunting. As it has been aforementioned, since this initiative was proposed, there have been different stakeholders that have opposed to it and it is an issue that usually appears in the media. Although the students who participate in this study did not live in Pyrenees, researchers thought that it was important talk with them about this reintroduction because it is a national debate, it usually appears on the media, and students probably will have to face with species reintroductions in the future and make decisions about them as citizens.

Once the classroom activity and the open-ended questionnaire have been designed, they were presented to an expert group composed by teachers and science education researchers who analyzed them and made suggestions which were then incorporated. In the second phase these instruments were applied.

2.1.3 The SSI Classroom Activity Designed

The activity designed lasted 2 h and was carried out in February 2009. It was divided in two sessions. At the first one, in order to introduce this SSI to students and know their previous ideas about bears' reintroduction, students read a news about a campaign started by a local organization with the aim of choosing a name for one of the cubs that Hvala (female bear released in 2006 from Slovenia) has given birth to. After this reading, students wrote down what did they know about bears' reintroduction in the Pyrenees before discussing all these information with the whole class. Once the issue has been introduced, the next step consisted of underlying the controversy related to it. According to this objective, students read another new that contains information about a human attack by a bear. In autumn of 2008, a boar hunter in the Vall d'Aran (Catalonia, Spain) was bitten by a bear and suffered minor injuries to his foot and hand. Meanwhile this fact has led to calls from the Aranese government for the removal of all bears from the range claiming that the "bear reintroduction experiment has failed"; the ecological organization Depana,[1] while lamenting the injuries to the man, lay the blame at poorly organized boar hunts and note that bears and boar hunting are perfectly compatible when managed properly, citing the example of the Cordillera Cantábrica. It is important to notice that Depana was the same organization that carried out the campaign to choose a name for one of the Hvalas' cubs and that Hvala was the bear which has injured the boar hunter. When students already knew this information, to gain deeper insights into

[1]Depana is an ecological organization founded in 1976 with the aim of supporting environment and sustainable development guarantying biodiversity conservation. In 1979 this organization join the IUCN (The International Union for Conservation of Nature), the world's oldest and largest global environmental organization.

the controversy, after talking about both read news, students discussed in pairs about the different viewpoints supported by all the stakeholders implied in this issue (e.g., as politicians, scientists, hunters, farmers, or environment organizations). They read arguments supported by pro-bear and anti-bear movements about socio-economic and ecological aspects related to the reintroduction. Students discussed the different viewpoints associated with the *subject of diet* (in the Pyrenees, shepherds tend to regard the bear as being carnivore, but the study of this diet shows his polyphagy and his taste for a variety of foods), *the guarantee of the survival of the bear population in the Pyrenees* (ecological justifications for the reintroduction), *the authorities and political campaigns associated with the creation of public sympathy regarding the reintroduction*, and the *socioeconomic consequences* (the financial helps for the shepherds to compensate any harm caused by bears, the influence of the image of the bear in tourists).

Finally, after discussing this information with the whole class, students read and discussed scientific information about bears and their ecological requirements, taking into account that we will analyze this information if students transfer these concepts to their decision making. Then, students were asked to respond individually to a set of open-ended questions and wrote down their answers. The question of the open-ended questionnaire that will be analyzed in this study is the following:

– Do you think that bear reintroduction should be done in other place where people would not be hurt by bears? (Justify your opinion.)

In the second session students had to write an argumentative text presenting their personal opinion about bears' reintroduction after following a guide and discussing it in groups before explaining what they have learned in this lesson and which changes would be proposed in order to improve the classroom activity development.

2.2 *Data Analysis*

According to other similar studies and to our research aims, qualitative methods combined with quantitative parameters of analyses were employed. The first phase of the analysis consisted of categories' definition following an inductive-deductive method. As a first step of inductive analysis of the data (Lincoln and Guba 1985), written responses to questionnaires were read in order to make preliminary notes regarding patterns emerging from the data. As a second step, the emergent categories were compared with the categories obtained in other studies and then they were used to classify the arguments given by the students in their responses. The second phase of the analysis was analyzing the frequency of the categories considering qualitative indicators and quantitative measures described in Wu and Tsai (2007) as will be explained in the next point. All this process was made by an independent examination of data done by investigators and the support of Atlas.ti.

3 Results and Discussion

In answer to the question *Do you think that bear reintroduction should be done in other place where people would not be hurt by bears? Justify your opinions*, students justified their opinions generating arguments from different perspectives. After defining categories following an inductive-deductive method, in this study, as is shown in Table 5.1, learners generated their arguments referring to various perspectives such as "social-oriented," "ecological-oriented," and "moral-oriented" perspectives.

Social-oriented arguments were based on the welfare of the society or human sympathy and an example was *I think bears should be in another place to protect flocks.* In the analysis, we considered that it was important to point out that there were students who referred to anthropomorphism talking about bears as if they were people and, in their arguments, referred to bears' welfare. An example of this kind of argument was *Bears should be in a territory where they can't be afraid and live safely.* On the other hand, there were students who justified their opinions with the same arguments but focusing their view on humans: *I think bears should be in a*

Table 5.1 Categories of perspectives from which students made their arguments to base their decisions about bears' reintroduction

Category	Keywords	Exemplar
"Social-oriented" arguments	Welfare or safety	
Focus on human	Human's welfare or safety, economic aspects	*In my opinion bears should be out of the Pyrenees to prevent bears from eating sheeps*
		Bears should be in another place because if they are in the Pyrenees, people will be afraid of them
Focus on bears	Bears' anthropomorphism, bears' welfare or safety	*Bears should be in another place because if they continue in the Pyrenees, they will be afraid of people*
		I think bears will be better in another place because they will be safer and live quietly
"Ecological-oriented" arguments	Habitats, extinction, ecosystems, food chains	*I think bears should continue in the Pyrenees because it's their habitat*
		Bears' reintroduction should continue to avoid its extinction
"Moral-oriented" arguments	Right to live, behaviors or beliefs considered to be good or bad	*I think that bears should be in a different zone because if they hurt people they have to be punished*
		Politicians should take into account that all species have rights and Earth is our planet

Table 5.2 Percentage of arguments of each category that were expressed by the students, considering all of them, considering students who agreed with the reintroduction, and considering students who disagreed with this initiative

	% of arguments of each category that were expressed by		
Category	All students	Anti-bear students	Pro-bear students
"Social-oriented" arguments	67	86.2	46.6
Focus on human	36.8	52.5	20.1
Focus on bears	30.2	33.7	26.5
"Ecological-oriented" arguments	12.3	3.7	18.3
"Moral-oriented" arguments	20.7	10.1	35.1

territory where people can't be afraid of them. As a result of this fact, we distinguished two subcategories in the "social-oriented" category depending on whether students focused on talking about humans' or bears' welfare.

In contrast, ecological-oriented arguments referred to the interactions established between species and the environment where they live as it can be seen in this example, *I think bears should continue in the Pyrenees because it's their habitat.* It has to be considered that successful biological conservation management programs like reintroductions also depend on an understanding of the biology of the organisms concerned and how they interact with their surrounding environment (Grace and Ratcliffe 2002).

Finally, moral-oriented arguments were related to attitudes, facts, behaviors, or thoughts that were considered good or bad, for example, *I think that bears should be in other zone because if they hurt people they have to be punished.*

These results were consistent with Wu and Tsai (2007) analysis because in their framework they determinate that in an SSI, learners may generate their arguments from different perspectives such as social, ecological, scientific, or technological. Nevertheless, there were some differences that we would explain later on. In our study, economic-oriented arguments were included in social-oriented perspective since we considered that economic safety is part of society's welfare. On the other hand, science-oriented arguments were represented by ecological-oriented, while technology-oriented arguments were not represented because students did not identify a technological component in this SSI. It is important to note that, although moral-oriented arguments were not distinguished in Wu and Tsai study, moral perspective is recognized in other studies like Fensham (2002) and Albe (2007). As SSI are of a contentious nature, they may be analyzed according to different perspectives, they do not lead to simple conclusions, and often, they involve a moral or ethical dimension (Sadler et al. 2004).

Regarding the use of these perspectives, it is interesting to analyze whether the perspectives referred by the students were the same for the students supporting the reintroduction of the bears in the Pyrenees (pro-bears students) and the ones against it (anti-bear students) and the possible differences between male and female students. As it can be seen in Table 5.2, the results of this analysis show that there

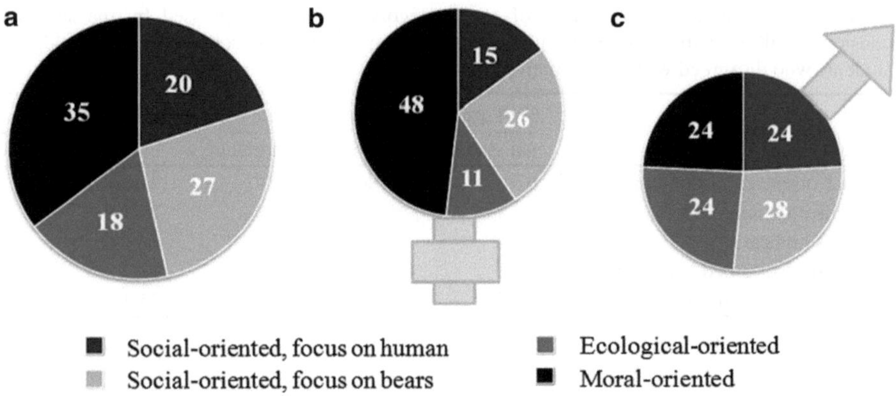

Fig. 5.1 Percentage of arguments of each category given by pro-bears students considering (**a**) all of them, (**b**) female students, and (**c**) male students

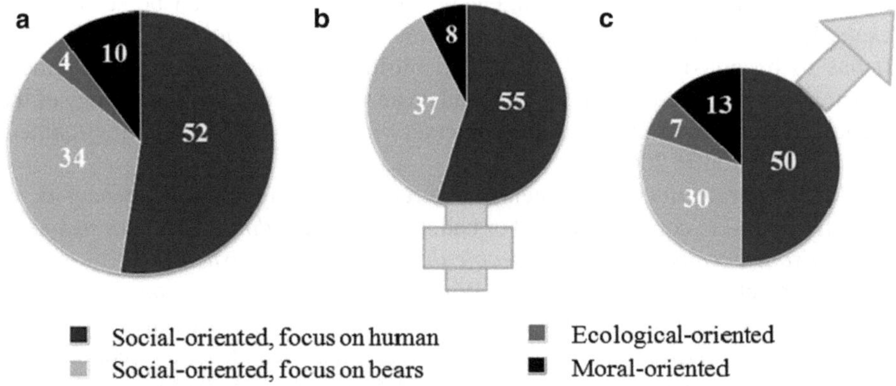

Fig. 5.2 Percentage of arguments of each category given by anti-bears students considering (**a**) all of them, (**b**) female students, and (**c**) male students

are differences in the kind of perspectives referred by the students depending on their viewpoint of the reintroduction.

In general, in this study, students who were against bears' reintroduction tended to refer to social-oriented arguments, mainly focusing on human aspects, while moral-oriented and, especially, ecological-oriented arguments were only mentioned by few students of this group. On the contrary, students who agreed with this initiative, although they also referred to the social perspective, they highlighted ecological-oriented and, especially, moral-oriented arguments to justify their opinions about the reintroduction. Despite these being global trends, there are differences between male and female students that can be seen in Figs. 5.1 and 5.2.

Table 5.3 Mean of arguments expressed by students considering all of them, considering students who agreed with the reintroduction, and considering students who disagreed with this initiative

Mean of arguments expressed to justify their opinion by		
All students	Pro-bear students	Anti-bear students
1.27	1.22	1.31

In the case of pro-bears students, both female and male students referred to all the perspectives defined although there were differences between the percentages related to each category of perspectives. It is important to notice that male students referred equally to the various perspectives; meanwhile about half of the female students referred to the moral perspective. In both groups, social-oriented arguments focused on the bear were expressed by the students. The same occurs with anti-bears students, since both female and male students also remarked this perspective in their arguments. In this case, one important difference that has to be highlighted is that, while male students also underlined ecological-oriented arguments, this perspective was not referred in the arguments of female students.

With respect to the arguments expressed by the students, it has been detected that, in spite of these differences on the percentage of arguments that referred to each perspective, we noted that the same arguments were used in order to defend opposite opinions. To clarify this, we want to comment the following example: the argument *bears should be in a territory where people can't be afraid of them* was used for a pro-bear student in order to justify the need of continuing reintroducing bears in the Pyrenees, while an anti-bear student used it to justify the need of rejecting this initiative and carrying it out in another place.

Finally, another interesting aspect of our study is to analyze how many arguments were provided by learners. As it is shown in Table 5.3, in this research, students utilized, on average, only one argument to justify their opinion.

If we analyze the differences between pro-bear and anti-bear students, we can only note that, although the differences are not significant, in general, anti-bear students expressed more arguments than pro-bear students. The reason of this trend is that, when students who were against this initiative justified their opinions, they usually tended to refer to social-oriented arguments focusing to both human and bears. These results concur with Fleming's study (1986a, b) where the author found that most subjects in his study tended to approach SSI with social concerns and differ from the results obtained in Sadler et al. (2004) and Yang and Anderson (2003) where learners were oriented to reason from multiple perspectives. Despite this difference, in this research, ecological-oriented considerations were the less proposed as occurred in the studies cited above. This trend might be due to the students' insufficient abilities to make the connections between what they had learned in science classrooms and the SSI they encountered in daily life (Wu and Tsai 2007). On the contrary, Kolsto (2001) argued that students' knowledge acquired in science classroom can serve as tools for their informal reasoning and decision making on controversial issues.

4 Conclusions and Implications

In the present research, students generated their arguments on bears' reintroduction from different perspectives such as social-oriented, ecological-oriented, and moral-oriented. Despite this fact, the results of this study suggest that students had difficulties to consider all these different perspectives related to an SSI when they had to make a decision, so it is suggested that they were not able to reason from multiple perspectives. Moreover, ecological-oriented arguments were the less used although students had studied ecology before carrying out this activity and they had analyzed and discussed scientific information about bears' and its ecological requirements. As a result, this study shows that students also had difficulties to use scientific knowledge when they make an SSI decision.

Taking this into account, it would be important to carry out activities to promote the analysis of the controversy from different viewpoints and to facilitate the transference of scientific knowledge to SSI decisions. To illustrate this idea in the development of our classroom activity, we would design a set of new exercises in order to (a) emphasize the science and values that each stakeholder implied in bears' reintroduction draws upon to justify their ideas, (b) help students to understand that they will need to take into account all this information concerning the stakeholders to be able to justify their opinions, and (c) guide the instructors to highlight the relationship between the SSI and the concepts covered in class or share their viewpoints as long as they explain how they use facts to come up with their view or resolution.

In conclusion, further research will be needed in order to improve our understanding of decision making related to SSI. In particular, some studies can be conducted to examine deeply how students make decisions in these contexts and to describe which teaching strategies could be used to enhance students' learning transfer and to engage them in scientific practices in order to help them understand the nature of scientific knowledge.

Acknowledgments This chapter is based on work supported by Spanish MCYT grant EDU-2009-13890-C02-02, Spanish MCYT grant EDU-2012-38022-C02-02, and Catalan PRI 2009SGR1543. The authors thank the teachers and students involved in the research.

References

Acar, O., Turkmen, L., & Roychoudhury, A. (2010). Student difficulties in socio-scientific argumentation and decision-making research findings: Crossing the borders of two research lines. *International Journal of Science Education, 32*(9), 1191. doi:10.1080/09500690902991805.

Aikenhead, G. S. (1985). Collective decision making in the social context of science. *Science Education, 69*(4), 453–475. doi:10.1002/sce.3730690403.

Aikenhead, G. S. (2005). *Science education for everyday life: Evidence-based practice*. New York: Teachers College Press. ISBN 0807746347.

Albe, V. (2007). When scientific knowledge, daily life experience, epistemological and social considerations intersect: Students' argumentation in group discussions on a socio-scientific issue. *Research in Science Education, 38*(1), 67–90. doi:10.1007/s11165-007-9040-2.

Albe, V. (2008). Students' positions and considerations of scientific evidence about a controversial socioscientific issue. *Science Education, 17*(8–9), 805–827. doi:10.1007/s11191-007-9086-6.

Barab, A., Sadler, T. D., Heiselt, C., Hickey, D., & Zuiker, S. (2006). Relating narrative, inquiry, and inscriptions: Supporting consequential play. *Journal of Science Education and Technology, 16*(1), 59–82. doi:10.1007/s10956-006-9033-3. Springer.

Ekborg, M., Idelan, M., & Malmberg, C. (2009). Science for life-a conceptual framework for construction and analysis of socio-scientific cases. *Nordina, 5*, 35–46. Retrieved from https://www.journals.uio.no/index.php/nordina/article/view/277

Fensham, P. J. (2002). Time to change drivers for scientific literacy. *Canadian Journal of Science, Mathematics, and Technology Education, 2*(2), 207–213.

Fleming, R. (1986a). Adolescent reasoning in socio-scientific issues, part I: Social cognition. *Journal of Research in Science Teaching, 23*, 677–687.

Fleming, R. (1986b). Adolescent reasoning in socio-scientific issues, part II: Nonsocial cognition. *Journal of Research in Science Teaching, 23*, 689–698.

Grace, M., & Ratcliffe, M. (2002). The science and values that young people draw upon to make decisions about biological conservation issues. *International Journal of Science Education, 24*(11), 1157–1169. doi:10.1080/09500690210134848.

Irwin, A., & Wynne, G. (1996). *Misunderstanding science*. Cambridge: Cambridge University Press.

Jimenez-Aleixandre, M. P., Rodrigues, A. B., & Duschl, R. (2000). "Doing the lesson" or doing science": Argument in high school genetics. *Science Education, 84*(6), 757–792.

Kolsto, S. D. (2001). Scientific literacy for citizenship: Tools for dealing with the science dimension of controversial socioscientific issues. *Science Education, 85*(3), 291–310. doi:10.1002/sce.1011.

Kortland, K. (2001). *A problem posing approach to teaching decision-making about the waste issue*. Utrecht: Cdb Press.

Lewis, J., Leach, J., & Wood-Robinson, C. (1999). Attitude des jeunes face a` la technologie ge´nique. In L. Simonneaux (Ed.), *Les biotechnologies a l'ecole*. Dijon: Educagri editions.

Lincoln, Y. S., & Guba, E. G. (1985). *Naturalistic inquiry* (Vol. 23, p. 416). Newbury Park: Sage. ISBN 978-0803924314.

Molinatti, G., Girault, Y., & Hammond, C. (2010). High school students debate the use of embryonic stem cells: The influence of context of decision-making. *International Journal of Science Education, 32*(16), 2235–2251.

Nuangchalerm, P. (2010). Engaging students to perceive nature of science through socioscientific issues-based instruction. *European Journal of Social Sciences, 13*(1), 34–37.

Osborne, J. F., Erduran, S., & Simon, S. (2004). Enhancing the quality of argument in school science. *Journal of Research in Science Teaching, 41*(10), 994–1020.

Patronis, T., Potari, D., & Spiliotopoulou, V. (1999). Students' argumentation in decision-making on a socio-scientific issue: Implications for teaching. *International Journal of Science Education, 21*, 745–754.

Ratcliffe, M. (1997). Pupil decision-making about socio-scientific issues within the science curriculum. *International Journal of Science Education, 19*, 167–182.

Reis, P., & Galvao, C. (2009). Teaching controversial socio-scientific issues in biology and geology classes: A case study. *Electronic Journal of Science Education, 13*(1), 1–24.

Ryder, J. (2002). School science education for citizenship: Strategies for teaching about the epistemology of science. *Journal of Curriculum Studies, 34*(6), 637–658. Taylor & Francis.

Sadler, T. D. (2004). Informal reasoning regarding socioscientific issues: A critical review of research. *Journal of Research in Science Teaching, 41*(5), 513–536. doi:10.1002/tea.20009.

Sadler, T. D., & Dawson, V. (2012). Socio-scientific issues in science education: Contexts for the promotion of key learning outcomes. In B. J. Fraser, K. Tobin, & C. J. McRobbie (Eds.), *Second international handbook of science education*. Dordrecht/Heidelberg/London/New York: Springer.

Sadler, T. D., & Zeidler, D. (2005a). The significance of content knowledge for informal reasoning regarding socioscientific issues: Applying genetics knowledge to genetic engineering issues. *Science Education, 89*(1), 71–93. doi:10.1002/sce.20023.

Sadler, T. D., & Zeidler, D. (2005b). Patterns of informal reasoning in the context of socioscientific decision making. *Journal of Research in Science Teaching, 42*(1), 112–138. doi:10.1002/tea.20042.

Sadler, T. D., Chambers, W. F., & Zeidler, D. (2004). Student conceptualizations of the nature of science in response to a socioscientific issue. *International Journal of Science Education, 26*(4), 387. doi:10.1080/0950069032000119456.

Simmoneaux, L. (2001). Role-play or debate to promote students' argumentation and justification on an issue in animal transgenesis. *International Journal of Science Education, 23*(9), 903–928.

Wu, Y.-T., & Tsai, C.-C. (2007). High school Students' informal reasoning on a socio-scientific issue: Qualitative and quantitative analyses. *International Journal of Science Education, 29*(9), 1163. doi:10.1080/09500690601083375.

Yang, F.-Y., & Anderson, O. R. (2003). Senior high school students' preference and reasoning modes about nuclear energy use. *International Journal of Science Education, 25*(2), 221–244. doi:10.1080/09500690210126739.

Zeidler, D., Walker, K. A., Ackett, W. A., Simmons, M. L. (2002). Tangled up in views: Beliefs in the nature of science and responses to socioscientific dilemmas. *Science Education, 86*(3), 343–367. Salem, Mass.: WG Whitman, 1929-. doi:10.1002/sce.10025.

Zeidler, D. L., Sadler, T. D., Simmons, M. L., & Howes, E. V. (2005). Beyond STS: A research-based framework for socioscientific issues education. *Science Education, 89*, 357–377. doi:10.1002/sce.20048.

Zimmerman, C., Bisanz, G. L., & Bisanz, J. (1999, March). *Science at the supermarket: What's in print, experts' advice, and students' need to know*. Paper presented at the annual meeting of the National Association for Research in Science Teaching (NARST), Boston.

Zohar, A., & Nemet, F. (2002). Fostering students' knowledge and argumentation skills through dilemmas in human genetics. *Journal of Research in Science Teaching, 39*(1), 35–62. doi:10.1002/tea.10008. Wiley Online Library.

Chapter 6
Learning About the Role and Function of Science in Public Debate as an Essential Component of Scientific Literacy

Ingo Eilks, Jan Alexis Nielsen, and Avi Hofstein

1 Introduction

Science and technology are essential components for every modern society. They offer the base from which to promote economic growth and welfare. Developments and decisions concerning science and technology are also essential for protecting the environment for coming generations and thus enabling a sustainable development into our future (Burmeister et al. 2012). Therefore, society is continuously driven to make decisions about science and technology – in particular about their application and use with a view to their consequences on local, as well as regional, national, and global levels. In a democratic society, every citizen is thought to contribute to respective debates and decisions, even if the citizen is not an expert in science or technology. That is why science education should provide a respective basic knowledge and understanding for all students, but it should also offer a framework to learn about the use of science in societal debate (Bauer 2009; Hofstein et al. 2011; Millar and Osborne 1998; Sjöström 2013; Ryder 2001).

Learning about the use of science in societal debates should enable learners to become future responsible citizens. They should learn to cope with their life individually within the society they live in but also to participate actively in societal discourse concerning socio-scientific issues (Roth and Lee 2004, Hofstein et al. 2011; Sjöström 2013). Science education should more thoroughly consider its role in

I. Eilks (✉)
University of Bremen, Bremen, Germany
e-mail: ingo.eilks@uni-bremen.de

J.A. Nielsen
University of Copenhagen, Copenhagen, Denmark
e-mail: janielsen@ind.ku.dk

A. Hofstein
Weizmann Institute of Science, Rehovot, Israel
e-mail: avi.hofstein@weizmann.ac.il

C. Bruguière et al. (eds.), *Topics and Trends in Current Science Education: 9th ESERA Conference Selected Contributions*, Contributions from Science Education Research 1,
DOI 10.1007/978-94-007-7281-6_6, © Springer Science+Business Media Dordrecht 2014

the foreground of what Activity Theory or *Allgemeinbildung* have to offer as guidance for any kind of education in general and for science education in particular (van Aalsvoort 2004; Elmose and Roth 2005; Sjöström 2013; Stuckey et al. 2013). From the perspective of these theories, science education should more thoroughly contribute to "education through science," allowing the learner to develop general skills that are important for a responsible life in present and future society rather than structuring science instruction as rote learning of science facts and theories detached from their role for individual and societal decisions in the means of only "science through education" (Holbrook and Rannikmae 2007). Also all modern perspectives that set scientific literacy as the overarching aim of science education encompass a societal dimension and the aim of preparing the learner for a self-determined life in society (Bybee 1997; Roberts 2007). Students should learn about the interaction of science with the environment and society to achieve competencies to shape their society in future for a more sustainable development (Burmeister et al. 2012). Anyhow, learning about the interrelationship between science and society and about dealing with socio-scientific issues within society in all its complexity seems still to be not sufficiently implemented in many countries, e.g., the cases of Israel, Germany, and the United States as discussed in a recent paper by Hofstein et al. (2011).

The socio-scientific issue movement in science education has emerged from different curricular approaches, especially the Science-Technology-Society framework (e.g., Solomon and Aikenhead 1994; Yager and Lutz 1995). In recent years, this movement has suggested that students need to learn more intensively about how science is interacting with society and its related problems (Sadler 2004). Socio-scientific issues proved to offer a valuable framework for science education aiming at both scientific knowledge as well as more general educational skills and thus contributing to multidimensional scientific literacy (Sadler 2011).

This chapter includes itself in the discussion about a more societal orientation of science education as suggested by Hofstein et al. (2011). This chapter does not attempt to review the broad picture of any aspects related to the socio-scientific-issues-based science teaching movement. The special focus of this chapter is communication. Communication – in particular argumentative communication – is the essential mediator of discourse and debate in society (Nielsen 2011). Debate in society takes place in public discussions, parliaments, the Internet, other mass media, or personal communication. Debate is influenced by communications in newspapers, digital media, advertisings, or publications by interest groups and political stakeholders. The individual as responsible citizen has to deal with this kind of information to make up her or his mind and to become able to participate in the debate – both with the presented information as well as with way the information is presented to him (Hofstein et al. 2011).

This chapter reflects the issue of communication in societal debate from different perspectives. For developing and implementing societal-oriented science teaching, the content and context as well as aligned pedagogies are necessary. This is why we will first provide a framework for selecting suitable topics for more thoroughly reflecting on societal debate in science education. Then two approaches of a more general understanding how science and scientific information are dealt with for the purpose of public debate will be presented to lead to pedagogies and assessment on debate about socio-scientific issues in the classroom. The chapter finishes with the idea of mimicking authentic societal practices as a concrete instructional strategy for learning about the handling of science and scientific information within society.

2 Suitable Topics for Learning About Science-Based Communications in Societal Debate

It is suggested that a thorough orientation on learning about how science is handled in societal debate for contributing to multidimensional scientific literacy for all students will have an influence on the selection of content and its related instructional approaches to science education. Several authors suggest using the potential of context-based science curricula, not only to more effectively learning science content but also to learn about the interdisciplinary connectedness of science (Gilbert 2006; Fensham 2009). However, e.g., Sadler (2004, 2011), Marks and Eilks (2009), or Holbrook and Rannikmae (2007) criticized the, often poor, links of science teaching with society – even as is the case in some of the context-based science curricula (Gilbert 2006). They suggest that teaching/learning materials should encompass a more radical socio-scientific-issues-based approach. Also from the theoretical approaches of *Allgemeinbildung* or Activity Theory, a more thorough societal orientation in selecting topics for science education is advocated (e.g., Elmose and Roth 2005; Hofstein et al. 2011; Holbrook and Rannikmae 2007; Marks and Eilks 2009; Sjöström 2013; van Aalsvoort 2004). A typical example, as suggested in Hofstein et al. (2011), is climate change. Climate change is recognized worldwide as one of the central problems with which all societies are faced and to which they must respond. Understanding climate change has the potential for a lot of science learning, e.g., about the physics and chemistry of the atmosphere or about the relationship of using different technologies and their effects on global warming or to react to it. Thus, the area of climate change has much potential to enhance students' knowledge in science through utilizing an issue that is definitely relevant to and essential for their future. However, this topic also has the potential for students to learn about how such an issue is handled within society, and one can introduce the interplay of science with economics, politics, as well as cultural beliefs and values. In addition one can also debate about how potential actions on climate change take place in society and how scientific information is used in this realm (Feierabend and Eilks 2010).

A second example might illustrate also that the selection of a suitable context for learning about society's use of science and scientific information is often context bound to the society in which the students live. The example shows the value of letting concurrent societal issues determine the selection of the teaching content, rather than the structure of the discipline, and it shows how this can be operated in the practice of science teaching.

In the project of implementing chemical industry issues into science teaching in Israel since the 1980s several attempts were made to incorporate socio-scientific issues into the chemistry curriculum (Hofstein and Kesner 2006). The goals were to (a) demonstrate the application of basic chemical principles and concepts in industrial chemistry; (b) demonstrate the importance and relevance of the chemical industry to the students personally and to the society in which they live (and operate); (c) develop a basic knowledge of the technological, economic, and environmental factors involved in the establishment and operation of the chemical industry; (d) investigate some of the specific problems faced by a local chemical industry, such as the location of an industrial plant, the supply of raw materials, labor, ways of taking care of the environment, and its related economic aspects; (e) present the special language used in industry regarding basic technology, economy, and industrial processes; (f) show the differences between a laboratory and an industrial process and the scaling-up steps needed for industrial production design; (g) demonstrate the dynamic nature of industry, i.e., changes that chemical industries undergo continuously; and (h) provide information regarding economic, environmental, and technological problems which face the chemical industry as well as to demonstrate the connection between the chemical industry and social and political issues. To get relevance and authenticity, the selection of examples was not driven by the structure of the discipline. The examples were created around those issues relevant for the society in the case of Israel and around (or together with) the chemical enterprises available within the country.

The program was developed and implemented over a period of almost 25 years. A range of topics, teaching units, and materials were cyclically developed and tested. Accompanying research evidenced the feasibility and effect of the materials including attitude and interest in industrial related issues and perception of the classroom learning environment using various pedagogical approaches. An important component was the students' perception regarding the relevance of the activities to their current life and as future citizens (e.g., Kesner et al. 1997; Hofstein and Kesner 2006). One example in this project concerned the issue of establishing an industrial plant for the production of Bromine and its related products in the Dead Sea region of Israel. This example is highly relevant for the chemical industry in Israel. But it is also controversial for society: Establishing industrial plants in the lesser developed Negev region in Israel has a notable impact on the regional and national economy of the state of Israel. Above all, the development is controversial since the Dead Sea is rather a fragile ecosystem and chemical industry and regional tourism businesses might come into competition.

These examples proved that topics like the example on bromine from the Dead Sea enabled discussion on real situations, actual dilemmas, and how science influences our daily life. Such real-life situations provide the learner with an opportunity for science-based debates and decision-making processes connecting the benefits of the modern world, the standards of life with the price one pays for them, and how society is debating and deciding about it. Students are asked to take into consideration scientific and its related technological applications as well as geographical, political, economic, and other societal related issues. Both from the teachers' and students' point of view, this innovation inserted in the curriculum were found to be effective and successful. Teachers reported positively on the learning environments in which the students were involved in various activities related to the different socio-scientific issues. Students were assessed using both qualitative as well as quantitative research methods. The results show that chemistry industry topics can serve as a platform for implementation of varied-type student-centered learning methods and pedagogical techniques. Students have the possibility of presenting their personal views, but at the same time they must consider scientific facts and thus start to learn about the handling of science-related information within public or political debate (Hofstein and Kesner 2006).

A related framework for developing socio-scientific-issues-based chemistry education was conducted in Germany by Marks and Eilks (2009). In this framework, Marks and Eilks started developing criteria for the selection of suitable topics to promote learning about how science and science-related information is handled in society. As above and as also suggested, e.g., by Sadler (2004), also Marks and Eilks characterized the most fruitful topics for learning about science in society are authentic societal debates, with personal relevance to the student, and being of a controversial character. As their underlying goal for science education was described to contribute to education for responsible citizenship, they placed an emphasis on having students learn about the actions and intentions of stakeholders, politicians, media, or advertisers who often use (very often wrongly and persuasively) science-based information in their actions and also in their decisions. Students should become informed about the use and character of the information they are faced with and how it is selected for use within public communication. Thus Marks and Eilks (2009) suggested a more thorough consideration of respective contexts as an initial point for science teaching representing authentic societal debate through having varied and contradictory opinions available in everyday-life media. To give criteria for selection and proof of respective topics they suggested the following: (1) authenticity, (2) relevance, (3) being undetermined in a societal respect, (4) allows for open discussion, and (5) deals with a question based on science and technology (see also Hofstein et al. 2011). These criteria are outlined in more detail in Table 6.1 giving also suggestions for verifying them. These five criteria also mirror the elements identified in Part 1 of this chapter. For example, the connection of the context with the students' life and its relevance can be found in the criteria of authenticity and relevance. The interdisciplinary connectedness of science can be related to the criterion of the societal indeterminateness of a topic and its potential consequences for the individual and the society in which they operate. Finally, the rationale regarding the issues themselves and its bases in science and technology offer rich possibilities to challenge science learning, as already suggested by Sadler et al. (2011).

Now that we have criteria for selecting content and context to learn about science-related public debate, the question is as to how to understand and model the use of communication in this realm. This issue can be reflected from two perspectives: the way the individual is handling respective information and the way society is doing so. Both perspectives will be discussed in the following two sections based on until now rarely used philosophical models for understanding communication in science education.

Table 6.1 Criteria for topics suitable for learning about authentic societal debate and the handling of scientific information by the public

Criterion	Requirement	Proof
Authenticity	The issue is authentic, because it is currently being discussed by society	The issue is checked for real presence in the authentic everyday media (newspapers, magazines, TV, advertising, Internet, etc.)
Relevance	The topic is relevant, because any societal decision on it will impact directly the life of the students, today or in future	Scenarios are outlined to whether societal decision between the different options for action/opinions on the topic will impact the students' life today or in future, e.g., concerning restrictions to personal consumption or options for acting
Undetermined in a societal respect	Societal evaluation is open and allows for principally different points of view	The issue is checked to see if different, controversial viewpoints are represented in the public debate as documented in statements by special interest groups, politicians, media reports, etc.
Allows for open discussion	The issue allows for open discussions and debating	Thought experiments are used to test the points of view present in public debate, as taking one of the perspectives in the classroom discussion could insult any of the individuals (e.g., on ethnical, religious, or socio-economic reasons) or push the spokesman to the fringes of society
Based on science and technology	This topic concerns itself with a techno-scientific query, which is based on science and technology	Discourse in the media is analyzed whether scientific concepts are addressed and either explicitly or implicitly used for argumentation

3 Understanding the Individual's Use of Scientific Information in Societal Debates

From the above outlined justification, it is clear that enabling students to use science and scientific information for participating in societal debate became a prominent focus in science education research. One of the dominant rhetorics behind this claim in the socio-scientific issues movement was that such activities have the potential to allow students to operationalize their science knowledge using argumentation to "*formulate positions, and provide supporting evidence*" (Sadler 2004, p. 515) – thus fostering the ability to use science as "evidence" and use them for "evidence-based" decisions on such issues (Sadler and Donnelly 2006; Zeidler et al. 2005). As a consequence, several studies made an attempt to investigate and describe the skills of individuals, especially students, when being asked to argue about socio-scientific issues, e.g., based on theories of argumentation patterns (e.g., Albe 2008; Kolstø 2006; or Sadler and Donnelly 2006).

A more recent approach contributed to our understanding in the foreground of some new theoretical base – normative pragmatics – to answer the question what argumentative roles invocations of science content have in students' group discussions about a controversial socio-scientific issue, and what effects on the dialectics of the discussion such invocations have (Nielsen 2010, 2012a, b). The theoretical approach applied in the studies by Nielsen elucidates the question on the theoretical basis of a generic framework within argumentation theory and philosophy (Jacobs 2000). Normative pragmatics concerns the study of the practical significance of linguistic performances in argumentative interactions – where such performances have a normative dimension (Blair 2006). In normative pragmatics, argumentation is understood as a way of managing disagreement, in which discussants use language in order to influence each other's decisions (Goodwin 2001). Discussants avert to strategies in attempts to make their interlocutors do something (e.g., acknowledge their standpoint, provide more reasons, clarify what they said before) through the design of messages that have specific contents (what is being said?) and designs (how is it being said?). The goal of normative pragmatic analysis is to identify such "*strategies as strategies* [*and*] *explain how an arguer's utterance of some words can be expected to accomplish things like the imposition of probative burdens*" (Goodwin 2001, p. 9).

Based on normative pragmatics, Nielsen (2012a, b) investigated secondary schools students' skills in argumentation on the issues of gene therapy, an archetypical issue for socio-scientific debate. The aim was to arrive at an informed interpretation of the argumentative role of science talk turns in the overall dialectics of the discussion. One of the key findings in the analysis was that students typically invoked science in attempts to frame the issue in a way that would be beneficial for their position. At times, the invoked science content had information relevance in the sense that it conveyed scientific information that potentially could be used in future argumentation. But invoked science content typically had pragmatic relevance in the sense that it was actually used in order to "*justify or refute a contested standpoint*" (Jacobs and Jackson 1992, p. 162).

It is suggested that, on the one hand, the students in Nielsen's (2012a, b) study drew on science in order to articulate and identify salient sub-issues in the process of discussing the overall issue of gene therapy. In such cases, factual information was used to signal to interlocutors that a particular sub-issue may potentially be an object of moral discussion. Such usages of science had the function of establishing the factual background on the basis of ensuing socio-scientific negotiations. Though episodes of this character could be co-opted by an individual student at a later point in the discussion, they represented the factual background in a sufficiently transparent fashion for others to draw on them as well. Such transparent and pseudo-neutral exchanges clearly contribute to the critical quality of a discussion. On the other hand, science content often played a role when students introduced, or framed, a sub-issue in a way that was beneficial to their position. The "matter-of-fact" lens offered by invoking science made it appear that there is an intuitive position to the sub-issue, or it at least challenges the interlocutor with an increased burden of proof. This can be, and was indeed, an effective argumentative strategy. The speakers were able to frame the sub-issue in a manner which was favorable for their position but to also cloud that it could be important for the group to deliberate whether that specific way of framing the sub-issue was important. In that sense a student could use science to unconsciously embed decision criteria that could favor his or her position. Ideally, however, a socio-scientific debate should contain explicit discussion of these decision criteria. The theoretical approach presented here might offer a chance to better understand such deficiencies and thus contribute for enabling a higher critical quality of students' socio-scientific discussions.

As having had the focus from the individual's use of science-related information in public debate in this section, another view regarding the issue may start from use of respective information in public debate by society at large. This point of view focuses where and how science appears in public debate and how the non-expert citizen is confronted to it.

4 Modeling the Society's Use of Scientific Information in Societal Debates

To consider a perspective on society's use of science-related information, one has to be aware that it is very probable that the majority of students in compulsory secondary science classes will not embark upon the career of becoming a scientist. Only a very small fraction of our students from the compulsory phase of school science classes will ever come into contact with authentic science, since authentic science can only be found in research institutes, scientific publications, or respective conferences. Access to such knowledge is limited to scientists by reasons of grasp to the original sources of information. Also, only sufficiently educated scientists are able to read the texts with their formal scientific language, technical terms, and symbols. Most science-related information the vast majority of our students will come into contact with are not directly taken from the scientific domains. Starting from academic handbook and textbook literature, the reader starts only indirectly dealing with original science (Bauer 2009). Even more, from popular science magazines or school textbooks, the reader is no longer dealing with original scientific information at all. In almost all everyday situations and societal decision-making bodies, there is nearly no direct contact to real science and authentic scientific information; sometimes the information is not even prepared by scientists who are experts in the particular area in question.

In everyday life and in the context of societal debates, there are rarely provisions of authentic scientific information. To use a model developed by Fleck (1935), as discussed in Bauer (2009), scientific information can be understood to be in the core of concentric spheres, similar to a planetary system of a center and circulating peripheries. In this model, a marked-off field of science activity is surrounded by exoteric domains starting with academic handbooks and textbooks, via a field of the public understanding of science, towards society (Fig. 6.1). Bauer describes that nearly all nonacademic domains from the society and education have only small

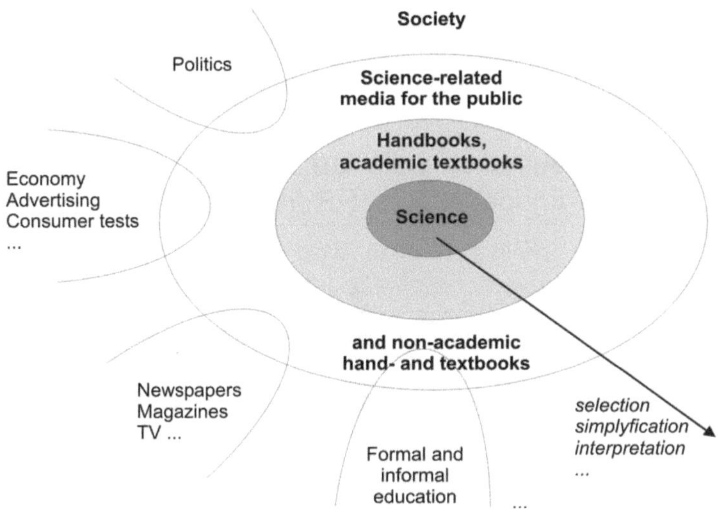

Fig. 6.1 The science-society-relationship with regard to types of information sources (Based on Bauer 2009)

overlap with the field of public understanding of science and have no overlap to the domains of science itself and the respective academic literature. Bauer describes gradients of simplification, iconicity, concreteness, certainty of judgment, and controversial reception coming from one of the domains to another. This gradient is clearly driven by the purpose and audience the information is prepared for.

Based on the model illustrated in Fig. 6.1, one can understand why science-related information used for public purposes appears only after several steps of transformation. Information in public use is presented to the consumer by special interest groups, politicians, journalists, or advertisers. This information is no longer authentically scientific, whether the source is TV, radio, newspapers, brochures, or advertisements. Every citizen is confronted with this kind of altered information, or better named filtered information (Hofstein et al. 2011). Everyone has to deal with it to make up her or his mind. In other words, for every nonscientist, the way suitable information for understanding a socio-scientific issue takes is very long and indirect. In a lot of single steps the original information is processed and filtered from one domain to another but also within the domains. This is done by individuals or groups through processes of selecting, simplifying, and interpreting the information in each of these steps.

The indirect link of the scientific information to society can be understood by means of valid models of information transfer. Belkin (1984) described information transfer as a dynamic interaction between three components: the user, the knowledge resource, and the intermediary mechanism. In his model the user initiates the information transfer because of a problem, goal, or intention. The intermediary mechanism mediates between the user's intention and interest and the source of

Fig. 6.2 The doubled filtering process of scientific information transfer

knowledge. In technology such intermediary mechanisms might be only simple transfer of devices. In our case, the intermediary mechanisms are very complex.

Bringing together the ideas of the model based on Fleck and the idea of models of information transfer, one can ask how the transfer works and what the intermediary mechanisms are. Each step of the information transfer from science to the general citizen is carried out by an individual person, be it a scientific journalist, news agent, author, editor, etc. As a result, the final product regularly is substantially different from the original, be it on reasons for comprehensibility or the purpose of the intended information use. The information is filtered by the individual foreknowledge, interests, and linguistic competencies of the writer, and also each person being involved along the filtering process. The further we move towards everyday life, the more the information most probably is filtered and altered. The further one travels from the domain of science, the greater the chance that the involved persons do not possess comprehensive subject matter knowledge necessary for securing reliability of the information transfer. As a result, the interaction with science-related information in everyday settings does not need only a simple evaluation of the pertinent scientific facts. Frequently, it is just as – if not more – important to understand which pathway the transmission of information followed and which interests have played a role in the transfer of this information. It is suggested that many different constitutes play a part in the process. It goes without saying that such groups as lobbyists, special interest groups, advertisers, and politicians are following their own agendas. Yet journalists also possess interests, even if they are as relatively clear as the wish to bring an attractive journalistic product to the readers. It has already been mentioned that the subject matter qualifications of the author are most often not clearly recognizable. But the filtering process is by no means determined solely by the knowledge-based scientific competences of the author. Of equal importance are such factors as personal interests and the competency to represent and deliver information in a solid, understandable, and objective manner (or at the least to intentionally and self-consciously craft such representations in a fashion which mirrors reality as accurate as possible). In this regard, a model of a double-filter mechanism is suggested for understanding information transfer in society and to teach about it as it is represented in Fig. 6.2. Understanding this double filtration is directly related to understanding how to (re)act towards such information. This is true, e.g., in the case of determining the credibility of an information source. This includes working out various strategies, e.g., self-consciously contrasting different pieces of information quite possibly stemming from sources with diametrically opposed interests.

The described alignment between science and society in this model is not unidirectional. Already Fleck in 1935 explained that aside from the influence of science on society, there is also an influence of society on science. He illustrated, e.g., this influence by ethical guidelines set from within society for research or by political decisions regarding which areas of research will be funded. However, there is also a more individual reaction towards science that can occur. If the individual is not able to cope with the information presented to him in public and if the public has no skills as to how to deal with the filtered information coming towards the individual citizen in daily life, there is always the danger that the individual too easily is misled by biased information. In other words, general criticism towards science and research emerges, and the information derived from science by those who lack evidence will be considered of equal importance for societal decision making. It is suggested that in order to avoid such and similar biased decisions, science educators must ensure that the learners will have opportunities to study about the interrelations of science and society. Part 5 of this chapter will provide insights into potential pedagogies.

5 Pedagogies to Learn About Individual's and Society's Handling of Scientific Information

Over the period of 10 years now, Marks and Eilks (2009) have been advocating a scheme for structuring socio-scientific issues based on science teaching, called the socio-critical and problem-oriented approach to science teaching. The approach is aligned with ideas developed by Sadler (2004) and also with the recent curriculum movement of socio-scientific-issues-based science education (Sadler 2011). The approach operates a 5-step model for structuring lesson plans about authentic and controversial socio-scientific issues in science classes. Lessons according to this approach always start with authentic media introducing a controversy from society and allowing for the analysis of the problem. In a second phase the essential science theory is learned to provide background knowledge. In the third step the learning is reflected upon, as to whether the learned content was of potential to support decision making within the controversial options for action – and in which way science was not able to give clear answers to one of the initially outlined questions, e.g., concerning decisions based on ethical, economic, or social reasons. In phase four the different points of view documented in society are analyzed and contrasted. Finally the ideas surrounding how society comes to decisions about such a socio-scientific question and which role science and scientific information is playing in this process are reflected upon.

In recent years, this model was refined alongside different curriculum development projects based on participatory action research (e.g., Marks and Eilks 2009). In particular, the phases of contrasting the different opinions and reflecting the societal mechanisms of discourse and decision making underwent continuous development. In this framework, the idea of mimicking a variety of authentic societal practices was developed. This idea offers a broad spectrum of suitable pedagogies for learning regarding the handling of science-related information by individuals or by society at large. When learners mimic authentic societal practices, they imitate the process of individuals or professional groups from the society how they are dealing with science-related information in a socio-scientific controversy. Originally the idea was connected to role-playing and business game exercises (e.g., Marks et al. 2008). However, the potential yield of role playing and business games is limited because it only can be applied to mimic a limited variety of scenarios of information handling within society. Role playing also is highly demanding because of the required time and students' skills in argumentation. Especially younger, novice students often cannot present their arguments in an eloquent manner and carry out their reasoning in a somewhat haphazard and non-integrated fashion. This is especially the case when a topic has only an indirect link to their everyday lives (Marks et al. 2008). When this occurs, students have difficulties with the complexity of the role, because they do not possess the necessary background and experiences. Because of this, the role players often cannot react flexibly enough and tend to stick to a "script" with regard to their assigned position. However, role-playing exercises can be generally quite successful whenever the learners bring a large amount of personal experience with them, e.g., if the scenario of the role play is quite common, if the roles are taken from the students life, and if the topic has direct meaning to them (Marks et al. 2008).

As said previously, role playing is often a demanding exercise and has only limited potential applications for mimicking societal practices of information transfer and use. That is why we suggest to look for more societal practices of information transfer and use to be mimicked in science education for learning about socio-scientific information handling within society. From these societal practices of information handling, alternative pedagogies can be derived with lesser demands on students' argumentation skills by giving the student clearer structures and graded learning materials. Such methods were inspired, e.g., by the working of a journalist where the students are asked to write a news report on the controversial topic, as if they were working in the editorial pool of newspaper or TV news magazine (Marks and Eilks 2010). Students receive their information through different sets of quotes ("news tickers") from sources of information provided by advertisings, pressure groups, political stakeholders, etc. For example, "news tickers" can be made up by using a Google search for the topic, e.g., musk fragrances, in combination with one more term, e.g., consumer protection, innovative product development, ecological impacts, or health (Marks et al. 2010). From the first 20 items found, one A4 page of quotes is taken and provided to the student, acting as a journalist in a local newspaper or TV company as he would start his search for any new and unfamiliar topic. Thus, within each of the various news tickers given to different groups of students, a set of short messages are offered which talk about the topic with respect to a different perspective on the issue. When starting to create their own news, most students become instantly aware about the weak base of information that the news tickers provided. Some asked for further information and searched the internet by themselves. These students tried to give a more balanced presentation but also became aware of not knowing whether every journalist would behave the same. Even these students at the end recognized the different points of view such a topic can have. It became clear that this is typical for much of the news written and presented in the media. In asking two groups of students starting from the same news ticker, it also became clearly apparent how big the influence of the single "journalist" is for the quality, presentation, and credibility of the final product. In researching the method it was clear that this method offers a good route for promoting wide-reaching self-reflection and critical thinking regarding various aspects of communication with respect to science in the everyday life. It offers an opportunity for opening dialogues with and among learners about socially relevant, scientific topics and how science-related information is used for public discussion and decision making regarding these topics. It promotes understanding of the role individuals play in such an information transfer process, be it either as the provider of the news ticker or the journalist selecting and transforming the information into the article in the foreground of his prior knowledge, skills, and interests (Marks and Eilks 2010; Marks et al. 2010). There are much more authentic societal practices that can be used in this way. Other recently developed examples mimic, e.g., the work of consumer test agencies (Burmeister and Eilks 2012) or advertisers (Stuckey et al. 2012). Also analyzing dialogues from everyday life or puzzling them based on given quotes of different potential persons involved (students, parents, salesman, consultant) proofed to be very appropriate for reflecting information use in public communication (Eilks et al. 2012).

In respective evaluation case studies (e.g., Burmeister and Eilks 2012; Marks and Eilks 2010), these instructional techniques proved to be very motivating. These results are similar to those discussed by Sadler et al. (2011) and Hofstein and Kesner (2006). The lesson plans allowed the students learning about different examples of socio-scientific information handling by individuals and society at large. The students' feedback gives initial evidence that they became more reflective and critical about them (Eastwood et al. 2011). Some of the students explicitly recognized having learned that the knowledge about the information transfer – the intermediary mechanism in the theory of Belkin (1984) – can have a bigger influence on how the information appears to the reader/listener than the information itself might have. From the experience of handling information as professionals do in society, the students begin to acknowledge the importance of knowing about the intermediary mechanisms and to reflect upon the source of science-related information in public discourse. The students start acknowledging that it might be even more important to learn about the intermediary mechanism than to be able to proof the reliability of the information on their own (Eilks 2002).

6 Conclusions

Clearly, communication is important for discussion and debate regarding socio-scientific issues. While accepting that the major goal of science education is to contribute to the preparation of the younger generation for active participation in society, science education should more thoroughly focus on the promotion of respective skills. From the studies presented in this chapter, it is clear that forming a comprehensive theory for understanding communication in societal discussions is still in an emerging stage. This chapter suggests different new contributions to the debate about understanding the use of science-related information in societal discussions, namely, normative pragmatics (Jacobs 2000) and the theory of the connection from science to society inspired by Fleck (1935). Both from empirical studies and research-based curriculum development, this chapter suggests new ideas for reflection on societal questions and on their potential to become a motivating component for the science classroom. In this chapter, we made an attempt to present several theoretical models and findings to better reflect on the role scientific information plays or should play on societal debate. From theoretical models concrete suggestions for classroom practice were derived illustrated by concrete teaching examples from the countries of Israel and Germany. The examples clearly illustrate that both the empirically derived models like the one based on the application of normative pragmatics as well as the one derived from the philosophy of science based on the works of Fleck have potential to enrich the discourse in science education. In addition, they are valuable in that they not only lead to a better understanding of students' learning and achievement, they also provide valuable ideas for innovating science teaching as well as provide very concrete suggestions for effective classroom science teaching practices.

References

Albe, V. (2008). When scientific knowledge, daily life experience, epistemological and social considerations intersect: Students' argumentation in group discussions on a socio-scientific issue. *Research in Science Education, 38*, 67–90.

Bauer, M. W. (2009). The evolution of public understanding of science – discourse and comparative evidence. *Science Technology Society, 14*, 221–240.

Belkin, N. J. (1984). Cognitive models and information transfer. *Social Science Information Studies, 4*, 111–129.

Blair, J. A. (2006). Pragma-dialectics and pragma-dialectics. In P. Houtlosser & A. van Rees (Eds.), *Considering pragma-dialectics*. Mahwah: Lawrence Erlbaum.

Burmeister, M., & Eilks, I. (2012). An example of learning about plastics and their evaluation as a contribution to education for sustainable development in secondary school chemistry teaching. *Chemical Education Research and Practice, 13*, 93–102.

Burmeister, M., Rauch, F., & Eilks, I. (2012). Education for Sustainable Development (ESD) and secondary chemistry education. *Chemical Education Research and Practice, 13*, 59–68.

Bybee, R. W. (1997). Toward an understanding of scientific literacy. In W. Gräber & C. Bolte (Eds.), *Scientific literacy – an international symposium*. Kiel: IPN.

Eastwood, J. L., Schlegel, W. M., & Cook, K. L. (2011). Effects of an interdisciplinary program on students' reasoning with socioscientific issues and perceptions of their learning experience. In T. D. Sadler (Ed.), *Socio-scientific issues in the classroom*. Dordrecht: Springer.

Eilks, I. (2002). Teaching 'Biodiesel': A sociocritical and problem-oriented approach to chemistry teaching, and students' first views on it. *Chemical Education Research and Practice, 3*, 67–75.

Eilks, I., Belova, N., Von Döhlen, M., Burmeister, M., & Stuckey, M. (2012). Kommunizieren und Bewerten lernen für den Alltag am Beispiel der Energydrinks. *Der Mathematische und Naturwissenchaftliche Unterricht, 65*, 480–486.

Elmose, S., & Roth, W. M. (2005). Allgemeinbildung: Readiness for living in a risk society. *Journal Current Studies, 37*, 11–34.

Feierabend, T., & Eilks, I. (2010). Raising students' perception of the relevance of science teaching and promoting communication and evaluation capabilities using authentic and controversial socio-scientific issues in the framework of climate change. *Science Education International, 21*, 176–196.

Fensham, P. J. (2009). Real world contexts in PISA science: Implications for context-based science education. *Journal of Research in Science Teaching, 46*, 884–896.

Fleck, L. (1935). *Entstehung und Entwicklung einer wissenschaftlichen Tatsache. English translation 1979: Genesis and development of a scientific fact*. Chicago: University of Chicago.

Gilbert, J. K. (2006). On the nature of context in chemical education. *International Journal of Science Education, 28*, 957–976.

Goodwin, J. (2001). One question, two answers. In H. V. Hansen, C. W. Tindale, J. A. Blair, R. H. Johnson, & R. C. Pinto (Eds.), *Argumentation and its implications*. Windsor: Ontario Society for the Study of Argument.

Hofstein, A., & Kesner, M. (2006). Industrial chemistry and school chemistry: Making chemistry studies more relevant. *International Journal of Science Education, 28*, 1017–1039.

Hofstein, A., Eilks, I., & Bybee, R. (2011). Societal issues and their importance for contemporary science education: A pedagogical justification and the state of the art in Israel, Germany and the USA. *International Journal of Science and Mathematics Education, 9*, 1459–1483.

Holbrook, J., & Rannikmäe, M. (2007). The nature of science education for enhancing scientific literacy. *International Journal of Science Education, 29*, 1347–1362.

Jacobs, S. (2000). Rhetoric and dialectic from the standpoint of normative pragmatics. *Argumentation, 14*, 261–286.

Jacobs, S., & Jackson, S. (1992). Relevance and digressions in argumentative discussion: A pragmatic approach. *Argumentation, 6*, 161–176.

Kesner, M., Hofstein, A., & Ben-Zvi, R. (1997). Student and teacher perceptions of industrial chemistry case studies. *International Journal of Science Education, 19*, 725–738.

Kolstø, S. D. (2006). Patterns in students' argumentation confronted with a risk-focused socio-scientific issue. *International Journal of Science Education, 28*, 1689–1716.

Marks, R., & Eilks, I. (2009). Promoting scientific literacy using a socio-critical and problem-oriented approach to chemistry teaching: Concept, examples, experiences. *International Journal Environment Science Education, 4*, 231–245.

Marks, R., & Eilks, I. (2010). The development of a chemistry lesson plan on shower gels and musk fragrances following a socio-critical and problem-oriented approach – a project of participatory action research. *Chemical Education Research and Practice, 11*, 129–141.

Marks, R., Bertram, S., & Eilks, I. (2008). Learning chemistry and beyond with a lesson plan on "potato crisps", which follows a socio-critical and problem-oriented approach to chemistry lessons – a case study. *Chemical Education Research and Practice, 9*, 267–276.

Marks, R., Otten, J., & Eilks, I. (2010). Writing news spots about chemistry – a way to promote students' competencies in communication and evaluation. *School Science Review, 92*(339), 99–108.

Millar, R., & Osborne, J. (1998). *Beyond 2000: Science education for the future*. London: King's College.

Nielsen, J. A. (2010). Functional roles of science in socio-scientific discussions. In I. Eilks & B. Ralle (Eds.), *Contemporary science education – implications from science education research about orientations, strategies and assessment*. Aachen: Shaker.

Nielsen, J. A. (2011). Dialectical features of students' argumentation: A critical review of argumentation studies in science education. *Research in Science Education, 43*, 371–393.
Nielsen, J. A. (2012a). Arguing from Nature: The role of 'nature' in students' argumentations on a socio-scientific issue. *International Journal of Science Education, 34*, 723–744.
Nielsen, J. A. (2012b). Science in discussions: An analysis of the use of science content in socioscientific discussions. *Science Education, 96*, 428–456.
Roberts, D. A. (2007). Scientific literacy/science literacy. In S. K. Abell & N. G. Lederman (Eds.), *Handbook of research in science education*. Mahwah: Lawrence Erlbaum.
Roth, W. M., & Lee, S. (2004). Science education as/for participation in the community. *Science Education, 88*, 263–291.
Ryder, J. (2001). Identifying science understanding for functional scientific literacy. *Studies in Science Education, 36*, 1–44.
Sadler, T. D. (2004). Informal reasoning regarding socioscientific issues: A critical review of research. *Journal of Research in Science Teaching, 41*, 513–536.
Sadler, T. D. (2011). *Socio-scientific issues in the classroom*. Dordrecht: Springer.
Sadler, T. D., & Donnelly, L. (2006). Socioscientific argumentation: The effects of content knowledge and morality. *International Journal of Science Education, 28*, 1463–1488.
Sadler, T. D., Klostermann, M. L., & Topcu, M. S. (2011). Learning science content and socio-scientific reasoning through classroom explorations of climate change. In T. D. Sadler (Ed.), *Socio-scientific issues in the classroom*. Dordrecht: Springer.
Sjöström, J. (2013). Towards bildung-oriented chemistry education. *Science and Education, 22*, 1873–1890.
Solomon, J., & Aikenhead, G. (Eds.). (1994). *STS education: International perspectives on reform*. New York: Teachers College Press.
Stuckey, M., Lippel, M., & Eilks, I. (2012). Sweet chemistry: Learning about natural and artificial sweetening substances and advertising in chemistry lessons. *Chemistry in Action*, 36–43.
Stuckey, M., Mamlok-Naaman, R., Hofstein, A., & Eilks, I. (2013). The meaning of 'relevance' in science education and its implications for the chemistry curriculum. *Studies in Science Education, 34*, 1–34.
Van Aalsvoort, J. (2004). Activity theory as a tool to address the problem of chemistry's lack of relevance in secondary school chemistry education. *International Journal of Science Education, 26*, 1635–1651.
Yager, R. E., & Lutz, M. V. (1995). STS to enhance total curriculum. *School Science and Mathematics, 95*, 28–35.
Zeidler, D. L., Sadler, T. D., Simmons, M. L., & Howes, E. (2005). Beyond STS: A research-based framework for socioscientific issues education. *Science Education, 89*, 357–377.

Chapter 7
Exploring Secondary Students' Arguments in the Context of Socio-scientific Issues

Fatih Çağlayan Mercan, Buket Yakmacı-Güzel, and Füsun Akarsu

1 Introduction

The commonly accepted goal for science education is to help all citizens become scientifically literate (NRC 1996; OECD 2000). Scientific literacy aims democratizing science so that citizens are equipped with knowledge and skills that are necessary for critically engaging in and making informed decisions about scientific, technological, and societal issues (Kolstø 2001; Oulton et al. 2004). Many decisions that await citizens such as genetically modified food, energy choices, and managing natural resources are called "socio-scientific issues" (SSI). SSI provides a unique opportunity for employing scientific literacy and therefore is important for citizenship education. Because decisions about SSI often result in taking political action, which may be determined by the votes of citizens and by citizens involvement in other forms of democratic participation, as well as individual or communal action, which may involve changes in lifestyles.

The outcome of the decisions about SSI is related to citizens' attitudes towards taking individual and political action. Accordingly, one line of research investigated students' attitudes towards taking pro-environmental action, especially about SSI related to the environment. Secondary school students' views on the usefulness of various specific actions at reducing global warming and their willingness to take these actions has been investigated in Australia (Boyes et al. 2009), in Spain (Rodríguez et al. 2010), in India (Chhokar et al. 2011), in Oman (Ambusaidi et al. 2012),

F.Ç. Mercan (✉) • B. Yakmacı-Güzel • F. Akarsu
Faculty of Education, Department of Secondary School Science and Mathematics Education, Bogazici University, Istanbul, Turkey
e-mail: fatih.mercan@boun.edu.tr; yakmacib@boun.edu.tr; akarsu@boun.edu.tr

C. Bruguière et al. (eds.), *Topics and Trends in Current Science Education: 9th ESERA Conference Selected Contributions*, Contributions from Science Education Research 1,
DOI 10.1007/978-94-007-7281-6_7,

and in Turkey (Kılınç et al. 2011). The results of this line of research showed that regardless of the country, there was a discrepancy between students' beliefs in usefulness of political actions and their willingness to undertake these political actions. For example, although most of the students believed in the usefulness of voting for international cooperation on environmental problems, fewer students were actually willing to vote for international cooperation. This discrepancy was more evident for Spain and Australia and less evident for India, Turkey, and Oman, implying that there might be differences in attitudes towards political pro-environmental action across different countries. More importantly, this line of research does not provide an explanation for the apparent discrepancy between belief in usefulness of political action and willingness to undertake political action, neither within nor among different countries. Understanding reasoning about SSI may help explain the discrepancy between the beliefs about usefulness of political action and willingness to undertake it.

Reasoning about SSI, often conceptualized as informal reasoning, is complex because SSI are often controversial, ill-structured, and lack clear-cut solutions (Sadler 2004). Informal reasoning is not purely rational in which relevant scientific evidence necessarily leads to an appropriate conclusion; instead, in the context of SSI, "reasonable people reasonably disagree" (King and Kitchener 2004, p. 5). The disagreements exist because reasoning about SSI involves making value judgments in which the kinds of evidence that people draw on and the standards of justification they employ vary. Informal reasoning involves moral considerations including personal experiences, beliefs, and values (Kuhn 1991; Sadler and Zeidler 2004). Informal reasoning also involves social considerations including economic, political, legal, and religious factors (Barab et al. 2007). Despite its complexity, reasoning about SSI does not necessarily require a huge amount of scientific knowledge (Ekborg 2008; Lewis and Leach 2006); however, it involves epistemic considerations including the role of scientists and uncertainty involved in producing and evaluating scientific knowledge (Albe 2008; Sadler et al. 2004).

The discrepancy between students' beliefs in usefulness of pro-environmental political action and their willingness to undertake political action may be explained by the complexity of informal reasoning. Based on the nature of informal reasoning, it is reasonable to expect that value judgments and personal considerations are influenced by the society and hence vary from one cultural context to another. Then, it is necessary to understand how students in specific cultures reason about SSI related to the environment. However, much of the science education research about informal reasoning has been conducted in Western Europe (Albe 2008; Ekborg 2008) and the USA (Barab et al. 2007; Sadler and Zeidler 2004). Few studies have explored informal reasoning in the context of SSI in a developing country such as Turkey.

The purpose of this study was to explore Turkish secondary school students' views on SSI and the reasons for holding those views in the context of responding to SSI dilemmas and to describe students' explanations for their beliefs in usefulness of political action and their willingness to undertake political action.

2 Methods

Until 2010 students in Turkey made the transition from primary to secondary education based on the results of a series of centralized exams called Level Determination Exams (or SBS) that students took at the end of grades 6, 7, and 8. For each age cohort, about one million students took the SBS and were placed to various secondary education institutions according to their ranks in the exam (MoNE 2007).

The school that this study was conducted was an academically selective school in Istanbul and accepted students whose ranks were above the top 5 % at the SBS for the year 2010 (MoNE 2010). The school's achievement in transition to higher education is high; 80 % of the graduates of 2010 were placed in a higher education institution. In the school the cohort of grade 10 (age 16) students had about 120 students in four classes. The teachers and administrators in the school stated that the classes at grade 10 were similar in terms of achievement, which was taken as an indicator for the homogeneity of the group in addition to the students' SBS scores. Two of the four classes were selected, from which 36 students volunteered to participate in the study. The reason for selecting high achieving students was the anticipation that these students would have some background knowledge on the SSI and would be articulate about their reasoning. Because the students in this school were high achievers based on their SBS scores, the results should be interpreted with caution for generalizing to all students in Turkey.

To attain in-depth understanding, a qualitative study was designed in which 36 students (22 male, 14 female) at grade 10 (age 16) responded in writing to three SSI dilemmas and then participated in a group interview. The SSI dilemmas were about energy crisis, global warming, and water taxation (see Appendix). The participants were given 90 min to respond in writing to two of the three dilemmas. Each dilemma was completed by 24 students. Afterwards, the participants were grouped with respect to the dilemmas they have responded to, and nine groups of four participants were formed for interviewing. The group interviews were conducted with the dual purposes of giving the participants a chance to clarify their written responses and to interact with the other participants. Each group interview was conducted in a quiet room at the students' school, lasted about 40 min, audio and video recorded, and was transcribed verbatim for further analysis. In all of the group interviews, two of the researchers co-mediated the discussions. During the group interviews the participants were given their written responses, then for each dilemma they summarized their ideas, and then the participants talked to each other, discussing, asking questions, challenging, or supporting others' arguments. The researchers also contributed to the discussions by asking clarifying questions when appropriate.

The data analyses involved cyclic processes of data reduction, data display, and conclusion drawing/verification, which are characteristics of qualitative analysis (Miles and Huberman 1994). All transcripts were imported to the qualitative analysis software ATLAS.ti to manage and organize the data as well as to keep track of the analytic progress. Data was reduced by coding in which the bulk of the dataset was

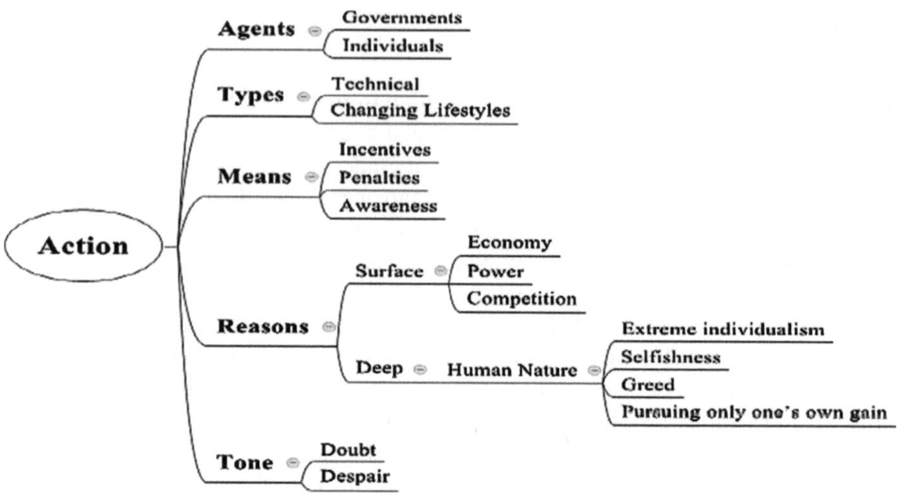

Fig. 7.1 Argument themes emerged in the context of the SSI dilemmas

condensed into analyzable units by creating categories from the data, utilizing constant comparative method (Glaser and Strauss 1967). The unit of analysis ranged from one to several sentences. Coding did not start with a theoretical framework and literature review was intentionally delayed to avoid being influenced by a particular theoretical perspective. The first and second author independently read the transcripts and marked sections of text with descriptive codes such as "selfishness" and "competition among countries." The first round of coding also involved context codes such as "energy crisis dilemma" to identify which dilemma was being discussed in particular segments of text, as well as identity codes such "student 1" to mark which student was talking in which part of the text. After completing open coding, the two researchers compared and discussed the open codes, identified overlapping codes, and merged similar codes under common names. The third researcher was involved in the axial coding stage of the analysis to provide external audit. In this stage, the most frequent codes were structured under more general categories. For example, "human nature" category was constructed to include the code "selfishness," and "deep reasons" category was constructed to subsume the category "human nature." "Action is necessary" was a code that encompassed all transcripts, and upon discussions "action" was judged as the organizing concept for the entire analysis. Then a map of the categories was constructed (shown in Fig. 7.1) with the consensus of the three researchers based on convergence, agreement, and coverage, which was used as the criteria for validity of the analysis (Gee and Green 1998). Afterwards, the transcripts were selectively coded to verify that the categories matched the data. Finally, a table showing each student's responses across the dilemmas and categories was constructed and used to count the number of students in each category for each dilemma.

3 Findings

As the result of the data analysis, the following themes emerged: action, agents of, types of, means of, and reasons for action, and the tone or attitudes towards action (see Fig. 7.1).

3.1 Action

Most of the students stated that they did not possess adequate scientific knowledge about the SSI dilemmas. Nevertheless, the students responded to the SSI dilemmas either positively, signifying that the dilemmas could be resolved, or negatively, suggesting that the dilemmas could not be resolved. In either case, resolution or lack of it was determined solely by acting or failing to act. For instance, according to the students the resolution of the global warming problem depended solely on the governments deciding to reduce carbon dioxide emissions and acting resolutely on these decisions. However, according to most of the students, such a decision was not likely and it will not be acted on. For example, a student said:

> Here it says that they will reduce the gas emissions by 30 %. This may look like a big ratio but I do not think it is enough. An assembly is gathered but they act like they are there for fun and sightseeing, returning home without reaching any serious decisions. Many experts got together but they came back without achieving anything. Because their governments do not allow it, it doesn't fit with their interests. (G6.S33)[1]

For the energy crisis dilemma, there were four positive and 20 negative responses; for the global warming dilemma, there were six positive and 18 negative responses; and for the water taxation dilemma, there were 22 positive and two negative responses. The students appear to think that water taxation can be resolved by taking action, whereas they seem less hopeful for resolving energy crisis and global warming.

3.2 Agents of Action

According to the students, the required actions for resolving the SSI dilemmas would have to be taken by governments or individuals. The majority of the students stated that individual actors could not change the world; it was mainly governments' responsibility to take action for the SSI, which they often failed. For instance, some students argued that regulating water consumption was beyond the individual,

[1] G6 represents group number 6 and S33 represents student number 33.

individual efforts would be too limited, and the government must act and intervene with water consumption by passing a water tax bill. For example, a student said:

> The water issue is not something that a person can solve by himself. Will my altruism really work? Maybe I will do the self-sacrifice but nobody else will? Maybe he will think about the short term. So there must be something that will force the people… The government can do this… By introducing a tax for instance. (G2.S09)

The students who did not support the water tax and stated that individuals may have some influence on reduction of water consumption, nevertheless, expressed that the water consumption issue cannot be resolved solely by individual efforts but by the government's action, acting to educate the citizens, launching advertisement campaigns for water conservation, or encouraging the citizens to adopt new technologies that reduce water consumption by offering incentives. For example, a student said:

> I wrote that the water tax is really unnecessary. For instance they tried to create awareness about water issues by broadcasting advertisements saying "don't leave the water running when you brush your teeth" or "when you shave and wash your cars use buckets instead of hoses". If the entire 72 million people do these maybe we can reduce water consumption. Still probably the authorities will have to do the campaigns for the awareness advertisements. (G7.S16)

For the energy crisis dilemma, 21 students said that the governments are the agents for the actions to be taken and three students said that individuals could be the actors; for the global warming dilemma, 22 students said that the actors are the governments and two students said that the actors are individuals; and for the water taxation dilemma, 15 students said that the governments are the actors and eight said the individuals are. The water taxation was apparently different than the other two dilemmas in terms of eliciting responses indicating that individuals may be actors for resolving SSI.

3.3 *Types of Action*

According to the students, there could be two types of action for resolving the SSI dilemmas: interpreting the problem as a technical one, it could be developing new technology or perpetuating existing technologies; or interpreting the problem essentially as an ethical one, it could be changing people's ways of living. For instance, some students stated that the energy crisis could be resolved by developing new energy technologies that provide clean and abundant power. For example, a student said:

> Since we need energy so desperately then we should produce more energy. This is about scaling up energy resources. For instance, in Turkey deserted areas those are not suitable for agriculture but get a lot of sunlight which may be used for power production. There was a desert like this, there are mirrors in a vast area that turn towards the sun as the sun moves in the sky, they reflect the sunlight to a steam turbine. (G4.S53)

Other students who thought that the energy crisis is an essentially ethical problem and not a technical one argued that the resolution is independent of the amount of available energy and resolution can only happen through fair distribution of energy, only if people are thoughtful about not only themselves but others as well and wanted to share the resources they have with others in a just way. According to these students, the resolution was only possible by a profound change in ethical considerations and with it adopting a new lifestyle based on sharing and responsibility towards all people on Earth. However, these students did not find this likely; the countries that are rich in energy resources would not agree to share their resources with other countries. For example, a student said:

> Energy resources should be distributed equally. That should be the aim; each country should get what it needs. Let's live in peace and share. But it just doesn't work that way. The most powerful always wins. I don't think that energy resources will ever be distributed equally. Somebody will just violate it. (G2.S08)

For the energy crisis dilemma six students interpreted the problem as a technical one and 18 as an ethical problem, for the global warming dilemma eight students interpreted it as a technical problem and 16 as an ethical issue, and for the water taxation dilemma 15 students interpreted it as a technical problem and nine as an ethical problem. The water taxation dilemma appears to be different than the other two dilemmas in terms of being interpreted as a technical problem.

3.4 *Means of Action*

The students suggested three ways as means for action including providing incentives, enforcing penalties, and creating awareness in relation to both types (technical and lifestyle changes) of action.

3.4.1 Incentives

According to the students, offering incentives is one way to implement both perpetuating new technologies and changing lifestyles. The incentives could be in the form of tax reduction or exemption, low or no interest loans, or some other benefit for the people to adopt new technologies or change their lifestyles. For example, when talking about the water taxation, a student expressed that adopting drip irrigation technologies in farming would reduce water consumption and the farmers could be encouraged for adopting drip irrigation by offering them a loan. She said:

> I did not support the tax… Instead of taxing people they can be made more aware: For example in agriculture drip irrigation technologies can be encouraged. I don't know if it is expensive, but even so the government can offer incentives to farmers such as giving them the cost upfront as a loan and then having the farmers pay back with their profits, as a way for reducing water consumption. (G9.S25)

3.4.2 Penalties

According to the students, a second way to help people adopt new technologies or change lifestyles is enforcing penalties when people do not comply with the proposed actions. The penalties are often stated as monetary fines, but some students also offered further sanctions including imprisonment, depending on the issue at hand. One of the common measures that the students offered as a means of action in the context of the water taxation issue was introducing a fine for overconsumption. For example, a student stated:

S03: Well maybe a limitation (on water usage) can be introduced and those who go over that limit may be fined. We can put a quota for those who do unnecessary consumption.
S43: Like the 4 GB quota (for the broadband internet).
S03: That would be good. Only those who go over the limit would be fined. (G9)

3.4.3 Awareness

A third way that the students offered as a means for action was creating awareness in the public. According to these students, explanations must be provided to the public for the proposed actions including the rationale behind the actions and the possible benefits for the people. Someone would have to go and explain the action to the stakeholders. For instance, when discussing the water taxation issue some of the students said:

S45: Well they have to go to every plant and every production facility and explain the tax very clearly. What other way can there be?
S18: It's all about creating awareness.
S45: Awareness could be created, the reasons could be shown. I mean if someone does not know why the tax was introduced at all why would he support it? He would object to it. (G8)

Some of the students offered education and especially voluntary participation in education as a way of creating awareness. By reflecting on their own experiences, some students said that they have changed the ways they behaved after participating in educational events such as conferences. However, some students objected to the possibility of changing people with education, according to these students, it was mainly a person's conscience that guided her behavior of complying with the proposed measures for resolving the SSI. For example, when talking about water taxation issue the following discussion took place with one of the groups:

S05: I remember that some teens went to the villages and educated the farmers on drip irrigation. Those teens went there completely voluntarily and because they did not aim for a personal gain they might have been able to touch the farmers. So if we launch such a campaign based on voluntary participation we might be more successful in reaching to the consciences of the people. They have helped the farmers understand the concepts (of drip irrigation and water conservation).

S47: Well I think that conscience is abstract and education is concrete. Before the advertisements on TV I was leaving the faucet running when I was brushing my teeth. OK It was a bit of acting without a conscience. But after participating in the conferences given in our old school I stopped doing that. If this is about conscience I am not a person with one about these issues.

S05: I think both are important. Obviously you were not completely ignorant because you have changed yourself. And there is one more issue, I am paying for this, why wouldn't I do it? So they don't see this as a problem that affects all of us.

S12: I think this is about people and their conscience and not education. Right now even education is not taken seriously. We see many people with good education doing really bad things. I don't think people care about education and this cannot be resolved by education. (G3)

For the energy crisis dilemma, 13 students offered incentives, 13 offered penalties, and none of the students offered raising awareness as means for action to resolve the issue. For the global warming dilemma, seven students offered incentives, 14 offered penalties, and two offered raising awareness as means of actions. For the water taxation dilemma, four students offered incentives, 14 offered penalties, and 16 offered raising awareness as the means of action (the total number of students that offered different means may be more than 24 because a student might have offered more than one means of action). The water taxation dilemma is different from the other dilemmas because it appears to have elicited raising awareness as the primary means of action.

3.5 *Reasons for Action or No Action*

According to the students, there were two classes of reasons, surface and deep reasons, for the governments and individuals taking or failing to take action. "Surface reasons" refer to the arguments for actions at societal, governmental, or intergovernmental levels. Fear of an economic crisis, the race for power, and competition for resources are stated by the students and were interpreted as the surface reasons. For example, on the surface reasons behind the lack of taking action for global warming a student said:

> The developing countries are actually acting rationally when you look from their perspective, they should not listen to it (the decision to reduce carbon dioxide emissions). Because when these countries reduce the carbon dioxide emissions they will fall behind. So they may put themselves in a situation like the African countries... But the United States is more developed. The reduction in carbon dioxide emissions will affect the developed countries most because at the moment they are emitting more carbon dioxide than others. If they reduce the emissions to the proposed levels, their power will be lessened, and America is a country who wants to keep the power, first it will obtain the economic power and then it will dominate the world. As long as it acts like this (reduce carbon dioxide emission) its economic power will be reduced and dominating the world will be out of question. (G9.S03)

"Deep reasons," the underlying reasons, stem from immutable traits of human nature, essentially articulated as extreme individualism, manifested by selfishness, greed, and pursuing only one's own gain. Human nature does not allow for action not in favor of one's own gain, unless the peril is immediate. Therefore, the SSI dilemmas essentially cannot be resolved, unless individuals and their human nature are regulated by a higher authority such as the government. For example, on deep reasons underlying the global warming issue during a group discussion, some students said:

Interviewer: The second question is about global warming. What did you say about it?
S31: This one is also selfishness. The very same thing with the energy crisis.
S17: Because people are the same. Everyone only thinks about himself. People shouldn't be selfish, they should not only think of themselves but also of the future generations. Or they should think about the human race. But it just isn't like that, people are selfish. If they keep acting like this there won't be a solution. We can never solve this problem.
S31: Yeah they put their own interests before everything else.
S17: I mean how long they have been living like this, for centuries. So they will never change. (G4)

For the energy crisis dilemma, nine students stated surface reasons, whereas 15 stated deep reasons; for the global warming dilemma, ten students stated surface and 14 stated deep reasons; and for the water taxation dilemma, 14 stated surface reasons and ten stated deep reasons. The water taxation seems to be somewhat different than the other two dilemmas; most of the students stated surface reasons for this dilemma.

3.6 *Tone or Attitude Towards Action*

The general tone of the students' writing and talk on the SSI dilemmas was marked by either doubt or despair. The students did not seem to believe that the governments and individuals were sincere in the efforts for taking action to resolve the SSI dilemmas. The students thought, individuals' decisions and actions could not contribute sufficiently to resolve the SSI dilemmas and the governments always had hidden agendas; therefore, these students expressed that they doubted that SSI could ever be resolved. For example, when discussing water taxation about developing awareness of water conservation among the citizens by education some students said:

S05: I said that education is a must; this issue can be resolved by education.
Interviewer: Can it be resolved by education? Can we solve it by educating people?
S05: This is something that is absolutely left to people's conscience.
S13: I don't think that a person would give up his interests even with education. So I don't think it can be solved by education.
S05: As I said before, as a developed country, if you can give that awareness to people whether it is water conservation or anything like that, people will get that awareness. But for a person who doesn't care about anything nothing can be done.

Interviewer: Can we give that general vision by education?

S13: I don't think that education would work, interests… I mean as long as a person doesn't have it in him, regardless of how much education you provide… Pardon me but even if you put a golden saddle on a donkey, the donkey is still a donkey; it has to be within the person. I mean if I have a personal gain that is in conflict with this (water conservation) I probably wouldn't do anything about this. So this is not that can be resolved with education.

S05: Then we have to push the people, if you change the direction of the interest.

S13: If we push them then they will have to do it, makes sense. (G3)

Similarly, some of the students appeared to have completely lost faith for even the possibility of resolving the SSI dilemmas. According to them, governments and individuals were self-serving, greedy, did not genuinely care about, and lacked the will for resolving these issues. Out of despair, these students suggested that a war would be the solution. For example, during a group discussion about energy crisis, some students said:

Interviewer: What if we assume that there has been a war, a really big war. The reason for the war would probably be energy in the first place. Let's say that eventually the energy resources were distributed with respect to the population of the countries. What would happen then?

S06: Another war would break out.

S30: Yes.

Interviewer: Why?

S06: Because the developed countries won't be content with it. With their historical exploitative personalities they will want to do the same thing.

Interviewer: Is there any possibility that they will change for the better?

S30: If people were actually humans then they might have changed for the better.

S10: But I don't think that is likely. (G1)

For the energy crisis dilemma, nine students' attitudes were classified as doubt and 14 as despair, and one student's attitude did not show any evidence of doubt or despair. For the global warming dilemma, eight students' responses were classified as doubt, ten as despair, and six students' responses did not show any evidence of doubt or despair. For the water taxation dilemma, only four students' tone was classified as doubt, three as despair, and 17 students' tone did not show any evidence of doubt or despair. The water taxation dilemma was different than the other two dilemmas in the tone expressed by the students; most of the students did not demonstrate doubt or despair for this issue.

4 Discussion

The purpose of this study was to explore the views of Turkish secondary school students on three SSI dilemmas and their reasons for holding those views. Previous research showed that students' beliefs in usefulness of political action and their willingness to undertake political action were not aligned in the case of global warming (Ambusaidi et al. 2012; Boyes et al. 2009; Chhokar et al. 2011;

Kılınç et al. 2011; Rodríguez et al. 2010), but this misalignment remains unexplained. The results of this study offer insights into why the students believed or did not believe in the usefulness of taking pro-environmental political action, and why they were willing or not willing to take these actions.

Most of the students expressed that political action was necessary to resolve the SSI; however, especially for the issues of energy crisis and global warming, they also expressed that the necessary political actions were not likely to be undertaken. The students explained the lack of taking political action with surface reasons, such as fear of an economic crisis, the race for power, and the competition for resources among countries. However, according to the students, the deep reasons underlying the failure to take necessary action has to do with the immutable traits of human nature. The students viewed human nature as essentially extremely individualistic; people only pursue their own gain. Therefore, one reason for the discrepancy between the belief in usefulness of pro-environmental political action and the willingness to undertake political action may be the beliefs about human nature. In other words, according to the students, although it is useful to take political action, human nature will not allow it because the political action is likely to conflict with the individuals' or governments' self-interests.

Most of the students considered particularly the energy crisis and global warming dilemmas not as technical but as ethical problems, a result in line with the results of previous studies (Kuhn 1991; Sadler and Zeidler 2004) that informal reasoning involves moral considerations. For the students, developing ethical conduct among people is not likely to happen spontaneously. Based on the description of human nature as extremely individualistic, the students expressed a need for regulating human nature by resorting to a higher authority; supporting the findings of previous research that show informal reasoning also involves legal, economic, and political considerations (Barab et al. 2007). The students stated that people will not sacrifice their own gain; but the authorities, for example, the government, may introduce incentives or penalties to regulate people's behavior related to SSI. For the water taxation issue, some students also expressed that education and launching campaigns for raising awareness might also help people change towards taking pro-environmental actions, which is again the responsibility of the authorities.

The students' tone or attitude towards the possibility of resolving energy crisis and global warming issues was not optimistic. For the students, individuals and governments are not sincere in their efforts related to resolving these SSI, leading to doubt; human nature will not change, leading to despair about the possibility of solving the SSI dilemmas. Such a pessimistic point of view from teenagers is alarming. The students cannot possibly imagine themselves as active agents, who can change the state of affairs in the world, hence are not only indifferent about taking pro-environmental action but they consider it futile. Perceiving that all solution possibilities are in the hands of governments and believing that governments cannot be trusted appear to lead them adopt a stance of paralysis.

The results show that the energy crisis and global warming dilemmas seem similar, whereas water taxation dilemma is different from the other dilemmas. The first two dilemmas mostly involve governments as actors, are treated as ethical problems, and are related to deep reasons that lead to doubt and despair; the third dilemma mostly involve individuals as the actors, is treated as a technical problem, and is related to surface reasons. One explanation of these differences might be related to the pervasiveness of impact: energy crisis and global warming are global issues, whereas water taxation is relatively a local one. Similarly the time span of the consequences of energy crisis and global warming is far into the future, whereas for the water taxation, the consequences are likely to be faced sooner. The implication is such characteristics of the SSI posed in research may require attention.

These results may be specific to academically talented Turkish secondary students and may have a cultural explanation. One such explanation is that Turkey is becoming a modern democracy, but the relative positions of the individual and the state are still quite behind the modern democracies, which may account for the students' perception of themselves as passive rather than active agents on issues that involve pro-environmental action. A second cultural explanation might be that the perceived corruption index of Turkey is high, with a rank of 61 among 183 countries (Transparency International 2011), which may have led the students not to trust the government and in fact to trust no governments. A third explanation may be the perception of the Western countries by the Turkish students, who seem to think that although the Western countries have great power on international decisions, they act with hidden agendas and pursue only their own benefit.

On the positive side, perhaps not on an individual basis, but collectively the students' arguments on the SSI dilemmas seem quite sophisticated. Although the sample consisted of high achieving students, it is striking that these 16 year olds tackled major philosophical questions about ethics, human nature, individualism versus collectivism, and the problems facing the world. The students had limited scientific knowledge on the SSI dilemmas and there is reason to doubt that they had any philosophical training. Yet, the students were able to capture and argue on these ethical and philosophical issues and their relations to the SSI dilemmas. Previous research (Ekborg 2008; Lewis and Leach 2006) already showed that students do not need huge amounts of scientific knowledge to discuss SSI; the current study confirms these results and extends it to philosophical and ethical knowledge; students, in the very least the academically talented, probably do not need huge amounts of philosophical knowledge to engage in SSI discussions. The results also imply that it may be meaningful to focus not solely on epistemology but also on ethics and politics. However, studies with representative samples are required to generalize these results to all students in Turkey, and similar studies in other countries are necessary.

Acknowledgments This study was supported by the Bogazici University Research Fund with project number 1862.

Appendix: The Dilemmas for the Writing Task

Dilemma 1: Energy Crisis

The year is 2030. The Third World War, which was caused by international disputes over energy resources, is over. The United Nations has decided to distribute energy resources differently. According to the resolution, no matter where energy resources are found, they belong to all nations. An international committee is being formed to discuss the distribution of these resources.

Plan: As a member of the committee, write a proposal for the new energy distribution.

Related Letters: Write two letters to the committee, one from a boy in Saudi Arabia and the other from someone in Israel.

Dilemma 2: Global Warming

There is a general consensus that climate change has a devastating effect in various parts of the globe. The Fourth Assessment Report of the Intergovernmental Panel on Climate Change (IPCC), which was completed in 2007, stated that there will be more frequent warm spells, heat waves, heavy rainfall, droughts, and tropical cyclones; extreme high tides; and accelerated ice loses in the Arctic in the future if the current levels of carbon dioxide emissions are not significantly reduced and the increase of the average global temperature is not halted beyond 2°C.

Recent scientific publications of climate change processes and impacts show an increased rate of greenhouse gas accumulation in the lower atmosphere and an accelerating sea level rise. The hope to limit the global temperature increase to about 2°C, which one expected a decade ago, has vanished; today there is growing scientific evidence that temperature rises could be up to 3–4 C. Although there is an increasing awareness for the risk to our social and economic well-being, the long-term impact of climate change on the biology and health of plant and animal species including our species seems not to be fully appreciated by the political institutions and decision makers.

The world climate conference in Copenhagen in December 2009 was a chance for developed, emerging, and developing countries of this globe to find a binding solution for all to this global problem.

The target was the reduction of 20 % from the starting point of 1990. While the European Union tried to push for a binding agreement, the emerging countries like India, Brazil, and especially China in an alliance with the USA saw their economic growth and development at risk. (The EU would have favored a 30 % reduction.) They made minor concessions to keep the process of discussion alive but avoided the joint signature of a binding agreement. The developing countries, the hardest affected group by the extreme weather situations, decreased food yields, and increased rates of infectious diseases, were left frustrated behind with bearing the bulk of consequences and with hardly any economic power to establish a sustainable future, while the EU had to realize that its influence on the global level is limited.

The United Nations have to build the institutional level for the future negotiations and agreements. The next climate conference will held in Bonn in 2010, to finish what was supposed to be achieved in Copenhagen. Here the EU can be the driving force for a global sustainable development and a mediator between the diverged interest groups. (The German government would go as far as reducing the emission as much as 40 % – not without critical comments from their own industry lobby.)

As an independent member of the UN, write a proposal for a new global climate change program. Take a stand in form of a short written comment from the Pacific Islands, Brazil, India, China, the USA, and the EU.

Dilemma 3: Water Taxation

In order to regulate water use and assure water sustainability, several measures are being offered by the government. Currently irrigation and industry takes a considerable percentage of all water usage in Turkey. One measure is to introduce a tax on water for irrigation and industrial use. This tax is expected to push agriculture and industry to act conservatively on water usage. However, this tax would inevitably be projected onto all consumers as increased prices on food and other goods. If the tax is not put into effect in the following decades, water shortage issues are expected. If the tax is put into effect, agriculture and industry is expected to move towards better technology that helps conserve water use.

Are you in favor of this tax or against it? Discuss why you support or why you oppose this tax with your reasons.

References

Albe, V. (2008). When scientific knowledge, daily life experience, epistemological and social considerations intersect: Students' argumentation in group discussions on a socio-scientific issue. *Research in Science Education, 38*(1), 67–90. doi:10.1007/s11165-007-9040-2.

Ambusaidi, A., Boyes, E., Stanisstreet, M., & Taylor, N. (2012). Omani students' views about global warming: Beliefs about actions and willingness to act. *International Research in Geographical and Environmental Education, 21*(1), 21–39. doi:10.1080/10382046.2012.639154.

Barab, S. A., Sadler, T. D., Heiselt, C., Hickey, D., & Zuiker, S. (2007). Relating narrative, inquiry, and inscriptions: Supporting consequential play. *Journal of Science Education and Technology, 16*(1), 59–82.

Boyes, E., Skamp, K., & Stanisstreet, M. (2009). Australian secondary students' views about global warming: Beliefs about actions, and willingness to act. *Research in Science Education, 39*(5), 661–680. doi:10.1007/s11165-008-9098-5.

Chhokar, K., Dua, S., Taylor, N., Boyes, E., & Stanisstreet, M. (2011). Indian secondary students views about global warming: Beliefs about the usefulness of actions and willingness to act. *International Journal of Science and Mathematics Education, 9*(5), 1167–1188. doi:10.1007/s10763-010-9254-z.

Ekborg, M. (2008). Opinion building on a socio-scientific issue: The case of genetically modified plants. *Journal of Biological Education, 42*(2), 60–65. doi:10.1080/00219266.2008.9656112.

Gee, J. P., & Green, J. L. (1998). Discourse analysis, learning, and social practice: A methodological study. *Review of Research in Education, 23*, 119–169.

Glaser, B., & Strauss, A. (1967). *The discovery of grounded theory: Strategies for qualitative research*. New Jersey: Aldine Transaction.

Kılınç, A., Boyes, E., & Stanisstreet, M. (2011). Turkish school students and global warming: Beliefs and willingness to act. *Eurasia Journal of Mathematics, Science & Technology Education, 7*(2), 121–134.

King, P. M., & Kitchener, K. S. (2004). Reflective judgment: Theory and research on the development of epistemic assumptions through adulthood. *Educational Psychologist, 39*(1), 5–18. doi:10.1207/s15326985ep3901_2.

Kolstø, S. D. (2001). Scientific literacy for citizenship: Tools for dealing with the science dimension of controversial socioscientific issues. *Science Education, 85*(3), 291–310. doi:10.1002/sce.1011.

Kuhn, D. (1991). *The skills of argument*. Cambridge: Cambridge University Press.

Lewis, J., & Leach, J. (2006). Discussion of socio-scientific issues: The role of science knowledge. *International Journal of Science Education, 28*(11), 1267–1287. doi:10.1080/09500690500439348.

Miles, M. B., & Huberman, A. M. (1994). *Qualitative data analysis: An expanded sourcebook*. Thousand Oaks: Sage.

MoNE. (2007). *Ortaöğretim kurumlarına geçiş yönergesi*. [Instructions for Secondary School Transitions]. http://mevzuat.meb.gov.tr/html/2602_1.html. Accessed 10 May 2011.

MoNe. (2010). 2010 *Ortaöğretim kurumlarına geçiş sistemi – I. Yerleştirme taban tavan puan listesi*. [2010 System for transition to secondary schools – First Placement Minimum and Maximum Scores List]. http://oges.meb.gov.tr/docs/2010_OGES_1_YerlesTavanTabanPuani.pdf. Accessed 10 May 2011.

NRC. (1996). *National science education standards: Observe, interact, change, learn*. Washington, DC: National Academy Press.

OECD. (2000). *Measuring student knowledge and skills: The PISA 2000 assessment of reading, mathematical and scientific literacy*. Paris, France: OECD Publishing, doi: 10.1787/9789264181564-en

Oulton, C., Dillon, J., & Grace, M. M. (2004). Reconceptualizing the teaching of controversial issues. *International Journal of Science Education, 26*(4), 411–423. doi:10.1080/0950069032000072746.

Rodríguez, M., Boyes, E., & Stanisstreet, M. (2010). Spanish secondary students' willingness to undertake specific actions to combat global warming: Can environmental education help? *Psyecology: Revista Bilingüe de Psicología Ambiental, 1*(1), 73–89. doi:10.1174/217119710790709496.

Sadler, T. D. (2004). Informal reasoning regarding socioscientific issues: A critical review of research. *Journal of Research in Science Teaching, 41*(5), 513–536.

Sadler, T. D., & Zeidler, D. L. (2004). The morality of socioscientific issues: Construal and resolution of genetic engineering dilemmas. *Science Education, 88*(1), 4–27. doi:10.1002/sce.10101.

Sadler, T. D., Chambers, F. W., & Zeidler, D. L. (2004). Student conceptualizations of the nature of science in response to a socioscientific issue. *International Journal of Science Education, 26*(4), 387–409. doi:10.1080/0950069032000119456.

Transparency International. (2011). *Corruption perceptions index 2011*. http://archive.transparency.org/content/download/64426/1030807. Accessed 15 May 2012.

Chapter 8
Teachers' Beliefs on Science-Technology-Society (STS) and Nature of Science (NOS): Strengths, Weaknesses, and Teaching Practice

Ángel Vázquez–Alonso, María-Antonia Manassero–Mas, Antonio García–Carmona, and Antoni Bennàssar-Roig

1 Introduction

The nature of science (NOS) represents an evolving field of meta-knowledge about science as a way of knowing, which has been constructed over many years from the interdisciplinary reflections of philosophers, sociologists, technologists, historians of science, scientists, and science educators. The NOS literature still maintains a kind of uneasy mixture of philosophical (epistemology), historical, social, and technological issues, which essentially involve the relationships of science, technology, and society (STS), the influence of society on science and technology (ST), the influence of ST on society, the relationships between science and technology, and the internal sociology of ST communities (values, principles, methods, commitments, etc.). These interdisciplinary roots and the usually evolving character of science have led to different visions, so that scholars do not agree on any shared specific definition of this complex field (Deng et al. 2011; Lederman 2007).

Á. Vázquez–Alonso (✉)
Department of Applied Pedagogy and Educational Psychology,
University of Balearic Islands, Spain
e-mail: angel.vazquez@uib.es

M.-A. Manassero–Mas
Department of Psychology, University of Balearic Islands, Spain
e-mail: ma.manassero@uib.es

A. García–Carmona
Departmento de Didáctica de las Ciencias, Universidad de Sevilla, Spain
e-mail: garcia-carmona@us.es

A. Bennàssar-Roig
Department of Biology, University of Balearic Islands, Spain
e-mail: abennassar@uib.es

C. Bruguière et al. (eds.), *Topics and Trends in Current Science Education: 9th ESERA Conference Selected Contributions*, Contributions from Science Education Research 1,
DOI 10.1007/978-94-007-7281-6_8,

However, the need in science education to construct a NOS curriculum that is at once authentic, practical, interesting, and viable for students has been driving a trend towards consensus about NOS, as is manifest in both the scholarly literature and many school science curricula. Indeed, the literature exhibits a consensual reductionist tendency to consider that the philosophical issues (epistemology of science) form the real core of NOS as against the remaining social, technological, and historical issues (Bartholomew et al. 2004; Millar 2005; Tsai and Liu 2005).

Other authors claim that NOS could be viewed from a broader perspective, not just reduced to the epistemology of science as the former studies suggest. Rather, a constellation of social, technological, and historical issues that are collectively labelled relationships of science with technology and society (STS). For instance, research on socio-scientific issues stresses the importance of social and ethical issues for science education (Sadler 2011; Zeidler et al. 2005). Aikenhead and Ryan (1992) suggest a complete and integrated framework of STS issues in nine dimensions, in which epistemology appears as one of these dimensions, although the whole framework is needed to account for the complexity of the STS orientation which involves the concepts, impacts, and solutions of science and technology in society (social, environmental, economic, cultural, ethical, etc.). Recently, Matthews (2012) has also argued for a broader set of consensual NOS features than the seven tenets of Lederman, so as to more faithfully reflect the multifaceted and contentious character of the NOS area. A recent inventory of specific consensual STS-NOS issues, drawn from different empirical studies, sets up today sound evidences in support of a broader STS orientation for the traditional NOS area (Vázquez and Manassero 2012).

Whether the approach will be to broaden STS or to reduce the epistemological orientation in NOS research is likely to depend on the relative importance assigned to STS relationships as against epistemological issues, since both of them could be deemed knowledge about science. Indeed, even the main advocates of the epistemological orientation for NOS acknowledge the STS orientation as a component of NOS under a diversity of generic formulations (e.g., scientific knowledge is "… socially and culturally embedded" – Lederman 2007, p. 833). Since much of current research on students' and teachers' beliefs has been performed under the umbrella of NOS, the present study will also come under this label, even though it assumes the broader STS orientation. Indeed, the instrument implemented here involves a broad mix of issues concerning epistemological tenets, the scientific community, and the interactions of science with society and technology (Vázquez et al. 2012).

The main relation connecting science teachers and STS-NOS issues is that the latter are today widely accepted as an indisputable target of science education for the attainment of authentic scientific and technological literacy for all (Rudolph 2000; DeBoer 2000; Millar 2006). Thus, NOS needs to be incorporated into the school science curriculum (e.g., Abd-El-Khalick and Lederman 2000; Lederman 1992), and indeed many current school science curricula already include the development of issues of NOS (AAAS 1993; Millar and Osborne 1998).

Consequently, research on teacher thinking on NOS is of interest for science education, although reviews of the last decade's literature (García-Carmona et al. 2011;

Lederman 2007; Liang et al. 2009; Tsai 2007) unfortunately reflect the inadequate comprehension of NOS and the naivety of the visions teachers hold on the topic. Teachers' understanding of NOS deviates from the modern views of science that the history, philosophy, and sociology (HPS) of science and technology studies have established, without forgetting their contentious and debatable status. Further, teachers' understanding shows scarce progress, as it is quite similar to that observed in previous years (e.g., Abd-El-Khalick and Lederman 2000; Lederman 1992; Nott and Wellington 1993). Teachers' beliefs lie close to traditional, positivist (logical empiricist), and idealistic views of S&T. They are similar to the McComas' list of myths (1996) and overly contrary to the list of consensual ideas on science drawn from the HPS modern views (Bartholomew et al. 2004; Lederman 2007; Vázquez and Manassero 2012). Teachers usually understand science as being either a static body of knowledge (thus, true and unchangeable) or a process of discovering what is out there, not as a human process of inventing explanations and theories (e.g., see a recent review in Lederman 2007).

Teachers' NOS beliefs are central for science education, because they decisively influence their classroom teaching practices. For instance, teachers who believe that science is an accumulation of knowledge tend to do experiments by following the textbook instructions and getting the right answers. In contrast, teachers who believe that science indeed changes are more likely to encourage students' discussions (Smith and Scharmann 1999). Furthermore, some studies have found that teachers with adequate NOS understanding do not necessarily automatically teach NOS issues in their classrooms (Lederman 2007). Although there are many factors involved in the decisions to teach NOS, in general, teachers lack the relevant pedagogical skills to do so effectively. Research has shown that some teaching practices are keys for curriculum NOS issues to be suitably addressed: planning, designing, and evaluating the NOS content; giving learners explicit access to appropriate NOS concepts; and providing general reflection and coherence between NOS tenets and the representations of science and technology in the classroom (Lederman 2007).

Until recently, scarce research on teachers' beliefs about NOS has focused on non-Anglo-Saxon, in-service teachers, and far less on Iberian teachers. Obviously, it is quite difficult generalizing about the complex reasons of this difference, though it can be ascertained that Iberian countries introduced STS-NOS contents, both in curricula and in science teacher training, much delayed in relation to Anglo countries. For instance, Spanish school science curricula explicitly presented STS-NOS contents from secondary to high school under the Education Act of 2006, though isolated proposals can also be documented earlier; the situation of science teacher training is even worse, as STS-NOS contents are not explicitly enacted, thus depending on the planning of teacher training implemented by each university. The present communication draws on data from an international cooperative investigation (Iberian-American Project of Evaluation of Attitudes Related to Science, Technology and Society – Spanish acronym PIEARCTS) that diagnoses STS-NOS beliefs across seven Iberian-American Spanish and Portuguese-speaking countries.

In particular, it presents the results of the application of a new methodological approach to assessing the STS-NOS beliefs of a large teacher sample consisting of preservice and in-service high-school science teachers. The STS-NOS beliefs are examined from a broad perspective that includes both external (relationships of science with technology and society) and internal aspects of science (its philosophy and sociology). The study is a search for empirical answers to the following questions: What are the strengths and weaknesses of secondary science teachers' beliefs on themes of STS-NOS? Do in-service science teachers understand STS-NOS issues better than their preservice counterparts?

2 Method

2.1 *Sample*

A master's degree is the official initial training for secondary science teachers in Spain. Their enrolment requires in turn the previous completion of a degree in science (e.g., life science, chemistry, physics, earth science, and engineering). The master's program involves both general pedagogy and specific didactics adapted to the specialization of the entrance degree.

The participants in the present study were 613 high-school science teachers – about one-third (32 %) in-service teachers, and the other two-thirds (68 %) preservice teachers. Their ages ranged from 23 to 63 years, and the sample split approximately into equal halves by gender (52 % men).

The preservice teachers were university graduates in physics, chemistry, biology, etc., who were following a master's program in science education to be accredited as prospective teachers (without educational experience). The in-service teachers were practicing teachers with variable practical experience in teaching science in the classroom. Since both the pre- and the in-service science teachers had received similar initial training in a pre-Bologna higher education system, overall they differed only in teaching experience.

2.2 *Instrument*

The Questionnaire of Opinions on Science, Technology and Society (Spanish acronym, COCTS) is an adaptation to the Spanish language and culture of the Views on Science-Technology-Society (VOSTS) pool (Aikenhead and Ryan 1992) and Teacher's Belief about Science-Technology-Society (TBA-STS) items (Rubba et al. 1996). In particular, COCTS is a pool of 100 multiple-choice items, which inherits the credit of the empirically developed VOSTS and TBA-STS items as one of the best pencil-and-paper instruments to evaluate STS-NOS beliefs, as the empirical development of the items guarantees the item validity (see the argumentation on this point in Aikenhead and Ryan 1992) and Lederman et al. (1998, p. 610) consider the VOSTS a valid and reliable instrument for investigating conceptions on the nature

Table 8.1 Labels of the items included in the two questionnaire forms (Form 1 and Form 2) across the structural dimensions of STS-NOS issues. A short description of the item's issue follows each key number

Item groups/dimensions (key)	Form 1 items (key/issue)	Form 2 items (key/issue)
(a) Definition of S&T (1)	F1_10111 science	F2_10211 technology
	F1_10411 interdependence	F2_10421 interdependence quality of life
(b) STS interactions (3)	F1_30111 STS interaction	
Influence of society in S&T (2)	F1_20141 country's government policies	F2_20211 industry
	F1_20411 ethics	F2_20511 educational institutions
Influence of S&T in society (4, 5)	F1_40161 social responsibility contamination	F2_40131 social responsibility information
	F1_40221 moral decisions	F2_40211 social decisions
	F1_40531 life welfare	F2_40421 application to daily life
		F2_50111 union two cultures
Internal sociology of science (6, 7, 8)	F1_60111 motivations	F2_60521 gender equity
	F1_60611 women's under-representation	F2_70211 scientific decisions
	F1_70231 consensus decisions	F2_70711 national influences
	F1_80131 advantages for society	
(c) Epistemology (9)	F1_90211 scientific models	F2_90111 observations
	F1_90411 tentativeness	F2_90311 classification schemes
	F1_90621 scientific method	F2_90521 role of assumptions
		F2_91011 epistemological status

of science. All the items have a common multiple-choice format: the stem presents a specific STS-NOS issue, which is followed by variable number of sentences (from 4 to 12; mean around 7), each labelled with a letter A, B, C, etc. Each sentence states a different rationale position (belief) on the stem issue, using nontechnical, familiar, and simple language (Vázquez et al. 2005, 2006).

The nine-dimensional structure of the original VOSTS is used here to classify the items into three main groups (leftmost column of Table 8.1), which correspond to the two underlying component fields of STS-NOS issues, i.e., STS interactions (groups a and b) and the epistemology of scientific knowledge (group c). Each item is labelled by a five-digit number. The first digit identifies the dimension, 1–8 for the different aspects of STS (internal sociology of science, etc.) and 9 for epistemology; the second pair of digits corresponds to themes; and the third pair of digits to the sub-themes of each item (science, technology, etc.). Each statement within an item is identified by the set of five digits corresponding to the item it belongs to, plus the letter that represents the position of the statement within the item (see detailed examples in Tables 8.2, 8.3, and 8.4). The main innovation of COCTS is the categorization of each sentence into a scheme of three categories (Adequate, Plausible, Naïve) by a panel of 16 experts on STS issues through a complex process of discovering the majority consensus (further details have been presented elsewhere: Vázquez and Manassero 1999; Vázquez et al. 2005, 2006). Moreover, some statements include the coding _C_ inserted before the tag number, which means the statement represents an idea that achieved the experts' consensus (over two-third of experts agreed on the category they assigned to the statement). The research team consensually selected 30 items that were distributed into two 15-item research booklets (F1 and F2; see Table 8.1) to provide a balanced coverage of the different dimension issues and for practicality of application (answering time under the person's fatigue threshold, average 50 min; easy self-administration; etc.). Each participant anonymously answered one randomly assigned form (F1 or F2), either on a pencil-and-paper brochure or on a web page; roughly half of the participants answered Form 1 (309), and the other half Form 2 (304).

2.3 Procedures

The respondents are not compelled to select any sentence. Rather, they are asked to rate their agreement (1, total disagreement; 9, total agreement) on each sentence of the items and are provided with alternative ways to not rate a sentence (not know, not understand, or leave it blank). These sentence scores are transformed into a homogeneous invariant normalized statement index in the interval [−1, +1] through a scaling procedure that takes into account the category of each statement (Adequate, Plausible, Naïve) assigned by the experts (further details in Vázquez et al. 2005, 2006).

For instance, an appropriate statement (_A_) expresses an adequate view on the issue, so that the scaling procedure assigns the index score +1 to total agreement (9) and −1 to total disagreement (1) and proportionally for the in-between scores; a naïve statement (_N_) expresses a view that is neither adequate nor plausible, so that the scaling assigns a scoring index that is the inverse of that of the appropriate statements; a plausible statement (_P_) assigns the +1 scoring index to the middle direct score (5) and −1 to the two extremes (1, 9) and proportionally for the in-between scores. Thus, the index meaning is invariant across all sentences: the higher (lower) the index score, the higher (lower) the belief's correctness according to current knowledge according to the history, philosophy, and sociology of S&T (Vázquez and Manassero 1999; Vázquez et al. 2006).

Thus, the closer to the maximum positive value (+1) an index is, the more informed (close to experts' conceptions on NOS) the respondent's view is; while the closer to the negative value (−1) the index is, the more misinformed (detached from current NOS conceptions) the respondent's view is. As misinformed conceptions are associated to the lowest negative values of the index, and the informed ones are

Table 8.2 Text of item 90621 (scientific method), displaying the text of sentences (center), the sentence labels (left column), the category assigned to each sentence (second column left), and the mean indices for the whole sample of the sentences, the three categories, and the whole item (right)

			Mean indices		
Variable	Category	Item	Sentence	Category	Item
		90621 The best scientists are those who follow the steps of the scientific method			
F1_C_90621A_N_	Naïve	A. The scientific method ensures valid, clear, logical, and accurate results. Thus, most scientists will follow the steps of the scientific method	–0.2422 (.5574)	–0.2142 (.4886)	
F1__90621B_N_	Naïve	B. The scientific method should work well for most scientists; based on what we learned in school	–0.1869 (.4936)		
F1_C_90621C_A_	Appropriate	C. The scientific method is useful in many instances, but it does not ensure results. Thus, the best scientists will also use originality and creativity	0.4854 (.4671)	0.4854 (.4671)	0.0400 (.2820)
F1__90621D_P_	Plausible	D. The best scientists are those who use any method that might get favorable results (including the method of imagination and creativity)	–0.0862 (.6506)	–0.1416 (.5034)	
F1__90621E_P_	Plausible	E. Many scientific discoveries were made by accident, and not by sticking to the scientific method	–0.1935 (.6470)		

Note the letters (A, B, C, D, etc.) that follow the question number and serve to identify each sentence within the question as well as the following set _A_, _P_, or _N_, which inform on the sentence category (Appropriate, Plausible, or Naïve)

Table 8.3 Complete text of item 10421 that displays the classification of its sentences as weaknesses (italics), strengths (bold), or medium, according to their mean indices for the teachers

10421 In order to improve the quality of life in our country, it would be better to spend money on technological research RATHER THAN scientific research	Mean (SD)	Category
A. Invest in technological research because it will improve production, economic growth, and unemployment. These are far more important than anything that scientific research has to offer	*.5332 (.4741)*	*N*
Invest in both:		
B. Because there is really no difference between science and technology	–.0348 (.6441)	P
C. Because scientific knowledge is needed to make technological advances	*–.4702 (.5414)*	*P*
D. Because they interact and complement each other equally. Technology gives as much to science as science gives to technology	*.5861 (.4347)*	*A*
E. Because each in its own way brings advantages to society. For example, science brings medical and environmental advances, while technology brings improved conveniences and efficiency	–.0712 (.5912)	P
F. Invest in scientific research – that is, medical or environmental research – because these are more important than making better appliances, computers, or other products of technological research	.1752 (.5198)	N
G. Invest in scientific research because it improves the quality of life (for example, medical cures, answers to pollution, and increased knowledge). Technological research, on the other hand, has worsened the quality of life (for example, atomic bombs, pollution, automation, etc.)	*.3750 (.5534)*	*N*
H. Invest in neither. The quality of life will not improve with advances in science and technology, but will improve with investments in other sectors of society (for example, social welfare, education, job creation programs, the fine arts, foreign aid, etc.)	*.8007 (.4002)*	*N*

represented by the highest positive values of the index, for brevity, they are simply referred to as "positive" or "negative" leaving out any biased meaning of these words.

The sentence's indices are the basis for further computations and statistics. For instance, a mean category index can be computed by averaging the sentences' indices belonging to each category; in turn, the average of the three category indices produces the weighted item global index, which represents the quantitative assessment of the overall conception about the item issue. This scheme describes the teacher's NOS thinking through 115 invariant indices for each form (about 100 sentence and 15 item indices – leaving aside category indices), thus being able to produce detailed profiles for persons, groups, or the overall sample (Table 8.2).

The indices allow the use of inferential statistics for hypothesis testing, which can be applied to compare groups (preservice vs. in-service teachers), or to set up cutting points to identify the strengths (highest positive), weaknesses (lowest negative), and

Table 8.4 Descriptive statistics of the 30 items (Form 1 and Form 2) for the whole sample of teachers (n valid cases, average item index and standard deviation)

Items	n	M	SD
F1_10111 science	309	0.275[a]	0.203
F1_10411 interdependence	308	0.231	0.278
F2_10211 technology	303	−0.050	0.239
F2_10421 interdependence quality of life	302	0.288[a]	0.222
F1_20141 country's government policies	307	0.283[a]	0.214
F1_20411 ethics	306	−0.094	0.303
F2_20211 industry	300	0.209	0.270
F2_20511 educational institutions	298	0.219	0.255
F1_30111 STS interaction	302	0.484[a]	0.293
F1_40161 social responsibility contamination	306	0.469[a]	0.207
F1_40221 moral decisions	304	0.232[a]	0.219
F1_40531 life welfare	306	0.010	0.372
F2_40131 social responsibility information	297	0.229	0.244
F2_40211 social decisions	295	0.046	0.293
F2_40421 application to daily life	293	−0.115	0.259
F2_50111 union two cultures	293	0.287	0.297
F1_60111 motivations	303	0.026	0.240
F1_60611 women's under-representation	304	0.218	0.289
F2_60521 gender equity	292	0.271[a]	0.214
F1_70231 consensus decisions	297	0.031	0.313
F2_70211 scientific decisions	286	0.042	0.284
F2_70711 national influences	288	0.073	0.298
F1_80131 advantages for society	304	0.095	0.242
F1_90211 scientific models	291	0.081	0.311
F1_90411 tentativeness	299	0.143	0.308
F1_90621 scientific method	295	0.040	0.282
F2_90111 observations	288	−0.174	0.402
F2_90311 classification schemes	292	0.139	0.255
F2_90521 role of assumptions	290	0.187	0.321
F2_91011 epistemological status	291	0.092	0.295

[a]Items whose mean indices are larger than one standard deviation

neutral or medium beliefs (Vázquez et al. 2006). The criteria for relevance were the statistical significance ($p<0.01$) and the effect size of the differences (differences measured in standard deviation units; $d>0.30$).

3 Results

The diagnostic method evaluates each item through a set of variables that includes the sentence indices, the three category average indices, and the overall item index (average of the three categories). The grand means of the sentence indices for the whole teacher sample were slightly different for Form 1 (m = 0.1862; SD = 0.5456) and Form 2 (m = 0.0809; SD = 0.5531); both are positive, but quite close to the null value. This result can be interpreted as an indicator of neutral, although somewhat positive, beliefs towards STS-NOS issues; some positive indices compensate the negative ones to produce this approximately neutral value for the overall mean of the entire sample, which suggests that informed beliefs coexist with other poorly informed beliefs.

The following example presents the classification of the sentences of item 10421 as weakness, strength, or medium, according to their mean index scores.

The Table 8.3 illustrates some common qualitative features detected through the new method in the results for all items: on each issue (item), the teachers hold at the same time strengths and weaknesses. In the example of Table 8.3, several sentences represent positive beliefs (A, D, G, H), while the remaining sentences represent uninformed beliefs, either clear weakness (C) or beliefs that do not attain a sufficiently high level to meet what the teachers need for quality teaching of NOS (B, E, F). Overall, this incomplete understanding of STS-NOS issues means that teachers do not master enough the issues to cope with their teaching in the classroom.

3.1 Teachers' Strengths and Weaknesses

The distribution of the average item indices is displayed in Table 8.4 for the items of both forms (F1 and F2). The two forms show asymmetrical distributions among positive and negative for item indices, where most of the items (26) are located in the positive area (F1 has only one item with negative mean index scores).

The items with the highest/lowest indices (more than one standard deviation above/under zero) identify the teachers' strengths and weaknesses. The list of F1 and F2 items which represent teachers' strengths is the following: F1_30111 STS interaction, F1_40221 moral decisions, F1_40161 social responsibility contamination, F1_20141 country's government policies, F1_10111 science, F2_10421 interdependence quality of life, and F2_60521 gender equity.

A few items (4) had negative scores on both forms (F1 and F2), though their mean indices did not attain low enough scores to be considered symmetric to the positive items (lower than one standard deviation under zero). Nonetheless, the items with negative indices, which represent a teacher's weaknesses, are the following: F1_20411 ethics, F2_40421 application to daily life, and F2_90111 observations.

From the perspective of teaching science, a good science teacher should have sufficient understanding of STS-NOS issues. Applying the criterion of relevant positive indices ($d > 0.30$ over zero) to the average item indices, one can conclude that approximately half of the items reflect the teachers having sufficient understanding of STS-NOS. Complementarily they need significant training on approximately another half of STS-NOS issues.

Similarly, the same overall analysis could be applied to the sentence variables with the highest positive and lowest negative indices to identify specific strengths and weaknesses. Each of the hundred sentences represents a specific belief on NOS, and the whole set conforms the spectrum of science teacher's thinking on STS-NOS. Due to space constraints, it is not possible to reproduce this extensive data here; instead, some main features of the index distributions are described.

The index distribution spectrum is slightly skewed towards positive scores since the number of sentences attaining positive indices is twice the sentences attaining negative indices. However, about a quarter of the sentences (28 %) surpass the relevance threshold ($d > 0.30$) of high positive mean indices over zero, so that these can be considered the teachers' relevant adequate conceptions on STS-NOS. The remaining sentences (72 %), whether positive or negative, represent sentences with insufficient scores to allow considering teachers to be sufficiently prepared, as these beliefs reflect. Overall, teachers did not attain a satisfactory level of understanding for over two-thirds of the sentences.

Again, the sentence mean indices allow identifying the positions with the highest positive indices (the best informed beliefs, or strengths) and the lowest negative indices (the worst informed beliefs, or weaknesses) to be identified.

Most sentences with the highest positive indices belong to the Appropriate (_A_) or Naïve (_N_) category sentences, while also noteworthy is the absence of Plausible (_P_) sentences. On the contrary, most sentences displaying to the lowest indices correspond to the Plausible or Naïve categories. Next paragraph shows comments on specific sentences for some selected items.

3.2 *Qualitative Analysis of Teacher Thinking*

The qualitative analysis of the sentence content allows one to examine the teachers' thinking in greater depth. For instance, the sentence F2_C_40211D_A_social decisions gets the highest item index and poses the issue of whether scientists and engineers should be the only ones to make decisions on the scientific affairs of our country. The teachers attained a high index in agreeing with the Appropriate sentence which states that the decision should be taken on a shared basis (the opinions of scientists and engineers, other specialists, and informed citizens should be taken into account in decisions that affect our society). Among the set of lowest index sentences, one can find the Naïve sentence F1_C_90621A_N_scientific method (Table 8.2), which states that the scientific method ensures valid, clear, logical, and accurate results (thus, most scientists will follow the steps of the scientific method). Teachers did not recognize the naïvety of this sentence as they attained a very low index on it. The two previous mentioned sentences are both examples of appropriate and naïve ideas often held by teachers and reported in the literature as myths of science (i.e., McComas 1996).

Qualitative analysis can be extended to each item to better understand teacher thinking on each topic, which is exemplified below through two items to meet the space constraints (Table 8.5).

Table 8.5 Mean indices, valid cases, standard deviation, and texts of items 40211 and 90311 for qualitative analysis

Items and sentences			Valid n	M	SD
40211 Scientists and engineers should be the ones to decide what types of energy our country will use in the future (for example, nuclear, hydro, solar, or coal burning) because scientists and engineers are the people who know the facts best					
Scientists and engineers should decide:					
F2_C_40211A_N_social decisions	A.	Because they have the training and facts which give them a better understanding of the issue	293	–0.064	0.574
F2_C_40211B_N_social decisions	B.	Because they have the knowledge and can make better decisions than government bureaucrats or private companies, both of whom have vested interests	292	–0.150	0.573
F2__40211C_P_social decisions	C.	Because they have the training and facts which give them a better understanding; BUT the public should be involved – either informed or consulted	294	–0.122	0.592
F2_C_40211D_A_social decisions	D.	The decision should be *made equally*; viewpoints of scientists and engineers, other specialists, and the informed public should all be considered in decisions which affect our society	295	0.638	0.406
F2__40211E_P_social decisions	E.	The *government* should decide because the issue is basically a political one; BUT scientists and engineers should give advice	295	0.034	0.607
F2__40211F_A_social decisions	F.	The *public* should decide because the decision affects everyone; BUT scientists and engineers should give advice	293	–0.035	0.581
F2__40211G_P_social decisions	G.	The *public* should decide because the public serves as a check on the scientists and engineers. Scientists and engineers have idealistic and narrow views on the issue and thus pay little attention to consequences	293	–0.077	0.641
F2__40211H_P_social decisions	H.	It depends of the type of decision; it is not the same thing to decide on the nuclear disarmament or on a baby. In some cases the scientists could make the decision, but in other, the citizens or the stakeholders should make it	292	–0.050	0.676

90311 When scientists classify something (for example, a plant according to its species, an element according to the periodic table, energy according to its source, or a star according to its size), scientists are classifying nature according to the way nature really is; any other way would simply be wrong					
F2_C_90311A_I_ classification schemes	A.	Classifications match the way nature really is, since scientists have proven them over many years of work	287	0.084	0.498
F2_C_90311B_I_ classification schemes	B.	Classifications match the way nature really is, since scientists use observable characteristics when they classify	287	−0.046	0.517
F2__90311C_P_ classification schemes	C.	Scientists classify nature in the most simple and logical way, but their way is not necessarily the only way	290	−0.190	0.585
F2__90311D_A_ classification schemes	D.	There are many ways to classify nature, but agreeing on one universal system allows scientists to avoid confusion in their work	290	0.653	0.411
F2_C_90311E_A_ classification schemes	E.	There could be other correct ways to classify nature, because science is liable to change and new discoveries may lead to different classifications	292	0.642	0.388
F2_C_90311F_A_ classification schemes	F.	Nobody knows the way nature really is. Scientists classify nature according to their perceptions or theories. Science is never exact, and nature is so diverse. Thus, scientists could correctly use more than one classification scheme	289	0.420	0.552

The item (40211, scientists and engineers should be the ones to decide...) raises the issue on the most appropriate way to make social decisions about issues related to science and technology (in this case, regarding energy decisions). Again the profile of teachers is low (around zero) along almost all statements, although teachers broadly guess right the most appropriate sentence (D, ...all (must) be considered in decisions which affect our society). Responses to this item exemplify very well the inconsistent and weak understanding of STS-NOS subjects by teachers: when teachers mostly recognize the equal participation of different stakeholders in decisions, just by applying simple logic, teachers should be able to reject statements that recognize either the participation of only one part or unequal participation in decisions.

The item (90311) on the epistemological status of scientific classifications shows a clear positive profile of teachers' understanding, as the three appropriate sentences (D, E, F) exhibit very high average indices. Unexpectedly, the other three sentences do not reach high indices, as it would be enough for teachers implementing elementary logical deduction from their major adherence to the appropriate sentences. For example, adherence to the possibility of different ways of classifying (D) and its change over time (E) must lead logically to major rejection of the sentences A and B (the only way is presumed nature really is) or C (classifications are simple and logical), but this is not the case (low mean indices). This feature suggests that the high commitment of teachers to the adequate complex, instrumental, and evolutionary nature of classifications (sentences D, E, F) is not the consequence of teachers' sound epistemological thought, but likely the simple reminder of textbooks, which usually display different classification schemes on the same issue (animals and plants, the soil's geological ages, chemicals, heavenly bodies, etc.), which also show change over time (E).

3.3 Differences Between Preservice and In-Service Teachers

The present methodological approach allows the application of inferential statistics in hypothesis testing. In particular, the simplest form of testing is group comparison, which is applied here to compare the STS-NOS beliefs of pre- and in-service teachers. To determine which variables might display significant differences between groups, a two-way between-groups analysis of variance was conducted to explore the impact of teaching practice on teachers' STS-NOS understanding, as measured by the 300 variables of items, categories, and sentences (dependent variables) (Figs. 8.1 and 8.2).

Overall, few variables (19 for the Form 1, and 24 for the Form 2 of the total variables involved in the two forms) displayed a statistically significant major effect ($p<0.01$; $d>0.30$) between in-service and preservice teachers. However, the differences did not have the same sign – some were positive (in-service teachers scoring higher than preservice teachers), and others were negative (in-service teachers scoring lower than preservice teachers) (Fig. 8.2). Thus, relevant differences not only were relatively scarce, but also they did not display the same trend (favoring one group over the other) across all the significant variables.

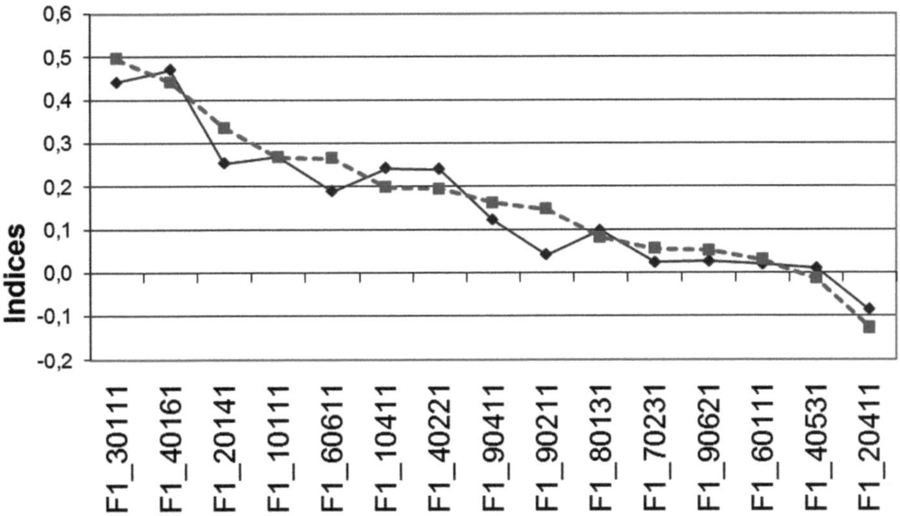

Fig. 8.1 Average item indices for in-service (*dotted line*) and preservice (*continuous line*) teachers (Form 1)

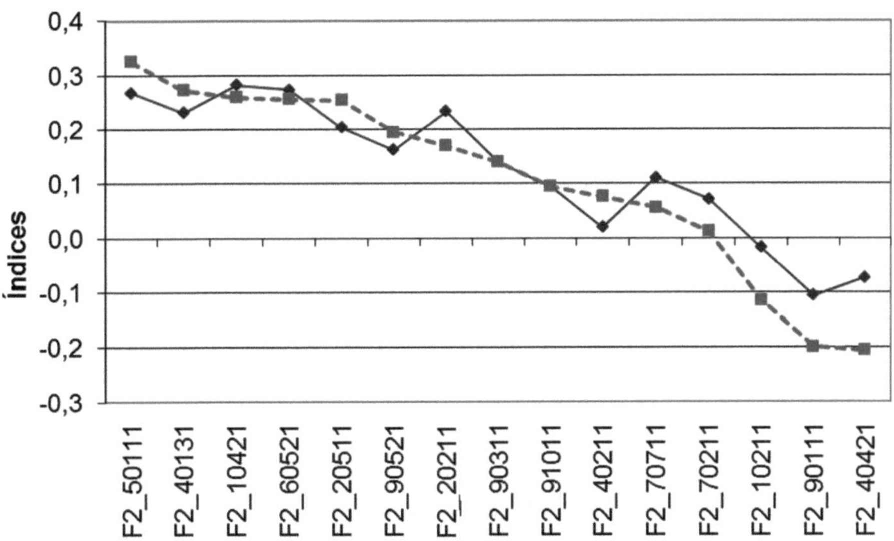

Fig. 8.2 Average item indices for in-service (*dotted line*) and preservice (*continuous line*) teachers (Form 2)

In sum, the results do not support the hypothesis that the practice of science teaching improves teachers STS-NOS beliefs. This means that improving STS-NOS understanding cannot rely on implicit practices, but instead might require explicit teaching of STS-NOS contents.

4 Conclusions and Implications

The new instrument (each sentence was categorized by a panel of experts) and methodological approach (new answering model and invariant standardized indices) described above improves the quality, validity, and reliability of the original instruments and avoids the usual objections against written questionnaires, such as participants' forced answers (choosing just one sentence) and the like. Moreover, the new approach helps to identify such qualitative features as profiles, strengths, and weaknesses in teachers' understanding of STS-NOS issues; further, the normalization of the scores and the standardization of the analyses allow standardized application and hypothesis testing statistics, such as between-group comparisons or comparisons among researchers. The method is also applicable to rapid and straightforward assessments of large representative samples, overcoming the limitations of case studies or open questionnaires to small samples, to really identify teachers' beliefs that are representative of teachers, either the strongest ones (those that closely match the experts' current understanding) or the weakest ones (those that are contrary to the experts' understanding).

The assessment instrument and method fit the requirements suggested by Allchin (2011) for appropriately assessing functional STS-NOS understanding: authentic context; well-informed analysis; adaptability to diagnostic, formative, or summative evaluation; adaptability to single, mass, local, or large-scale comparative use; and respect for relevant stakeholders. For instance, the research project PIEARCTS is an example of large-scale applications across seven countries and involving over 16,000 valid responses (Bennássar et al. 2010; Manassero et al. 2010).

The results provide a global and detailed picture of teachers' STS-NOS beliefs that is more complex than usually described (Lederman 2007): the teachers have inappropriate beliefs coexisting with appropriate ones across the entire range of STS-NOS topics. Many teachers' STS-NOS beliefs (about two-thirds of the beliefs examined) do not attain the high level required for quality teaching of STS-NOS. As the Spanish teachers involved in this study have not received specific STS-NOS training, their profiles of beliefs could be explained through a mixture of factors from informal education, such as personal reflection along years of experience on science teaching, shared exchanges with colleagues, and the reinforcement of distorted images of science conveyed by the media, the textbooks, etc. (Hodson 2009).

It is usually agreed that years of teaching improve a teacher's pedagogical content knowledge, so that the in-service teachers should have attained higher scores than the preservice teachers (Abd-El-Khalick and Lederman 2000; Lederman 2007). This would now seem to be a rather optimistic view since the comparisons between the pre- and the in-service teachers revealed scarce relevant differences. Indeed, the differences in favor of preservice teachers balance out those in favor of the in-service teachers, and the two groups seemed more similar than different. Spanish science teachers, whether in-service or preservice, have not been trained institutionally in STS-NOS issues (control variable), so that the present results provide evidence that science teaching practice by itself does not contribute to refining teachers' beliefs.

Accordingly, it should not be expected that teaching experience may (implicitly) contribute to training science teachers in STS-NOS. Complementarily the results are coherent with current compelling evidence in favor of explicitly teaching STS-NOS in order to attain effectiveness (Lederman 2007).

Nevertheless, the teachers' appropriate beliefs could be educationally invaluable because they could serve as pedagogical hooks in teacher training and in the planning of a teaching STS-NOS curriculum. Overall, the qualitative analysis, exemplified here through the mean indices for the overall teacher sample, would also be valuable for the diagnosis of individual teachers: superficial traits, weakness, and lack of logical consistency would also be identified with greater strength. At the time, teacher's awareness of the weakness of thought and subsequent cognitive conflict between adequate and naive ideas can be a basic hint for achieving conceptual change in the understanding of STS-NOS through teacher training.

All in all, the numerous inappropriate, or simply insufficiently appropriate, beliefs found in the science teachers' thinking, together with the lack of effectiveness of teaching experience to improve this situation, point to the necessity of designing and implementing both pre- and in-service teacher education programs that involve the participants in an explicit and reflexive analysis of STS-NOS topics (Hanuscin et al. 2006). The aim ought not to be for teachers to become philosophers of science, or to be more knowledgeable about STS-NOS, but for them to gain in competence teachers to teach STS-NOS issues effectively in the science classroom.

Acknowledgments Research Project SEJ2007-67090/EDUC funded by a grant of the Ministry of Education (Spain).

References

AAAS American Association for the Advancement of Science. (1993). *Project 2061. Benchmarks for science literacy*. New York: Oxford University Press.

Abd-El-Khalick, F., & Lederman, N. G. (2000). Improving science teachers' conceptions of nature of science: A critical review of the literature. *International Journal of Science Education, 22*, 665–701.

Aikenhead, G. S., & Ryan, A. G. (1992). The development of a new instrument: "Views on science-technology-society" (VOSTS). *Science Education, 76*, 477–491.

Allchin, D. (2011). Evaluating knowledge of the nature of (whole) science. *Science Education, 95*, 518–542.

Bartholomew, H., Osborne, J., & Ratcliffe, M. (2004). Teaching students "ideas-about-science": Five dimensions of effective practice. *Science Education, 88*, 655–682.

Bennássar-Roig, A., Vázquez-Alonso, Á., Manassero-Mas, M. A., García-Carmona, A. (Coor) (2010). Ciencia tecnología y sociedad en Iberoamérica: Una evaluación de la comprensión de la naturaleza de ciencia y tecnología [Science Technology and Society in Latin America: An assessment of the understanding of the nature of science and technology]. Madrid: OEI

DeBoer, G. E. (2000). Scientific literacy: Another look at its historical and contemporary meanings and its relationship to science education reform. *Journal of Research in Science Teaching, 37*, 582–601.

Deng, F., Chen, D.-T., Tsai, C.-C., & Chai, C.-S. (2011). Students' views of the nature of science: A critical review of research. *Science Education, 95*, 961–999.

García-Carmona, A., Vázquez, Á., & Manassero, M. A. (2011). Estado actual y perspectivas de la enseñanza de la naturaleza de la ciencia: una revisión de las creencias y obstáculos del profesorado [Current status and prospects for teaching the nature of science: a review of teacher beliefs and obstacles]. *Enseñanza de las Ciencias, 29*, 403–412.

Hanuscin, D. L., Akerson, V. L., & Phillipson-Mower, T. (2006). Integrating nature of science instruction into a physical science content course for preservice elementary teachers: NOS views of teaching assistants. *Science Education, 90*, 912–935.

Hodson, D. (2009). *Teaching and learning about science: Language theories methods history traditions and value*. Rotterdam: Sense Publishers.

Lederman, N. G. (1992). Students' and teachers' conceptions of the nature of science: A review of research. *Journal of Research in Science Teaching, 29*, 331–359.

Lederman, N. G. (2007). Nature of science: Past present and future. In S. K. Abell & N. G. Lederman (Eds.), *Handbook of research on science education* (pp. 831–879). Mahwah: Lawrence Erlbaum Associates.

Lederman, N. G., Wade, P. D., & Bell, R. L. (1998). Assessing understanding of the nature of science: A historical perspective. In W. F. McComas (Ed.), *The nature of science in science education: Rationales and strategies*. Dordrecht: Kluwer Academic.

Liang, L. L., Chen, S., Chen, X., Kaya, O. N., Adams, A. D., Macklin, M., & Ebenezer, J. (2009). Pre-service teachers' views about nature of scientific knowledge development: An international collaborative study. *International Journal of Science and Mathematics Education, 7*, 987–1012.

Manassero, M. A., Vázquez, A., & Bennassar, A. (2010). *Latin-American project of evaluation of attitudes related to science technology and society (PIEARCTS): Comparison of teacher thinking*. In B. Lazar & R. Reinhardt (Eds.), Proceedings of the XIV IOSTE Symposium Socio-cultural and Human values in Science Education (CD). Ljubljana: Institute for Innovation and Development of University.

Matthews, M. R. (2012). Changing the focus: From nature of science (NOS) to features of science (FOS). In M. S. Khine (Ed.), *Advances in nature of science research concepts and methodologies*. Heidelberg: Springer Dordrecht.

McComas, W. (1996). Ten myths of science: Reexamining what we think we know about the nature of science. *School Science and Mathematics, 96*, 10–16.

Millar, R. (2005). Contextualised science courses: Where next? In P. Nentwig & D. Waddington (Eds.), *Making it relevant context based learning of science*. New York: Waxmann Münster.

Millar, R. (2006). Twenty first century science: Insights from the design and implementation of a scientific literacy approach in school science. *International Journal of Science Education, 28*, 1499–1521.

Millar, R., & Osborne, J. (1998). *Beyond 2000: Science education for the future*. London: King's College.

Nott, M., & Wellington, J. (1993). Science education notes: Your nature of science profile – an activity for science teachers. *School Science Review, 75*, 109–112.

Rubba, P. A., Schoneweg, C. S., & Harkness, W. J. (1996). A new scoring procedure for the views on science-technology-society instrument. *International Journal of Science Education, 18*, 387–400.

Rudolph, J. L. (2000). Reconsidering the 'nature of science' as a curriculum component. *Journal of Curriculum Studies, 32*, 403–419.

Sadler, T. D. (Ed.). (2011). *Socio-scientific issues in the classroom: Teaching learning and research*. Dordrecht: Springer.

Smith, M. U., & Scharmann, L. C. (1999). Defining versus describing the nature of science: A pragmatic analysis of classroom teachers and science education. *Science Education, 83*, 493–509.

Tsai, C.-C. (2007). Teachers' scientific epistemological views: The coherence with instruction and students' views. *Science Education, 91*, 222–243.

Tsai, C. C., & Liu, S. Y. (2005). Developing a multi-dimensional instrument for assessing students' epistemological views toward science international. *Journal of Science Education, 27*, 1621–1638.

Vázquez, A., & Manassero, M. A. (1999). Response and scoring models for the 'views on science technology-society' instrument. *International Journal of Science Education, 21*, 231–247.
Vázquez, Á., & Manassero, M. A. (2012). La selección de contenidos para enseñar naturaleza de la ciencia y tecnología (parte 1): Una revisión de las aportaciones de la investigación didáctica [The selection of content to teach nature of science and technology (part 1): A review of the contributions of educational research]. Revista Eureka sobre Enseñanza y Divulgación de las Ciencias 9:2–31
Vázquez, A., Manassero, M. A., & Acevedo, J. A. (2005). Quantitative analysis of complex multiple-choice items in science technology and society: Item scaling. Revista Electrónica de Investigación Educativa 7, Retrieved from http://redieuabcmx/vol7no1/contents-vazquezhtml.
Vázquez, A., Manassero, M. A., & Acevedo, J. A. (2006). An analysis of complex multiple-choice science-technology-society items: Methodological development and preliminary results. *Science Education, 90*, 681–706.
Vázquez, Á., García-Carmona, A., Manassero, M. A., & Bennàssar, A. (2012). Spanish secondary-school science teachers' beliefs about science-technology-society (STS) issues. *Science & Education*. doi: 10.1007/s11191-012-9440-1.
Zeidler, D. L., Sadler, T. D., Simmons, M. L., & Howes, E. V. (2005). Beyond STS: A research-based framework for socioscientific issues education. *Science Education, 89*, 357–377.

Part III
Teachers' Practices and Teachers Professional Development

Chapter 9
Professional Learning of Science Teachers

Jan H. Van Driel

1 Introduction

Professional learning of teachers, both pre-service and in-service, is a complex process. Only recently research has begun to demonstrate that what and how teachers learn from teacher education and professional development programmes has an impact on whether and how they change their knowledge and practice (Desimone et al. 2002; Fishman et al. 2003). Studies on teachers' professional learning have shown that high-quality professional development programmes must entail a form of inquiry (Little 2001; Lotter et al. 2006) that enables (pre-service and in-service) teachers to actively construct knowledge through practice and reflection (Guskey 1986, 2002; Schön 1983). Moreover, it has become clear that multiple strategies are necessary to effectively promote teacher learning. Several review studies revealed that for strategies aimed at promoting professional learning of teachers to be successful, the following elements are important: (a) an explicit focus on teachers' initial knowledge, beliefs and concerns, (b) opportunities for teachers to experiment in their own practice, (c) collegial co-operation or exchange among teachers and (d) sufficient time for changes to occur (e.g. Bell and Gilbert 1996; Garet et al. 2001; Hawley and Valli 1999; Hewson 2007; Van Veen et al. 2010).

Many professional development programmes, however, have been found lacking with respect to stimulating teacher learning (Ball and Cohen 1999; Little 2001), since they neglect the knowledge, beliefs and attitudes that these teachers bring into the programme (Van Driel et al. 2001) and also ignore the context in which teachers work (Kennedy 2010). Furthermore, many professional learning programmes also fail to take into account existing knowledge about how teachers learn (cf. Ball and Cohen 1999; Borko 2004).

J.H. Van Driel (✉)
Graduate School of Teaching, ICLON – Leiden University,
P.O. Box 905, 2300 AX Leiden, The Netherlands
e-mail: Driel@ICLON.Leidenuniv.nl

C. Bruguière et al. (eds.), *Topics and Trends in Current Science Education: 9th ESERA Conference Selected Contributions*, Contributions from Science Education Research 1,
DOI 10.1007/978-94-007-7281-6_9, © Springer Science+Business Media Dordrecht 2014

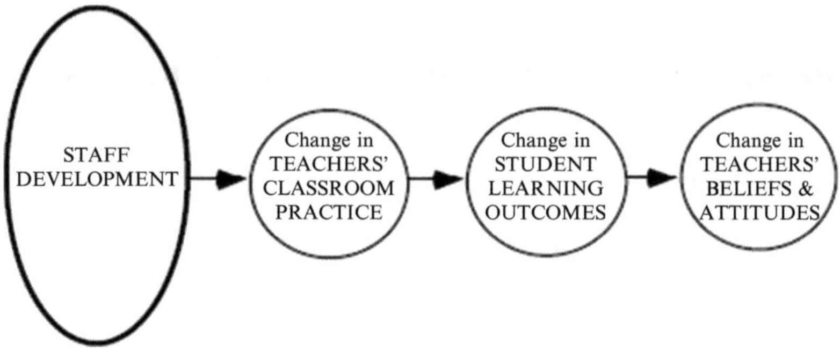

Fig. 9.1 A model of teacher change (Guskey 1986)

In this chapter, I will first discuss models of professional learning from the literature, selecting one that will serve as a framework for the following sections. Using this framework, I will then concentrate on the learning of pre-service science teachers, in the context of initial science teacher education, in particular by discussing a case study from my own work. Next, I will use the same framework to focus on professional learning of in-service science teachers, in the context of programmes of continuing professional development, discussing another case study from my own research.

2 Models for Teacher Professional Learning

There is a general agreement in the educational research community about the importance of teachers' professional learning as one of the ways to improve education. However, there is no consensus about how such a process occurs and how it can be analysed and promoted. This may be because it was only in the last decades that the nature and development of teachers' knowledge started to be understood by educational researchers (Munby et al. 2001).

A major question in teacher learning literature relates to the issue of whether and how changes in knowledge, beliefs and attitudes relate to changes in teacher practice (Wubbels 1992; Richardson and Placier 2001). For a long time, it has been widely assumed that when teachers change their knowledge, beliefs and attitudes on, for example, new instructional methods, their teaching practice will improve and accordingly result in better student outcomes. Since the middle of the 1980s, ideas about teacher change have been more focused on learning through reflection on one's own practice (Guskey 1986, 2002; Korthagen et al. 2001). Guskey (1986), for example, proposed a linear model of teacher change, assuming that a professional development programme causes changes in teachers' practice, which in turn lead to changes in students' learning and therefore result in changes in teachers' knowledge, beliefs and attitudes (see Fig. 9.1).

The facilitating process here is reflection. Other researchers, however, cautioned that teacher learning is not a linear process, but covers a complex system of processes in which teachers are engaged in active and meaningful learning (Borko 2004, Clarke and Hollingsworth 2002; Desimone et al. 2002). In a review study, Borko (2004) proposed a non-linear model in which the programme, the teachers, the facilitators and the context in which the professional development occurs are key elements in a professional development system. Borko states that the relations between these elements have been investigated in various studies. These studies focused on explaining factors found in each element, but were not explicit about what the precise relations are between these elements or how exactly the elements are related, thus leaving the nature of actual teacher growth processes vague.

According to Sprinthall et al. (1996), there are three main types of models for explaining teachers' development: the craft, the expert and the interactive models. The first model advocates the view that teachers develop as a result of becoming experienced teachers. In this case, knowledge emerges from classroom experiences. However, the model does not make clear how teachers produce new meanings from their experiences nor why some teachers only reproduce the same experience many times without learning from it. The expert model is focused on teachers being taught what and how to do by experts. As discussed by Clarke and Hollingsworth (2002), for a long time, changes in teachers' knowledge have been assumed to be the results of 'training', that is, of something that is done to teachers and in which they are relatively passive participants. Typically, the outcomes of such changes are generally 'measured' at the end of the training. In order to be effective, however, many researchers have recognised that such programmes should involve meaningful learning activities. This is the basis of what Sprinthall et al. (1996) characterised as the interactive model. Although programmes which are characterised as interactive present particularities concerned with how teachers' learning is both analysed and supported, they assume the occurrence of a process of changing prior knowledge with some help from experts. Within this perspective, one of the models proposed for teachers' professional growth is the Interconnected Model of Teacher Professional Growth (IMTPG; Clarke and Hollingsworth 2002). Using empirical data on which to base their findings, this model is made up of four different domains: (1) the personal domain (PD), which is concerned with teachers' knowledge, beliefs and attitudes; (2) the external domain (ED), which is associated with external sources of information or stimuli; (3) the domain of practice (DP) which involves professional experimentation; and (4) the domain of consequence (DC), which is comprised of salient outcomes related to classroom practice (see Fig. 9.2).

According to this model, teachers' professional learning is represented by changes in these four domains, through the mediating processes of 'reflection' and 'enactment' (represented as arrows linking the domains). The authors explain:

> The term 'enactment' was chosen to distinguish the translation of a belief or a pedagogical model into action from simply 'acting', on the grounds that acting occurs in the domain of practice and each action represents the enactment of something a teacher knows, believes or has experienced. (Clarke and Hollingsworth 2002, p. 951)

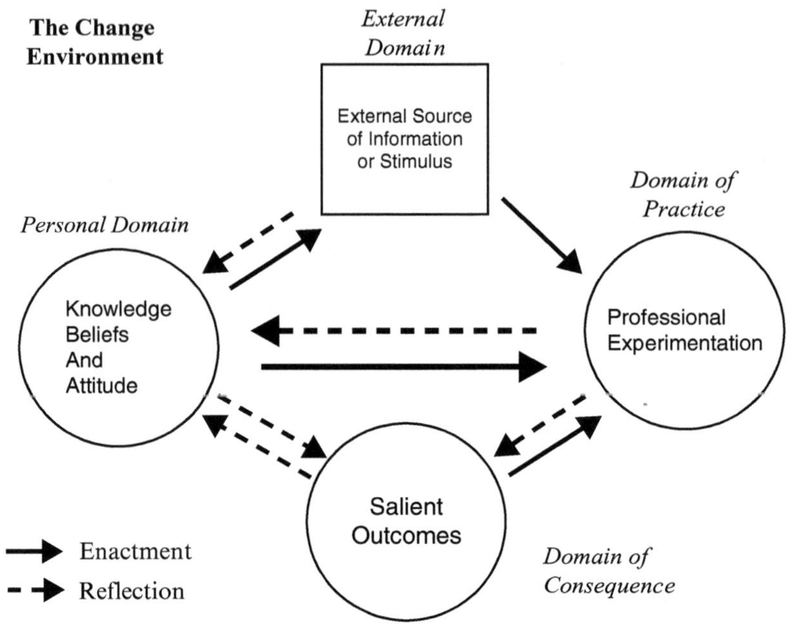

Fig. 9.2 The interconnected model of teacher professional growth (Clarke and Hollingsworth 2002, p. 951)

The term 'reflection' refers to 'a set of mental activities to construct or reconstruct experiences, problems, knowledge or insights' (Zwart et al. 2007, p. 169), for example, when a science teacher realises that an analogy to explain the model of an atom enables the students to visualise the model so that they understand the differences between the protons and the electrons. Within their model, Clarke and Hollingsworth distinguish different types of pathways for teacher learning, that is, a 'change sequence' or a 'growth network'. Change sequences occur when a change in one domain leads to a change in another, supported by enactive or reflective links; a growth network refers to a more complex and ongoing change process in more than one domain. The multiplicity of possible pathways between the domains reflects the non-linearity and the complexity of teachers' professional development. In terms of the development of teachers' professional knowledge, the authors assert that

> teacher growth becomes a process of the construction of a variety of knowledge types (content knowledge, pedagogical knowledge and pedagogical content knowledge) by individual teachers in response to their participation in the experiences provided by the professional development programme and through their participation in the classroom. (Clarke and Hollingsworth 2002, p. 955)

In the next section, the Interconnected Model of Teacher Professional Growth will be used as a framework to study the professional learning of pre-service science teachers.

3 Pre-service Science Teachers' Learning

3.1 The Curriculum for Science Teacher Education

Most programmes for science teacher preparation around the world recognise the importance of subject matter knowledge, general pedagogical knowledge and pedagogical content knowledge in preparing high-quality science teachers (Van Driel and Abell 2010). Thus, these programmes include recognisable and common components to build these knowledge bases: liberal arts courses, science content courses, general pedagogy courses (e.g. educational psychology, classroom management, educational history and philosophy) and subject-specific teaching and learning courses (called 'methods courses' in the USA and an equivalent of 'subject matter for teaching' in Europe, e.g. 'Fachdidaktik' in Germany or 'didactique disciplinaire' in France). Most programmes also value the authority of learning from experience (Russell and Martin 2007) and thus include significant supervised field experiences in school classrooms. Programmes vary widely in terms of the relative emphasis placed on these components, on programme length and on academic level. As an example, in the Netherlands, a 4-year programme at college level ('Hogeschool', or University of Applied Sciences) prepares science teachers for the lower grades of secondary education (students aged 11–14), whereas science teacher preparation for upper secondary education (students aged 15–18) is a 1-year programme either within or after a 5-year master's programme at university. Recently, countries suffering from science teacher shortages, such as the USA, the UK and the Netherlands, have seen the advent of relatively short post-baccalaureate alternative certification programmes that attract mid-career individuals into the teaching profession.

The enactment of these components is influenced by the philosophical orientation of the teacher preparation programme. Researchers have proposed different ways to categorise the purposes and goals for science education. For example, Roberts (1988) defined various curriculum emphases in science education, including providing a solid foundation in science, preparing students for the next grade level, science skills development and helping students generate everyday explanations. Such classification schemes can help us think about the curriculum of science teacher education as well. For example, science teacher preparation programmes of the past focused on preparing teachers as technicians capable of implementing specific strategies (e.g. wait time, cooperative groups). Recent views of science teacher education recognise the importance of challenging future teachers' prior knowledge and beliefs and helping them see viable alternatives to transmission-oriented science teaching. Russell and Martin (2007) called this orientation to science teacher education 'teaching for conceptual change'. Science teacher education curricula that follow a conceptual change orientation are aimed at helping future teachers reflect on practice and make decisions grounded in student learning.

3.2 An Empirical Study on Pre-service Teachers' Learning

Whereas the 'technical' approach to science teacher education referred to above often resulted in teachers strictly following their textbooks and focusing on 'teaching for the test', recent programmes for science teacher education often aim at preparing science teachers to design and test curricular materials. An example of a study in the context of such a programme was conducted in the Netherlands (Justi and Van Driel 2006). In this study, the Interconnected Model of Teacher Professional Growth (IMTPG) was chosen as the framework for the design of a part of the teacher education programme, which was aimed at the development of pre-service teachers' knowledge and practice about models and modelling in science. In this context, the pre-service teachers conducted a project during which they developed and tested curricular materials focusing on models and modelling, in particular, about states of matter and phase transitions and about the particulate nature of matter. In this study, the IMTPG was also used as a framework for analysing the data. The study was guided by the following questions:

1. How does the 'external domain' contribute to the development of pre-service teachers' content knowledge, curricular knowledge and pedagogical content knowledge (PCK) on models and modelling?
2. How do specific aspects of pre-service teachers' content knowledge, curricular knowledge and PCK on models and modelling change when they participate in the project?
3. How do pre-service teachers' changes manifest themselves in their classes?

3.2.1 Method

Five science teachers, who were following the 1-year postgraduate teacher education programme at Leiden University, the Netherlands, voluntarily participated in this research. Three were male and two were female. Before entering this programme, the participants had obtained master's degrees in chemistry or physics. The programme prepared them to teach at the level of upper secondary education (students aged 15–18). Below, the teachers are identified by the use of codes (T1, T2, etc.).

In order to characterise teachers' initial knowledge on models and modelling, they answered the written questionnaire VOMM C (Views on Models and Modelling, version C; Justi and Gilbert 2003) and were interviewed. They then took part in four meetings of 3 h each that were held over a period of 6 weeks. During these meetings, they were involved in learning activities concerned with all the main aspects that might be part of their knowledge on models and modelling (e.g. the nature and uses of models in science, the production and use of various sorts of teaching models [two-dimensional, three-dimensional and computerised models], the nature and importance of the modelling process in science and in science education). These activities were explicitly related to teachers' practice (for instance, by analysing pictures – two-dimensional models – provided by textbooks), to involve them in thinking about new aspects from their existing knowledge, to consider them as learners during the discussion and, simultaneously, to ask them to reflect on their teaching practices. This approach aimed to contribute to making their learning more meaningful (Borko and Putnam 1996).

Next, the teachers chose one of the aspects of models and modelling that was discussed in the meetings as the basis of a research and development project that they conducted in their own classes. For this purpose, they designed a lesson series that they conducted in a particular class. To support their project, an extra meeting focusing on methodological aspects was organised. Before the teachers actually conducted their project, they were interviewed for a second time. As part of their projects, the teachers collected data in their own classes (for instance, three-dimensional or written material produced by their students, video recordings of classroom discussions) that they analysed before writing reflective research reports about their projects. A third and final interview occurred after the presentation of their research reports during a final group meeting.

After everything that was said or written by each of the teachers during their participation in this project was transcribed or copied, the process of analysis occurred in distinct phases:

1. The categorisation of all the data collected for each teacher. This meant that their knowledge expressed in each of the data sources was identified for each of the various knowledge aspects mentioned in the research questions.
2. The analysis of each teacher's personal development, part A. This meant the reorganisation of all the data previously categorised in order to show the content knowledge, the curricular knowledge and the PCK expressed by each teacher in each of the four domains of the IMTPG.
3. The analysis of each teacher's personal development, part B. For each category of teachers' knowledge, this meant the characterisation of the relationships between the four domains of the IMTPG.
4. The representation of the relationships according to the pictorial representation of the IMTPG. For each teacher, the relationships established between the four domains in relation to each of the knowledge aspects were represented in a summarised picture of the IMTPG, which was then classified in terms of either a change sequence or a growth network (see below).

In order to assure the internal validity of the data analysis, some of the phases were conducted independently by the two researchers (Cohen et al. 2000). The results obtained by each of them were compared. Whenever there was a difference in the initial categorisation, it was discussed in order to reach agreement.

3.2.2 Results

In analysing the pictorial representations of the IMTPG, the complexity of these representations was assessed. A change sequence was characterised by the

Table 9.1 Identification of the types of teachers' change for each of the aspects (*CS* change sequence, *GN* growth network)

Aspect		T1	T2	T3	T4	T5
Content knowledge	Models	GN	GN	GN	GN	GN
	Modelling process	GN	GN	CS	GN	GN
Curricular knowledge	Curricular models	–	CS	CS	CS	–
	Introduction of modelling activities	GN	GN	GN	GN	CS
PCK	Teaching models – purpose of their use	GN	CS	CS	CS	CS
	Teaching models – production	GN	GN	GN	GN	CS
	Teaching models – use	GN	GN	GN	GN	CS
	Conducting of modelling activities	GN	GN	GN	GN	GN
	Students' ideas about models and modelling	GN	GN	GN	GN	GN

establishment of one or two relationships between different domains for a given aspect of teachers' knowledge, what we interpreted as a superficial change in teachers' knowledge. On the other hand, when the pictorial representation of the IMTPG of a given aspect consisted of more than two relationships between different domains, thus meaning more complex changes in teachers' knowledge, it was identified as a growth network. The classification as change sequences or growth networks is presented in Table 9.1.

Table 9.1 shows that only 11 (24 %) of the representations can be identified as change sequences. Most were concerned with teachers' ideas about curricular models or with teachers' PCK on the purposes of using teaching models. Nearly all other representations (32 out of 45, that is, 71 %) were identified as growth networks, with several levels of complexity. For instance, the development of T4's content knowledge about models did not include any relationship with the domain of consequences, while the development of the same knowledge by T2 included relationships originating from all the domains, making evident the improvement of her personal domain.

In addition, the pictorial representations of the IMTPG were analysed in terms of the types of relationships established between different domains. Sometimes, reflective relationships dominated a teacher's growth network – as occurred in the development of T1's PCK about the production of teaching models (Fig. 9.3).

T1 commented on the production of teaching models in her classes and reflected on this experience at different levels, e.g. by emphasising aspects that she had never paid attention to before and by considering students' current outcomes. However, as the focus of her research project was on the building of models by students, it was not possible to make other enactment relationships evident. In other cases, the teacher's growth networks were mainly built from enactment relationships – as occurred in the development of T3's PCK about the use of teaching models (Fig. 9.4). From the ideas expressed by T3 at different times, it became clear that this occurred because, in his research project, he analysed how students understood different teaching models for a given phenomenon, and from his results, he was able to both propose how he would change the activity for the following academic year and think about similar activities that could be developed for the teaching of other scientific ideas.

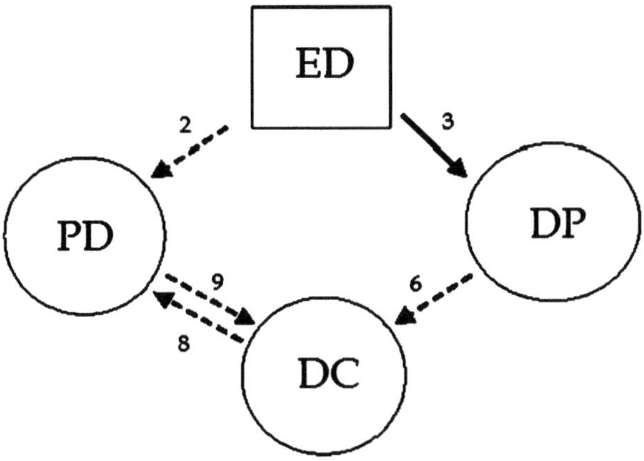

Fig. 9.3 The development of T1's PCK about the production of teaching models. Meaning of the numbers at the *arrows*: *2*: Reflection on activities in the external domain, leading to changes in a teacher's knowledge. *3*: Enactment of input from the external domain in a teacher's practice. *6*: Activities of students and teacher in practice leading to certain outcomes. *8*: Reflection of outcomes, leading to changes in a teacher's knowledge. *9*: Using a teacher's knowledge to reflect on certain outcomes of teaching

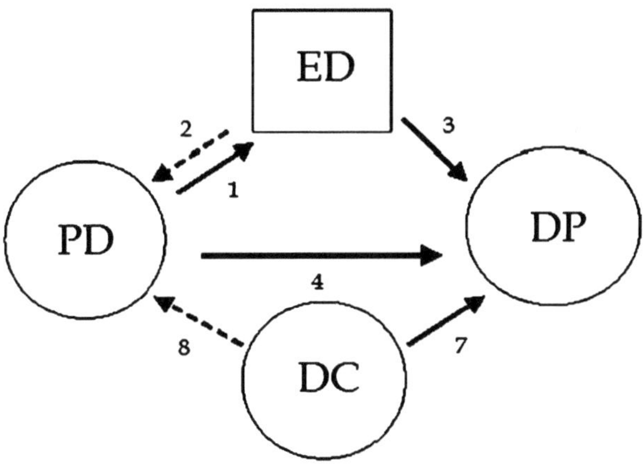

Fig. 9.4 The development of T3's PCK about the use of teaching models. Meaning of the numbers at the arrows: *1*: Using a teacher's knowledge to react on input given in the external domain. *2*: Reflection on activities in the external domain, leading to changes in a teacher's knowledge. *3*: Enactment of input from the external domain in teachers' practice. *4*: Enactment of a teacher's knowledge in his or her practice. *7*: Enacting on certain outcomes to make modifications in a teachers' practice. *8*: Reflection of outcomes, leading to changes in a teacher's knowledge

3.2.3 Conclusion and Discussion

On the basis of the findings, it was concluded that the use of the IMTPG was fruitful and productive. As for the design of the professional learning project, the meetings (the external domain) were carefully organised in such a way that these were connected both with the teachers' initial ideas (the personal domain) and with their teaching practice (domain of practice). In particular, the IMTPG informed the decision to organise the activities in the domain of practice in the form of an action research project, which stimulated pre-service teachers to experiment, in their practices, with some of the ideas that were discussed during the meetings. Asking the teachers to write a report about their action research project and to discuss their reports helped to establish connections between the domain of practice and their personal domains. As for the analysis of the teachers' knowledge growth, the use of the IMTPG was crucial. Without the IMTPG, all aspects of teachers' knowledge could have been categorised, but a framework would have been lacking to monitor and understand the development of these knowledge aspects. By identifying relationships between the four domains, the IMTPG made it possible to understand each teacher's development in a detailed way. Moreover, data analysis with the IMTPG was essential in order to support the characterisation of differences in the development of the knowledge across different teachers. It also supported the emergence of differences in the development of distinct aspects of the knowledge of a given teacher. In other words, this made possible the characterisation of the teachers' knowledge development as an idiosyncratic process. Overall, it was concluded that in the present project, the IMTPG helped to organise and to discuss data in a way that favoured the discussion of the research questions.

In the following section, I will seek to demonstrate how the IMTPG can also be useful to frame the professional learning of in-service science teachers.

4 In-Service Science Teachers' Learning and Curriculum Reform

Professional learning of in-service teachers is often, although not always, related to educational reform or innovation. Often, the question is how to involve teachers in these reform efforts so that the chances of a successful innovation are enhanced. For science education, in particular, 'ever since the birth of the science curricular reform movement in the late 1950s, a large portion of science teacher education has been connected in some way to attempts to introduce curricular change' (Anderson and Mitchener 1994, p. 36). Traditionally, this process consisted, roughly, of the following steps (Van Driel et al. 2001):

1. The core elements of the innovation were defined by curriculum developers or policymakers.
2. A description was made of the teaching behaviour expected of teachers who would loyally implement the innovation or of the skills teachers should acquire.
3. A series of training sessions or supervision activities were designed, aimed at developing the desired teaching behaviour (cf. Joyce and Showers 1980). In particular, 'single shot interventions', like in-service workshops, were used to achieve this aim.
4. Usually, the implementation was not adopted by the teachers in the manner intended, or initially observed changes in the teachers' behaviour did not persist.

Of course, not every reform effort in the past followed this scheme. There have been many attempts to improve on this outline (cf. Loucks-Horsley et al. 2003), but on the whole, it can be concluded that the role of teachers in the context of curriculum change usually has been perceived as 'executing' the innovative ideas of others (policymakers, curriculum designers, researchers and the like). Ball and Cohen (1999) have argued that the role of the government should be limited to establishing a framework for reforms (e.g. by setting standards and providing useful tools, like curricular materials). The reform of actual practice, however, should be in the hands of the professional sector.

In the research literature, there is a growing consensus that educational reform efforts are doomed to fail if the emphasis is on developing specific teaching skills, unless the teachers' cognitions, including their beliefs, intentions and attitudes, are taken into account (Haney et al. 1996). Reforms call for radical changes in teachers' knowledge and beliefs about subject matter, teaching, children and learning. The implementation of reforms can therefore be seen as essentially a matter of teacher learning (Ball and Cohen 1999). However, many authors have pointed out that teachers' ideas about subject matter, teaching and learning do not change easily or rapidly. There are various reasons why teachers' cognitions are usually stable and why innovative ideas are not easily applied in their teaching practice. First, teachers do not tend to risk changing their own practice, which is rooted in practical knowledge built up over the course of their careers. Over the years, this knowledge has proven workable in a satisfying way. Rather, teachers tend to change their practice in a tinkering manner, picking up new materials and techniques here and there and incorporating these in their existing practice (Thompson and Zeuli 1999). Others, such as Kennedy (2010), have explained how reform efforts tend to ignore the practicalities of working in schools and classrooms and therefore often lack ecological validity (Doyle and Ponder 1977). Finally, although experience contributes to an increase in a teacher's practical knowledge, at the same time, the variety within this knowledge tends to decrease. This phenomenon is known as knowledge concentration: Professionals gradually feel more at home in an area that becomes smaller (Bereiter and Scardamalia 1993). Consequently, it becomes more and more difficult for someone to move into an area with which he or she is not familiar. For these reasons, innovators often tend to consider teachers' practical knowledge conservative (cf. Tom and Valli 1990). However, as it is the expression of what teachers really know and do, teachers' practical knowledge is a relevant source for innovators when preparing educational reform, in particular when designing professional development programmes aimed at implementing such reform.

4.1 An Empirical Study of In-Service Teachers' Learning

Although numerous studies have focused on the development of teachers' knowledge (Beijaard et al. 2000), teachers' individual professional learning processes have not been studied extensively (Zwart et al. 2007; Hashweh 2003; Wilson and Berne 1999).

In a recent study (Wongsopawiro 2012), the aim was to understand what and how individual teachers learn from taking part in a professional development action research programme, specifically with respect to the development of their pedagogical content knowledge (PCK). Regarding PCK, Kind (2009) argued that studies on professional development programmes are needed in order to gain a deeper understanding of whether and how such programmes affect individual PCK development. Wongsopawiro (2012) studies this development using Clarke and Hollingsworth's IMTPG model. The research focused on identifying possible pathways of change that indicate the development of science teachers' pedagogical content knowledge, in the context of participating in a professional development programme that incorporated conducting an action research project in their classrooms. The following research question was central to the study: What are the possible pathways that lead to changes in science teachers' pedagogical content knowledge in a professional development programme?

To answer the research question, the following sub-questions were formulated:

1. What pathways of change can be identified among the participants of a professional development programme using the IMTPG model?
2. Which of the identified pathways are related to the development of science teachers' pedagogical content knowledge?
3. Which specific elements of the professional development programme contribute to development in the teachers' pedagogical content knowledge?

4.1.1 Method

The study was conducted in the context of a 1-year professional development programme called the Mathematics and Science Partnership (MSP) programme, which aimed at increasing teachers' professional knowledge. In this programme, teachers were encouraged to use action research as part of a professional development tool by which to improve their classroom performance. The MSP programme started with a 2-week summer session in which teachers were introduced to action research. In the first week, the teachers created an action research plan in which they selected a topic from their curriculum and considered materials and strategies to teach this topic. They attended presentations from university staff on various science and mathematics topics and best practices in education. In the second week, the teachers continued working on their plan, doing literature research in order to deepen their understanding of the subject and to find successful instructional strategies on the topic in question. The teachers were asked to reflect upon their earlier teaching of this topic and to provide reasons why they now intended to use different instructional methods. They developed research questions and identified methods by which to assess their projects. After creating lesson plans and teaching materials, they conducted their action research programme in the following school year. During that year, they had four meetings with the university staff. The academic staff acted as facilitators and peers (i.e. school colleagues) as critical friends in this professional development programme (Ponte et al. 2004).

Table 9.2 Demographics of the in-service teachers participating in the study

Teacher	Name (fictitious)	Years of experience	Subject taught	Grade level
1	Betsy	12	Deserts	8th
2	Josh	7	Atomic theory	5th
3	Carlene	8	Rocks and minerals	8th
4	Dana	17	The human body	4th
5	Diane	22	Cell structure	7th/8th
6	Donna	21	Volcanoes	7th
7	Matt	28	Photosynthesis and respiration	7th
8	Norma	3	Cell structure	7th
9	Rhonda	26	Bats	7th
10	Shania	21	Cell structure	6th
11	Stephanie	10	The human body systems	7th
12	Trisha	2	Earthquakes	4th

Twelve in-service science teachers from middle and high schools in the Midwest region of the USA volunteered to participate in this study. Their schools were located in small rural communities (Table 9.2). All participating teachers were present at the 2-week summer programme and the four follow-up sessions during the school year 2005–2006. Three teachers were male, nine were female. The subjects that they taught were biology (e.g. cell structure, human body) and earth science (e.g. volcanoes, earthquakes), and in order to understand the complex pathways between the domains for each PCK component (Magnusson et al. 1999), three data sources were used: (1) the teachers' action research reports, (2) the teachers' reflective journals about their professional learning processes and (3) a semi-structured interview. During the MSP programme, the teachers worked on their action research reports. As the programme continued, the teachers were able to build upon this document and make revisions. In this way, they gradually compiled their report, which also included an overview of their lesson plans and of products made by students that they collected during the year. During the entire programme, all teachers kept a personal electronic journal in which they reflected on their personal progress. Teachers were asked to reflect on the presentations by the university staff and the workshop activities during the summer course, as well as on their findings in the classroom and their action research project. At the end of the year, the teachers submitted this journal together with their action research report, as part of the evaluation process. Finally, they were interviewed about these documents, in particular about what they had learned from their action research project.

Data analysis followed a procedure similar to that used by Justi and van Driel (2006) (see Sect. 3.2.1). First, the data sources were explored looking for statements that were considered indicators of change in PCK. Next, these changes were examined to determine relationships between the different domains of the IMTPG. Then, pictorial representations (pictograms) were constructed for the development of each PCK component, showing relationships between the domains of the IMTPG. One pictogram for each PCK component per teacher was thus constructed, resulting in

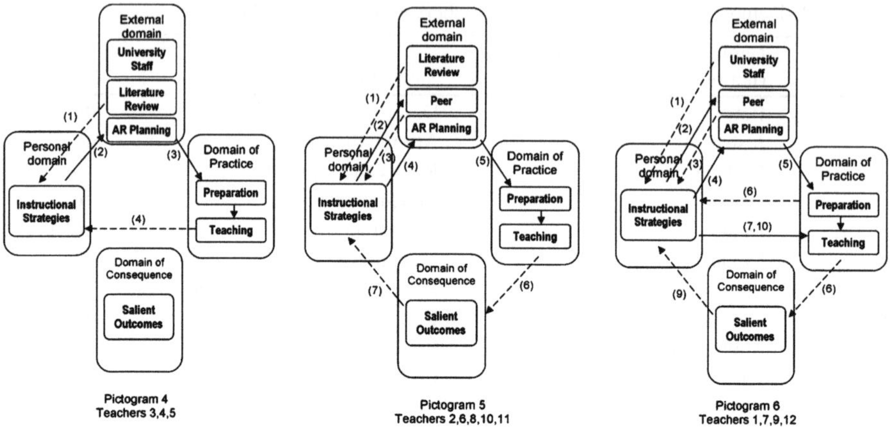

Fig. 9.5 Pictorial representations of development of knowledge of instructional strategies

48 pictograms. In accordance with the work of Zwart et al. (2007), the 48 pictograms were studied in order to identify particular pathways on the basis of the common entry points (starts), the sequences of changes and the end points.

4.1.2 Results

We found three different pathways of change for each PCK component. In this section, we discuss an example of each pathway by explaining how they were constructed and how they differed from each other. For this purpose, we focus on the PCK component *knowledge of instructional strategies*. Where necessary, we will use statements from the teachers' journals to explain the typical enactments and reflections associated with each of the pathways.

Data analyses for the PCK component *knowledge of instructional strategies* show pictograms with similar entry points but with three different pathways leading to three distinctly different learning outcomes (see Fig. 9.5).

All entry points are in the external domain, where teachers reviewed the literature. The participants used the literature extensively to search for appropriate instructional strategies for their lessons. Some teachers discussed their instructional strategies with their peers (pictograms 5 and 6), and others did not (pictogram 4). After planning (arrow 2), preparing (arrow 3) and conducting their lessons, pictogram 4 teachers reflected on their lessons (arrow 4). An example from Dana (teacher 4):

> I used experiments while studying the human body because I wanted my students to have as many experiences as possible. I think that they do learn better by providing different evidence themselves, not just out of a book. (pictogram 4, arrow 4; source: teacher interview)

Pictogram 5 teachers reflected on their classroom practice (arrow 6) and their classroom outcomes (arrow 7). An example of arrows 6 and 7 is as follows: After Shania (teacher 10) taught her sixth grade class about volcanoes, she told us that her students did not learn that much when they were taught in the traditional way. Now, she was convinced that her students did learn something:

> Now they remembered something... throughout their school life, anything that has to do with cells will come back to them and I think that alone makes a lot of difference. (pictogram 5, arrows 6 and 7; source: teacher interview)

Pictogram 6 teachers continuously reflected on their instructional strategies: after presentations from the university staff (arrow 1), after consulting peers (arrow 3), after preparing lesson plans (arrow 6), and after teaching (arrow 8). Furthermore, after these teachers reflected on their classroom outcomes (arrow 9), they acted on it in order to change their classroom teaching (arrow 10). Matt's (teacher 7) example of arrows 9 and 10:

> Through using them [micro-based computer labs], I was forced to reflect on how these types of labs work with seventh graders. I saw how they impacted the learning in my room as we reviewed video tapes of students doing microcomputer-based labs (arrow 10; source: action research report)...We also did a study last year on our pond. And it had all kinds of little spin-offs, where we wanted to go with it... So the second time I did it [the micro-based computer labs], it was actually better than the first. (arrow 11; source: teacher interview)

4.1.3 Conclusion and Discussion

Although we found different pathways for each teacher, we were able to categorise these pathways, based on similar entry points, similar domains and similar ending points. We found two distinct pathways that lead to changes in PCK: pathways that include the domain of consequence (DC; see pictograms 5 and 6) and pathways without the domain of consequence (pathways in pictogram 4). We consider pathways *without* the DC to reflect 'simple growth networks', whereas pathways *including* the DC can be seen as more 'complex growth networks'. When closely examining those pathways showing a 'simple growth network', we did find changes in the different domains; however, the teachers did not demonstrate whether they learned from their classroom actions. For example, Dana (teacher 4) reflected on her knowledge of instructional strategies after preparing lesson plans, but failed to reflect on how her students perceived this new way of teaching (see pictogram 4). In the pathways with a 'complex growth network', the teachers reflected on their students' learning (a change in the domain of consequence) and were able to specify what they learned from their students. For example, Matt (teacher 7) reflected on the teaching strategy used in his classroom on the basis of student feedback and was able to argue whether the instructional strategy was effective or not (see pictogram 6). In our study, we found that teachers with a more 'complex growth network' indicated obvious changes in their pedagogical content knowledge. Teachers with a 'simple growth network' did show change, for example, in cognition, but it is doubtful whether this change affected their teaching. These findings show that reflections on classroom outcomes were important for the PCK development of these in-service teachers.

Investigation of the different entry points led us to conclude that changes in the external domain often induced major changes in the PCK found in the personal domain. 41 of the 48 entry points were located in the external domain. Fourteen entry points were linked to the university staff, 17 entry points were found when teachers used their literature review and ten were prompted by teachers participating in peer discussions. Furthermore, we noted that the university staff contributed most in helping participants define science curricula and in constructing knowledge of student understanding. The literature review and peer discussions were used extensively in the search for instructional strategies and assessment methods. It should also be noted that teachers valued the use of the educational and science literature reviews to improve their teaching. When teachers studied the literature, they were able to adapt their instructions more to current recommendations from this literature (e.g. pictograms 4 and 6). This tallies with the findings of Rhine (1998), who asserted that resources on educational research can be crucial for in-service teachers as a 'lifelong resource' for lesson planning. Teachers in this study used the literature to find information on science subjects and to learn about effective ways to teach these subjects. Then when they discussed their findings from the literature with peers, this helped them reflect on this new-found knowledge, providing a deeper understanding of their PCK (e.g. pictogram 6). In general, we found that teachers who conducted a literature review and participated in peer discussions acquired a better understanding of the use of instructional strategies and assessment methods, such as the use of micro-based computer labs to increase students' science skills and the use of students' journals to assess their students' knowledge. In the planning of professional development programmes, therefore, teachers' reading of educational research literature should not be underestimated, since it creates opportunities to construct new knowledge.

5 Final Remarks

Reviewing the findings of the studies in the previous sections, it may be concluded that science teachers' professional learning PCK may effectively be supported by providing opportunities to experiment with new teaching approaches in their classroom and to reflect on their experiences, both individually and collectively. This approach acknowledges that teachers, as professionals, working individually at different schools, hold the key to improving the effectiveness of science education (Bell and Gilbert 1996). In particular, working with the IMTPG as an analytical tool proved to be helpful, giving more insight into the processes involved in professional learning. The IMTPG appears to help to make the often tacit and implicit change pathways explicit, and, furthermore, it makes it possible to indicate powerful elements within professional learning programmes. However, it should be noted that more research evidence is needed to support the claims made above. Ultimately, we need research that demonstrates how professional development programmes contribute to changes in teachers' professional knowledge and their practice, in a way that enhances student learning and appreciation of science (cf. Yoon et al. 2007; Desimone 2009).

Acknowledgments The author wishes to thank Dr. Rosária Justi and Dr. Dirk Wongsopawiro for allowing me to use the results of their studies and Dr. Amanda Berry for her help with the preparation of my plenary lecture and this chapter.

References

Anderson, R. D.,& Mitchener, C. P. (1994). Research on science teacher education. In D. Gabel (Ed.), *Handbook of research on science teaching and learning* (pp. 3–44). New York: National Association for Research in Science Teaching and National Science Teachers Association/ Macmillan Publisher.

Ball, D. M., & Cohen, D. (1999). Developing practice, developing practitioners: Toward a practice-based theory of professional development. In L. Darling-Hammond & G. Sykes (Eds.), *Teaching as the learning profession: Handbook of policy and practice* (pp. 3–32). San Francisco: Jossey-Bass.

Beijaard, D., Verloop, N., Wubbels, T., & Feiman-Nemser, S. (2000). The professional development of teachers. In R. J. Simons, J. van de Linden, & T. Duffy (Eds.), *New learning*. Dordrecht/ Boston/London: Kluwer Academic.

Bell, B., & Gilbert, J. K. (1996). *Teacher development: A model from science education*. London: Falmer Press.

Bereiter, C., & Scardamalia, M. (1993). *Surpassing ourselves: An inquiry into the nature and implications of expertise*. Chicago: Open Court.

Borko, H. (2004). Professional development and teacher learning: Mapping the terrain. *Educational Researcher, 33*(8), 3–15.

Borko, H., & Putnam, R. (1996). Learning to teach. In D. C. Berliner & R. C. Calfee (Eds.), *Handbook of educational psychology* (pp. 673–708). New York: Macmillan.

Clarke, D., & Hollingsworth, H. (2002). Elaborating a model of teacher professional growth. *Teaching and Teacher Education, 18*, 947–967.

Cohen, L., Manion, L., & Morrison, K. (2000). *Research methods in education* (5th ed.). London: Routledge Falmer.

Desimone, L. M. (2009). Improving impact studies of teachers' professional development: Toward better conceptualizations and measures. *Educational Researcher, 38*, 181–199.

Desimone, L., Porter, A. C., Garet, M., Yoon, K. S., & Birman, B. (2002). Does professional development change teachers' instruction? Results from a three-year study. *Educational Evaluation and Policy Analysis, 24*, 81–112.

Doyle, W., & Ponder, G. A. (1977). The practicality ethic in teacher decision-making. *Interchange, 3*, 1–25.

Fishman, B. J., Marx, R. W., Best, S., & Tal, R. T. (2003). Linking teacher and student learning to improve professional development in systemic reform. *Teaching and Teacher Education, 19*, 643–658.

Garet, M., Porter, A., Desimone, L., Birman, B., & Yoon, K. S. (2001). What makes professional development effective? Results from a national sample of teachers. *American Education Research Journal, 38*, 915–945.

Guskey, T. R. (1986). Staff development and the process of teacher change. *Educational Researcher, 15*(5), 5–12.

Guskey, T. R. (2002). Professional development and teacher change. *Teachers and Teaching: Theory and Practice, 8*, 381–391.

Haney, J. J., Czerniak, C. M., & Lumpe, A. T. (1996). Teacher beliefs and intentions regarding the implementation of science education reform strands. *Journal of Research in Science Teaching, 33*, 971–993.

Hashweh, M. Z. (2003). Teacher accommodative change. *Teaching and Teacher Education, 19*, 421–434.

Hawley, W., & Valli, L. (1999). The essentials of effective professional development: A new consensus. In L. Darling-Hammond & G. Sykes (Eds.), *Teaching as the learning profession. Handbook of policy and practice* (pp. 127–150). San Francisco: Jossey-Bass.

Hewson, P. W. (2007). Teacher professional development in science. In S. Abell & N. Lederman (Eds.), *Handbook of research on science education* (pp. 1179–1203). Mahwah: Lawrence Erlbaum Associates.

Joyce, B., & Showers, B. (1980). Improving in-service training: The message of research. *Educational Leadership, 37*, 379–385.

Justi, R., & Gilbert, J. K. (2003). *Investigating teachers' ideas about models and modelling: Some issues of authenticity*. In: Paper presented at the fourth international conference of the European Science Education Research Association, Noordwijkerhout, 19–23 Aug 2003.

Justi, R., & Van Driel, J. H. (2006). The use of the IMTPG as a framework for understanding the development of science teachers' knowledge on models and modelling. *Teaching and Teacher Education, 22*, 437–450.

Kennedy, M. M. (2010). Attribution error and the quest for teacher quality. *Educational Researcher, 39*, 591–598.

Kind, V. (2009). Pedagogical content knowledge in science education: Perspectives and potential for progress. *Studies in Science Education, 45*(2), 169–204.

Korthagen, F. A. J., Kessels, J., Koster, B., Lagerwerf, B., & Wubbels, T. (2001). *Linking practice and theory: The pedagogy of realistic teacher education*. Mahwah: Lawrence Erlbaum Associates.

Little, J. W. (2001). Professional development in pursuit of school reform. In A. Lieberman & L. Miller (Eds.), *Teachers caught in the action: Professional development that matters* (pp. 28–44). New York: Teachers College Press.

Lotter, C., Harwood, W. S., & Bonner, J. J. (2006). Overcoming a learning bottleneck: Inquiry professional development for secondary science teachers. *Journal of Science Teacher Education, 17*, 185–216.

Loucks-Horsley, S., Love, N., Stiles, K. E., Mundry, S., & Hewson, P. W. (2003). *Designing professional development for teachers of science and mathematics* (2nd ed.). Thousand Oaks: Corwin.

Magnusson, S., Krajcik, J., & Borko, H. (1999). Nature, sources and development of pedagogical content knowledge. In J. Gess-Newsome & N. G. Lederman (Eds.), *Examining pedagogical content knowledge* (pp. 95–132). Dordrecht: Kluwer Academic.

Munby, H., Russell, T., & Martin, A. K. (2001). Teachers' knowledge and how it develops. In V. Richardson (Ed.), *Handbook of research on teaching* (4th ed., pp. 877–904). Washington, DC: American Educational Research Association.

Ponte, P., Ax, J., Beijaard, D., & Wubbels, T. (2004). Teachers' development of professional knowledge through action research and the facilitation of this by teacher educators. *Teaching and Teacher Education, 20*, 571–588.

Rhine, S. (1998). The role of research and teachers' knowledge base in professional development. *Educational Researcher, 27*(5), 27–31.

Richardson, V., & Placier, P. (2001). Teacher change. In V. Richardson (Ed.), *Handbook of research on teaching* (4th ed., pp. 905–947). Washington, DC: American Educational Research Association.

Roberts, D. A. (1988). What counts as science education? In P. J. Fensham (Ed.), *Development and dilemmas in science education* (pp. 27–54). London: Palmer Press.

Russell, T., & Martin, A. K. (2007). Learning to teach science. In S. Abell & N. Lederman (Eds.), *Handbook of research on science education* (pp. 1151–1176). Mahwah: Lawrence Erlbaum Associates.

Schön, D. (1983). *The reflective practitioner: How professionals think.* New York: Basic Books.
Sprinthall, N. A., Reiman, A. J., & Thies-Sprinthall, L. (1996). Teacher professional development. In J. Sikula, T. J. Buttery, & E. Guyton (Eds.), *Handbook of research on teacher education* (2nd ed., pp. 666–703). New York: Macmillan.
Thompson, C. L., & Zeuli, J. S. (1999). The frame and the tapestry: Standards-based reform and professional development. In L. Darling-Hammond & G. Sykes (Eds.), *Teaching as the learning profession. Handbook of policy and practice* (pp. 341–375). San Francisco: Jossey-Bass.
Tom, A. R., & Valli, L. (1990). Professional knowledge for teachers. In W. R. Houston (Ed.), *Handbook of research on teacher education* (pp. 372–392). New York: Macmillan.
Van Driel, J. H., & Abell, S. K. (2010). Science teacher education. In B. McGraw, P. L. Peterson, & E. Baker (Eds.), *Third international encyclopedia of education* (pp. 712–718). Amsterdam: Elsevier.
Van Driel, J. H., Beijaard, D., & Verloop, N. (2001). Professional development and reform in science education: The role of teachers' practical knowledge. *Journal of Research in Science Teaching, 38*, 137–158.
Van Veen, K., Zwart, R., Meirink, J., & Verloop, N. (2010). *Professional development of teachers: A review study on the effective characteristics of professional interventions of teachers. (Professionele ontwikkeling van leraren: een reviewstudie naar effectieve kenmerken van professionaliseringsinterventies van leraren).* Leiden: ICLON/Expertisecentrum: Leren van Docenten.
Wilson, S. M., & Berne, J. (1999). Teacher learning and the acquisition of professional knowledge: A review of research on contemporary professional development. *Review of Research in Education, 24*, 173–209.
Wongsopawiro, D. (2012). *Examining science teachers' pedagogical content knowledge in the context of a professional development programme.* Doctoral dissertation, Leiden University.
Wubbels, T. (1992). Taking account of student teachers' preconceptions. *Teaching and Teacher Education, 8*(2), 137–150.
Yoon, K. S., Duncan, T., Lee, S. W. Y., Scarloss, B., & Shapley, K. (2007). *Reviewing the evidence on how teacher professional development affects student achievement* (Issues and answers report, REL 2007, Vol. 033). Washington, DC: US Department of Education.
Zwart, R. C., Wubbels, T., Bergen, T. C. M., & Bolhuis, S. (2007). Experienced teacher learning within the context of reciprocal peer coaching. *Teachers and Teaching: Theory and Practice, 13*(2), 165–187.

Chapter 10
Nanoeducation: Zooming into Teacher Professional Development Programmes in Nanoscience and Technology

Ron Blonder, Ilka Parchmann, Sevil Akaygun, and Virginie Albe

1 Introduction and Literature Review

Nanotechnology is concerned with the ability to create materials, devices and systems with fundamentally new properties and functions by working at the atomic, molecular and supramolecular levels (Roco 2001). These new properties are utilised for developing new electronic, magnetic, optoelectronic, medical, alternative energy and other applications. Roco (2003) described the importance of education for the future development of this field: 'One of the "grand challenges" for nanotechnology is education, which is looming as a bottleneck for the development of the field' (Roco 2003, p. 1247).

From science education research literature, four strands of research can be identified (Hingant and Albe 2010): reflections prior to curriculum development on nanosciences and nanotechnologies, studies on students' conceptualisations of nano-related concepts, use of haptic tools to teach nanosciences and nanotechnologies, and professional development for secondary school teachers. In this chapter, we focus on the latter.

R. Blonder (✉)
Weizmann Institute of Science, Rehovot, Israel
e-mail: ron.blonder@weizmann.ac.il

I. Parchmann
Leibniz-Institute for Science and Mathematics Education (IPN), Kiel, Germany
e-mail: parchmann@ipn.uni-kiel.de

S. Akaygun
Bogazici University, Istanbul, Turkey
e-mail: sevil.akaygun@boun.edu.tr

V. Albe
STEF, Ecole Normale Supérieure de Cachan, Cachan, France
e-mail: virginie.albe@stef.ens-cachan.fr

C. Bruguière et al. (eds.), *Topics and Trends in Current Science Education: 9th ESERA Conference Selected Contributions*, Contributions from Science Education Research 1,
DOI 10.1007/978-94-007-7281-6_10, © Springer Science+Business Media Dordrecht 2014

When examining the field of nanoeducation, educators have been faced with a problem: on one hand, most nanoscience researchers received training in a discipline other than nanoscience and therefore have difficulties in teaching nanoscience; on the other hand, high school teachers have difficulties in updating their knowledge on advanced topics such as nanoscience and will naturally have difficulties in teaching content with which they are not acquainted. Programmes therefore need to address the content knowledge, the pedagogical content knowledge as well as attitudes and beliefs, as described in the well-known model by Shulman (Shulman 1987) and later adaptations.

Several difficulties of (future) teachers have been described in the literature. Hundred and nine teachers from Australia filled in a ten-item questionnaire that mapped their knowledge in nanotechnology (Kumar 2007). The study revealed a lack of understanding of the underlying physical scale of nanoscience and nanotechnology and the etymology of the term 'nano' among pre-service teachers. In a recent study, pre- and in-service teachers' conceptions of spatial scale were compared (Jones et al. 2011). Ninety-two in-service and 134 pre-service teachers from Australia, Taiwan and the United States participated in the study. Jones et al. (2011) found that there was an underlying common dimension to understanding scale and that accuracy of understanding scale was not related to professional teaching experience. Australian and Taiwanese teachers were significantly more accurate than the US group. The Australian participants scored significantly higher than the US and the Taiwanese participants, and that was related to their learning/not learning scales in their formal learning experience and informal learning due to common hobbies in the different countries. The authors claimed that bridging the gap between existing knowledge of school science teachers and the knowledge and pedagogy required to teach nanotechnology should be the heart of professional development programmes created for teachers.

Some authors have described the hurdles which may dissuade teachers from integrating 'nanos' into their lessons: difficulties in taking into account an interdisciplinary approach and reluctance to deal with subjects with which they have not been acquainted during their initial training and on which they may be at a loss to answer pupils' questions (Schank et al. 2007). The few professional development programmes in nanoscience offer first insights into promising approaches to overcome such barriers.

Blonder (2011) describes a comprehensive course for chemistry teachers. The course was designed to make chemistry teachers nanoliterate by providing them with the basic principles of nanoscience, enabling them to independently study any subject in nanoscience and elevating their enthusiasm for chemistry. An evaluation of teachers' knowledge and self-efficacy was performed throughout and after the course period. Posttest results showed that the participants achieved average scores quite close to the full answer in relation to nanoscience topics. The teachers were also interviewed and stated that the course provided them with the basic tools and knowledge to become nanoliterate.

Tomasik et al. (2009) designed an online professional development course on nanoscience and nanotechnology for in-service teachers. The course included eight weekly sessions and a final project, in which each participant created a nanoscience module for his or her classroom that was peer-reviewed by other participants. They followed teachers' knowledge by analysing their weekly quiz responses. It was found that teachers achieved significant learning gains during the 8 week course. Teachers found the virtual course environment to be as effective as face-to-face interactions that are usually used in other professional development programmes for teachers. A support for these results was obtained by Nichol and Hutchinson (2010). They developed videoconferencing combined with a desktop sharing tool as a platform to conduct a course including nanotechnology topics that matched with concepts in chemistry or physics. Using distance-learning technologies, they aimed to expand the impact of the programme and to reach more teachers than in the previous years. Surveys, feedback and follow-up workshops have shown that lesson plans and concept questions were developed and group discussions have been fostered, even for teachers participating in remote locations. They found that the outcomes from the synchronous videoconferencing format were similar to those of the face-to-face course. It is critical to note that the target audience for the teachers' courses is their students.

Daly and Bryan (2010) examined how teachers, who participated in a nano-education professional development programme, used models in their lessons to facilitate learning nanotechnology concepts and phenomena. The researchers focused on the ways teachers used models for different goals in their nanotechnology lessons and also used the teachers' written lesson plans, the questions on model use and their reflections on the lesson. They found that the teachers used models for four purposes: as a tool for visualisation, as products of student design, as representation for student critique and as a means for investigation. Understanding the reasons for teachers choosing to use models only as visual tools in nanotechnology instruction is particularly important, since it is a newly developed educational area. Moreover, the way models are used in teaching nanotechnology is important also in informal education, where models play an important role in transferring concepts.

The last example named here investigates the role of the instructors. Drane et al. (2009) developed a nanoscience module and compared the teaching of a nanotechnology professor and a graduate student who differ in their experience of nanotechnology research and teaching. They found that as a consequence of the short course involving the nanoscience module, students' knowledge and interest in nanotechnology increased in both groups. However, students indicated that the nanotechnology professor used different teaching strategies. They concluded that more extensive and detailed training by the module developer focusing not only on nanotechnology content but also on the pedagogical elements is necessary to ensure a 'faithful transfer' of the nanoscience module.

Alongside the content and specific methods of nanoscience, there is also a need to include science and society issues in courses on nanoscience, both for students and teachers. During the last decade, new types of courses have appeared in science and technology universities, first in the United States and now reaching Europe and France, associating a call for an interdisciplinary approach with a strong convergence of science and industry and new ways of integrating social and/or human sciences in scientific curricula. Intended for science students, such 'science and society' courses rely on (and participate in the construction of) specific ways of describing science, technology and the social world, generally saturated with political values. The integration of these courses in scientific education, linked with a strong effort to specify new institutional organisations of scientific practice and education (in particular through political support of emerging fields like nanoscience, synthetic biology or the cognitive sciences), plays an important role in the acculturation of future scientists and engineers to 'good' scientific practices and discourses.

2 Further Exploration of the Field: Four Approaches to Investigate and to Teach Nanoscience to (Future) Teachers

Following this comprehensive summary of research and developmental approaches regarding teacher education in the area of nanoscience, we will now lay out four different courses and the accompanying research studies. These four approaches form a unity in the aim of enhancing teachers' backgrounds in nanoscience and nanotechnology and convey diversity in the methodologies used. The first project aims at developing instruments and gaining further knowledge on the goals of and preconditions for introducing nanoscience into teacher professional development programmes. The Model of Educational Reconstruction (MER) has been used to achieve this. The second project reports on the framework and the first empirical results of a pre-service teacher training course approaching the challenges named in Sect. 1 and identified also in the first project from a phenomenographic point of view. The third and fourth projects describe the structures and results of two in-service teacher training courses, one focusing specifically on the use of models and the second one on teachers visiting real science labs (see the literature review as well). The questions leading the four projects, which had not been developed in coherence but have first been combined in the ESERA conference symposium, can be summarised as follows:

- How can we analyse and make use of the expectations and knowledge of (future) teachers and experts to develop professional development programmes in the area of nanoscience?
- Which elements of professional development programmes seem to be promising according to exploratory research results?

2.1 Analysing and Including Future Teachers and Experts' Knowledge and Expectations in Professional Development Programmes

The design of teacher training courses demands knowledge about the goals aimed for in teaching a topic, the pre-knowledge of the audience to be dealt with and the approaches that have been evaluated as promising for use in educational settings

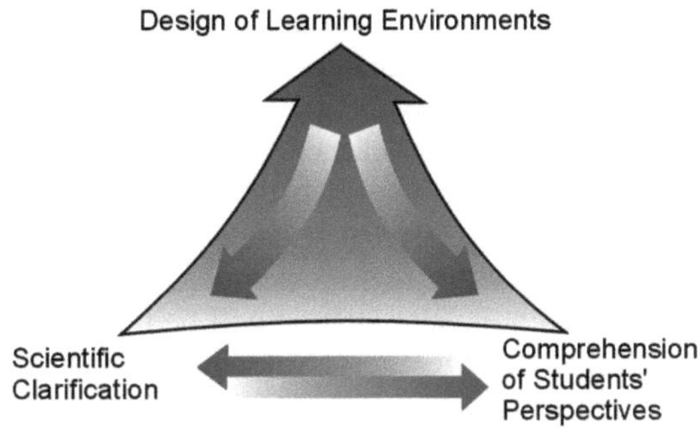

Fig. 10.1 The model of educational reconstruction

(see the literature review). Building on the studies described in Sect. 1 of this chapter, this project describes a model that can be applied for both the analyses of goals and preconditions and the development of professional development programmes explicitly starting from those preconditions.[1] The Model of Educational Reconstruction (Duit et al. 2012; Parchmann and Komorek 2008) was used in this project as a framework to combine content analyses, empirical research and the design of educational settings (see Fig. 10.1).

The scientific clarification aimed at enlarging the knowledge on goals and structures for teaching nanoscience at school level (Stevens et al. 2009, and Sect. 1 of this chapter). Next to socio-scientific issues such as risk assessment (Englander and Kim 2011), the explanations of structure-property relations are a basic scientific concept for which nano perspectives can be included to set a focus on the importance of size and scale. Traditional school curricula focus on macroscopic properties first and lead to explanations on the (sub)microscopic level. These explanations begin with simple particle models, leading to more and more sophisticated models about atoms, bonds and resulting structures. The inclusion of nanostructures can set the focus on properties related to a specific scale between single atoms and macroscopic properties. This is regarded as a promising perspective to bridge the well-documented difficulty for students in relating macroscopic properties to submicroscopic models. One of the problems is, for example, that students directly explain properties of substances by properties of atoms. The perspective on nanostructures points out the necessity to find explanations in between, as single atoms can obviously not be the reason for different properties in different sizes and scales of substances – they are always the same.

In addition to the literature review and the theoretical educational considerations, university physics and chemistry professors have been (and currently still are) interviewed. The interview guideline includes questions concerning their beliefs about the specific worth of investigating and of teaching examples and explanations of nanosciences at university and at school level. The research experts were also asked to give possible models or explanations for important techniques in nanoscience,

[1] Researchers involved: Katrin Bock, Stefanie Herzog and Ilka Parchmann

such as scanning tunnel microscopy (STM) or atomic force microscopy (AFM, see also Sect. 2.4) on different levels, which could be used in teacher training workshops. Exemplary results have been published nationally to enhance the discussion between teachers (Parchmann et al. 2010). The experts use analogies like a record player to explain the functioning of an STM, for example. They also give suggestions for different steps of explanation, leading to the quantum mechanical concept only at the highest step.

The scientific clarification will be iteratively combined with empirical analyses of the potential learners. Student teachers were chosen as the first target group to analyse the learners' perspectives, according to the Model of Educational Reconstruction. Two pilot questionnaire studies have been carried out to investigate pre-service teachers' self-estimated knowledge, their expectations about teaching nanoscience at secondary schools and their beliefs about nanostructures and about techniques such as STM and AFM. Currently, additional interviews, analogous to the expert interviews, are in progress.

Based on answers in an open questionnaire, a Likert scale instrument was developed and answered by students at three different universities (N = 96).

A factor analysis led to four different scales with acceptable reliabilities:

- Experiences and interest in further knowledge (e.g. 'I wish I had learned more about nanoscience at university'; four items, $\alpha = 0.727$)
- Visions and potential use of nanoscience (e.g. 'Nanoscience will help to optimise many tools in the future'; five items, $\alpha = 0.716$)
- Expectations on students' knowledge ('I think nanoscience can help students to optimise their understanding of models'; four items, $\alpha = 0.714$)
- Expectations on students' interest ('I think nanoscience will be very interesting for students'; three items, $\alpha = 0.613$)

On a Likert scale from 1 (I do not agree) to 4 (I agree), the agreement on interest in further knowledge and experience was high (mean value: 3.13). A positive mean value was also found for the visions, referring to positive expectations for future nano developments (3.14). The expectations of the students' knowledge and interest in the topic showed mean values of 2.5.

Correlation analyses showed that the student teachers' own experiences and interests correlate significantly with their positive visions of nanoscience and their expectations on the students' knowledge gain, but not on the expectations of the students' interests. Their judgment seems to be more independent from the teachers estimated own knowledge base.

The analyses of exemplary items confirm the findings reported in the literature. Student teachers are interested in gaining further knowledge and understanding, but they do not feel confident yet about teaching nanoscience topics (see Fig. 10.2).

Indeed, the student teachers' knowledge is also insufficient in different areas, as described in some of the studies mentioned in Sect. 1 of this chapter. The open answers in the first pilot study offer starting points for the development of learning tasks. For example, 68 % of the students did not give any answer on structure-property relations or named wrong influences such as 'carrying functional groups'.

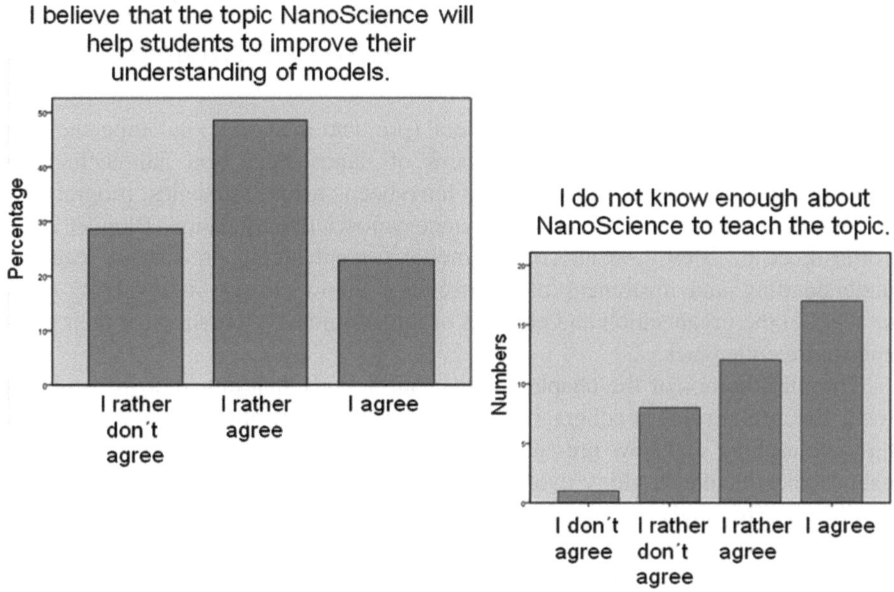

Fig. 10.2 Results of two exemplary items; in the first figure, the option 'I don't agree' was not chosen

Less than 10 % named the surface structure as important. In contrast to those results, the students gave various reasons why teaching nanoscience could have positive effects on their teaching and the students: the relevance for daily life and the future, an alternative and interdisciplinary approach to teaching the basic concept of structure-property relations and a possibility to show abstract concepts by images of structures (using STM or AFM).

The next step in this project will be the application of the empirical results and further content analyses to develop professional development programmes both for pre-service and in-service teachers. The Model of Education Reconstruction highlights the importance of incorporating not only the scientific knowledge but also the preconditions of the learners, in this case the (missing) knowledge foundation and the expectations or beliefs of the (future) teachers. The interviews of the experts have been seen to offer additional insights into topics, models and stepped explanations that could be used in teacher workshops. The investigation of the student teachers has revealed several results of the literature review but has also shown new results, such as the correlations between the teachers' own beliefs and their expectations towards their students. Last but not least, the resulting questionnaire itself can now be used in other studies to analyse the starting conditions in each professional development programme. The Model of Educational Reconstruction could then be applied to discuss goals and empirical findings for students and to develop a course for secondary education within the group of teachers, as partly adapted in the following section of this chapter.

2.2 An Alternative Method for Incorporating Nanoscience in Science Teachers' (Pre- and In-Service) Training

Recent research conveyed science teachers' (pre- and in-service) incompetencies in conceptual understanding and awareness of nanoscience and nanotechnology (Kumar 2007; Jones et al. 2011). There have been various strategies, programmes or courses developed for teachers to enhance nanoscience education (Blonder 2011; Tomasik et al. 2009). An alternative model for enhancing pre-service teachers' understanding and awareness of nanoscience could consist in involving them *actively* in the organisation and conduct of a nanoscience workshop for their peers and future colleagues.

The third author of the chapter carried out a study that aimed at investigating what the pre-service teachers initially know and feel about nanoscience and nanotechnology and how pre-service teachers' understanding and awareness of nanoscience/nanotechnology evolved as they became involved in a nanoscience workshop. The study was designed qualitatively in terms of phenomenography (Marton 1994), which allowed more in-depth investigation and understanding of the situation and people's experiences (the other studies, presented in Sect. 2, applied different methodologies). Triangulation of the data was done by open-ended questionnaires, one-on-one interviews and participant observations.

The participants in the study were nine pre-service teachers, who were majoring in chemistry (N = 2), in physics (N = 3) and in mathematics (N = 4) education in their junior and senior years. They voluntarily prepared, organised and participated in 1, 2 or 3 'nanoscience/nanotechnology workshops' for pre- or in-service teachers. The study got started while the students were taking a class called 'Science, Technology and Society (STS)'. First, the announcement was made regarding a future 'nanoscience/nanotechnology workshop' in which volunteer students would be preparing, organising and presenting certain topics such as introduction, history, instruments, classroom activities, applications and ethics. Nine out of a class of 26 students voluntarily decided to take part in this workshop. Next, individual interviews with each volunteer pre-service teacher were conducted to better understand their conceptualisation and awareness of nanoscience and nanotechnology. After the interviews, students were given 3 weeks to prepare a PowerPoint presentation on their topic by working in pairs. In the following semester, the groups were called upon to take part in a workshop setting where they presented their sections to a new group of pre-service teachers. Then the groups presented, discussed and revised their sections after two weekly scheduled meetings. The pre-service teachers were also responsible for organising the setting, documents and materials needed for the workshop. The 'nanoscience/nanotechnology workshop' for the peer pre-service teachers took 2 h, and it was video recorded for a better understanding of pre-service teachers' understanding and awareness. One week after the workshop, the presenters got together to discuss their experiences and impressions, and there was also feedback from the participants. At the end of that semester, three of the pre-service teachers graduated. The remaining six of them agreed to participate in a summer 'nanoscience/nanotechnology workshop' for in-service elementary teachers who were visiting university for a summer school. Finally, four of the pre-service teachers participated in the third 2-h-long 'nanoscience/nanotechnology workshop' in the following semester for another group of pre-service teachers. This time, the workshop was designed by a graduate student, and the volunteer pre-service teachers worked as group mentors who led small group discussions for various activities instead of presenting certain parts. After the third workshop was completed, post-interviews were conducted with all the pre-service teachers including the ones who had graduated earlier. In summary, three pre-service teachers participated in only one workshop, three of them participated in two out of three workshops and three pre-service teachers participated in all three workshops by preparing, organising, presenting and leading group discussions. Being actively involved in the nanoscience/nanotechnology workshops by taking on various roles had some effect on pre-service teachers' understanding of nanoscience. The four different selected cases are discussed below.

Pre-service teacher 'Z' was majoring in teaching mathematics and graduated after the first workshop. In her initial interview, she said that she had never heard anything about nanoscience/nanotechnology until she was taking the STS course. In the post-interview, she remembered all the fundamental concepts, applications, benefits and the risks. She said that she started to talk about nanoscience/nanotechnology to the people around her. She said that she gave examples from nanoscience while she was teaching logic to her 9th grade mathematics students. She emphasised that she developed awareness not only in nanoscience and nanotechnology but in science and technology in general because she said that she got excited by the acceptance of a Turkish group to the European Organization for Nuclear Research (CERN) to work in the research team.

Pre-service teacher 'N' was majoring in teaching chemistry and graduated after participating in the first workshop. In her pre-interview, she said that she had heard about nanoscience/nanotechnology, but she said that she did not have any conceptual understanding. In the post-interview, she said that as she got involved in the workshop, she started to read more about nanoscience to learn in detail. Most impressively, she said after graduating she decided to pursue her graduate studies in chemistry emphasising nanoscience, and she added that she wished to learn it when she was in high school.

Pre-service teacher 'S' was majoring in teaching mathematics and participated in two of the three workshops as organiser and presenter. Initially she viewed nanoscience as 'dream science' and then changed this view to 'science coming from everyday life'. As she got involved in the workshop, she refined her understanding of the fundamental concepts of nanoscience. In the post-interview, she said that she had started to inform people around her and had even thought of giving them some ideas of some possible new products containing nanotechnology – 'curtains storing solar energy'. In addition, she said that she had started to think about the risks as she had never thought before and had become more sceptical about nanotechnology.

Pre-service teacher 'R' was majoring in teaching mathematics and participated in all three of the workshops as organiser, presenter and mentor. From the beginning of the first workshop to the end of the third one, she refined her conceptual understanding of nanoscience as she started to use terminology appropriately and listed new areas of application. The things that she wanted to learn about nanoscience changed from 'the relation of nano to mathematics' (her major) to 'chemistry of nanoscience'; in other words she wanted to learn more in depth. Her understanding of the risks of nanotechnology changed from 'use in war' to 'not being informed about the risks'. When she was asked how nanoscience should be integrated in teaching, her perspective changed from 'using videos and activities' to 'raising awareness', as from being more learning oriented to being awareness oriented. Finally, in taking part in all the workshops, her goal changed from learning the topic to thinking big about nanoscience and improving herself in every aspect of nanoscience and increasing her awareness.

In conclusion, this study helped to identify how the understanding and awareness of pre-service teachers evolved as they got actively involved in the preparation, organisation and presentation and mentoring stages of the nanoscience/nanotechnology workshops. In spite of their different backgrounds, all the pre-service teachers who took part in the workshop developed better conceptual understanding and awareness of nanoscience and nanotechnology. In the long run, they started to value other scientific activities, such as the acceptance of a Turkish team to CERN. It can be claimed that they may have greater motivation and interest to follow scientific activities not only in nanosciences but also in other disciplines. It could be suggested that the model be used as a training model for nanoeducation given to pre-service teachers. As the issues such as risks, benefits and ethics, regarding nanoscience and nanotechnologies, and the possible activities such as organising workshops are incorporated into the course 'Science, Technology and Society (STS)' for pre-service teachers, their motivation to learn, reflect and react could be increased, and in turn, they would become more scientifically literate citizens.

2.3 How to Address Societal Issues in Science Teachers' (Pre- and In-Service) Training

The fourth author of the chapter described a summer school programme elaborated by a group of teachers and researchers from research institutions[2] and from scientists' and teachers' unions[3] with the aim of promoting scientific research, informing

[2] National Institute of Nuclear and Particle Physics (IN2P3), part of CNRS (National Center for Scientific Research, a government-funded research organisation, under the administrative authority of France's Ministry of Research); French Atomic Energy Commission, leader in research, development and innovation, with the objective of ensuring that the nuclear deterrent remains effective in the future.

[3] The French Physics Society, an association that aims to promote physics and physicists; Physics and Chemistry Teachers' Union, a teachers' union involved in a 2 year undergraduate programme specific to France, leading to a nationwide competitive examination for admission into one of the 'Grandes Ecoles', the leading schools in engineering, management and research – the programme includes high-level courses in mathematics, physics, chemistry, computer and engineering sciences as well as the humanities (foreign languages and philosophy).

teachers of recent scientific developments and providing links between schools and research centres in France. The 2010 E2phy summer school was the tenth edition of a series of yearly events aimed at science teachers from secondary and higher education. The 2010 summer school focused on the physics of the nanoworld. It included lectures on advanced domains in nanosciences, the social impacts of nanosciences (ethics, health, toxicity, law, regulation, jurisdiction, etc.) given by French researchers well known in their respective fields, lab work, visits to research centres and inclusive special events devoted to discussions on research in nanoscience and their societal impacts. Information concerning the knowledge possessed by the science teachers attending the summer school as regards nanoscience and nanotechnologies and their intentions to teach nanoscience and nanotechnology was collected via pre- and posttest questionnaires. The questionnaires were structured into two parts: a first part on nanoscience and nanotechnologies and a second part on nanoeducation. The first part consisted of six open questions, which were identical for the pre- and posttests: *According to you, what are nanoscience and nanotechnologies? What are the characteristics or specificities of nanoscience and nanotechnologies? What do you think of the stakes, interests and strategies of research in nanoscience and nanotechnologies? How much have you heard about nanoscience and nanotechnology developments raising debates, particularly on health and environment effects? According to you, what is debated? What definitions of nanoscience and nanotechnologies do you give or would you give in class?* The second part of the questionnaire consisted of four questions, which were identical for the pre- and posttests: *What do you teach in nanosciences? What curriculum contents do you think could be linked to nanosciences? With what forms? To what ends?* There were three additional open questions for the posttest to assess the summer school: *What did you gain from the summer school? What do you think you can integrate into your teaching? On the contrary, what do you consider as ill suited to your teaching?* 125 questionnaires were collected for the pretest and 47 for the posttest. All the answers were transcribed and coded with categories that emerged from three rounds of analysis. As a first step, two coders elaborated categories of answers independently and then confronted their categorisations to build consensus categories (second step). As a third step, the previously elaborated categories were revised through another round of coding. Sphinx software was used for the statistical analysis.

It was found that a large majority of teachers, both in pre- and posttests, consider that nanoscience and nanotechnologies are knowledge relative to the nanoscale, and for some teachers it is a new domain with specific applications. The characteristics or specificities of nanoscience and nanotechnologies from the teachers' viewpoints also mostly concern the nanoscale then technological developments, and a link with quantum mechanics is underlined in the pretest. In the posttest, teachers also focused on new properties specific to this scale, raising technological difficulties and debates. Stakes and interests are focused in teachers' responses to the pretest on medicine, electronics and information technologies within the context of international competitiveness and on economic and social registers in the posttest. Teachers were mostly aware of debates raised by nanoscience and nanotechnologies, both in the pre- and posttests. They underlined the absence of epidemiological results in the pretest and mostly health and interactions with living organisms, and the environment to a lesser extent, in the posttest. Definitions of nanoscience and nanotechnologies to give in class are mostly focused on the nanoscale size and linked to knowledge on atoms and molecules for both the pre- and posttests and on the idea of a new domain with specific applications and physical properties in the pretest (more rarely in the posttest for the latter). A majority of the teachers declared that at that time they did not teach nanoscience and cited curriculum contents that could be linked to nanosciences: basic chemistry, optics and medicine applications being mentioned the most often (both in the pre- and posttests) and electronics and size and scale in the posttest. The forms of teaching envisaged are specific educative activities dedicated to the exploration and documentation of a new scientific domain, in both the pre- and posttests. Specific lessons in the science class are also mentioned in the pretest with the purposes being to show science in the making and to develop engagement in science and understanding of scientific concepts. In the posttest, collaboration with research labs, debates and video presentations are cited, with the purposes being to show science in the making, to develop engagement in school science and understanding of scientific concepts, to contribute to information literacy and to citizenship education and to learn how to debate. In their responses to the three additional open questions for the posttest, the teachers expressed the view that from the summer school they gained knowledge on nanoscience and nanotechnologies and on nanoscience and nanotechnologies research, ideas for pedagogical activities and personal reflections. They considered that they can integrate information and knowledge on nanoscience into their teaching. The pre-/posttest comparison of teachers' knowledge showed that nanoscience and nanotechnologies mostly defined minima by the nanoscale size in the pretest were also apprehended with specific physical properties and applications in the posttest. This result on teachers' knowledge improvement through training programmes converges with the literature (Blonder 2011; Tomasik et al. 2009). Teachers tend to favour specifically designed teaching activities to integrate nanoscience and nanotechnologies in class.

2.4 The Use of a Simple Teaching Model in a Professional Development Programme in Nanotechnology

The first author of this chapter explored the influence of explicitly treating PCK by a teaching model that was used to introduce in-service chemistry teachers to one of the most used instruments in nanoscience: the atomic force microscope (AFM). The development of the nanosciences was strongly dependent on the development of characterisation methods at the nanoscale level and especially on the scanning probe microscopy methods. The atomic force microscope, first demonstrated in 1986, quickly became the workhorse of nanoscience, providing nanometre-scale imaging, local mechanical and electronic probing and manipulation (Binnig et al. 1986). The extensive use of the AFM in research triggered its incorporation into teaching labs in colleges, undergraduate studies and even high schools. However, in order to teach nanochemistry, chemistry teachers should learn this new topic, understand it and feel confident enough to teach it in school, i.e. CK in not enough. Blonder (2010) presented the use of a simple teaching model in nanotechnology and explored its influence on chemistry teachers with different backgrounds in nanoscience (nano-novices and nano-experts). This model was built of metal corks, a magnet and a used CD and therefore considered as a simple teaching model. The goal of this research was to learn how the AFM teaching model influenced teachers' understanding of the AFM and their attitudes about using this teaching model with their students.

Two groups of teachers participated in the research (each contained seven experienced chemistry teachers): Group 1 consisted of nanoliterate teachers who participated in a special programme for enhancing chemistry teachers' content knowledge and had already learned about the AFM. Group 2 consisted of nano-novice teachers. The data that we were interested in referred to the main goal of the study, namely, determining the influencing of the teaching model on teachers' knowledge and attitudes towards using the models in class. The data consisted of a constructivist tool, namely, Personal Meaning Mapping (PMM) (Falk and Dierking 2000), and teacher lab reports and were analysed according to Chi (1997). The analysis of the PMMs and the lab reports was done according to Chi: quantification of qualitative data. Each teacher's pre-knowledge and attitudes about AFM were compared with his or her post-experiment knowledge and attitudes, as reflected by the pre- and post-PMM. Learning was measured by assessing change across four semi-independent dimensions that emerged from the data. For the statistical analyses, we used the Wilcoxon signed-rank test for small samples as well as the Spearman correlation test. In addition to statistics, we examined the content and the essence of the new concepts that the teachers had learned after using the teaching model. We attempted to ascertain what influenced their teaching attitudes towards using the model and teaching nanochemistry in their classes.

It was found that previous knowledge was significantly different between the two groups (as was expected from their background), as reflected from the vocabulary mentioned, the conceptions and the misconceptions. The Wilcoxon signed-rank test for nonparametric variables was applied to compare the pre-PMMs of the two groups. The teachers in Group 1 had organised previous knowledge regarding the AFM, whereas the teachers in Group 2 were novices with respect to nanotechnology. Overall, there was evidence that teachers from both groups learned, as manifested by a significant change regarding the four PMM learning dimensions. The quantitative analyses show that the AFM teaching model is a useful teaching tool that increased the teachers' knowledge about the AFM. In addition to statistics, we examined what new concepts the teachers had learned, after using the teaching model, and what misconceptions had disappeared. In the remarks section of the lab report, all the teachers wrote that they enjoyed working with the teaching model and that it helped them to better understand the way that AFM works. They also wrote that they intended to use it in their class when teaching about nanochemistry; as one of the teachers wrote: 'Such microscopes are not frequently found in high schools, and are not available for use. Because of this, my knowledge of the subject was theoretical and very narrow, it was very difficult for me to understand the way AFM works, and even more difficult to imagine how to simplify it for my students. Now, after using the teaching model, my conceptions have changed, I have found a simple way that has helped me to better understand it and that would help me to explain it to my students'. This research served as a valuable opportunity to present a subject from the frontiers of chemistry research to high school chemistry teachers. The usage of a simple teaching model allows one to teach advanced content, which otherwise would not be part of the teachers' education. The ability to use a simple teaching model to explain a modern microscopy technique provides a platform to introduce the AFM to high school students.

3 Conclusions and Discussion

The four described projects all developed and analysed opportunities to present nanotechnology, a subject from the frontiers of research, to high school chemistry and physics in- and pre-service teachers. The understanding of the teachers' pre-knowledge and attitudes is an important stage for the design of an effective programme for teachers. The Model of Educational Reconstruction combines teachers' perspectives and experts' knowledge to develop a coherent educational programme that takes into consideration scientific parameters as well as science education (Sect. 2.1). The second project has been identified as promising approaches for the professional development of pre-service teachers regarding knowledge and attitudes (Sect. 2.2). Section 2.3 has shown in-service teachers' knowledge on nanoscience is focused on the small size. The forms of teaching envisaged are specific educative activities dedicated to the exploration and documentation of a new scientific domain with the purposes being to show science in the making, to develop engagement in school science and understanding of scientific concepts, to contribute to information literacy and to citizenship education and to learn how to debate. Section 2.4 has shown the ability to use a simple teaching model to explain a modern microscopy technique provides a platform to introduce relevant techniques such as the AFM to high school students and support the development of teachers' PCK in this area. Participating in different nanotechnology courses enhanced teachers' knowledge (Blonder 2011). These results converge with the literature showing that teachers tend to privilege an extension of the science curriculum on nanoscience and nanotechnology (Daly and Bryan 2010) with reasons based on relevance, student motivation and content knowledge (Nichol and Hutchinson 2010; see also the studies reported). In addition, including nanoethics and SSIs related to nanotechnology was proven to be a vehicle for introducing nanotechnology into professional development of pre- and in-service teachers. As the teachers develop understanding and awareness of nanoscience, they could also discuss the topic with their students.

Regarding the leading questions, the four projects have all contributed to the analyses of goals and preconditions, proving methods and instruments that could also be used in future professional development programmes in other countries. Aspects identified as promising for such programmes are the following: connecting nanoscience to socio-scientific and ethical issues, using simple teaching models in the professional development that can be transferred to school teaching and introducing leading instrumentation in the field as part of the teachers' course.

The authors will continue to develop and to investigate such programmes both for pre-service and in-service teachers in the nano area. Further interests will focus on the teachers' implementation of nano modules in different school curricula but also on the co-operation between school and out-of-school learning facilities, such as science centres and laboratories.

References

Binnig, G., Quate, C. F., & Gerber, C. (1986). Atomic force microscope. *Physical Review Letters, 56*, 930–933.

Blonder, R. (2010). The influence of a teaching model in nanotechnology on chemistry teachers' knowledge and their teaching attitudes. *Journal of Nano Education, 2*, 67–75.

Blonder, R. (2011). The story of nanomaterials in modern technology: An advanced course for chemistry teachers. *Journal of Chemical Education, 88*(1), 49–52.

Chi, M. T. H. (1997). Quantifying qualitative analyses of verbal data: A practical guide. *The Journal of the Learning Sciences, 6*(3), 271–315.

Daly, S., & Bryan, L. A. (2010). Model use choices of secondary teachers in nanoscale science and engineering education. *Journal of Nano Education, 2*, 76–90.

Drane, D., Swarat, S., Light, G., Hersam, M., & Mason, T. (2009). An evaluation of the efficacy and transferability of a nanoscience module. *Journal of Nano Education, 1*, 8–14.

Duit, R., Gropengießer, H., Kattmann, U., Komorek, M., & Parchmann, I. (2012). The model of educational reconstruction – A framework for improving teaching and learning science. In J. Dillon & D. Jorde (Eds.), *The world handbook of science education – Handbook of research in Europe* (pp. 13–37). Rotterdam: Sense Publisher.

Englander, O., & Kim, A. (2011) Nanocore at the FAMU-FSU College of Engineering: Program overview and unique assessment approach. In: *Proceedings of the ASME 2011 International Mechanical Engineering Congress & Exposition,* Denver.

Falk, J. H., & Dierking, L. D. (2000). *Visitor experiences and the making of meaning*. Walnut Creek: AltaMira Press.

Hingant, B., & Albe, V. (2010). Nanosciences and nanotechnologies learning and teaching in secondary education: A review of literature. *Studies in Science Education, 46*(2), 121–152.

Jones, M. G., Paechter, M., Yen, C.-F., Gardner, G., Taylor, A., & Tretter, T. R. (2011). Teachers' concepts of spatial scale: An international comparison. *International Journal of Science Education*, 1–21. doi: 10.1080/09500693.2011.610382.

Kumar, D. D. (2007). Nanoscale science and technology in teaching. *Australian Journal of Education in Chemistry, 68*, 20–22.

Marton, F. (1994). Phenomenography. In T. Husen & T. N. Postlethwaite (Eds.), *The international encyclopedia of education* (pp. 4424–4429). Oxford: Pergamon.

Nichol, C. A., & Hutchinson, J. S. (2010). Professional development for teachers in nanotechnology using distance learning technologies. *Journal of Nano Education, 2*, 37–47.

Parchmann, I., & Komorek, M. (2008). The model of educational reconstruction – A research model for the investigation of students' and teachers' conceptual ideas. In B. Ralle & I. Eilks (Eds.), *Promoting successful science education – The worth of science education research* (pp. 169–181). Aachen: Shaker Verlag.

Parchmann, I., Lienau, C., Klüner, T., Drögemüller, S., & Al-Shamery, K. (2010). Kann man atome sehen? – eine reflexion aus sicht verschiedener wissenschaften. *Chemkon, 17*(2), 59–65. doi:10.1002/ckon.201010112.

Roco, M. C. (2001). From vision to the implementation of the U.S. National Nanotechnology Initiative. *Journal of Nanoparticle Research, 3*, 5–11.

Roco, M. C. (2003). Converging science and technology at the nanoscale: Opportunities for education and training. *Nature Biotechnology, 21*, 1247–1249.

Schank, P., Krajcik, J., & Yunker, M. (2007). *Nanoethics: The ethical and social implications of nanotechnology. Chapter: Can nanoscience be a catalyst for education reform?* (pp. 277–289). Hobeken: Wiley.

Shulman, L. S. (1987). Knowledge and teaching – Foundations of the new reform. *Harvard Education Review, 57*, 1–22.

Stevens, S., Sutherland, L. M., & Krajcik, J. S. (2009). *The big ideas of nanoscale science and engineering: A guidebook for secondary teachers*. Arlington: NSTA Press.

Tomasik, J., Jin, S., Hamers, R., & Moore, J. (2009). Design and initial evaluation of an online nanoscience course for teachers. *Journal of Nano Education, 1*, 48–69.

Chapter 11
Education for Sustainable Development: An International Survey on Teachers' Conceptions

Pierre Clément and Silvia Caravita

1 Implementation of ESD Around the World

1.1 A General State of Implementation

Education for a Sustainable Development (ESD) is more and more implemented all around the world since the Brundtland report (1987), the Rio Summit and the Agenda 21 (1992). ESD is promoted in most of the national politics of education and curricula (UNESCO 2009a, b), even if some countries still avoid to use the word "development" arguing, as in Brazil[1] or in Columbia,[2] a possible confusion between "development" and "growth", and still using the appellation "Environmental Education" already dedicated, in these countries, to the three classical perspectives of ESD (social, economical and environmental). Nevertheless, "Development" is not necessarily "Growth"; it can be a transformation, a metamorphosis of our societies and environments to keep living in the future. Clément and Caravita (2011) use a metaphor: the development of the frog – the final adult frog is smaller than the last steps of development of its initial tadpole. The resources of our planet are too limited to support an international extension of the today level of consumption of US for all the countries; in the most developed countries, the sustainable development probably implies a "sustainable decrease" (Latouche 2006). The notion of sustainability is central in ESD. In Australia, ESD is called "Environmental Education for Sustainability".[3]

[1] Governo federal Brasil (2008). *Educação Ambiental no Brasil*, (March 2008), p. 9.

[2] Torres, M. (2009). *La Educación Ambiental : una realidad en Colombia, 2009.*

[3] http://www.environment.gov.au/education/publications/sustainable-future.html

P. Clément (✉)
EA 4148 S2HEP, University of Lyon 1, ENS-Lyon, University of Lyon, Villeurbanne Cedex, France
e-mail: Pierre.Clement@univ-lyon1.fr

S. Caravita
IRPPS-CNR, Rome, Italy
e-mail: silvia.caravita@istc.cnr.it

C. Bruguière et al. (eds.), *Topics and Trends in Current Science Education: 9th ESERA Conference Selected Contributions*, Contributions from Science Education Research 1,
DOI 10.1007/978-94-007-7281-6_11, © Springer Science+Business Media Dordrecht 2014

In most of the countries, the previous "Environmental Education" is now called "Education for a Sustainable Development" (ESD), adding the social and economical perspectives to the classical environmental/ecological ones.

Nevertheless even when the national policy clearly claims a priority for ESD, the implementation in the school practice is not always so clear, as shown by Clément and Caravita (2011). They analysed the implementation of ESD in 13 countries all around the world. ESD is today present in all their curricula and for several matters of secondary schools but more often in biology and geography. In some countries as Cameroon, but also Lebanon or Senegal, ESD is still now absent in several textbooks, and teachers are reticent to introduce new active pedagogy and to take into account the three dimensions of ESD (social, economical and environmental). There are other differences among the 13 analysed countries. For instance, ESD is more focused on the economical challenges (and also on religious education) in Malaysia,[4] while it is more centred on values in Australia.[5]

Challenging features are implicit for teachers involved in ESD: knowledge contents that are not always well stabilised, awareness of the different facets of an environmental problem (mostly outside the teachers' mono-disciplinary training), importance of teaching critical choosing rather than choices, relevance of being aware of values implied in each choice, development of projects that deal with problems having no a priori known solutions, openness to facilitate discussions with respect of diverging opinions, willingness to open their classrooms and even their school to interactions with external partners in their territory, etc. Sometimes, the values and practices underlying ESD can strongly differ from the teachers' values and practices. Therefore, when looking more closely to actual experiences, ESD objectives are not so well implemented correspondingly to the Ministry's official declarations (Clément and Caravita 2011).

That is why we focus this chapter on conceptions of teachers, having the perspective to understand what might be crucial in improving their training. Before presenting our results, we introduce just below two categories of teachers' problems when they are involved in EDS.

1.2 Multidisciplinary Approach and Teachers' Postures

An important problem of teachers when they are confronted with ESD is the necessity of a multidisciplinary approach to take into account the ecological, social and economical dimensions of ESD; their professional training usually deals with only one of these dimensions in most of the countries.

In a large inquiry recently done in Sweden (Borg and Gericke 2012), 3,229 upper secondary teachers from 223 schools (teaching science, or language, or social

[4] *Education for Sustainable Development (ESD) in Malaysia: An Overview*. By Dr. Siti Eshah Mokshein, Ministry of Education Malaysia (August 2004)

[5] http://www.environment.gov.au/education/publications/sustainable-future.html

science, or vocational and esthetical practices or other disciplines) answered an online questionnaire with mostly Likert-scale questions and some multiple choice questions. The results show that differences in the way ESD is understood are linked to the subject taught: e.g. natural science and social science teachers think to a lesser extent that economic growth is important, compared to language, vocational and esthetical – practical teachers. The ecological perspective is the perspective singled out by most teachers as integral to the concept of SD. This study points to the need for teachers' education to provide a holistic understanding of SD including all perspectives.

A similar trend concerning this diversity of teachers' conceptions of the multidisciplinary approach of ESD was also shown in France by Lange (2008) from an inquiry of 165 pre-service teachers (of biology – geology, or physics – chemistry, or history – geography). His results show significant differences among the disciplines and also different postures of teachers to introduce ESD.

The precedent references mainly concerned the teachers' knowledge and postures related to their difficulty to conceive a multidisciplinary approach of ESD. Another category of teachers' postures is related to their values and their capacity of making them explicit and debatable: their neutrality of engagement for a SD (Sustainable Development) when they are enrolled in ESD.

Qualitative approaches bring a deeper understanding of this kind of difficulty. Case studies of ESD are often descriptive. Some of them analyse the teachers' competences (UNECE 2010) and only few are focused on the teachers' more or less impartial postures (e.g. Kelly 1986; Gayford 2002; Cotton 2006; Sadler et al. 2006). Kelly (1986) defined three different postures that teachers assume when they are implied in ESD: (1) a *neutral impartiality* focused on only scientific contents in order to avoid to promote their own opinions, (2) an *interventionist educational aim* to make students aware of the environmental emergencies and (3) a *critical educational model* which does not avoid controversies about problems and promotes students' responsible decision-making. Analysing interviews of eight high school biology teachers involved in teaching global climate change in France, Urgelli and Simonneaux (2012) found the same three types of teachers' postures, the third one being more frequently adopted by teachers when several of them (of different scholar disciplines) are copresent in debates with students or when they collaborate in a project.

1.3 Values in ESD

The consensus related to the importance of ESD is hiding a diversity of environmental ethics, rooted in different values and philosophies of nature and environment (Larrère 1987; Quillot 2000; Clément 2004).

The values underlying ESD can be highlighted by making reference to the existing literature in educational and social studies (synthesis in Caravita et al. 2008). Several of these values are also underlying Education for All, Education to Citizenship and

Table 11.1 Categories of values pertaining to different dimensions of human practical and intellectual activities that are relevant to environmental education

1. *Ecological*: Referring to the maintenance of natural systems that require biodiversity; to the perceived quality of the environment (local and global); to the feeling of "inter-being" with the other living beings; to the acceptance of constraints for human action, population growth, etc.; to the awareness and acceptance of the planet as a limited pool of resources
2. *Aesthetic*: Referring to an appreciation of beauty and harmony through our senses, to the pleasure gained by this perception, to the value assigned to beauty relative to other environmental affordances
3. *Economic*: Referring to the exchange of goods and services among people and countries, to the definition of the value of resources (natural resources, labour, knowledge, technology), to the ownership of resources, to equity in the accessibility of goods, to the creation and distribution of profit, to the criteria for calculating the costs and benefits of human plans of actions
4. *Cultural*: Referring to the maintenance of the attitudes and the practices of social and cultural units (traditions, habits, knowledge), to the free circulation of information, to the accepted/promoted agencies of cultural changes, to the image of science
5. *Social*: Referring to the maintenance of the cohesiveness of the social environment, to the management and ruling of individual freedom and of interactions within a society, to attitudes about diversities (gender, sex, age, culture), to the evaluation of quality
6. *Political*: Referring to the ways of managing, ruling and controlling the interactions between individuals and society, humans and environment; the attainment and diffusion of human rights; the rights of minorities; the participation of citizens
7. *Ethical*: Referring to taking responsibility as users of resources, as consumers of affordances (goods, services, information), as citizens who vote, as living beings empowered by conscience; to equity, justice, respect and tolerance as objectives of individual and society actions
8. *Existential*: Referring to the ways of conceiving and managing one's own life and private sphere in relation to the environment where one lives; to the value assigned to the quality of life, to the person, to life itself (and to the other forms of life); to the role assigned to the spiritual dimension (religious, artistic, ideological) in one's own life; to the attitudes related to risk taking, to sacrifice enduring, with happiness seeking

Human Rights: e.g. for the equality of all the human beings independently to their gender, ethnic group, religion or sexual orientation (UNESCO 2009a, p. 9). Other values, more focused on Environment and Sustainability, are specific to ESD.

Caravita et al. (2008, pp. 102–103) give the following sociological definitions of values:

- systems of collective preferences that orient and justify the social actions of humans. (Huron 1994)
- an element of a shared symbolic system which serves as a criterion or standard for selection among the alternatives of orientation which are intrinsically open in a situation (Miceli and Castelfranchi 1989, p. 177).

Clément (2010) proposed a synthetic definition, calling values that which is the base of, or founds, a judgement. Depending on our values, we decide if something is good or bad, important or not, right or wrong, true or false, beautiful or ugly.

Caravita et al. (2008, p. 105) considered different ways of categorising values related to environmental issues, having in mind that the sense of a specific value (e.g. freedom, health) is a function of the ethical field in which it is embedded, because it is related to the system of values that features a particular ethic (Table 11.1).

After that, they presented the BIOHEAD-Citizen[6] methodology used to analyse teachers' conceptions, including their values, in several countries. We present below results coming from this methodology, using the BIOHEAD-Citizen questionnaire in more countries than those initially involved in this research project.

As justified by Caravita et al. (2008), our research is trying to identify eventual interactions between scientific knowledge and values in teachers' conceptions (as suggested by the KVP model: Clément 2006, 2010, analysing conceptions as possible interactions between scientific knowledge, K, values, V and social practices, P). These three components are clearly interacting in Environmental Education (Clément and Hovart 2000). If some values listed in Table 11.1 are only related to nature and environment, several of them are also universal values, related to human rights. In consequence, the results presented below are not only focused on the classical attitudes to the environment (ecocentric, biocentric, anthropocentric), formalised as two poles, utilisation and preservation, by Wiseman and Bogner 2003. They also deal with teachers' conceptions related to the fundamental human rights, as equality among all human beings. All these conceptions, and the values sustaining them, strongly vary among different sociocultural contexts. To face this question, we did an international survey implying different sociocultural contexts, to try to identify eventual interactions between the teachers' knowledge, their values and these contexts.

2 Teachers' Conceptions of Environment and Human Rights

2.1 Introduction and Questions of Research

Most of the research of Science Education dealing with Environmental Education is focused on the analysis of students' conceptions and conceptual changes and, more recently, on analysis of sequences of teaching, mainly socio-scientific issues (SSI). Little research was done at an international level, with nevertheless some exceptions as Schultz and Zeleny (1999) who compared students' conceptions (values and attitudes) in 14 countries (2,160 college students), mainly in North and South America. Schultz et al. (2000) analysed on the same sampling a relation between Judeo-Christian religious belief and attitudes of environmental concern. They showed a consistent pattern across countries, students who expressed more literal beliefs in the Bible scoring lower on ecocentric environmental concerns and higher on anthropocentric environmental concerns.

Our question of research is double:

1 Are there correlations between the teachers' conceptions related to the environment and those related to human rights?

[6] BIOHEAD-Citizen (2004–2008). *Biology, Health and Environmental Education for better Citizenship*, STREP CIT2-CT-2004-506015, E. C., Brussels, FP6, Priority 7

2 Are there differences among countries related to the teachers' conceptions of environment and of human rights?

2.2 Methodology

Our methodology was defined by the BIOHEAD-Citizen research project.[7]

The research project aimed at identifying the eventual interactions between scientific knowledge and values in teachers' conceptions. What are the teachers' conceptions related to Environment and Human Rights in a variety of different countries? More precisely, is there some correlation between their values related to the environment (as ecolocentric or anthropocentric ethics) and their values related to some human rights as equality among all the human beings, an equality independent to their gender, ethnic group, sexual orientation or religion?

The teachers' conceptions were analysed from their anonymous answers to a long questionnaire, elaborated and validated collectively during 2 years, using previous interviews, pilot test, etc. (Clément and Carvalho 2007). Twenty-seven questions were related to nature, the environment and environmental education; 15 questions dealt with the eventual biological justification of differences related to the gender, to ethnic groups or to sexual orientation; other questions concerned personal information, including social, political and religious opinions.

In each country, the sampling was well balanced: one third primary schools teachers, one third secondary schools teachers of biology and one third secondary schools teachers of language. In each sample, there were half of in-service teachers and half of pre-service teachers (end of their training). The countries were chosen from their diversity: 18 by the BIOHEAD-Citizen project (13 in all Europe, four in Arabic countries and 1 in sub-Saharan Africa). Since 2008, the research was extended to other countries, and we present here some of our results coming from 24 countries. New data have been collected in six other countries: two European ones (Denmark and Serbia), two belonging to sub-Saharan Africa (Burkina-Faso and Cameroun), Brazil and Australia. The number of the interviewed teachers in each of the 24 countries is reported below in Figs. 11.2, 11.3, 11.5, and 11.6 (total = 8,749).

The data were centralised in Lyon (France) and analysed with the help of statisticians, by classical statistics as well as by multivariate analyses (software "R", analyses justified in Munoz et al. 2009). We present here some of our results, coming from a between-class analysis differentiating the countries, and from a Co-Inertia analysis correlating teachers' conceptions on the environment and on human rights, both being included in ESD.

[7] "Biology, Health and Environmental Education for better Citizenship". This project was focused on six topics: evolution, human genetics, human brain, sex education, health education and ecology and environmental education. We were coordinating this last topic. For each topic, syllabuses and school textbooks were also analysed (Caravita et al. 2008, 2012, Caravita and Valente 2013; Clément et al. 2006, 2008; Carvalho et al. 2011; Quessada et al. 2008).

2.3 Results and Discussion

A between analysis (Fig. 11.1) identifies the answers which discriminate the countries most. A test of randomisation (Monte Carlo type: Fig. 11.1d) shows that the difference among the countries is very significant ($p < 0.0001$).

The following questions, located on the left of the horizontal axis in Fig. 11.1c, mostly discriminate the teachers' conceptions grouped by countries:

- A16: "Our planet has unlimited natural resources" (Fig. 11.2).
- A18: "Human beings are more important than other living beings" (Fig. 11.3).
- A17: "Society will continue to solve even the biggest environmental problems".
- A39: "Genetically modified plants are good for the environment because their cultivation will reduce the use of chemical pesticides (e.g. insecticides, herbicides)".
- Other anthropocentric conceptions: A32: "Humans have the right to change nature as they see fit". A4: "Nature is always able to restore itself". A54: "Only plants and animals of economical importance need to be protected".
- And also the three "sentimentocentric" conceptions: "Frogs" (A29), "Flies" (A45) and "Snails" (A10) "... are able to feel happiness".

The answers to these questions discriminate the teachers' conceptions into two groups along the horizontal axis (Fig. 11.1c): on the left the most anthropocentric and sentimentocentric conceptions are more numerous in Lebanon, Algeria, Tunisia, Morocco, Senegal and Burkina Faso (in this last country, they are a little less anthropocentric) and on the right these anthropocentric conceptions are less numerous in the European countries as well as in Brazil and Australia, with the exception of Lithuania (being in the middle of the two groups).

The amount of answers inside each of the 24 countries is illustrated more precisely by the histograms of the Fig. 11.2 (question 16) and three (question 18). Even if there are some little differences in the ranking of countries depending on the teachers' answers to these two questions, globally the teachers of the less developed countries display more anthropocentric conceptions, as minor awareness of the problem of the limit of resources in our planet. They are also less reticent to using GMO (genetically modified organisms) and more sentimentocentric, believing that most of the animals can feel sentiments.

These results are not contradictory with those of Schultz et al. (2000), because the most anthropocentric conceptions are found in the countries where the interviewed teachers said that they believe in God and practice religion. Nevertheless, most of them are Muslim in some countries (in North Africa or in Senegal) or Christian (in Cameroon). A deeper analysis of our results shows no difference among religions, only a correlation of anthropocentric conceptions with the degree of believing in God and practising religion. Nevertheless, this effect of religious opinion is totally linked to the effect of country: it disappears when we suppress the country effect (doing a specific multivariate analysis called PCAVOI: Sabatier et al. 1989), while this one does not disappear when we suppress the religion effect (from another PCAVOI).

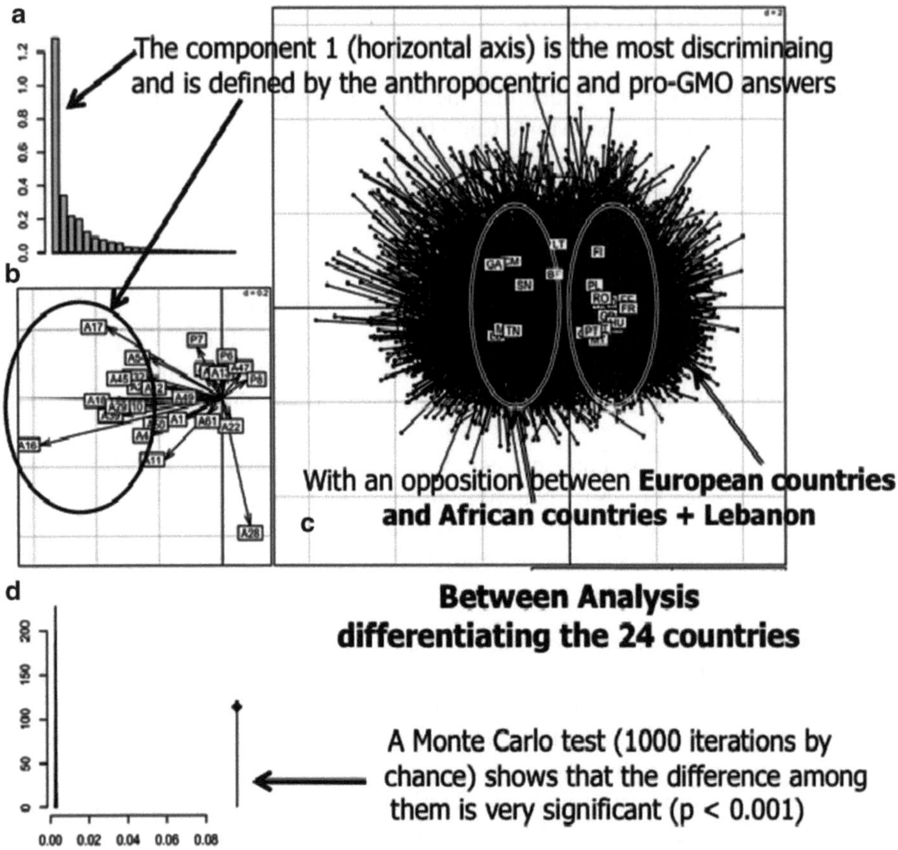

Fig. 11.1 Between-class analysis differentiating the 24 countries. See the text for explanations. In Fig. 11.1c, each point corresponds to one teacher's conceptions and is related to the centre of gravity of his/her country; Lithuania (LT) is just in the middle of the two groups of countries, not far from Burkina Faso (BF)

There is also a significant correlation between the degree of believing in God and the economical level of the country, independently to the Muslim or Christian religion. The most anthropocentric conceptions can be related with the importance of poverty in these countries and the urgent necessity of a better economical development. The implementation of ESD in these countries has to take into account this specificity and should not propose the same situations in the didactical projects as in developed countries where inverse problems are a priority (to avoid wasting and excessive consumption).

The sentimentocentric conceptions were firstly described as differentiating teachers' conceptions inside some countries: France, Portugal and Germany (Forissier 2003; Forissier and Clément 2003), Lebanon (Khalil et al. 2007), Algeria (Khammar et al. 2008), Morocco (Khzami et al. 2008) as well as Poland (Clément et al. 2011) or when comparing two countries as France and Australia (Quinn and

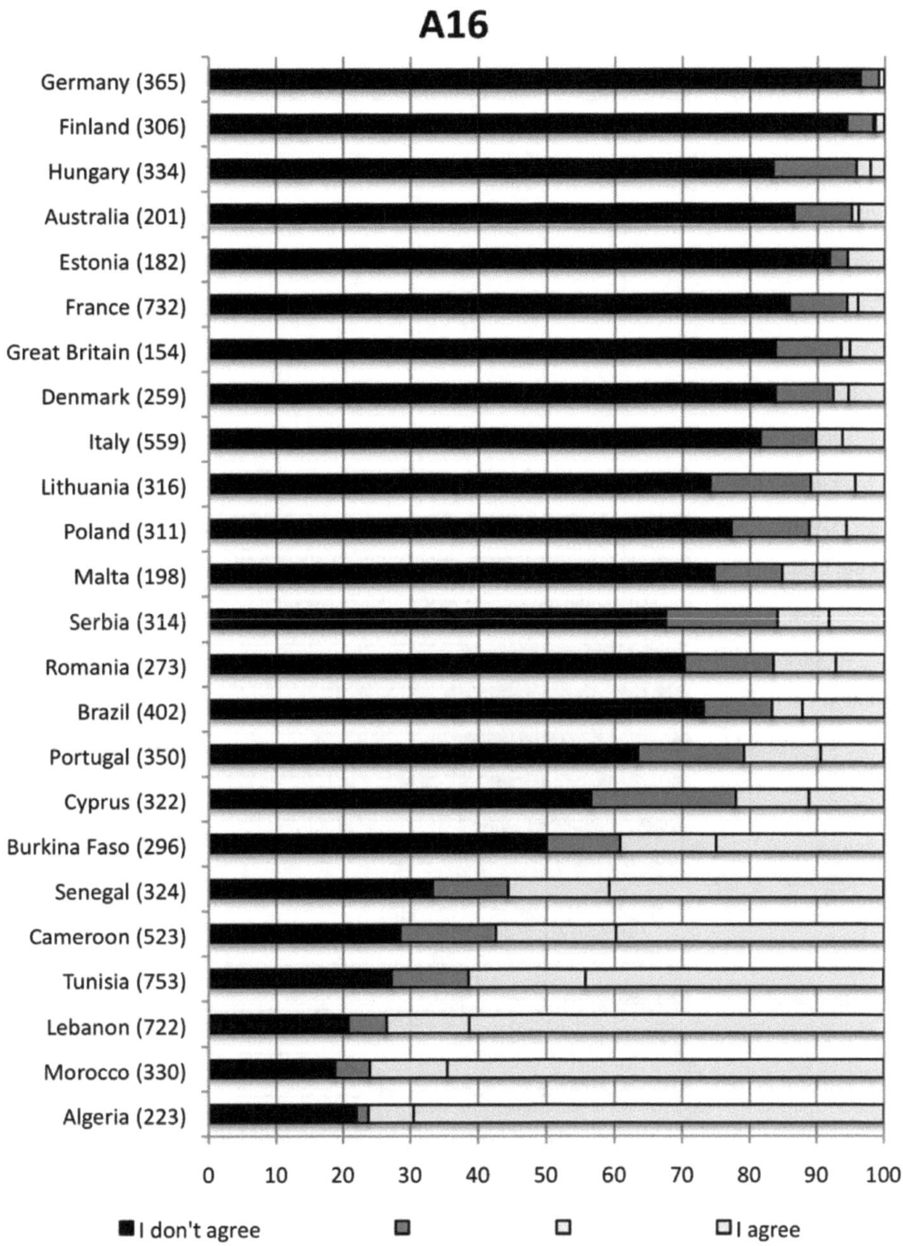

Fig. 11.2 Answers of the 8,749 teachers (grouped by country) to the question *A16* – "Our planet has unlimited natural resources"

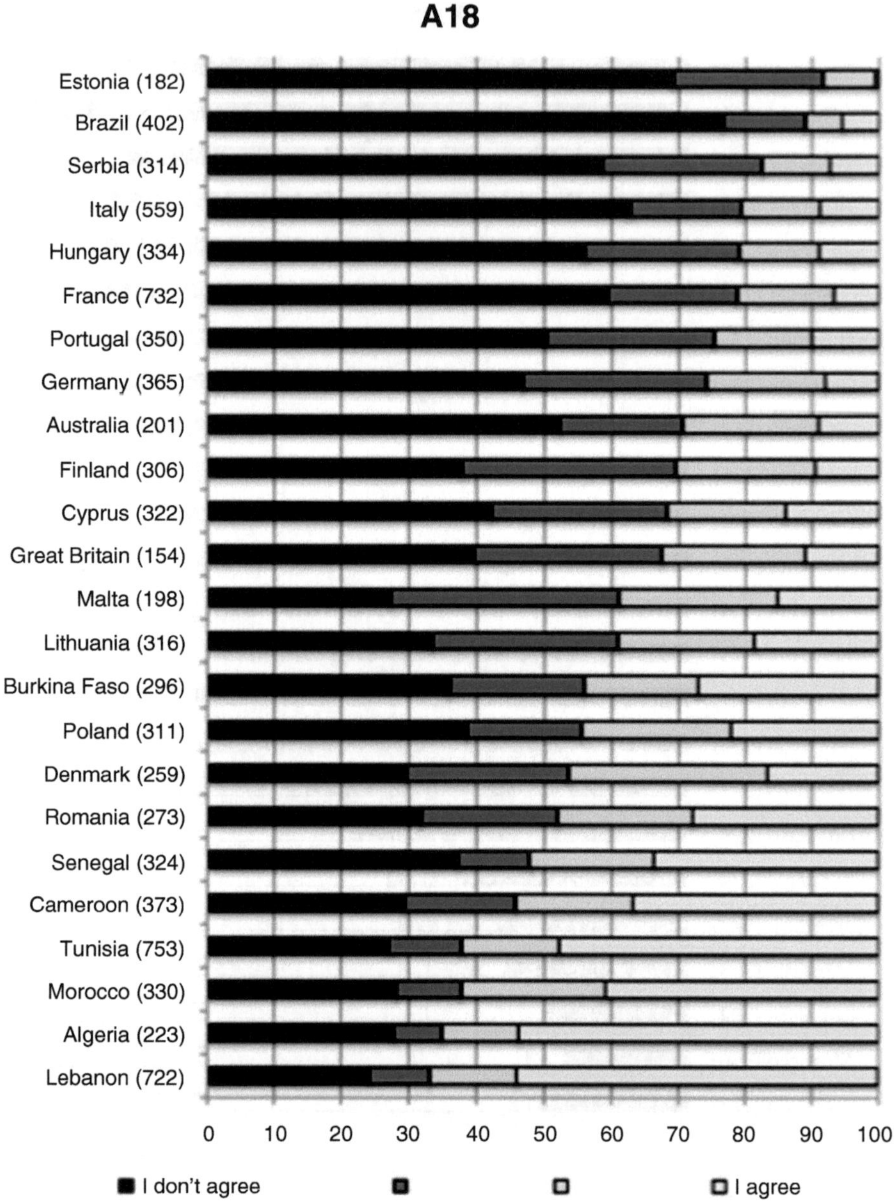

Fig. 11.3 Answers of the 8,749 teachers (grouped by country) to the question *A18* – Human beings are more important than other living beings

Clément 2012). Nevertheless, until now, these conceptions did not show differences among countries[8]: our results show the contrary (Fig. 11.1). Our hypothesis is a possible correlation of these sentimentocentric conceptions with a higher degree

[8] With the exception of biology teachers' conceptions differentiating France and Portugal (Forissier and Clément 2003)

of believing in God: we will do complementary analyses of our data to test this hypothesis.

Schultz and Zeleny (1999) did not observe any difference among the 14 countries where they analysed students' conceptions, in North and South America. A contrario, our results show very important differences among countries: while in Germany 99 % of the interviewed teachers answered they do not agree or rather do not agree with "Our planet has unlimited natural resources" (question A16, Fig. 11.2), this percentage is only 23 % in Algeria and intermediate amounts are observed in the 22 other countries, with this percentage <50 % in five countries.

This important result answers to the part of our second question of research, related to teachers' conceptions of environment. What about their conceptions of human rights, and our first question of research, related to possible correlations between teachers' conceptions on the environment and on human rights, two important dimensions of ESD?

To analyse the possible correlations between the teachers' conceptions related to the environment and to human rights, we did a Co-Inertia analysis from the two PCA related to these two sets of data (Fig. 11.4). The test of randomization (Monte Carlo: Fig. 11.4d) shows that the correlation between the two PCA is very significant ($p<0.0001$). It is supported by the first component (85 % of the total variance: Fig. 11.4a), represented by the horizontal axis of graphs 4b and 4c. On the left of this axis are the most anthropocentric conceptions (Fig. 11.4b: A16, A18, A39, A32, A54: see above for the formulation of these questions). They correspond to several conceptions related to human rights (Fig. 11.4c):

- The most racist conceptions agreeing with A35: "Ethnic groups are genetically different and that is why some are superior to others" (Fig. 11.5).
- The most sexist conceptions agreeing with A38: "It is for biological reasons that women more often than men take care of housekeeping" (Fig. 11.6); A25: "It is for biological reasons that women cannot hold positions of as high responsibility as men can"; A9: "Women are less intelligent than men are because their brains are smaller than men's brains are"; or disagreeing with A2: "In a modern society, men and women should have equal rights".
- The most homophobic conceptions disagreeing with A41: "Homosexual couples should have the same rights as heterosexual couples".
- And against immigration, agreeing with A26: "There are too many foreigners in my country: the government should limit immigration".

We also did another Co-Inertia analysis, from the PCA related to conceptions on the environment and another PCA related to the teachers' personal opinions (in the domains of politics, society, economy, religion, etc.: a total of 17 questions). The results show a very significant correlation between the most anthropocentric conceptions and a high degree of faith in God and practice of religion, political positions for a strong central power, against the separation between religion and politics, and between science and religion, and for a liberal economy.

The results of these two Co-Inertia analyses, done from the 8,749 teachers (here without taking into account any groups as countries), show mainly two opposite

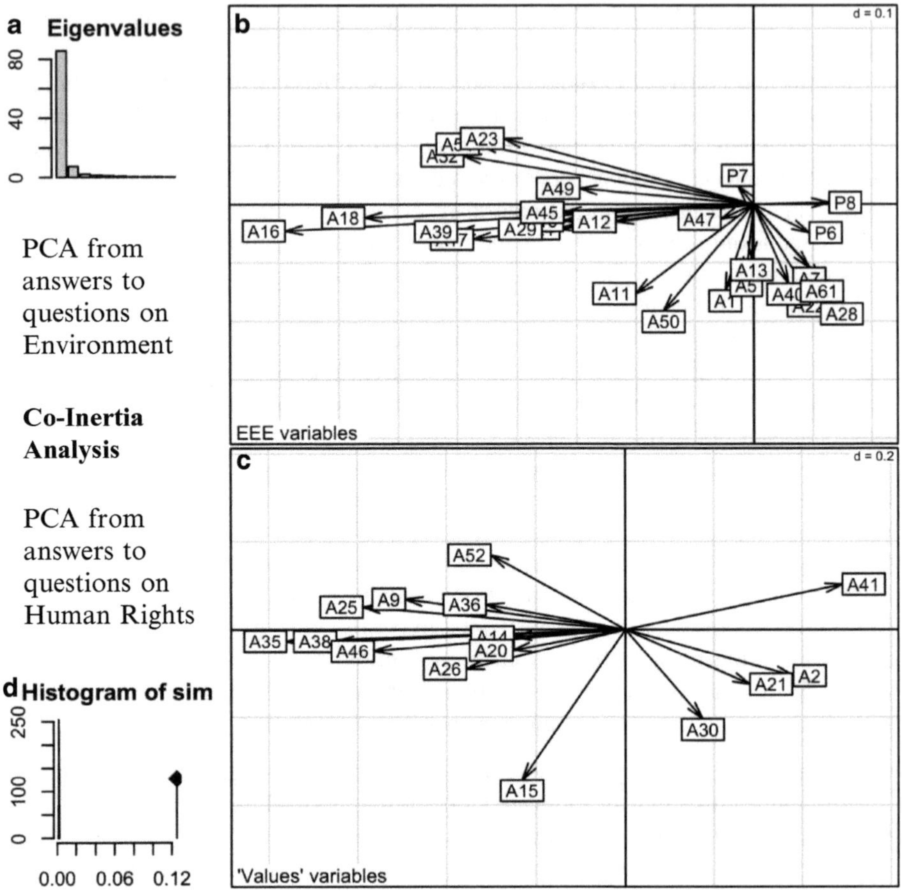

Fig. 11.4 Co-Inertia analysis, correlating teachers' conceptions on the environment (**b**) and on Human Rights (**c**): (**a**) Only the first component (85 % of the total variance) is explaining the correlation between the two PCA. It is represented by the horizontal axis of graphs (**b**) and (**c**). (**b**) and (**c**): see the text for explanations (the variables supporting the correlations). (**d**) The randomization test (Monte Carlo) shows that the co-inertia analysis is very significant ($p>0/0001$): the observed value (*on the right*) is far from the values coming from1,000 random essays (thin histogram *on the left*)

systems of conceptions, rooted into two opposite systems of values and often in outdated scientific knowledge (as the consequences of the brain size on intelligence):

(a) Conservative conceptions, justifying by (outdated) biological explanations the superiority of males, and of some ethnic groups, or the reject of homosexuality. These conceptions are linked to the most anthropocentric attitudes related to environment and to a high degree of faith and practice of religion.

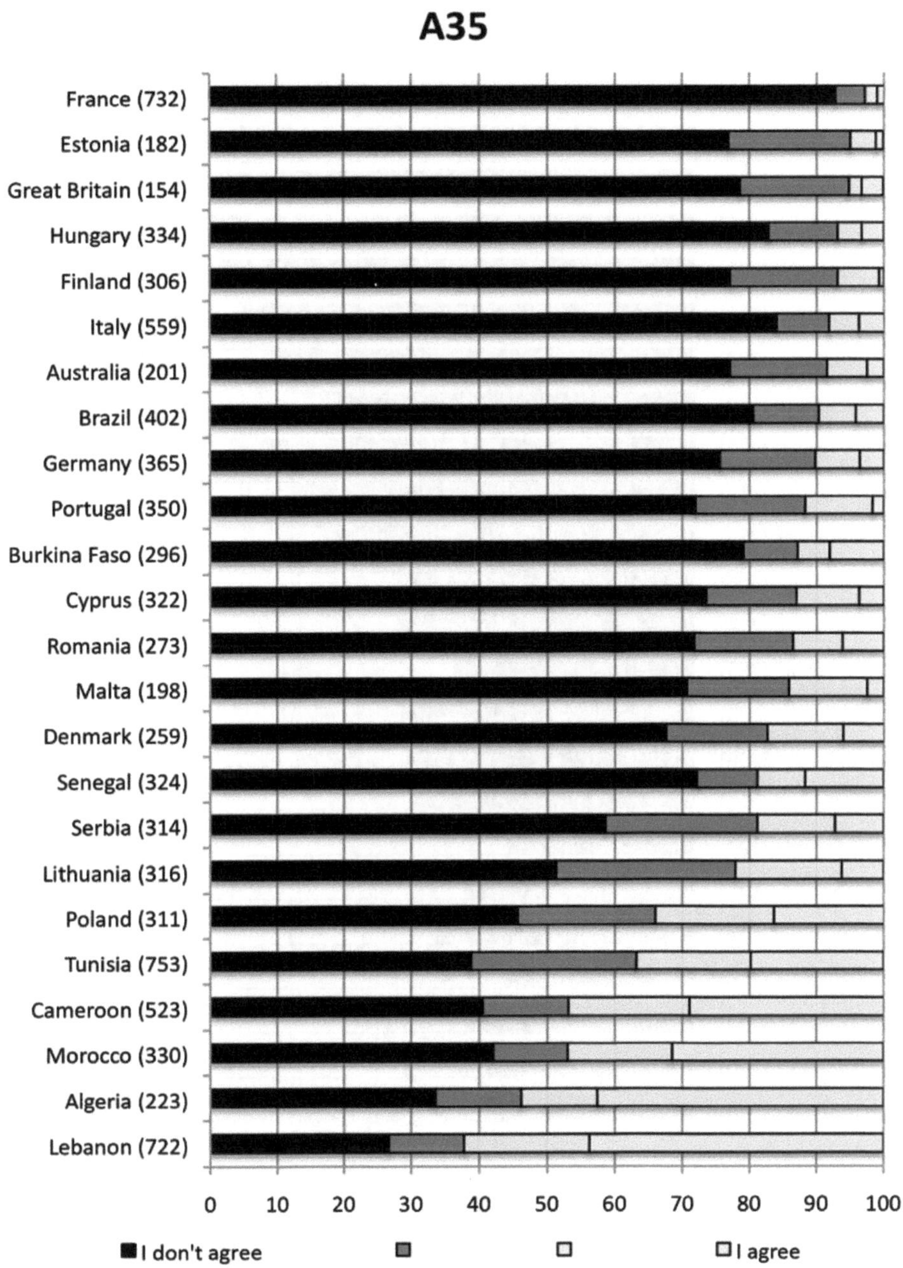

Fig. 11.5 Answers of the 8,749 teachers (grouped by country) to the question *A35 – Ethnic groups are genetically different and that is why some are superior to others*

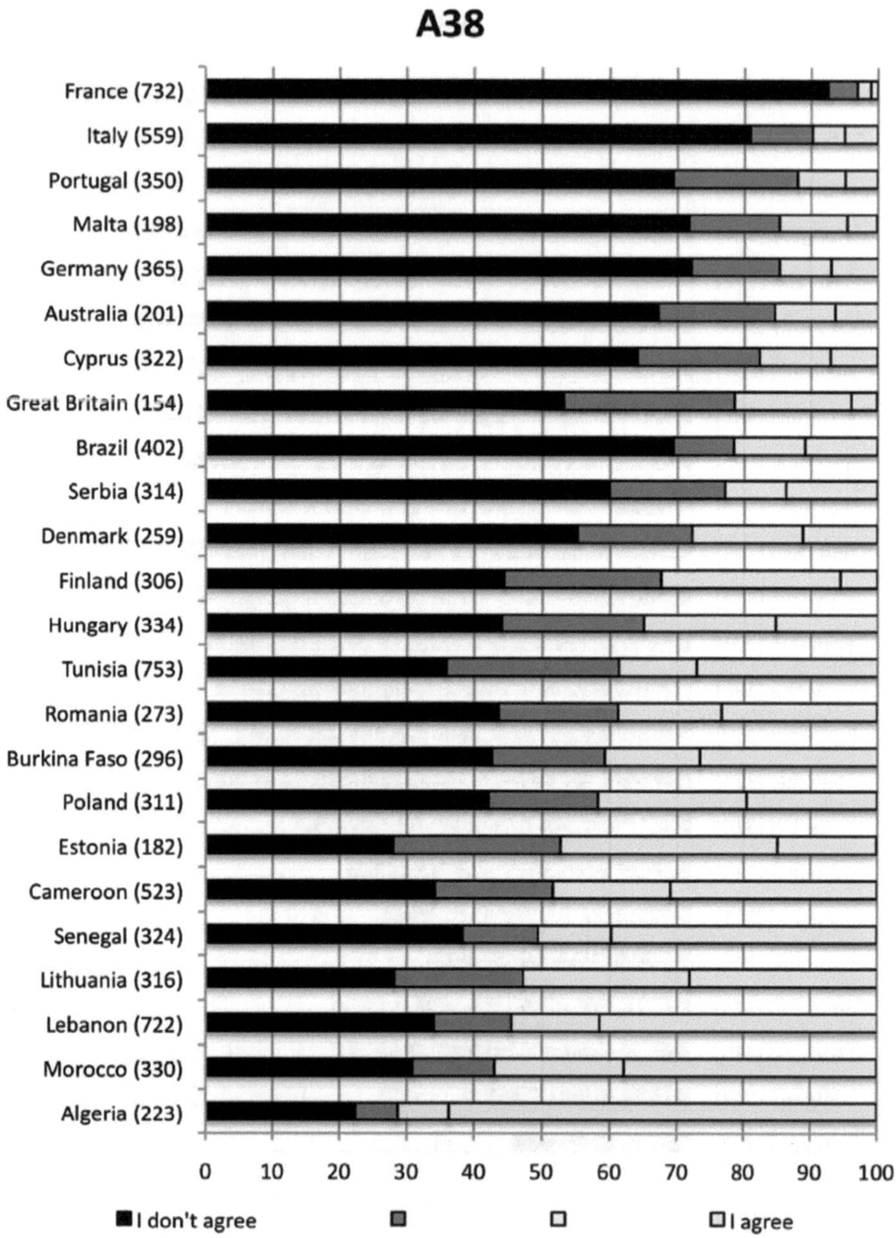

Fig. 11.6 Answers of the 8,749 teachers (grouped by country) to the question *A38* – "It is for biological reasons that women more often than men take care of housekeeping"

(b) Opposite conceptions, rejecting sexism, racism, homophobia, being less believing in God and practising religion, and having less anthropocentric and more ecolocentric conceptions.

The values promoted by ESD, listed in Table 11.1, are corresponding to the system of conceptions (b). Nevertheless, in several countries (as shown in Figs. 11.1, 11.2, 11.3, 11.5, and 11.6), a large part of teachers answered in coherence with the system of conceptions (a).

3 Conclusion

The biological justification of superiority of men or of some ethnic groups is today considered as an ideological and not scientific argument, as an interaction KV (knowledge and values) justifying some still actual social practices (P). These positions emerge from a large part of Christian or Muslim teachers in Arabic countries, in sub-Saharan Africa, but also in some countries of North and East Europe. They are correlated to anthropocentric attitudes, which are the contrary of the goals of Environmental Education, and they are themselves contrary to the promotion of universal Human Rights that are central in ESD.

Improving the training of teachers is therefore an urgent necessity in any country. Nevertheless, our results show that it will not be so easy to do in countries where a majority of teachers' conceptions are, till now, still contaminated by sexism, racism or homophobia, values rooted in deep social practices, and often believing in outdated scientific knowledge to justify their values (a clear illustration of the KVP model).

In summary, the present work shows that it is easier to change the curriculum and the syllabuses, to apparently focus them for ESD, including values related in Human Rights, as we have shown in the first part of our chapter, than to change the teachers' conceptions, as we have shown in the second part of this chapter. Teachers' conceptions, including their scientific knowledge, their values and their social practices, are rooted in the sociocultural context of their country, with important differences among countries.

In further publications, we will analyse more precisely how the sociocultural traditions are correlated, in each country, with values of ESD related to anthropocentrism as well as with universal human rights.

The attainment of a higher consistency between actualised knowledge, citizenship values and practices is probably only an elusive goal, which nevertheless helps us to reflect upon our own conceptions in any country.

An increased awareness of these factors and of their dynamics can enable educated citizens to be better prepared for the public arena. It should become part of teacher's education to enable them to consider the influence of conceptions on didactic method and in school manuals rather than letting them unconsciously impact their teaching and student discussions. Openness to discuss alternative conceptions is also connected with higher consciousness of their nature and of the arguments that sustain them.

Acknowledgements Thanks to all our colleagues of the 24 countries, who gathered data in their respective countries. For 18 of the 24, it was done with the help of the BIOHEAD-Citizen project (*Biology, Health and Environmental Education for better Citizenship*, STREP CIT2-CT-2004-506015, E. C., Brussels, FP6, Priority 7).

References

Borg, C. & Gericke, N. (2012). Teachers' understanding of sustainable development – discipline bound differences. In C. Bruguière, A. Tiberghien, & P. Clément (Eds.), *E-book ESERA 2011 conference, science learning and citizenship* (pp. 18–23), Part 8.

Brundtland. (1987). *The report of the Brundtland Commission our common future*. Oxford: University Press.

Caravita, S., & Valente, A. (2013, in press). Educational approach to environmental complexity in life sciences school manuals. An analysis across countries. Myint Swe Khine (Ed.), *Critical analysis of science textbooks: Evaluating instructional effectiveness*. Dordrecht/London: Springer.

Caravita, S., Valente, A., Luzi, D., Pace, P., Khalil, I., Berthou, G., Valanides, N., Kozan-Naumescu, A., & Clément, P. (2008). Construction and validation of textbook analysis grids for ecology and environmental education. *Science Education International, 19*(2), 97–116.

Caravita, S., Berthou-Gueydan, G., Agorram, B., & Clément, C. (2012). Environmental complexity and pollution in the life sciences manuals of six Mediterranean countries. In M. Boutan, B. Maurer, & H. Remaoun (Eds.), *La Méditerranée des méditerranéens à travers leurs manuels scolaires* (pp. 71–97). Paris: Harmattan.

Carvalho, G. S., Tracana, R. B., Skujiene, G., & Turcinaviciene, J. (2011). Trends in environmental education images of textbooks from Western and Eastern European countries and non-European countries. *International Journal of Science Education, 33*(18), 2587–2610.

Clément, P. (2004). Construction des umwelts et philosophies de la nature. In J. M. Exbrayat & P. Moreau (Eds.), *L'homme méditerranéen et son environnement* (pp. 93–106). co-éd. Soc. Linéenne Lyon & Univ. Catholique Lyon: Fac. Sc. & Fac. Philo.

Clément, P. (2006). Didactic transposition and the KVP model: Conceptions as interactions between scientific knowledge, values and social practices. In *Proceedings ESERA Summer School* (pp. 9–18), IEC, Braga.

Clément, P. (2010). Conceptions, représentations sociales et modèle KVP. *Skholê (Univ. de Provence, IUFM), 16*, 55–70.

Clément, P., & Caravita, S. (2011). Education pour le Développement Durable (EDD) et compétences des élèves dans l'enseignement secondaire. Report for UNESCO. www.ensi.org/Publications/Publications-reports/

Clément, P., & Carvalho, G. (2007). Biology, health and environmental education for better citizenship: Teachers' conceptions and textbook analysis in 19 countries. In *Proceedings WCCES XIII (World Council of Comparative Education Societies)* (15pp.), Sarajevo, CD-Rom.

Clément, P., & Hovart, S. (2000). Environmental education: Analysis of the didactic transposition and of the conceptions of teachers. In H. Bayerhuber & J. Mayer (Eds.), *State of the art of empirical research on environmental education* (pp. 77–90). Münster: Waxmann Verlag.

Clément, P., Mouehli, L., & Abrougui, M. (2006). Héréditarisme, béhaviorisme, constructivisme: Le système nerveux dans les manuels scolaires français et tunisiens. *Aster, 42*, 187–222.

Clément, P., Mouelhi, L., Kochkar, M., Valanides, N., Nisiforou, O., Thiaw, M. S., Ndiaye, V., Jeanbart, P., Horvarth, D., Ferreira, C., & Carvalho, G. (2008). Do the images of neuronal pathways in the human central nervous system show feed-back? A comparative study in fifteen countries. *Science Education International, 19*(2), 117–132.

Clément, P., Laurent, C., & Samonek, E. (2011b). Polish teachers' conceptions related to environment. *Annales Universitatis Paedagogicae Cracoviensis, Studia ad Didacticam Biologiae Pertinentia I, 86*, 104–115.

Cotton, D. R. E. (2006). Teaching controversial environmental issues: Neutrality and balance in the reality of the classroom. *Educational Research, 48*(2), 223–241.

Forissier, T. (2003). *Les valeurs implicites dans l'Education à l'Environnement. Analyse de la formation d'enseignants de SVT et des conceptions de futurs enseignants français, allemands et portugais*. Ph.D. thesis, Université Lyon 1, Lyon.

Forissier, T., & Clément, P. (2003). Les systèmes de valeurs d'enseignants du Secondaire sur la Nature et l'Environnement. Une analyse comparative en France, en Allemagne et au

Portugal. *Actes JIES* (A. Giordan, J. L. Martinand, & D. Raichvarg eds, Univ. Paris Sud), 25, (pp. 393–398).
Gayford, C. (2002). Controversial environmental issues: A case study for the professional development of science teachers. *International Journal of Science Education, 24*(11), 1191–1200.
Huron, P. (1994). *Les valeurs*. Lyon: Académie de Lyon.
Kelly, T. (1986). Discussing controversial issues: Four perspectives on the teacher's role. *Theory and Research in Social Education, 14*, 113–138.
Khalil, I., Clément, P., & Laurent, C. (2007). Anthropocentrées, écolocentrées ou sentimentocentrées: Les conceptions d'enseignants et futurs enseignants libanais sur la nature et l'environnement. *Feuilles Libanaises (Ligue Professeurs de l'Univ. Libanaise), 29*, 67–92.
Khammar, F., Clément, P., Laurent, C., & Remki, L. (2008). Les conceptions d'enseignants algériens sur la nature et l'éducation à l'environnement. In *Enjeux dans la rénovation de l'éducation à l'environnement et à la biologie* (pp. 95–111). Alexandrie: Presses Université Senghor, http://www.usenghor-francophonie.org/publications/ActesColloqueCBEavril2008.pdf
Khzami, S., Agorram, A., Selmaoui, S., Elabboudi, T., & Clément, P. (2008). Les systèmes de valeurs d'enseignants et de futurs enseignants marocains des sciences de la vie et d'arabe sur l'environnement. In *Enjeux dans la rénovation de l'éducation à l'environnement et à la biologie* (pp. 223–238). Alexandrie: Presses Université Senghor. http://www.usenghor-francophonie.org/publications/ActesColloqueCBEavril2008.pdf
Lange, J. M. (2008). L'éducation au développement durable au regard des spécialités enseignantes. *Aster, 46*, 123–154.
Larrère, C. (1987). *Les philosophies de l'environnement*. Paris: Presses universitaires de France.
Latouche, S. (2006). *Le pari de la décroissance*. Paris: Fayard.
Miceli, M., & Castelfranchi, C. (1989). A cognitive approach to values. *Journal for the Theory of Social Behavior, 19*, 169–193.
Munoz, F., Bogner, F., Clément, P., & Carvalho, G. S. (2009). Teachers' conceptions of nature and environment in 16 countries. *Journal of Environmental Psychology, 29*, 407–413.
Quessada, M. P., Clément, P., Oerke, B., & Valente, A. (2008). Human evolution in science textbooks from twelve different countries. *Science Education International, 19*(2), 147–162.
Quillot, R. (2000). *La nature*. Paris: Ellipses, Coll. Philo.
Quinn, F., & Clément, P. (2012). Primary schools teachers' conceptions of environment. A comparison between Australia and France. In C. Bruguière, A. Tiberghien, & P. Clément (Eds.), *Ebook ESERA 2011 conference, science learning and citizenship* (pp. 110–114), Part 8.
Sabatier, R., Lebreton, J. D., & Chessel, D. (1989). Principal component analysis with instrumental variables as a tool for modelling composition data. In R. Coppi & S. Bolasco (Eds.), *Multiway data analysis* (pp. 341–352). North-Holland: Elsevier Science Publishers B.V.
Sadler, T. D., Amirshokoohi, A., Kazempour, M., & Allspaw, K. M. (2006). Socioscience and ethics in science classrooms: Teacher perspectives and strategies. *Journal of Research in Science Teaching, 43*, 353–376.
Schultz, P. W., & Zelezny, L. (1999). Values as predictors of environmental attitudes: Evidence for consistency across 14 countries. *Journal of Environmental Psychology, 19*, 255–265.
Schultz, P. W., Zelezny, L., & Dalrymple, N. J. (2000). A multinational perspective on the relation between Judeo-Christian religious beliefs and attitudes of environmental concern. *Environment and Behavior, 32*(4), 576–591.
UNESCO. (2009a). Contextes et structures de l'Education pour le développement durable, DEDD 2005–2014, p. 9.
UNESCO. (2009b). The ESD Lens. Results of a review process by seven Southern Africa countries.
UNECE (2010). *Expert Group on Competencies in ESD*. Draft 2 (21 May 2010).
Urgelli & Simonneaux. (2012). Global warming: A case study of eight French teachers' involvement in education for sustainable development. In C. Bruguière, A. Tiberghien, & P. Clément (Eds.), *Ebook ESERA 2011 conference, science learning and citizenship* (pp. 140–144), Part 8.
Wiseman, M., & Bogner, F. X. (2003). A higher-order model of ecological values and its relationship to personality. *Personality and Individual Differences, 34*, 783–794.

Chapter 12
Learning to Teach Science as Inquiry: Developing an Evidence-Based Framework for Effective Teacher Professional Development

Barbara A. Crawford, Daniel K. Capps, Jan van Driel, Norman Lederman, Judith Lederman, Julie A. Luft, Sissy Wong, Aik Ling Tan, Shirley S.L. Lim, John Loughran, and Kathy Smith

1 Need for Empirical Evidence of Effective Professional Development for Inquiry

In this chapter, we present empirical evidence from studies focused on supporting teachers in engaging children in inquiry science learning. We provide promising and empirically based examples of professional development (PD) programmes for teachers related to using inquiry-based approaches. In the United States and other

B.A. Crawford (✉) • J.A. Luft
University of Georgia, Athens, USA
e-mail: barbarac@uga.edu; jaluft@uga.edu

D.K. Capps
University of Maine, Orono, USA
e-mail: dkc39@cornell.edu

J. van Driel
Graduate School of Teaching, ICLON – Leiden University, Leiden, The Netherlands
e-mail: Driel@ICLON.Leidenuniv.nl

N. Lederman • J. Lederman
Illinois Institute of Technology, Chicago, USA
e-mail: ledermann@iit.edu; ledermanj@iit.edu

S. Wong
University of Houston, Houston, USA
e-mail: sissy_wong@hotmail.com

A.L. Tan • S.S.L. Lim
National Institute of Education, Singapore, Singapore
e-mail: aikling.tan@nie.edu.sg; shirley.lim@nie.edu.sg

J. Loughran • K. Smith
Monash University, Glen Eira, Australia
e-mail: john.loughran@monash.edu; kathy.smith@monash.edu

C. Bruguière et al. (eds.), *Topics and Trends in Current Science Education: 9th ESERA Conference Selected Contributions*, Contributions from Science Education Research 1,
DOI 10.1007/978-94-007-7281-6_12, © Springer Science+Business Media Dordrecht 2014

countries, there is an increasing emphasis on teaching children how to engage in scientific questions, use data as evidence to develop explanations, create models, develop and defend arguments and communicate findings (NRC 2012). In the United States, the K-12 Framework (NRC 2012) strongly emphasises that students engage in designing and carrying out investigations, in order to learn about what scientists do and about what science is (NRC 2012). In countries other than the United States, there is an emphasis on using authentic scientific inquiry experiences and conducting investigations in classrooms (Hasson and Yarden 2012). Although science education reforms highlight the importance of inquiry-based instruction, the state of affairs is that teachers struggle to enact inquiry-based instruction in their classrooms. It appears that little has changed regarding how science is taught in the majority of classrooms in many countries. There are several issues related to effectively supporting teachers in using inquiry-based pedagogy. First, there is a great deal of confusion about what it means to teach science as inquiry across the spectrum of standards documents. Researchers, curriculum designers and teachers hold various views. One view of inquiry is the following: 'the diverse ways in which scientists study the natural world and propose explanations based on evidence derived from their work' (NRC 1996, p. 23) or more simply, 'the process by which scientific knowledge is developed' (Lederman 2004, p. 308). Second, although there have been a myriad of studies on various professional development programmes (Hewson 2007), a recent review of the literature revealed very few empirical studies focused *specifically* on science inquiry published in major peer-reviewed journals (Capps et al. 2012). There are publications that propose elements of professional development programmes in general, including setting of goals, planning, enacting, looking at outcomes and reflecting on the entire process (Loucks-Horsley et al. 2010). Other important features include aligning teachers' professional development goals and experiences with those people designing the professional development experiences, in addition to national, state and local goals. Goals of professional development include teacher growth in subject matter, pedagogical practices and attitudes towards reform-based teaching and learning. The ultimate goal is enhancement of students' knowledge of science concepts and principles, in addition to greater understanding of scientific practices and nature of science. The purpose of this chapter is to focus on professional development projects that show evidence of impacting teachers and students related to learning about science as inquiry.

This chapter summarises the evidence for several models of professional development. These models were presented during a symposium at the 2011 ESERA conference in Lyon, France. The researchers are from Europe, North America, Australia and Asia. Two main questions are addressed in each section: (1) What are the views and framework used in supporting teachers in carrying out inquiry in the science classroom? (2) What is the evidence for effective strategies of supporting teachers in learning to teach science as inquiry? In this chapter, we present the context of each PD programme, framework and summary of findings. Jan van Driel from Leiden University offers comments at the end of the chapter, pulling out themes and raising critical issues.

1.1 *Building Sustainable Capacity for Inquiry-Oriented Instruction*

Project SLICE, Science Learning through Inquiry, Content, and Engagement, is a programme developed by Norman Lederman and Judith Lederman, the Illinois Institute of Technology, United States. The primary focus of this ambitious systemic professional development initiative was to enhance US high school students' science achievement (as measured by standardised tests and internally developed assessments) in biology, chemistry and physics (grades 9–11) through the delivery of extensively revised curriculum and continuous on-site and off-site professional development support for teachers. A strong focus was placed on scientific inquiry and nature of science and the design of the project provided a natural context for the collaboration of science teachers, scientists and science educators.

Project SLICE has just completed its fifth year and currently involves all of the biology, chemistry and physics teachers in 23 high schools. There are currently 197 participant teachers and 27,852 students involved in the project. Schools are active in the project for a period of 3 years. All ninth grade science teachers in identified schools are involved in year 1, year 2 involves both ninth and tenth grade teachers and year 3 involves teachers spanning grades 9–11.

Project SLICE consists of three essential elements that are repeated, with some modification, during each of the 3 years of each school's engagement. These phases consist of (1) initial professional development for participating teachers, (2) monthly academic year professional development workshops (these are divided between the university and an informal education site) and (3) on-site academic year support from science coaches. A science coach is assigned to each school and this individual works closely with each of the teachers on a daily basis. Support ranged from observing lessons and providing feedback, to co-planning of lessons, to team teaching, or the coach actually modelling instruction for the teacher. In addition, the science coach helped coordinate science instruction by meeting with the science department as a whole each week. Science teachers in participating schools had a common planning time to facilitate this coordination. Participating schools and teachers also received all needed materials and revised or newly developed curriculum materials for the entire course taught.

Although the primary goal of this systemic initiative is to enhance high school students' science achievement, other important goals are:

- Increase students' scores on the ACT Exam
- Enhance high school students' understanding of and ability to do scientific inquiry
- Enhance high school students' understandings of nature of science
- Enhance in-service science teachers' understanding of and ability to do scientific inquiry
- Enhance in-service teachers' understandings about nature of science
- Enhance in-service science teachers' ability to teach inquiry, about inquiry and nature of science

- Enhance in-service teachers' ability to use informal education sites to improve science instruction and student achievement
- Develop leadership skills in participant teachers so that they may subsequently work with other teachers in their school districts.

1.1.1 Data Sources

Achievement scores were derived from standardised instruments developed for the project. These instruments went through strict content validation procedures using multiple groups of subject matter experts and educators. A level of agreement of 80 % or higher was achieved for each item on each of the resulting instruments. In terms of reliability, Kuder-Richardson (21) estimates exceeding 0.80 was achieved for each subject matter test. For scientific inquiry, the Views of Scientific Inquiry (VOSI) survey was used. For nature of science, the Views of Nature of Science (VNOS) was used. Each of these latter instruments has previously established validity and reliability. Finally, data from the standardised ACT Examination were used to assess knowledge and abilities. The results are presented below.

Science Achievement: During each of the first 5 years of the project, pre- and posttest data were collected on students' achievement. For each subject area, correlated t-tests ($\alpha = 0.05$) were used to verify that students exhibited significant gains in achievement. Significant improvement in test scores ($p < 0.05$) was exhibited in each of the subject matter areas. More impressive is that upon completion of the academic year, students were achieving at relatively high levels. That is, biology achievement reached 75 % for the first year, 76 % for year 2, 78 % for year 3, 79 % for year 4 and 81 % for year 5. For chemistry, the average achievement score was 84 % for year 1, 85 % for year 2, 85 % for year 3 and 86 % for year 4. The physics achievement level was 85 % for year 1, 86 % for year 2 and 88 % for year 3.

Understandings of Scientific Inquiry: Students and teachers were pre- and posttested on understandings of scientific inquiry during each year. If a student or teacher was part of the project for 5 years, they were assessed on their understandings for each of those years. Chi-square analyses ($\alpha = 0.05$) indicated significant improvements in each aspect of scientific inquiry addressed. Within the group of teachers, the greatest gains were shown with respect to understandings that there is no single scientific method and that scientists viewing the same data may arrive at different interpretations of the data. As expected, teachers assessed in multiple years showed consistent improvement from year to year. As with subject matter understandings, the final understandings exhibited by students and teachers are more impressive than the fact that significant changes occurred from pre- to posttests. That is, the 'final' understandings noted here are not commonly noted in student and teacher populations.

Understandings of Nature of Science: Teachers and students were assessed with respect to their understandings of nature of science as they were with scientific inquiry. Chi-square analyses ($\alpha = 0.05$) were used to note any changes in understandings from pretest to posttest. Significant changes were noted in all aspects of nature of science assessed within the group of teachers. As with subject matter knowledge and understandings of scientific inquiry, the 'final' understandings are more important than the significant changes from pre- to posttest.

Comparisons Across Years of Engagement: Since project SLICE is a multiple-year systemic change effort, it was assumed that teachers and students would become more proficient in knowledge and skills with additional years of engagement in the project. Although students involved for more than 1 year were taking different subject matter courses (e.g. biology, then chemistry, then physics), comparisons of subject matter improvement across years indicated that students' achievement levels increased from year to year. That is, students in the project for 3 years tended to achieve at a higher level in their physics course than their chemistry course and higher in their chemistry course than their biology course. The trend of increasing achievement levels from biology to chemistry to physics runs counter to students' typical performance in these different areas of science.

Analysis of Covariance (ANCOVA) was used to assess student performance in the same subject matter area for teachers that participated in the project for more than 1 year. The ANCOVA tests ($\alpha=0.05$) indicated significant differences across years with student achievement increasing with each additional year of teachers' experience.

1.1.2 Conclusions and Implications

The project has completed its fifth year and has involved a total of 23 high schools with instruction in biology, chemistry and physics and 197 teachers and 27,852 students. The project has been quite successful with respect to improvement in students' subject matter achievement, knowledge about scientific inquiry and knowledge about nature of science. Research shows that systemic change requires intensive, frequent and long-term interaction with schools, teachers and students and there is believed to be an accumulated effect over time. Students improved, on the one hand, as they benefit from the accumulated knowledge and perspectives provided by curriculum themes. They also benefit from the change in academic culture in the school as a very focused effort on systemic change is pursued.

1.2 Exposure to Authentic Science Inquiry in Shaping Teachers' Ideas About Science

In this section, Aik Ling Tan and Shirley S L Lim, the National Institute of Education, Singapore, present their model for enhancing teachers' views of teaching science as inquiry using authentic science experiences. Singapore is a young city state that has earned herself a reputation for commendable mathematics and science education reflected in good performances at international benchmarking studies such as Trends in Mathematics and Science Study (TIMSS) (Gonzales et al. 2004). However, nested within these achievements are questions about policies that shape science education and science teacher education. Science as inquiry has been touted to hold the promise of enabling learners of science to learn science in authentic ways, such that they become informed citizens who are scientifically literate. The science curriculum in Singapore underwent a major change in 2008 to reinvent itself – having science as inquiry as the core and guiding philosophy. The change requires teachers to teach science differently, shifting from a dominantly teacher-centred science teaching culture to one that required teachers to share autonomy with students in the learning process. Consequently, science teacher education in Singapore had to re-examine itself in order to (1) help prospective teachers to change their mind-sets about effective science teaching, (2) create opportunities for authentic science inquiry and (3) help prospective teachers develop sound pedagogical practices of science inquiry.

1.2.1 Recontextualising Science Teacher Education

With the emphasis of science as inquiry, prospective teachers need to (1) craft authentic science inquiry activities, (2) identify potential areas of need as students work with science inquiry tasks and (3) align classroom instructions with requirements of national examinations. As noted in the TIMSS 2003 International Science Report (Martin et al. 2004), only 41 % of grade 4 students in Singapore were taught science by teachers with university degrees that may or may not be science related. This percentage was lower than the international average of 52 % observed in participating countries in the study and implies that more than 50 % of Singaporean students learnt science from teachers without even a postsecondary specialisation in science. It is not surprising these teachers were concerned that their limited exposure to formal learning of science (learning science in formal institutions like schools and universities) would be inadequate for them to implement inquiry science effectively. Similarly, many of our prospective teachers (especially those trained to teach in elementary schools) also do not have formal science education at the high school level or beyond. Appreciating authentic science inquiry experience requires prospective teachers to develop sound knowledge and understanding of the nature of science, science inquiry and current content knowledge in science. One possible way to resolve the tension between a lack of authentic experience in science and the need to teach science as inquiry is through teacher education programmes (both pre-service as well as in-service) that provide the teachers with authentic experience in scientific inquiry. Research has delved into different aspects of teacher education programmes that are useful in helping teachers with authentic science teaching (e.g. Windschitl 2003). It is then logical to think that successful implementation requires teachers to understand and have experienced scientific inquiry themselves (Morrison 2008).

To equip prospective teachers with authentic inquiry experiences, a study was carried out in an institute of education that offers science degree courses with an education focus to pre-service as well as in-service (termed as returning teachers)

teachers. These prospective teachers, who are enrolled in this 4-year course, engage in and complete a science inquiry project under the mentorship of a scientist for 14 weeks. The framework that informs our programme is sociocultural perspective of learning (Vygotsky 1986; Wertsch 1998). Adopting a social view of learning means that higher order functions like logic and argumentation of scientific knowledge are a result of social interaction. Physical tools and language are used to facilitate the learning process by mediating the relationship between the learners and the world. Enculturation within an environment in which authentic science inquiry is carried is hence fundamental to prospective teachers learning about aspects of and how science as inquiry can be carried. As such, our prospective teachers work on a science research project under the mentorship of scientists, and in this 3-month project, they generate scientifically oriented questions, collect data, analyse data and finally present their results in the form of a final report and poster for assessment purposes. This programme enables the prospective teachers to be immersed in a science research environment, to speak the language of science and to experience the norms and practices of members of a scientific community. They use the tools and language that scientists used to negotiate and communicate knowledge. With these authentic inquiry experiences, we argue that prospective teachers will be equipped with the skills and knowledge to design learning environments and materials that adhere to and conform to authentic science inquiry. Further, they will be able to recreate and facilitate the social interactions required for scientific inquiry in the science classroom.

The prospective teachers worked on projects in different subdisciplines of science (physics, chemistry or biology), but the goals of scientific investigation component remain the same. This course aims to enable prospective teachers to learn (1) content of science by 'doing' science, (2) processes of science by engaging in scientific investigations and (3) processes of scientific inquiry that are necessary for learners of science to make sense of science. Their views about nature of science and inquiry-based science teaching were detailed after their projects. Data were collected from 86 participants through a 20-item questionnaire. Analysis of the data showed that while the prospective teachers in the study had predominantly contemporary ideas about nature of science and inquiry-based science teaching, they also had some naive ideas about inquiry-based science teaching. Results from this study suggest that an exposure to authentic science inquiry is evident in shaping prospective teachers' ideas about science. The prospective teachers in the study showed that they appreciated the fact that inquiry-based science teaching does not require teachers to have all the answers to the questions and that it emphasised the process rather than the product and it involves the skills of hypothesising, experimental design, data collection, data analysis and writing science reports. However, the prospective teachers in the study also showed that they believe that inquiry-based science teaching does not focus on the science content to be taught and that students need to have good laboratory skills to carry out inquiry. We hypothesise that being involved in different scientific investigations that led to the construction of different scientific knowledge, the prospective teachers are, therefore, of the view that the absolute content of science is secondary to the means of arriving at the conclusion.

Similarly, as a by-product of their heavy engagement in data collection during their course of research, they believe that good laboratory skills are essential for inquiry-based science teaching.

1.2.2 Results

The results of the study are encouraging for science educators who are struggling to equip teachers (both pre-service as well as in-service teachers) with current and relevant skills for effective science teaching. However, it is important to note that changing prospective teachers' knowledge is relatively simple as compared to changing teachers' beliefs (Prawat 1992) and eventual practice of science as inquiry. The constituents of beliefs are attitudes, confidence, motivations, self-concepts and self-esteem, and these are formed and expressed socially through interactions, communications and evaluations (Tobin et al. 1994). Beliefs are hence deep-seated ideas and changing teachers' beliefs about inquiry science will require more than one professional development course or teacher education course. A more recent study in Singapore by Talaue and Tan (2011) reported that besides having adequate and current scientific knowledge and knowledge about science as inquiry, teachers' intentions to enact science as inquiry are also strongly influenced by (1) subjective norms and expectations of the community, (2) their personal attitudes towards science as inquiry and (3) teachers' perception of the level of control that they have over their practice. These three factors suggest that effective science teacher education needs to go beyond acquisition of knowledge and skills to incorporate and build a mentoring framework for prospective teachers. Our findings suggest that it is reasonable to initiate the change process by increasing prospective teachers' knowledge in science, examining their existing ideas about science inquiry and then work collaboratively with them and their school community towards changing their beliefs and practice.

1.3 How Does a Science-Specific Induction Programme Impact a Teacher's Use of Inquiry?

Julie A. Luft, University of Georgia, and Sissy S. Wong, University of Houston, United States, describe in this section how different induction experiences impacted the instruction of beginning science teachers. In science education, pre-service teachers experience coursework that supports their learning of how to teach with a 'science as inquiry' approach (see National Research Council [NRC] 1996). Unfortunately, when they graduate from their pre-service programme, the only form of support that they have to teach in this manner is the teacher who is assigned to mentor them. This mentor may or may not be familiar with using inquiry in the classroom, or for that matter, they may not be familiar with science.

If newly qualified science teachers are going to enact inquiry in their classrooms, they will need adequate support in order to ensure their use or even attempt at inquiry instruction. In attempting to design such a programme, those who work with new teachers would learn that there are few studies in this field to guide the development of this type of programme. Instead, most studies focus on the design of a programme that supports the general practices of a new teacher and not the needs of a content specialist (Davis et al. 2006; Wang et al. 2008).

In order to address this void and to contribute to the discussion in this area, a study was designed in 2005 that explored how different induction experiences impacted the instruction of beginning science teachers. The programmes of interest included science-specific university programmes (SSUP) that worked with local districts to support secondary science teachers, electronic mentoring programmes (eMP) offered by universities and national associations to a large number of science teachers, general induction programmes (GP) offered by school districts that are developed for all of their teachers, and internship programmes (IP) that provided coursework towards certification while teaching. A complete description of the programmes, the study design and the methods of data collection and analysis can be found in Luft (2009) and Luft et al. (2011).

This paper reports on the part of this study that is concerned with the impact of the different induction programmes on the teaching of inquiry. It specifically covers the first 3 years of the study. During the first 2 years, the teachers experienced different induction programmes, and in the third year, the teachers were followed to determine the persistent forms of inquiry instruction.

1.3.1 Methods Participants and Programmes

Our data comes from 98 teachers, who came from five states in the United States. The teachers were located by contacting schools and districts and directors of science teacher education programmes in the Southwestern and Midwestern. Most of the beginning science teachers were female, Caucasian and first-career teachers. Most entered teaching with a bachelor's degree and worked at the high school level. They taught an average class size of 23 students, in an average school size of 1,406.

All of the teachers participated in one of four induction programmes during their first 2 years in the classroom. Briefly, the teachers in the SSUP were mentored by a science educator from a Southwestern or a Midwestern university, graduate students in science education, as well as experienced science teachers who were from the region. The beginning science teachers met with the mentors as a group, once a month, during the school year for 2 years. Teachers in the eMP were mentored by science educators and experienced science teachers through a Web-based programme. The other two programmes, GP and IP, had less of a focus on content-specific induction. The GP teachers were assigned a mentor by a school or district administrator. The interactions between the beginning teachers and their assigned mentors were dictated by the district and often observations and conversations around general teaching practices. The IP teachers were also assigned mentors, but the curriculum provided to these teachers was focused on supporting them as they pursued certification.

1.3.2 Data Collection and Analysis

The data in this study were classroom practices. The practices of teachers were captured in two ways – observations of classroom practice and interviews about classroom practice. The observations of teachers were conducted four times during each school year. During an observation, the classroom instruction was coded in 5-min increments, following the Collaboratives for Excellence in Teacher Preparation core evaluation classroom observation protocol (CETP-COP), which was piloted, field tested and refined to document the instruction of science and mathematics teachers by Lawrenz et al. (2002) and Appeldoorn (2004). This protocol was modified in order to document investigations that were guided (teacher provides some autonomy to students), directed (teachers provides little autonomy to students), verification (teacher provides no autonomy to students) or process/skill focused. Inter-rater consistency was established prior to observing classrooms.

Interviews about practice, which were digitally taped, were used to capture a week of classroom instruction and the school and classroom experiences of the new teachers. The practice protocol in this interview matched the observational protocol by Lawrenz et al. (2002). As the teacher reported his or her practices, the occurrence of the practice was indicated on a scoring sheet. Data collected from the classroom observations were totalled in minutes by specific practices over each year for each teacher, while data from the interviews were totalled by occurrence over the year for each teacher. An average was calculated for both scores. These two numbers were then multiplied to present a proportional amount of time that groups of teachers participated in specific practices. The final practice numbers were then compared to the practice mean in order to identify the prevailing practices between groups and years. In this study, the prevailing practices are reported.

1.3.3 Findings

Table 12.1 reports the use of different forms of inquiry among the teachers in the different induction programmes. The year in which each practice was prevailing is indicated by a number. Thus, guided inquiry (1, 3) indicates that the practice of guided inquiry was prevailing in year 1 and 3.

In this study, there are two important conclusions that emerge from the data. The first pertains to the importance of science teachers participating in a science induction programme. For the teachers in the eMP and the SSUP, they had a consistent use of laboratories and some forms of inquiry in the classroom. This suggests that science-specific induction programmes may be important. This second conclusion

Table 12.1 Use of inquiry among the different induction groups over 3 years

GP	eMP	SSUP	IP
Guided inquiry (1,3)	Guided inquiry (1–3)	Guided inquiry (1–3)	Demonstration (1)
Directed inquiry (1,2)	Directed inquiry (1–3)	Directed inquiry (1,2)	
Process/skill (3)	Process/skill (1–3)	Process/skill (1,3)	
		Verification (1–3)	

reinforces the importance of pre-service education in science. The teachers in the IP, who had little formal training in teaching science, enacted few inquiry or laboratory lessons.

1.4 Whole-School Approach to Developing Scientific Literacy

John Loughran and Kathy Smith (Australia), *Monash University*, report on a professional development model that uses an in-school inquiry approach. Fensham (2008) argued that 'scientific literacy does not have a fixed meaning or definition. Nor is it a single notion' (p. 28) while Roberts (2007) suggested two differing positions (Vision I and Vision II) of curriculum design as a way of conceptualising what scientific literacy might be. He described Vision I as scientist-centred with a strong focus on science content knowledge. On the other hand, he described Vision II as being more student-centred and context-driven. When reviewing Roberts' ideas, Aikenhead (2007, May) pointed out that Vision II 'seeks to enhance students' capacities to function as life-long, responsible, savvy participants in their everyday lives; lives increasingly influenced by science and technology' (p. 1).

It is not difficult to see how a Vision II view of the curriculum would be appealing to teachers as it moves well beyond ideas about the transmission of science facts and knowledge. However, not so easy to see are the implications inherent in implementing such a vision in science teaching in primary schools. Chasing the vision brings into question not only what is taught but also how it is taught and these issues were at the centre of this inquiry project.

1.4.1 Multidomain Approach

The research site for the study was a primary school that aimed to focus on scientific literacy through an in-school inquiry approach. The project set out to explore and clarify the nature of scientific literacy in terms of the practical implications for both curriculum and learning by implementing a multidomain inquiry planning approach. The multidomain approach involved curriculum planning designed to foster meaningful links across curriculum areas in order to enhance students' learning across subject areas, enabling them to draw on a range of skills and understandings. The multidomain approach centred on collegial planning around four unit topics (relationships, sustainability, technology, safety) for the year (as opposed to each subject area being distinct and separate) with each topic allocated to a specific school term. Each topic was covered simultaneously at all levels within the school at the same time.

The multidomain approach worked to develop learning connections across topics, explicitly linking learning in each unit across the year. It developed as teachers recognised that existing planning approaches had interrupted and disconnected student learning. Exploring the multidomain approach enabled teachers to find ways to 'keep learning alive' and encouraged students to see that their thinking was not merely connected or limited to only one unit. Teachers were not only encouraged to consider some of the key ideas around scientific literacy but also to engage in professional discussion and examine alternative approaches to planning and teaching to provide effective conditions to enhance students' scientific literacy.

The project systematically supported teachers in taking charge of their own professional learning. The project was based on a 'critical friend' being available to support the school-based professional learning project. The critical friend provided specific planning support for teachers across all levels within the school. Specifically, the role was about building the decision-making capacity of teachers and focusing their conversations on the development of key thinking and communication skills and science concepts to support students in developing their scientific literacy. Topics were conceptualised in different ways from the more traditional content-based approach. In collaboration with the Teaching and Learning Coordinator, over an extended period of time (3 years), the critical friend supported the classroom teachers. Through that support, teachers became more expert at articulating their pedagogical reasoning as they responded to changes in their students' learning. A vital element of support was in the provision of explicit structures for reflection to encourage teachers to develop their own understandings of 'that which matters' in science education.

Case writing (Shulman 1992) encouraged teachers to reflect on their practice and provided a structure for thinking differently about science teaching and learning. Teachers were fully supported in ways that they determined appropriate through writing workshops and weekend writing retreats designed to make explicit that participants were valued as professionals and respected as innovators and knowledge builders – not implementers of external, mandated teaching ideas. By implementing this approach to inquiry, teacher-led whole-school change was encouraged. Teachers had ownership of their ideas, control over the direction of inquiry and clarity about what scientific literacy meant for them – and the ways in which they might develop it with their students. The key feature accounting for the success of this project was the impact of collaboration. The multidomain planning approach depended on participants sharing their professional practice, their honest critiques of their own teaching and learning and a desire to work together to make a difference.

1.4.2 Evidence

The evidence of teachers' learning is documented in the book that captures participants' accounts of their professional learning: Scientific Literacy Under the Microscope (Loughran et al. 2011). Teachers' chapters offer insights into their professional knowledge through their everyday questioning of practice, their desire to continually enhance students' learning, and through an expectation that their professional growth would be driven from within – not mandated by others. The following excerpts are indicative of individuals' views of their work in this project:

> Challenges and excitement in science and the thought of empowering students to find answers and embark on new initiatives is at the heart of the goals I now set for myself ... I can see that I have come from being a teacher in charge of a limited science curriculum for students to a confident risk taker who allows students to lead the way within a co-operative learning environment in order to build up an evolving curriculum. (Howard 2011, p. 56)

> ... developing meaningful learning opportunities really needed careful planning and discussion. These conversations were supported by the head of Teaching and Learning and our critical friend. With their support we began to find opportunities to make connections between students' present learning and their learning from previous terms' work, and this became action that mattered for our planning and our students' learning. (Adams 2011, p. 63)

> By no means were we locked into specific unit understandings from the outset, but we were thinking of the skills to be developed over the course of a year and ways in which this could be achieved. This reflected a major shift in our thinking and attitudes towards planning units. (Walsh 2011, p. 97)

> ... I have drastically changed the way I teach and plan ... The level of professional dialogue at our school is remarkably different to previous years. I see the need for much talk with my colleagues and students about the direction in which our learning will travel. (France 2011, p. 108)

Through their accounts of inquiry (as documented in Scientific Literacy Under the Microscope), these teachers made the nature of educational change clear in new ways that made a difference in their school, their practice and their students' learning.

1.5 An Authentic Investigation of Fossils to Explore Foundational Concepts of Evolution

Barbara A. Crawford, the University of Georgia, and Daniel K. Capps, the University of Maine, United States, describe the context and findings for a PD model centred on an authentic Devonian fossil study. The context for the teacher PD model is a scientific investigation of authentic samples of fossils from the Devonian. Teachers and students collaborate with palaeontologists to gather and analyse data and develop explanations about past environments. The project is a collaboration of three main groups of professionals: science teachers of 5–9th grade (10–15 years) from across the United States, scientists at the Paleontological Research Institution in Ithaca, New York, and science education researchers at a university in the north-east United States, all focused on helping children learn what science is. We hypothesised that by engaging teachers themselves in the scientific study, combined with inquiry-based pedagogy and nature of science (NOS) and use of innovative materials, teachers can support their own students in using science practices. The science practices include an emphasis on collecting data (finding and measuring fossils in the rock samples) and using data as evidence to make sense of the fossil samples and then developing explanations of what the environment was like 380 million years ago. Our PD framework involves using an integrated approach (inquiry, NOS, and core science subject matter) to enhance teacher knowledge and support teachers in engaging children in doing science and understanding how scientists conduct scientific studies. The ultimate goal is to increase students' understandings and their interest in science, including those children from typically underrepresented groups. The approach makes use of an authentic investigation of fossils to explore foundational concepts of evolution and investigate biological responses to change in past environments. We suggest that the authentic fossil investigation could be a model for other kinds of investigations in other domains of science.

1.5.1 Views and Framework

We view learning as situated and value the importance of authentic activities in classrooms (Brown et al. 1989). The main purpose of the Fossil Finders PD project is to help teachers facilitate instruction for 5th through 9th grade (10–15 years) children in complex situations that involve learning about evolutionary concepts, NOS and inquiry. Our view of teaching science as inquiry always involves the learner engaged in trying to find answers to questions by using scientific practices. The learner asks and answers scientifically oriented questions about the natural world, gives priority to evidence in responding to questions, comes up with explanations using evidence, connects explanations to scientific knowledge and communicates and justifies explanations. The kinds of roles taken on by the classroom teacher highly influence the creation of an inquiry-based learning environment (e.g. Crawford 2000); therefore, attention to intense PD is important. Prior research indicates the need to combine opportunity for teacher PD with innovative curriculum materials (i.e. Borko 2004; Parke and Coble 1997). Our PD framework consists of immersing teachers in authentic science, combined with attention to essential features of scientific inquiry and explicit teaching about NOS. We situate teachers in the dual role of learner and collaborator. We integrate field experiences and pedagogy (moving back and forth from the field to the classroom to the museum) in order to help teachers understand how to support their students in carrying out an authentic palaeontological investigation. Palaeontology is the kind of science, in which scientists try to determine the nature of past environments by reconstructing it from fossil evidence. This kind of scientific thinking is different from most teachers' view of science, which involves mainly the experimental method using controls and variables in a laboratory setting. Central to our position is that children and teachers need to understand how scientists gather and use evidence to build theory.

1.5.2 Evidence for Effective Strategies

We used a mixed method approach and collected multiple forms of data of teachers' knowledge, views, practice and of their children's learning. For the quantitative data, we used a quasi-experimental design, in which we used comparison classrooms, pre- and posttested project teachers and comparison teachers, project classroom children and comparison classroom children. Data included (1) teacher and students' pre-/posttests of science content, views of inquiry and of NOS, as well as teachers' perceptions of their own learning; (2) interviews of teachers following the summer resident institute; (3) videotapes of all PD work sessions and fieldwork; and (4) videotaped teachers' lessons. We quantitatively scored the pre-/posttests and the evaluation questionnaire, and we transcribed and systematically interpreted post-institute interviews and created data displays to track changes (Miles and Huberman 1994). Analyses of the views and knowledge pre-/posttests and interview data revealed that our teacher participants deepened their subject matter knowledge and views of NOS and of scientific inquiry. Further, teachers appeared to gain confidence in teaching science. Participants demonstrated willingness and ability to use the designated instructional materials in their classrooms; and there was evidence of professional networking opportunities that were established and supported by the project. For example, the following is one project teacher's pre- and post-written responses to the question 'what does it mean to teach science as inquiry?'

Pre: 'Students have to discover the concept. That is we cannot give the students all the correct answers they have to discover the answers on their own. Students need to ask their own questions and then discover the "correct" answer.'	Post: 'Science as inquiry is that it should be taught through meaningful learning experiences. First the students need to be engaged. Once they are engaged we need to keep the excitement. We need the students to think critically about the concept at hand. We want them to make meaningful conclusions to what is being taught. We are not lecturing the kids and giving them the right answer. Instead they are engaged, exploring different concepts, and coming to their own understanding. This is different from traditional approaches because it requires students to think and not be passive. Science is not when students read books and write down notes, instead they work hard to make meaning. The students are not just being lectured about a cell, they are looking at a cell and making conclusions about them. Science needs to engage students to be critical thinkers not passive note takers.'

Related to views of inquiry and NOS, in the preinterviews, most participant and comparison teachers felt confident in their ability to teach science as inquiry, but their conceptions of inquiry *did not align* with ideas put forth by the NSES (1996). Analyses of the quantitative data indicated that participant teachers' post-questionnaire scores were significantly higher than comparison teacher scores for both the content knowledge and the NOS. Related to student learning, we are seeing very positive results related to children's knowledge of NOS (see Fig. 12.1).

Additionally, we have evidence that upper elementary students (ages 10–15) made significant gains in their understandings of inquiry and how scientists work. Analyses of student responses (of those teachers in the second cohort) showed that after exposure to teacher instruction utilising the Fossil Finders curriculum, students were better able to recognise the difference between inferences and observations and use fossil and geologic evidence to make inferences regarding the environment, the geologic timeline and changes to organisms.

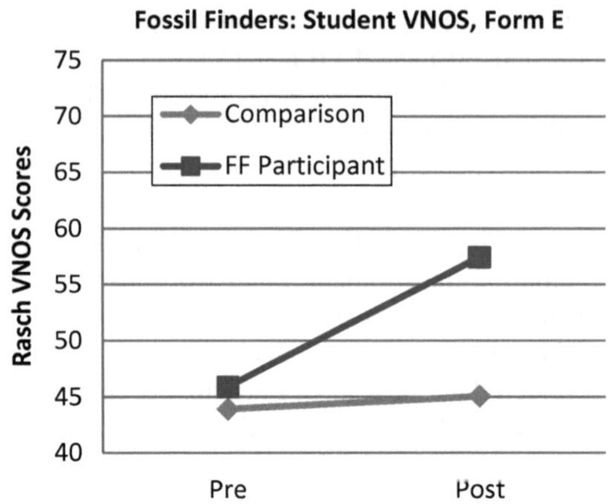

Fig. 12.1 Students' pre- and post-Rasch mean scores on 'VNOS Form E' by Fossil Finders participation, participant and comparison teacher, 2008–2009

1.6 Insights into Essential Elements

1.6.1 Comments and Summary

Jan van Driel, Leiden University, the Netherlands, provides the following insights. Research on teacher professional development has made it clear that multiple strategies are necessary to effectively promote teacher learning. Within successful strategies, the following elements are particularly important: (a) an explicit focus on teachers' initial knowledge, beliefs and concerns, (b) opportunities for teachers to experiment in their own practice, (c) collegial co-operation or exchange among teachers and (d) sufficient time for changes to occur (e.g. Bell and Gilbert 1996; Garet et al. 2001). In the studies presented in this chapter, all these elements can be recognised. However, in each study, emphases were different. Clearly, these differences are related to different contexts and different goals. For instance, in a situation where teachers never had a chance to experience authentic scientific inquiry as part of their prior education, it makes a lot of sense to incorporate authentic inquiry experiences in a programme aimed at learning to teach science as inquiry. In such a programme, research scientists obviously play an important role. Also, the design of a programme for beginning science teachers will need to pay attention to generic aspects, as any induction programme would do, in addition to a focus to teaching science as inquiry.

The studies in this chapter are also quite different in terms of the variables that were studied, in particular, the outcome measures. While some focused on changes in teachers' understandings, beliefs or views, others looked into teachers' practices. Only in the projects SLICE and Fossil Finders, student understanding was part of the study. These differences can be viewed in terms of levels of impact of teacher development (Guskey 2002) or in terms of the domains of teacher professional growth (Clarke and Hollingsworth 2002). Also, the studies differ in design.

Some are based on teachers' self-reports, or responses to interviews or questionnaires, while others applied pre- and posttests. None of the studies, however, had a (quasi-) experimental design. The above is not to say that one study is superior to another. Obviously, each study should focus on the goal of its respective professional development programme, and overall, this seems to be the case for each of the studies reported here. Taken together, these studies demonstrate that science teachers' professional learning is effectively supported by providing opportunities to experiment with new teaching approaches in their classroom, sometimes in combination with authentic experiences to learn science (i.e. scientific inquiry) and to reflect on these experiences, both individually and collectively. Such approaches acknowledge that teachers, as professionals, hold the key to improving the effectiveness of science education (Bell and Gilbert 1996).

1.6.2 Final Summary

In summary, the purpose of this chapter is to present evidence-based examples of teacher PD programmes related to inquiry, across the globe. Our ultimate goal is to work towards constructing an evidence-based framework for effective PD for inquiry. It is our hope that this chapter will inspire others to consider these ideas and findings, and continue this line of research, in order to contribute towards the ultimate goal of engaging children in inquiry and learning how to think.

References

Adams, D. J. (2011). *Effective learning in the life sciences: How students can achieve their full potential*. Chichester: Wiley.

Aikenhead, G. (2007). *Expanding the research agenda for scientific literacy*. Paper presented at the Linnaeus tercentenary symposium promoting scientific literacy, Science Education Research in Transaction, (pp. 28–29). Uppsala: Uppsala University, May 2007.

Appeldoorn, K. (2004). Developing and validating the collaboratives for excellence in teacher preparation (CETP) core evaluation classroom observation protocol (COP). Ph.D. dissertation, University of Minnesota, Minneapolis.

Bell, B., & Gilbert, J. K. (1996). *Teacher development: A model from science education*. London: Falmer Press.

Borko, H. (2004). Professional development and teacher learning: Mapping the terrain. *Educational Researcher, 33*(8), 3–15.

Brown, J. S., Collins, A., & Duguid, P. (1989). Situated cognition and the culture of learning. *Educational Researcher, 18*, 32–42.

Capps, G., Crawford, B. A., & Constas, M. (2012). A review of empirical literature on inquiry professional development: Alignment with best practices and a critique of the findings. *Journal of Science Teacher Education, 23*(3), 291–318.

Clarke, D., & Hollingsworth, H. (2002). Elaborating a model of teacher professional growth. *Teaching and Teacher Education, 18*, 947–967.

Crawford, B. A. (2000). Embracing the essence of inquiry: New roles for science teachers. *Journal of Research in Science Teaching, 37*, 916–937.

Davis, E. A., Petish, D., & Smithey, J. (2006). Challenges new science teachers face. *Review of Educational Research, 76*, 607–651. doi:10.3102/00346543076004607.

Fensham, P. J. (2008). *Science education policy making: Eleven emerging issues*. Paris: UNESCO.
France, A. (2011). Speaking about scientific literacy. In J. Loughran, K. Smith, & A. Berry (Eds.), *Scientific literacy under the microscope: A whole school approach to science teaching and learning* (pp. 101–112). Rotterdam: SENSE.
Garet, M., Porter, A., Desimone, L., Birman, B., & Yoon, K. S. (2001). What makes professional development effective? Results from a national sample of teachers. *American Education Research Journal, 38*, 915–945.
Gonzales, P., Guzman, J. C., Partelow, L., Pahike, E., Jocelyn, L., Kastberg, D., et al. (2004). *Highlights from the Trends in International Mathematics and Science Study (TIMSS) 2003*. http://nces.ed.gov/pubs2005/2005005.pdf. Accessed Aug 2012.
Guskey, T. R. (2002). Professional development and teacher change. *Teachers and Teaching: Theory and Practice, 8*, 381–391.
Hasson, E., & Yarden, A. (2012). Separating the research question from the laboratory techniques: Advancing high-school biology teachers' ability to ask research questions. *Journal of Research in Science Teaching, 49*(10), 1296–1320.
Hewson, P. W. (2007). Teacher professional development in science. In S. K. Abell & N. G. Lederman (Eds.), *Handbook of research on science education*. Mahwah: Lawrence Erlbaum Associates.
Howard, M. (2011). You don't have to have all the answers. In J. Loughran, K. Smith, & A. Berry (Eds.), *Scientific literacy under the microscope: A whole school approach to science teaching and learning* (pp. 47–58). Rotterdam: SENSE.
Lawrenz, F., Huffman, D., Appeldoorn, K., & Sun, T. (2002). *CETP core evaluation, classroom observation handbook*. Minneapolis: CAREI.
Lederman, N. G. (2004). Syntax of nature of science within inquiry and science instruction. In L. B. Flick & N. G. Lederman (Eds.), *Scientific inquiry and nature of science* (pp. 301–319). Dordrecht: Kluwer Academic.
Loucks-Horsley, S., Stiles, K., Mundry, S. E., Love, N. B., & Hewson, P. W. (2010). *Designing professional development for teachers of science and mathematics* (3rd ed.). Thousand Oaks: Corwin.
Loughran, J., Smith, K., & Berry, A. (2011). *Scientific literacy under the microscope: A whole school approach to science teaching and learning*. Rotterdam: SENSE.
Luft, J. A. (2009). Beginning secondary science teachers in different induction programs: The first year of teaching. *International Journal of Science Education, 31*(17), 2355–2384.
Luft, J. A., Firestone, J., Wong, S., Adams, K., Ortega, I., & Bang, E. J. (2011). Beginning secondary science teacher induction: A two-year mixed methods study. *Journal of Research in Science Teaching, 48*(10), 1199–1224.
Martin, M. O., Mullis, I. V. S., Gonzalez, E. J., & Chroslowski, S. J. (2004). *TIMSS 2003 international science report*. Retrieved 25 Mar 2010, from http://timss.bc.edu/timss2003i/scienceD.html
Miles, M., & Huberman, A. M. (1994). *Qualitative data analysis: An expanded sourcebook* (2nd ed.). Thousand Oaks: Sage.
Morrison, J. A. (2008). Individual inquiry investigations in an elementary science methods course. *Journal of Science Teacher Education, 19*, 117–134.
National Research Council. (1996). *National science education standards*. Washington, DC: National Academic Press.
National Research Council (NRC). (2012). *A Framework for K-12 science education: Practices, crosscutting concepts, and core ideas*. Washington, DC: The National Academies Press.
NSES (National Science Education Standards). (1996). *Observe, interact, change, learn*. Washington, DC: National Academic Press.
Parke, H., & Coble, C. (1997). Teachers designing curriculum as professional development: A model for transformational science teaching. *The Journal of Research in Science Teaching, 34*, 773–789.
Prawat, R. (1992). Teachers' beliefs about teaching and learning: A constructivist perspective. *American Journal of Education, 100*, 354–395.
Roberts, D. A. (2007). Scientific literacy/science literacy. In S. Abell & N. Lederman (Eds.), *Handbook of research on science education* (pp. 729–780). Mahwah: Lawrence Erlbaum Associates.

Shulman, J. H. (1992). *Case methods in teacher education*. New York: Teachers College Press.
Talaue, F., & Tan, A. -L. (2011). *Partnership for change towards science inquiry in elementary science classrooms: Collective responsibility of teachers and students*. Unpublished half year report for Office of Educational Research, National Institute of Education.
Tobin, K., Tippins, D. J., & Gallard, A. J. (1994). Research on instructional strategies for teaching science. In D. L. Gabel (Ed.), *Handbook on research on science teaching and learning* (pp. 45–93). New York: Macmillan.
Vygotsky, L. S. (1986). *Thought and language*. Cambridge: MIT Press.
Walsh, S. (2011). Busting the myths about science teaching. In J. Loughran, K. Smith, & A. Berry (Eds.), *Scientific literacy under the microscope: A whole school approach to science teaching and learning* (pp. 93–100). Rotterdam: SENSE.
Wang, J., Odell, S. J., & Schwille, S. A. (2008). Effects of teacher induction on beginning teachers' teaching: a critical review of the literature. *Journal of Teacher Education, 59*, 132–152.
Wertsch, J. (1998). *Mind as action*. New York: Oxford University Press.
Windschitl, M. (2003). Inquiry projects in science teacher education: What can investigative experiences reveal about teacher thinking and eventual classroom practice? *Science Education, 87*, 112–143.

Chapter 13
Weaving Relationships in a Teaching Sequence Using ICT: A Case Study in Optics at Lower Secondary School

Suzane El Hage and Christian Buty

1 Introduction

ICT-based tools have reached a high level of use in science teaching in developed countries (de Vries 2001). Despite the different level of use depending on the discipline, ICT in education has multiple uses. Among other authors, Pinto et al. (2010) present a typology of software used in science classrooms (hypermedia, simulator, computer animations, computational modelling tools, etc.). For the same level, same discipline and same chapter, there is a variety of software. The teacher then has to choose among this wide range of programs. However, most of the time, these programs are not accompanied by information resources, which could explain their use and their underlying learning hypotheses. This has been qualified as a 'great defect' by de Vries (2001).

The mastery of these technological tools requires much time and personal investment both in terms of their management and of their integration into instructional settings. Therefore, it is possible for some teachers to have a low level of information, leading them to believe that ICT determines knowledge. We consider that the ICT activity has no meaning in itself for students; it acquires its meaning when the different parts of the session or the sequence are linked. The teacher has a major role in building relationships between these parts. For example, El Hage et al. (2010) studied the teaching practice mobilized in a physics lesson about optics at grade 8.

S. El Hage (✉)
UMR ICAR, Université de Lyon 2, CNRS, ENS, Lyon, France
e-mail: suzan_hj@hotmail.com

C. Buty
EA ECP, Université de Lyon 2, ENS Lyon, Université de Saint-Etienne, Lyon, France
e-mail: christian.buty@univ-lyon2.fr

C. Bruguière et al. (eds.), *Topics and Trends in Current Science Education: 9th ESERA Conference Selected Contributions*, Contributions from Science Education Research 1,
DOI 10.1007/978-94-007-7281-6_13, © Springer Science+Business Media Dordrecht 2014

In this case, the teacher uses a simulator (Visiolab[1]) to teach additive synthesis. The teacher makes relationships between the results obtained by the simulator and the results obtained by an experiment.

A meta-analysis made by Hogarth et al. (2006) on 37 studies showed that most research studies on the ICT are interested in situations limited in time. As far as students' learning is taken into consideration, we consider on the contrary that the long-term effects, i.e. the coherence of ICT use regarding the other parts of the teaching sequence, must be analysed as well: what is the role of an ICT device in a teaching sequence? In particular, is it used to introduce new notions or to reinforce concepts that have already been studied?

Analysing classroom situations makes it necessary to combine a set of parameters, in order to sort out their complexity and their richness (Buty et al. 2012). Therefore, a single angle of analysis is not enough; the researcher must have a set of parameters. In our case, the theoretical tools that we shall use in this case study take into account the multimodal dimension in science teaching (gesture, discourse, writing, proxemics, etc.) because science discourse is multimodal (Lemke 2002). Therefore, our theoretical tools are divided among three elements: the modelling processes, the semiotic registers and the affordances or learning opportunities.

2 Theoretical Framework

2.1 Modelling Processes

Models and modelling are an essential dimension in science teaching. A lot of research presents modelling processes as the core of physics and chemistry (Gilbert 1993; Martinand 1992; Tiberghien 1994). The importance of modelling in learning physics is often reflected by the fact that the models are products of a scientific activity. Modelling approaches in physics offer the same perspectives for researchers and students, so the students can perceive the mode operation of the discipline. It is the reason why the number of research projects studying the transposition of this activity has increased within the past years. The power of modelling is due to the capacity to explain, interpret and predict phenomena in science and education.

For Tiberghien (1994), the process of modelling in the teaching of physics requires the establishment of links between two levels: the theories and the models and the description of the material world. She defines:

- The world of objects/events as knowledge elements that refer to the observable and material world
- The world of models/theories as qualitative or quantitative notions and procedures that correspond to the generalizable aspects of the studied situation

[1] http://www.intellego.fr/soutien-scolaire-4eme/aide-scolaire-physique/telecharger-logiciel-23-synthese-des-couleurs-et-couleurs-de-la-tv/14987

We need to be clearer concerning the distinction between model and theory. The distinction is based on two criteria: the application field and the complexity (Bachelard 1979; Robardet and Guillaud 1997). Theories are much larger, more complex and heavier than models; the model is a tool for studying a perfectly circumscribed local reality and takes into account only some aspects of the theory. Tiberghien (2000) included theories and models in the same world despite these differences, because the distinction between theories and models is very subtle for students in teaching at secondary school.

These notions can be applied both in the context of everyday life and in the context of science instruction. Of course, the nature of theory in one case and in the other is very different (Tiberghien 2000).

Very often, ICT-based tools represent models of empirical realities (computer simulation, animation, computational modelling tools). The dynamic images on the computer screen can be motivating or illuminating for many students and facilitate the relationship between reality and models or theories, if the teacher makes explicit to the student the difference between reality and the representations on the software.

For instance, in our case, the software used in physics (Cabri-Géomètre[2]) obeys the rules of physics; therefore these representations belong to the world of theories/models.

2.2 Semiotic Registers

Science classroom discourse is a particular one; it is multimodal in principle (Lemke 1998) and uses several kinds of representations. For instance, teaching optics consists in knowing the laws of geometrical optics and knowing how we make geometric constructions that translate those laws. In terms of semiotic register, it is necessary for students to know how to associate the laws (which are given in natural language or algebraic register) with the representations of objects that obey those laws. Let us consider a simple example: teaching optics consists in knowing the symbols of the physics materials, drawing the representations of converging lens, converging lens, light rays, light beams, screens and so on. In the line of Duval (1995), we define those representations as semiotic registers.

Duval (1995) defines the semiotic registers in the following way: 'they are the productions made by the use of signs (utterances in natural language, algebraic formulas, graphs, geometrical figures…)'.

Duval defines some semiotics registers in mathematics education. In the following, we shall adopt semiotic registers that are very often used in physics teaching:

- The register of natural language: it is the primary tool used before and/or during the acquisition of scientific vocabulary – this register is flexible, i.e. used in the world of objects/events, the world of theories/models and to make a link between the two worlds.

[2] www.cabri.com

- The algebraic register shows the relationship between different variables: mathematical expressions and laws.
- The symbolic register is a tool that allows the representation of variables by symbols – for example, the symbols (O to represent the optical centre of the lens) and the units of physical variables.
- The drawing is an 'exact representation of the shape of an object' (Davy and Doulin 1991).
- The schema is a representation of a referent. It can be non-figurative or figurative – a figurative schema refers to something sensible, visual and perceived, while a non-figurative schema designates a mental referent (Estivales 2003).

The move from one register to another is an important step because each semiotic register only embodies some elements of the same concept and not all of them. For many authors, learning a concept requires understanding how the information is expressed in a given register in particular and, namely, how to move from one register to another. Duval (1995) defines some types of operations in the semiotic register and distinguishes between the necessity to handle a concept in the same semiotic register, called 'treatment', and the move from one register to another, called 'conversion'.

Different studies (Duval 1998b; Szczygielski 2009) emphasize the difficulties for students in making a link between various semiotic registers. The ability to change the representation or to make links between the various registers is a part of the teacher's task to help students to attain this objective (Mortimer and Buty 2009). Missing links between different registers can lead students to a misunderstanding of the concept involved.

When using ICT in a science classroom, many modes of representation, many 'semiotic registers' are used in the classroom discourse; the dynamic graphical representations, which are the core of most ICT-based tools, must be articulated with natural language (words and gestures as well), or static graphical representations (e.g. drawings or schemas on the blackboard), or mathematical symbolism.

2.3 Affordance Versus Learning Opportunity

The psychologist James Gibson proposed the term 'affordance' in his ecological theory of perception (Gibson 1979). Gibson described affordances as what the environment offers the organism. He noted that an affordance depends not only on the environment but also on the interactions between the animal and its environment: 'Affordance implies the complementarity of the animal and the environment' (Gibson 1979).

In the field of ICT-based science education, Webb (2005) expresses a definition for the notion of affordance inherited from Gibson: 'In an ICT-supported learning environment, affordances are provided by interactions between the hardware, software, other resources, teachers and other students' (Webb 2005).

We consider it meaningful to introduce a distinction between two phases of the teacher's activity when implementing an ICT-based activity in the class: what the teacher has previously planned before the session and what the teacher performs during the session. This distinction is a consequence of our sociocultural point of view on science learning, which gives a key role to social interactions in the construction of knowledge.

When a teacher plans to use an ICT-based tool in a teaching session, she/he first chooses the software (or special functionalities of the software), depending on the school equipment, on the starting point of the students, on the teaching aims, etc. The teacher also foresees the way she/he will play the scene. We keep the term 'affordances' for that set of characteristics, which is planned before the session begins.

During the activity itself in the class, the interactions can widely modify what has been planned. When we analyse the classroom discourse, we find evidences of unpredicted suggestions by students, or unforeseen (correct or incorrect) procedures that they use, which deviates the knowledge flow in the classroom discourse and activity and which as a consequence can activate or inhibit previously organized affordances. We call 'learning opportunities' the affordances that have been effectively offered to the students during classroom interactions; students can grasp these opportunities or not; that is another question asking for different methodologies and is not the purpose of this article.

2.4 Links Between the Theoretical Tools

These three kinds of theoretical tools are relevant to our purpose in analysing the teacher's practice: to study the constructed relationships between the moments when the teacher uses ICT-based activities and other moments of the sequence. These kinds of activities:

- Can be at different modelling levels (the representation of the software belonging to the world of theories and models, for instance, and the experiment belonging to the world of objects and events)
- Mobilize different types of representations (graphic, drawing, static representation, dynamic representation) in the same modelling level or in different modelling levels – for instance, the teacher can say in the register of natural language that the light crosses a magnifying glass (world of objects/events) and then produce a representation of this fact in the world of theories and models using symbolic and graphic registers (he can draw light rays having a given direction and arrows to indicate the sense of light propagation – these rays concentrate in a point after crossing a converging lens)
- Can be planned before the activity, for example, the teacher could have planned how she/he will use a given software to offer different affordances to the students in order to construct relationships between different moments of the sequence

To conclude, what we have detailed above is closely linked to the teacher's practice. The teacher's role is critical (Pinto et al. 2010); if some software exists, it does not mean that the teacher is able to use it and to make links between the different semiotic registers and the modelling levels. Depending on the discourse of the teacher, she/he can offer (or not) many learning opportunities to the students or strengthen a learning opportunity, which had been offered during the previous sessions where she/he did not use the software.

3 Research Question

'ICTs as teaching resources are not considered a benefit in themselves' (Pinto et al. 2010). Taking into account the previous theoretical elements, the questions in this study are as follows: How does the use of an ICT-based tool allow the teacher to establish relationships between different moments in a teaching sequence? How does the use by the teacher of an ICT-based tool offer new learning opportunities (compared to the affordances) to students, in the flow of classroom discourse?

4 Methodology and Context

We recorded a male physics teacher at a lower secondary school in France, during a complete teaching sequence. The sequence consists of six activities, each session lasting one and a half hours. The sequence[3] is a research-based one, as it was elaborated in a joint group involving a researcher and several teachers. We are not interested in the elaboration of the sequence but we are interested in the behaviour of the teacher in classroom during the sequence.

Cabri is a dynamic geometry software belonging to the class of computational modelling tools following the typology of Pinto et al. (2010). It makes it possible to directly manipulate the representations. The objects are created according to mathematical knowledge (laws). These laws are preserved in all moving objects. Therefore, the representations that appear on the computer screen always obey the mathematical laws concerned.

The laws of geometrical optics can be implemented by geometrical constructions; Cabri is particularly relevant to the modelling of optical systems insofar as they obey the laws of geometrical optics.

This software has a specificity that involves several semiotic registers. Cabri can present graphs, natural language and symbols closely on the screen. For instance, the software enables the representations of the ray of light, the light beam that diverges at the same point by crossing a converging lens, to be built from a light source.

We have indexed our video data of the entire sequence in a table (script). We identify the sessions where the teacher used the Cabri software in the sequence, i.e. two times.

[3] See the current version at http://pegase.inrp.fr/theme.php?rubrique=1&id_theme=57

We looked in particular at the moments when the teacher used the software and we transcribed those moments. In the discourse of the teacher, we searched for all the indicators relating to the activities earlier in the sequence.

We present below (Table 13.1) the 'script' of the sequence including moments when the teacher used the software.

It was not planned to use Cabri-Géomètre in the sequence. That is why we have not completed the learning objective on the corresponding lines. One of the authors suggested that the teacher use Cabri files in this sequence, as they may facilitate the understanding of the modelling processes by the students.

5 Analysis

We analyse two moments when the teacher uses Cabri-Géomètre: on May 11 and May 25. In both cases, we present the analysis results according to each direction of our theoretical framework and, as a conclusion in each case, in terms of building relationships in the teaching sequence.

5.1 *Case 1: First Use of Cabri-Géomètre on May 11*

5.1.1 Analysis in Terms of Modelling Processes

- In this case, the teacher uses two files created using Cabri-Géomètre:

  First file: Cabri was used to model the convergence of light rays after passing through a converging lens. In this file, the teacher has the possibility to:
  - Choose and modify the focal length of the converging lens.
  - Choose if he wants to show to the students the convergence of one ray or of a light beam coming from an infinite distance.

  Second file: Cabri was used to model the divergence of light rays after passing through a diverging lens. In this file, the teacher has the possibility to:
  - Choose and modify the focal length of the diverging lens.
  - Choose if he wants to show to the students the divergence of one ray or of a light beam coming from an infinite distance.

These Cabri files and those used on May 25 obey the laws of optics, and the representation obtained by Cabri refers to the world of theories/models.

Nevertheless, there is a difference between the kind of representation adopted by the Cabri files and the conventional representational rules of optics: in Cabri files, rays are represented by lines without arrows, when optical rays carry arrows to indicate the direction of light propagation. Depending on the Cabri file used by the teacher, we can visualize graphic registers only or accompanied with natural language or symbolic register.

Table 13.1 Script of the sequence titled 'lens' (grade 8). The analysed use of Cabri files took place on May 11 and May 25 (bold characters)

Date of the session	Number of the activity	Learning objective, as indicated at the beginning of each activity	Content of the activity
2010_04_06	Activity 1	1. Recognize the different types of lenses	1. The students work in pairs to find which of four lenses can serve as a magnifying glass. The strategy consists in looking through the lenses
	Activity 2	2. Understand the effects of a converging lens in terms of optics	2. The teacher introduces the terms *diverging* and *converging* lenses
2010_05_11	*Cabri (addition)*		*Modelling a converging/diverging beam of light after passing through a converging/diverging lens*
	Activity 3	3. Show that the energy is concentrated at the focus of the converging lens for sources situated at the infinite	3. The students carry out experiments using a thermometer, a converging lens, a screen and a lamp. Students place the lamp as a light source followed by a converging lens. They measure the temperature of the light spot formed on a screen as they move it from the lens
2010_05_18	Activity 4	4. Show that some specified condition is necessary to obtain an image	4. The students send the light of the lamp on a slide with the letter P. The light passes through a fixed converging lens. Students move the screen (backwards and forwards from the lens) to obtain an image
	Activity 5	5. Understand the vision mechanism from the point of view of optics	5. The students establish the correspondence between the terms mentioned in the drawing of the eye (Fig. 13.1) and the following words in the text: diaphragm, converging lens and screen
2010_05-25	*Cabri (addition)*		*Correspondence between the elements of the eye in a schema and a drawing* *Convergence of the rays passing through a lens when they come from an object at a finite distance or an infinite one* *Explanation of the accommodation phenomenon*
	Activity 6	6. Corrective lenses for eye defects	6. Correction of eye defects

- Teacher's Discourse Analysis During the Session:
  We can notice through the discourse of the teacher that he has tried to create an explicit relationship between the two worlds of modelling processes: he utilizes the practical activity, consisting in finding which of four lenses can serve as a magnifying glass. He establishes a relation with the files created using Cabri-Géomètre. The teacher also tries to explain that the difference between the three converging lens is due to the focal length.

5.1.2 Analysis in Terms of Semiotic Representations

- The teacher makes a conversion between the natural language and the graphic register 'first, we will present a light beam [the teacher presses a button on Cabri, a presentation of several lines then appears on the screen]'. He has already explained that a light beam comprises a set of light rays. When he uses Cabri-Géomètre, the students have the possibility to visualize a graphic representation of a light beam; they can understand that a light beam is not composed only of four parallel light rays (Figs. 13.1 and 13.2) but much more (Fig. 13.4).
- The teacher makes a conversion between the drawing and the graphic representation. These two semiotic representations are for the same experiment in Activity 3 (see Table 13.1 for more details). During an interview with the teacher, he explains that Fig. 13.2 is the model of Fig. 13.1. He specifies the difference between the two figures and makes explicit his point of view: the teacher considers that Fig. 13.1 belongs to the world of objects/events and Fig. 13.2 to the world of theories/models. He explains the differences as follows:

  In Fig. 13.1, the arrows are placed on the end of the line; these representations do not represent light rays like Fig. 13.2 'the arrows of a light rays are presented on the line'.

  In Fig. 13.1, the light source is a light pocket, something belonging to everyday life; in Fig. 13.1, there is a space between the lines and the drawing of a magnifying glass.

In this case, the change in semiotic register (natural language, graphic) was used to explain the function of the magnifying glass (world of objects/events) by a physical model (Fig. 13.2).

5.1.3 Analysis in Terms of Learning Opportunities

- The convergence and divergence of light may appear to be stationary when students carry out an experiment with a light and a converging lens or when the teacher projects a slide to model the experiment (Fig. 13.2). The use of Cabri-Géomètre by the teacher allows students to observe the phenomenon in two phases: the first phase before the light rays cross the lens (Fig. 13.3) and the second phase when the light rays pass through the lens (Fig. 13.4). To switch

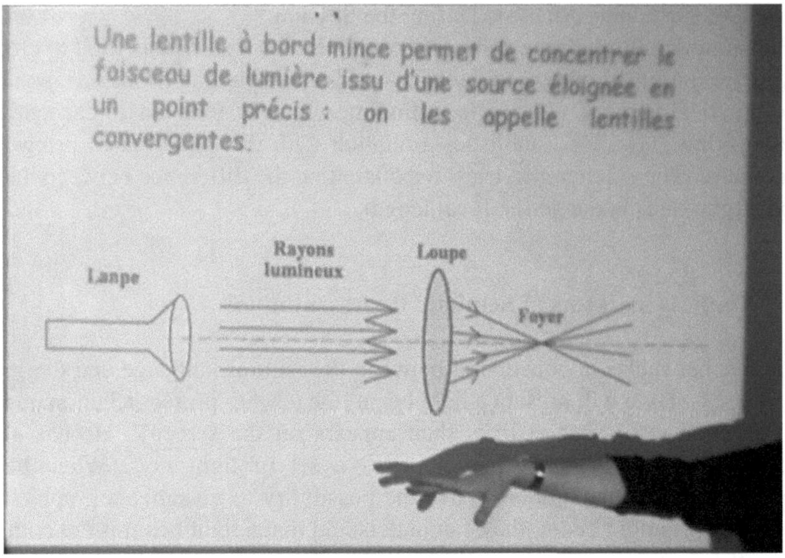

Fig. 13.1 Screenshot of the slide that the teacher uses to present the experiment, Activity 3, performed by the students

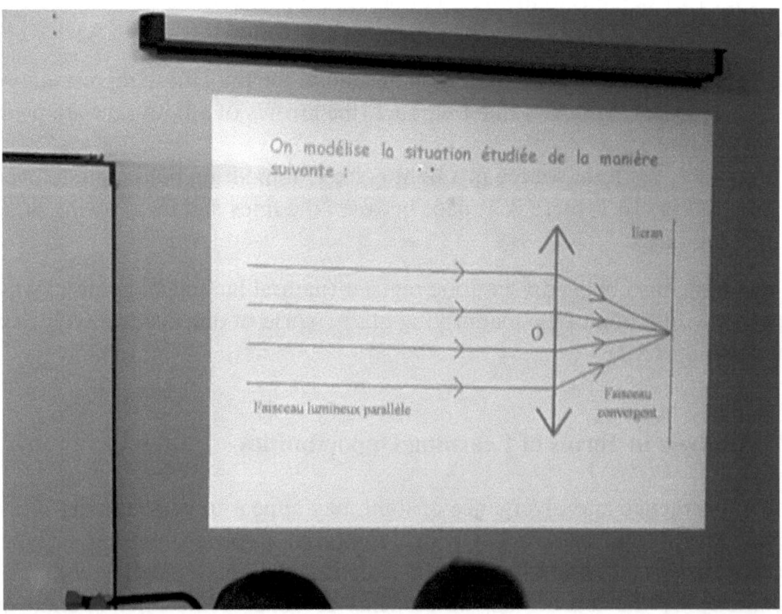

Fig. 13.2 Screenshot of a projected slide that the teacher uses to present a model of the same experiment, Activity 3, carried out by the students

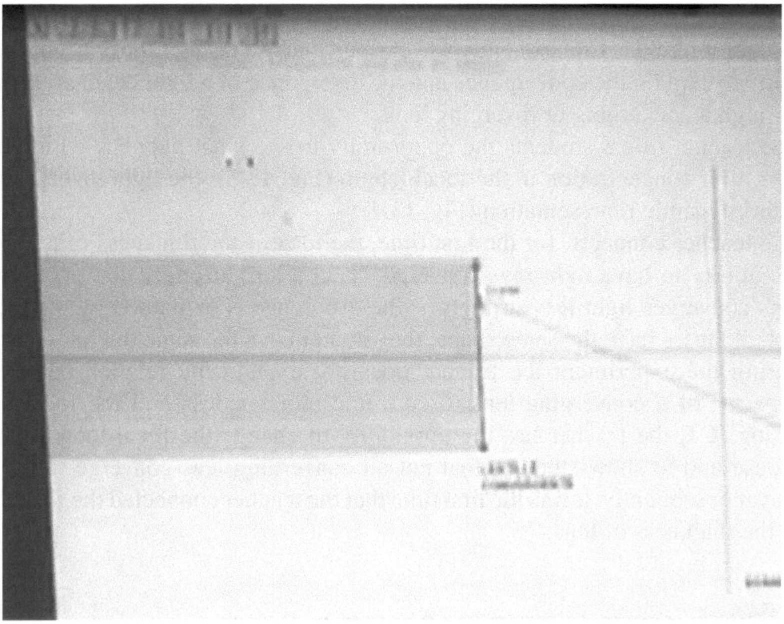

Fig. 13.3 Screenshot of the filmed video when the teacher uses Cabri; we visualize a light beam after crossing a converging lens

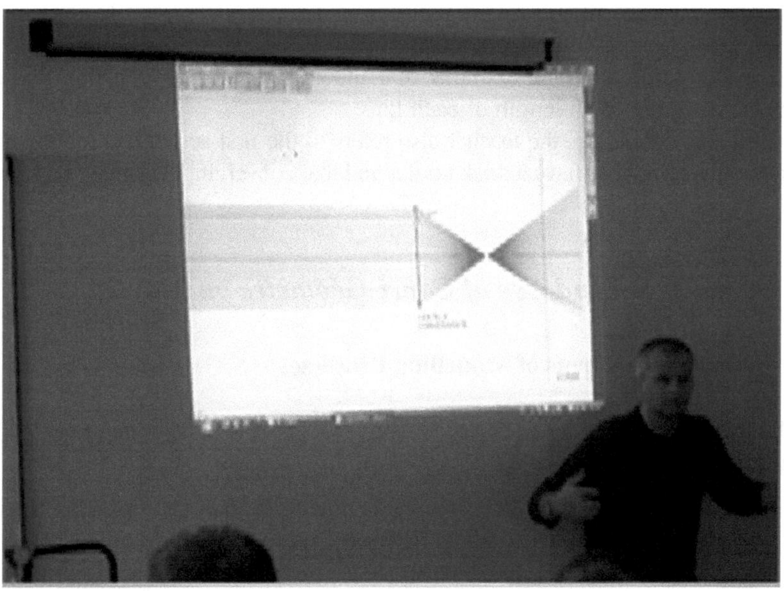

Fig. 13.4 Screenshot of the filmed video when the teacher uses Cabri-Géomètre; we visualize the deviate of the light rays after being concentrated in a single point

between the light rays before passing through the lens (Phase 1) to Phase 2, the teacher took 50 s. Consequently, we can say that the teacher introduced a temporality to explain the convergence and the divergence of a light beam after passing through a converging or diverging lens.

- The teacher offers students the opportunity to see what happens with the light rays after concentration in the focal length (Fig. 13.3); the light diverges on the Cabri dynamic representation (Fig. 13.4).[4]
- The teacher connects, for the first time, the idea of the thickness of a lens with the ability to have light rays converge: '[...] a lens strongly converges light, a lens converges light less strongly... the difference is explained by the fact that not all lenses have the same shape, they do not have the same thickness'. In fact, during the experiment the teacher does not explain the relation between the capacity of a converging lens to converge more or less and the focal length. Using ICT, the teacher has the possibility to change the focal length with the mouse and to show students that not all converging lens converge in the same way; consequently, it was the first time that the teacher connected the focal length to the thickness of lens.

5.1.4 Analysis in Terms of Constructed Relationships

The teacher referred to the previous activities performed 1 month before by the students ('you remember in activity 3', 'in activity 3, you find that', 'you have already carried out') several times during the 4 min where he uses ICT. The teacher recalled the experiment showing that all lenses are not converging lens. He asks, as a supplementary question, if all the converging lenses produce light ray convergence in the same way; he explains that the difference between the converging lenses is due to the focal length of each lens.

During this debriefing, the teacher also refers to the first activity, to remind the students that diverging lens have a thick border and that converging lens have a thin border.

5.2 *Case 2: Second Use of Cabri-Géomètre on May 25*

5.2.1 Analysis in Terms of Modelling Processes

- The teacher uses a third file created using Cabri-Géomètre, for a period of 5 min in this session:

  Cabri was used to model the convergence of light rays after passing through a converging lens. In this file (Fig. 13.6), the eye is represented by an open

[4] Cabri allows a 'direct manipulation'. It is a way to show students various proprieties of transformations immediately.

Fig. 13.5 Static representation of the eye

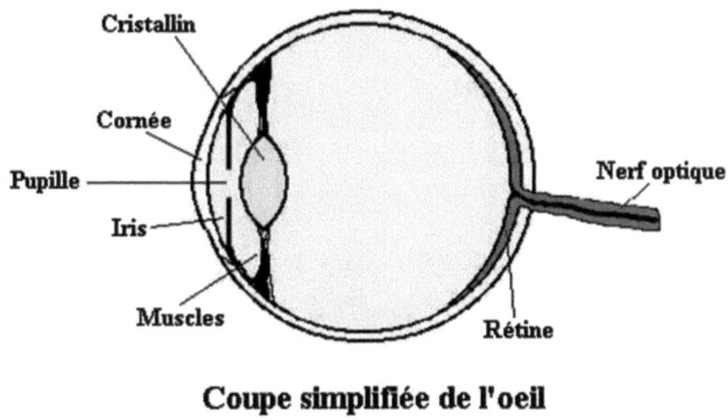

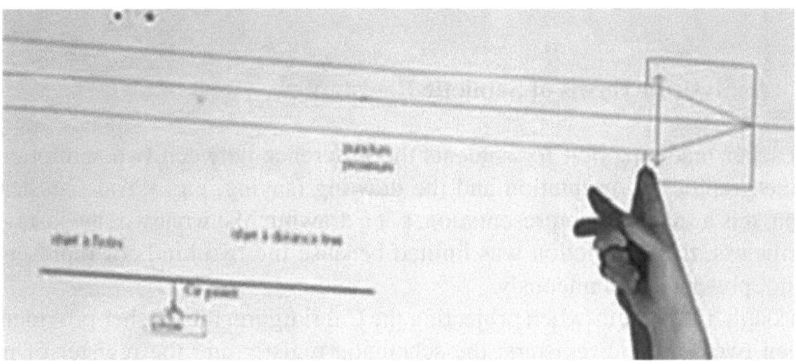

Fig. 13.6 Dynamic representation of the eye, from a physics viewpoint; representation obtained using Cabri software; case of an object at an infinite distance from the eye

square (hole square): a hole to represent the pupil, the back of the square to represent the retina and a converging lens to represent the crystalline lens.

The teacher has the possibility to choose and modify the location of the object (finite distance, infinite distance) relative to the lens.

- Teacher's Discourse Analysis during the Session:

The teacher aimed to establish a link between the two worlds using natural language. We base this interpretation on the observed identification of some elements of the world of objects and events (the crystalline lens, the pupil, the retina, all specified by the 'drawing', Fig. 13.5) to corresponding elements of the world of theories and models: the lens, the opening in the square and the screen on the bottom of the square (Fig. 13.6).

In his discourse, the teacher tried to explain a phenomenon of everyday life (accommodation) through a physical model. He suggested that the situation belonged to the world of objects/events for students ('Probably you noticed it when reading your textbook or a book, and suddenly you look at something far from you') and then produced his explanation.

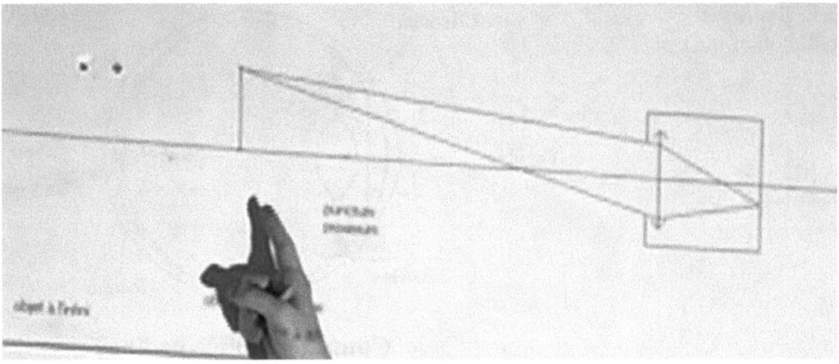

Fig. 13.7 Dynamic representation of the eye, from a physics viewpoint (representation obtained using Cabri software; case of an object at a finite distance from the eye)

5.2.2 Analysis in Terms of Semiotic Registers

The teacher made explicit for students the difference between two semiotic registers: the graphic representation and the drawing (saying, e.g., 'You see, here is a schema, it is a schematic representation, not a drawing like what we saw last week'). Nevertheless, this distinction was limited because the two kinds of representation were not present simultaneously.

Through his gestures when projecting the Cabri figure, the teacher provided links between two semiotic registers: the schematic register and the register of natural language. In fact, the teacher pointed out with his finger relevant elements on the representation; for the converging lens, for example, he said: 'Light rays as they pass through the lens, as they are all converging lens…' (Figs. 13.6 and 13.7).

As we can see, the change in semiotic registers is related to the modelling processes involving the point of view of everyday life (crystalline lens, pupils) and the world of theories/models (symbols of converging lens, light rays).

5.2.3 Analysis in Terms of Learning Opportunities

Using Cabri-Géomètre, the teacher introduced a temporality to explain the convergence of a light beam after passing through a converging lens. This was a learning opportunity already offered to the students in the first case (see Sect. 5.1.3) and strengthened in the discourse of the teacher when he uses the third Cabri file.

The use of a Cabri file offers students the opportunity to visualize, for the first time, the modelling of the image obtained by a converging lens in the case of an object at a finite distance from the lens (Fig. 13.3). The teacher gave here a new learning opportunity to the students, characterized by the fact that the image is formed on the retina of the eye, whatever the place of the object relative to the lens.

5.2.4 Analysis in Terms of Constructed Relationships

On many occasions during these 5 min, the teacher referred to the previous activities performed the week before by students ('you remember'): he recalled the experiment showing that the image through a converging lens is located in a special place behind the lens (Activity 4). But he also indicated that the situation is different for the image formation on the retina (the screen does not move), and that is his way of introducing the changes in the focal length of the crystalline, the accommodation phenomenon. Moreover he made the parallelism between Activity 5 (the schema of the eye) and the modelling of the eye in the Cabri environment.

6 Conclusion

We recall that our results are driven by a case study. We analysed the multimodal discourse of the teacher in some episodes involving Cabri in a whole class.

In these episodes, we can observe that the meaningful articulation between the semiotic registers, the modelling levels and the activities of the sequence enabled this experienced teacher to offer new learning opportunities to the students (such as the link between their everyday life experience and the accommodation phenomenon) and to strengthen the learning opportunities which had been offered in the previous sessions (such as the fact that the optical image is formed in a located area behind the lens).

Despite the limited use of Cabri files (two times only during the sequence and for a few minutes), we can conclude that the use of ICT has assisted the teacher in building bridges and creating links between different sessions and activities in the sequence. This link between experiments and the results obtained by software is not easy; it requires expertise and knowledge of the potential of the software used. We notice however that the links are always to the past and not to the future. We can question if the relationship only to the past depends on the fact that the teacher uses Cabri files specifically at the beginning of this session, when he needs to summarize the work done earlier before starting. Is it a way to remind students about what they have worked on earlier, by presenting these contents to them in a dynamic and different way? Why is the projection to the future absent?

The multiplication of case studies of this type, by varying the conditions (teacher, type of activity, type of ICT, etc.), can provide a basis for proposals for teacher training with videos on the one hand and, at the opening of research, on the effects of the ties on the understanding and learning of students on the other hand.

Acknowledgements We wish to express our gratitude to the teacher (Fabien) who agreed to be recorded and to the researcher (Karine Robinault) who facilitated our access to the experimental field.

References

Bachelard, S. (1979). Quelques aspects historiques des notions de modèle et de justification des modèles. In P. Delattre & M. Thellier (Eds.), *Élaboration et justification des modèles (Vol. 1)*. Paris: Maloine S.A.

Buty, C., Badreddine, Z., & Régnier, J. C. (2012). *Didactique des sciences et interactions dans la classe: quelques lignes directrices pour une analyse dynamique*. ENSAIO. http://150.164.116.248/seer/index.php/ensaio/article/view/929/801. Accessed 3 Dec 2012.

Davy, J., & Doulin, J. (1991). Schémas: Nature, fonction, valeur. *Cibles ENNA de Nantes, 25*, 9–18.

De Vries, E. (2001). Les logiciels d'apprentissage: Panoplie ou éventail? *Revue Française de Pédagogie, 137*, 105–116.

Duval, R. (1995). *Sémiosis et Pensée Humaine: Registres sémiotiques et apprentissages intellectuels*. Berne: Peter Lang.

Duval, R. (1998). Geometry from a cognitive point of view. In C. Mammana & V. Villani (Eds.), *Perspectives on the teaching of geometry for the 21st century*. Dordrecht: Kluwer Academic.

El Hage, S., Bécu-Robinault, K., & Buty, C. (2010). Consistency of an optics lesson including ICT at grade 8. In: Z. Zacharia, C. Constantinou, & M. Papaevripidou (Eds.), *Application of new technologies in science and education*. Proceedings of the 9th Conference on Computer-Based Learning in Science (CBLIS), Warsaw 2010. Poland.

Estivales, R. (2003). *Théorie générale de la schématisation 3*. Paris: l'Harmattan.

Gibson, J. (1979). *The ecological approach to visual perception*. Boston: Houghton Mifflin.

Gilbert, J. K. (1993). *Models and modelling in science education*. Hatfield: Association for Science Education.

Hogarth, S., Bennett, J., Lubben, F., Campbell, B., & Robinson, A. (2006). *ICT in science teaching: The effect of ICT teaching activities in science lessons on students' understanding of science ideas*. http://eppi.ioe.ac.uk/cms/LinkClick.aspx?fileticket=pGZPQL4IPG8%3d&tabid=945&mid=2149. Accessed 1 Nov 2011.

Lemke, J. L. (1998). Multiplying meaning: Visual and verbal semiotics in scientific texts. In J. R. Martin & R. Veed (Eds.), *Reading science*. London: Routledge.

Lemke, J. L. (2002). Enseñar todos los lenguajes de la ciencia: Palabras, símbolos, imagines y acciones. In M. Benlloch (Ed.), *La educación en ciencias: Ideas para mejorar su práctica* (pp. 159–185). Barcelona: Ediciones Paidós Ibérica.

Martinand, J. L. (1992). *Présentation. Enseignement et apprentissage de la modélisation en sciences*. INRP, LIREST.

Mortimer, E. F., & Buty, C. (2009). What does 'in the Infinite' mean? The difficulties with dealing with the representation of the 'Infinite' in a teaching sequence on optics. In C. Andersen, N. Scheuer, M. D. P. Pérez Echeverría, & E. V. Teubal (Eds.), *Representational systems and practices as learning tools*. Rotterdam: Sense Publishers.

Pinto Casulleras, R., Cousa Lagaron, D., & Hernandez Rodriguez, M. I. (2010). An inquiry-oriented approach for making the best use of ICT in the classroom. *eLearning Papers, 20*, 1–13.

Robardet, G., & Guillaud, J. C. (1997). *Éléments de didactique des sciences physiques*. Paris: Presses universitaires de France.

Szczygielski, C. (2009). Lecture et compréhension dans différents systèmes sémiotiques en électricité: Raisonner sur des schémas électrocinétiques ou électrotechniques et des montages électriques. *ASTER, 48*, 161–185.

Tiberghien, A. (1994). Modelling as a basis for analysing teaching-learning situations. *Learning and Instruction, 4*, 71–87.

Tiberghien, A. (2000). Designing teaching situations in secondary school. In R. Millar, J. Leach, & J. Osborne (Eds.), *Improving science education: The contribution of research*. Buckingham: Open University Press.

Webb, M. E. (2005). Affordances of ICT in science learning: Implications for an integrated pedagogy. *International Journal of Science Education, 27*(6), 705–735.

Chapter 14
Inquiry-Based Mathematics and Science Education Across Europe: A Synopsis of Various Approaches and Their Potentials

Katrin Engeln, Silke Mikelskis-Seifert, and Manfred Euler

1 Renewing Mathematics and Science Pedagogy on a European Level

In many developed nations, serious concerns are being raised about the status and the impact of science and mathematics education and the decrease of students' interest for the corresponding key subjects. Over the past decades, there has been a growing consensus that the lack of a basic education and of mathematical and scientific literacy will hinder young people to become active citizens. Furthermore, the present uptake of science and technology-related studies and professions is

COMPASS has been funded with support from the European Commission (503635-LLP-1-2009-1-DE-COMENIUS-CMP). This publication reflects the views only of the author, and the Commission cannot be held responsible for any use which may be made of the information contained therein.

PRIMAS has received funding from the European Union Seventh Framework Programme (FP7/2007–2013) under grant agreement n° 244380. This chapter reflects only the author's views and the European Union is not liable for any use that may be made of the information contained herein.

K. Engeln (✉)
Leibniz-Institute for Science and Mathematics Education,
Olshausenstraße 62, 24098 Kiel, Germany
e-mail: mikelskis@uni-potsdam.de; engeln@ipn.uni-kiel.de

S. Mikelskis-Seifert
Institute for Physics Education, University of Education Freiburg,
Kunzenweg 21, 79117 Freiburg, Germany
e-mail: s.mikelskis@web.de

M. Euler
Leibniz-Institute for Science and Mathematics Education,
Olshausenstraße 62, 24098 Kiel, Germany
e-mail: euler@ipn.uni-kiel.de

C. Bruguière et al. (eds.), *Topics and Trends in Current Science Education: 9th ESERA Conference Selected Contributions*, Contributions from Science Education Research 1,
DOI 10.1007/978-94-007-7281-6_14,

considered insufficient to keep up the pace of innovation and to react adequately to the economical, ecological and social challenges of a rapidly changing world. As a consequence, a variety of new educational programmes and projects to improve the quality of science and mathematics education at school and also at university were created on a regional, national and supranational level.

In the European context, reports by expert groups have identified the necessity of a renewed pedagogy in school that transforms the traditional mainly deductive teaching styles towards more appealing and cognitively activating forms of learning. Inquiry-based science education is identified as the method of choice to increase students' interest and achievement in science and mathematics (Rocard et al. 2007). Accordingly, the European Commission finances many projects that focus on various ways to foster inquiry-based approaches in mathematics and science education. In addition to improving school teaching, there are also increased efforts to transform university education by better linking research and teaching at the tertiary level (Healey 2005).

The present chapter will focus on findings from the EU projects PRIMAS and COMPASS. Both projects aim to support teachers in integrating and applying inquiry-based learning pedagogies in their mathematics and science classrooms. Whereas the project PRIMAS is still running (2010–2013), COMPASS had a lifespan from 2009 until 2011. Within the PRIMAS project, a baseline study was carried out to gain insight into the current IBL situation in Europe. This chapter presents results from that baseline survey. Supplementary results of a study are reported that evaluate the effects of using material being developed by COMPASS team members and teachers. Both studies focus on the status and the impact of IBL. They address disciplinary differences in the teaching and learning cultures within mathematics and science.

2 Views of Inquiry-Based Teaching and Learning

Inquiry-based science education has a long history and there are many approaches to teaching and learning science as inquiry (Barrow 2006; Prince and Felder 2007). In the USA the tradition of inquiry-based learning goes back to Dewey (1910), whereas in Germany, for example, inquiry learning was introduced within the reformist pedagogy starting in the 1920s (cf. Wagenschein 1962). In discussions on improving education, inquiry is used in different ways and contexts. Not only is inquiry-based learning used without clarifying connections and distinctions, terms such as *inquiry-based teaching*, *inquiry-based method*, *inquiry-based education* and *inquiry-based pedagogy* are also widespread. Moreover, inquiry is often conflated or used interchangeably with other terms that describe similar learning and teaching approaches such as *anchored instruction*, *hands-on*, *problem-based, project-based, student-centred, inductive* and *dialogic approaches* (Anderson 2002; Hayes 2002). For example, the Rocard report (2007) identifies problem-based learning as the method of choice for mathematics education and inquiry-based learning for science education.

In this chapter, inquiry-based learning (IBL) is used in a broad sense as it is in the projects PRIMAS and COMPASS. Even though there is no generally accepted clear definition of inquiry-based education, common core elements do exist. As a complement to conveying the content and concepts of mathematics and science which dominates traditional deductive teaching, inquiry-based education focuses on acquainting students in experience-based ways with the typical processes and methods of the science subjects. Students are supposed to take more responsibility for their own learning as they learn to work individually as well as in groups. They explore and discover, ask questions, identify problems, find solutions, create models, formulate hypothesis, devise experiments, design tools and systems, share, reflect and communicate their knowledge. To make learning meaningful, the learning context plays an important role. The covered topics are relevant to the students and their prior experience is adequately taken into account. Thus, by engaging students actively in the construction, evaluation and reflection of knowledge, inquiry-based education is thought to promote competencies that are relevant for lifelong learning and for a successful orientation in a complex world. On the teachers' part orchestrating and facilitating learning processes is a subtle skill that is critical to make IBL function well. The teacher as an expert for learning helps the students through modelling and coaching in particular by the use of questioning strategies (Barrow 2006; Colburn 2006; Hmelo-Silver 2004; Prince and Felder 2007). An appropriate balance of challenge and learning support has to be provided.

With this broad multifaceted conception, inquiry education goes along with many contemporary school improvement projects, which are largely based on constructivist ideas, learner-centred and oriented towards the development of higher-order understandings and skills and/or emphasise collaborative efforts by students in learning communities. Therefore, the spectrum of IBL conceptions is wide. For example, IBL can be differentiated according to type and the complexity of the problem, the degree of student-centred learning and also the order of problem and information presentation (Walker and Leary 2009). Staver and Bay (1987) distinguish between structured inquiry, guided inquiry and open inquiry.

In a more narrow sense, inquiry-based education, especially in science, refers to learning that takes place following the processes that are involved in scientific inquiry. Students are guided to pose questions, to devise experiments and investigations, to gather and analyze data and to construct evidence-based arguments.

> We describe inquiry instruction as engaging students in the intentional process of diagnosing problems, critiquing experiments, distinguishing alternatives, planning investigations, researching conjectures, searching for information from experts, and forming coherent arguments. (Linn et al. 2004, p. xvi)

> Inquiry science is a hands-on constructivist approach to science education. Students address teachers' and students' questions about natural phenomena or events by conducting scientific investigations in which they collaboratively develop plans, collect and explain evidence, connect the explanations to existing scientific knowledge, and communicate and justify the explanations. (Anderson 2002)

Although critical voices have been raised against minimal guidance instruction and pure discovery learning (Kirschner et al. 2006; Mayer 2004), there is evidence for the benefits of inquiry-based education from empirical research. Substantial quantitative improvements are found on various measures, including cognitive achievement, process skills and attitude towards science. Inquiry-based education proves indispensable especially with respect to achieving complex and comprehensive "higher-order" objectives such as understanding science principles, comprehension of scientific inquiry and applying science knowledge to personal and societal issues (Anderson 2002). In his meta-analyses Hattie (2009, pp. 209–210) concludes, "Overall, inquiry-based instruction was shown to produce transferable critical thinking skills as well as significant domain benefits, improved achievement and improved attitude towards the subject".

With respect to improving students' performance, the meta-analyses indicate only moderate effects of IBL as compared to various forms of activating instruction by the teacher (Hattie 2009, p. 243). These findings clearly point out that in spite of its obvious benefits, a widely accepted and successful implementation of inquiry-based mathematics and science education is far from trivial. Although many good examples for inquiry-based learning in mathematics and science education have been put forward by researchers, teacher educators and experienced teachers, changing the prevailing deductive teaching style is a highly challenging issue. Teachers' professional competences are of crucial importance for keeping a proper balance between instruction and autonomous construction in the teaching and learning of mathematics and science. "However, there's no such thing as a teacher-proof curriculum, and there are lots of times when inquiry-based instruction is less advantageous than other methods. It's up to you to find the right mix of inquiry and non-inquiry methods that engage your students in the learning of science" (Colburn 2000, p. 44).

3 Teachers' and Students' Views on Classroom Practices

The project PRIMAS like other projects funded within the FP7 framework focuses on various ways to foster inquiry-based approaches in mathematics and science education. Twelve European countries (Cyprus, Denmark, Germany, Hungary, Malta, the Netherlands, Norway, Romania, Spain, Slovak Republic, Switzerland, United Kingdom) are working together to develop, implement and promote inquiry-based learning in mathematics and science education. The project provides materials for direct use in class and for further professional development of teachers (PRIMAS 2010).

In view of the above findings on the widely varying views of IBL and its impact, it is crucial to collect information in order to clarify the present situation of mathematics and science teaching in the framework of the traditions of teaching and learning in each country. Therefore, within the project a baseline study was carried out to collect information on the status of IBL in the participating countries. Additionally, the baseline study is conceived as a reference line against which the impact of intervention through the project can be measured. In order to link the baseline study to findings from large scale assessments, the PRIMAS teacher questionnaire adapted items from the PISA 2006 student questionnaire (OECD 2009a). The PRIMAS findings can be compared with and validated by the data from the PISA 2006 study. Altogether, 925 teachers throughout Europe participated in the baseline survey. The following findings refer to data collected within the baseline study (Euler 2011).

Classroom practice depends on the culture, the subject and the personal beliefs of the teacher regarding the subject and teaching and learning in general. Each curriculum area and each traditional school subject embodies its own tradition of teaching, learning and assessment. Nevertheless, all over Europe there is a widespread belief that learning will improve through being more learner-centred and problem-oriented. It is not enough for students to know the facts; they also have to learn to deal with problems and to adapt to new situations (OECD 2009b, pp. 88–99). In order to probe into classroom practice, we report the results of five items used in both the PRIMAS baseline and the PISA 2006 study. These five items belong to scales referring to science teaching and learning (OECD 2009a, pp. 333–335). They provide insight into central facets of inquiry-based education in classroom teaching practices both from the teachers' and the learners' perspective:

- Students are given opportunities to *explain* their own *ideas* (ST34Q01).
- Students are allowed to design their *own experiments* (ST34Q08).
- The teacher uses science to help students understand the world *outside* school (ST34Q12).
- There is *class* debate or *discussion* (ST34Q13).
- Students do experiments by following the *instructions* of the teacher (ST34Q14).

Teachers' (PRIMAS) and students' (PISA) estimate the frequency of the described classroom teaching practice on a four-point scale, ranging from "never or hardly ever" to "in almost all lesson". The results show that the students' and teachers' perspectives are largely comparable (Fig. 14.1). However, a few exceptions can be highlighted. Science teachers (PRIMAS) and students referring to science (PISA) both report that designing "their own experiments" does not occur often in science lessons. Similarly, there are rather few occasions for doing experiments by following the "instructions" of the teacher, and opportunities for class "discussions" are not very frequent. On the other hand, in most lessons students have the possibility to explain their ideas. These results support the finding of the TALIS study that student-centred practices exist, but structuring practices, such as summarising former lessons, homework review and checking the exercise book, are the most frequently employed practices across all participating countries (OECD 2009b, p. 98). Interestingly, helping students understand the world "outside" school is estimated differently by PRIMAS teachers and PISA students. PRIMAS teachers report using science to help students understand the world "outside" school in most lessons, while PISA students report a lesser frequency. It needs further research to explain this phenomenon that indicates a different perception of the "world outside school" between students and teachers. One can assume that students and teachers as participants of the ongoing teaching and learning processes (as well as detached

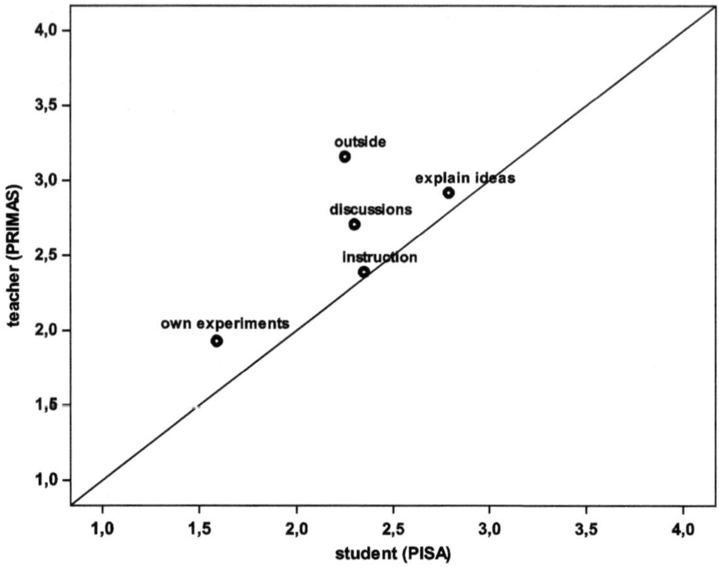

Fig. 14.1 Scatter plot of five items describing IBL-related elements of teaching practice. The student data are taken from PISA 06 and the teacher data from the PRIMAS baseline study (1: never or hardly ever, 2: in some lessons, 3: in most lessons, 4: in almost all lessons)

external observers) take up different perspectives and use different clues and value systems to describe classroom practice. Only limited research comparing the reliability and the differences of these different sources exists (Lüdtke et al. 2006). Clausen (2002, p. 186) concludes that teachers, students and observers have a different perception of classroom practice. Furthermore, he points out that teachers might miss a reference frame because they rarely have the chance to experience and assess classroom practice of their colleagues. Nevertheless, the teachers' perception is important because the teacher is the initiator for changing classroom practice.

In summary, student-centred teaching practice is already visible both in the students' and the teachers' statements, whereas there is a special need for strengthening practical activities in the daily classroom practices and to increase measures for engaging students in the design and evaluation of experiments. The reflective interplay of experimenting and modelling is at the heart of inquiry-based education and still needs to be better included into mathematics and science education. These descriptions of lesson practices are also supported by another result from the PRIMAS baseline study. Throughout Europe, teachers are positively oriented towards IBL, but inquiry-based learning has not yet found its way into daily teaching practice.

Another research question of the PRIMAS baseline study was to investigate if there are differences in the lesson patterns between the school subjects. The results reveal considerable differences not only between science and mathematics but also between the sciences. The five items from the preceding chapter and, additionally,

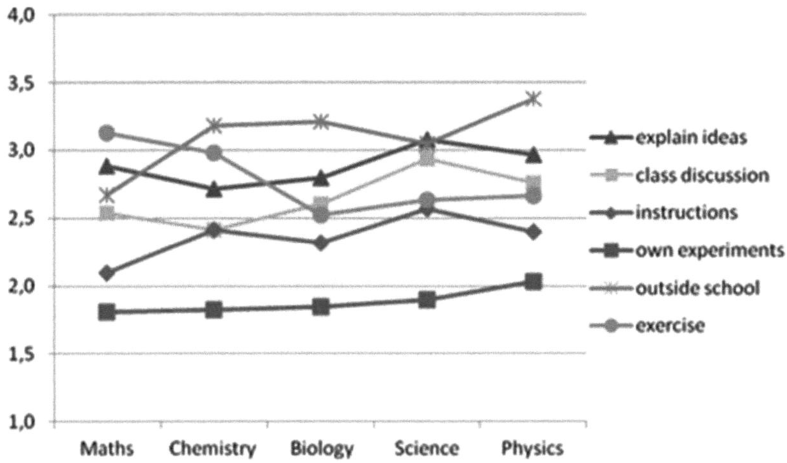

Fig. 14.2 Mean of five items describing IBL-related elements of teaching practice and the item "exercise" depending on the subject. The data is taken from the PRIMAS baseline study (n_maths=524, n_physics=97, n_biology=75, n_chemistry=47, n_science=81, 1: never or hardly ever, 2: in some lessons, 3: in most lessons, 4: in almost all lessons)

the item "My students learn through doing exercises" are analyzed with respect to the different subjects. As expected, mathematics teachers use their subject least to help students understand the world outside school. Furthermore, practical activities (students' own experiments, instructions) play a minor rule. Mathematics teachers report the highest amount of exercises (Fig. 14.2). Interestingly, significant differences within the sciences can also be detected. For example, chemistry teachers report a frequent use of exercises while physics teachers more strongly emphasise the use of their subject to help students understand the world outside school. Within the natural sciences, physics teachers implement IBL strategies that relate to experiments and to the world outside school more than their colleagues from biology and chemistry and general science teachers that teach science as an integrated subject.[1]

The results are supported by the research of Stodolsky and Grossman (1995). They believe that subject matter has a strong influence on actual teaching practices. In a study with nearly 400 teachers of five academic subjects from 16 highs schools in both California and Michigan, they showed differences in the conception of the subjects and the implication for teaching. Mathematics teachers see their subject as more defined, more sequential and more static than science teachers. Moreover, they report having less control over content and less choice over instructional instrument.

Therefore, all measures to promote inquiry-based education have to consider subject-specific differences in the degree of IBL already implemented in the subject area. The daily teaching practices as well as the conception of inquiry strategies

[1] It is not possible to make a reference to the PISA study, because within PISA the different science subjects are not treated separately.

depend on the subject. Consequently, professional development has to be tailored to meet these different needs.

The use of IBL in class not only depends on the subject area, it also reveals considerable cultural differences. The baseline study collected information concerning potential difficulties of practising IBL in the countries. The relevance of the different problems anticipated when implementing IBL depends on the cultural background. This finding can be anticipated because the school system including assessment practice and the syllabus as well as the available resources depend on the national and local policy. The problems teachers indicate with respect to the implementation and the routine use of IBL can be subsumed into three factors. The first factor subsumes *system restrictions (SYR)* of the school system like the curriculum, assessment practice and also class size. The second factor accounts for *classroom management* (*CLA*). The third factor stands for missing *resources* (*RES*) including adequate textbooks and computers but also for the need of continuing professional development (Colburn 2000; Walker 2007).

These three factors are also found in the answers of open question regarding difficulties that hinder the implementation of IBL:

> My major stumbling block is time. I cannot leave out whole chunks of the syllabus to implement IBL, therefore I am using an inquiry based approach but collaborative group work is at the bare minimum, e.g. once a month or less frequently". (teacher from Malta)

A member of the Hungarian PRIMAS team commented:

> .... The opinions here can be divided into two main clusters: (1) lack of PD programs about IBL, and even more frequently (2) lack of resources and equipment. Maybe people from the Western part of the river Elbe can hardly believe that even pencils, pens, cardboards and other very simple equipment is missing from the school or are bought by parents and brought to the school by the children.

Within Europe not only the relevance of each of these categories but also the overall problem of the implementation of IBL is rated differently. In all countries, actual classroom management is considered the least significant problem. A closer examination of the responses to the two other categories (system restrictions and resources) shows that teachers in Denmark, Germany and Norway see fewer problems with the implementation of IBL than in the other countries. Teachers in Spain, Slovakia, Cyprus and Romania have the greatest worries about implementation. In these countries system restrictions and lacking resources are seen as almost equally obstructive. With the remaining five countries, it is striking that in Switzerland and in Malta teachers regard system restriction clearly as more of a hindrance than resources (Fig. 14.3).

4 Roles of Materials to Develop Teachers' Practices Relating to Inquiry-Based Learning

Another European project focusing on changes in methods of teaching mathematics and sciences is COMPASS – Common Problem Solving Strategies as Links between Mathematics and Science. In COMPASS teaching materials are developed and

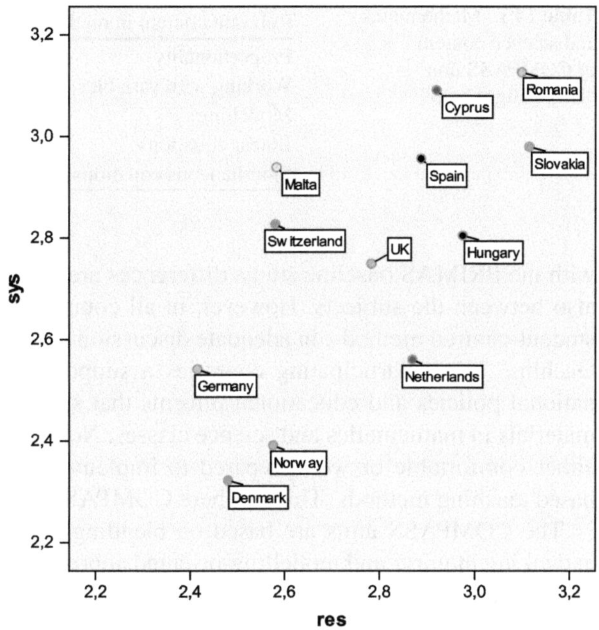

Fig. 14.3 Scatter plot of system-related problems versus resources-related problems (1: strongly disagree, 2: disagree, 3: agree, 4: strongly agree)

evaluated that emphasise inquiry-based learning through tasks which engage students in working with important concepts that bridge mathematics and science. COMPASS pays special attention to intertwining mathematics and the sciences and using socially relevant topics in its teaching modules. COMPASS comprises six institutions from six European countries (Germany, the Netherlands, Spain, Slovakia, Cyprus, United Kingdom). Across these countries, the project team developed teaching materials in collaboration with mathematics and science teachers from the EU to ensure the adaptability of the tasks in different nations. The cooperation between teachers and educators is viewed as "symbiotic". This means that teachers are seen as experts in the field of practice, whereas mathematics and science educators are familiar with recent research and the literature on new instructional methods. The role of the educators is to serve as coaches guiding the work of the "community of practice" composed of teachers and educators. A close cooperation between teachers and educators can be regarded as key to successful implementation of COMPASS materials in schools. Teachers test and use the COMPASS materials in their classroom and gain experience by changing classrooms when using COMPASS materials. One major research interest of the COMPASS team is the constraints and the supporting conditions for the implementation of COMPASS materials and therefore of inquiry-based learning strategies in class (COMPASS 2009).

Like the PRIMAS baseline study, an analysis at the beginning of the COMPASS project shows a heterogeneous picture of daily school practice concerning inquiry and interdisciplinary learning in the different European countries. In accordance

Table 14.1 Mathematics and science content of COMPASS unit "Ban of light bulb"

Relevant content in mathematics	Relevant content in science
Proportionality	Efficiency
Working with variables	Production of light
Modelling	Light spectrum
Linear equations	
Simultaneous equations	

with the PRIMAS baseline study, differences are visible between the countries and also between the subjects. However, in all countries there seems to be deficits in student-centred methods, in adequate discussion culture and in application-oriented teaching. In all participating countries a supportive framework exists regarding national policies and educational reforms that supports implementing COMPASS materials in mathematics and science classes. Nonetheless, the teachers did not feel either comfortable or well prepared to implement inquiry- and interdisciplinary-based teaching methods. This is where COMPASS applies.

The COMPASS units are based on blending science and mathematics content and taking inquiry- and modelling-oriented approaches into account. Everyday contexts which are relevant to the students and societal problems fostering the students' ability of critical argumentation play an important role. In the following, the unit *light bulb ban in European countries* is introduced as an example of a COMPASS unit. The key research question of this unit is: "Why have traditional light bulbs been abolished by the EU?" In Germany, as in many other European countries, conventional light bulbs are to be abolished. From 2009 to 2012 conventional light bulbs are be gradually withdrawn from the markets starting from high to low wattage. The COMPASS unit aims to provide students with opportunities to discover why the EU introduced this change while exploring relevant mathematical and scientific content (Tables 14.1 and 14.2).

To analyse the impact of the COMPASS unit's practice, three scales measuring facets of teaching practice were developed and tested. These scales show satisfactory reliabilities and were used in teachers' and students' questionnaires. The following table (Table 14.3) presents students' estimations (n = 150 students in Germany) concerning the perceived teaching practices in mathematics comparing the beginning and the end of the implementation process. In the scales "student-centred methods in mathematics" and "application-oriented teaching in mathematics", there are significant increases. The same effect was found throughout the entire COMPASS sample.

Comparing mathematics and science classes, significant differences become visible in Germany. In accordance with the results of the PRIMAS baseline study, the mean values of the scales "student-centred methods" and "application-oriented teaching" are on a higher level in science than in mathematics classes at the beginning of the implementation. Teaching practices in science tend to be more student-centred and application-oriented as opposed to the teaching in mathematics.

COMPASS materials were well received by the participating teachers. Both science and mathematics teachers report a considerable impact of COMPASS

Table 14.2 The task plan of COMPASS unit "Ban of light bulb"

Number/subject	Content
1	*Introduction to the context*
M/P	Why were light bulbs banned? Why was there a run on light bulbs?
2	*What are the differences between light bulbs and energy-saving bulbs from a physical point of view?*
P	Look into the construction of a light bulb and an energy-saving light bulb and how they work
	Which physical principles play an important role? Explain and compare the construction and workings of a light bulbs using models and conducting experiments
	Possible extension: Investigate the light spectrum of each light. Look into the impact of lights on humans
3	*What is the difference in efficiency between light bulbs and energy-saving lights?*
P	Investigate the efficiency of a light bulb and an energy-saving light. Look into information about efficiency and compare the various lights to each other. Investigate the light intensity of the various types of lights using, for example, the grease spot photometer
4	*How much money can you save by using energy-saving lights?*
M	Calculate the lighting costs for an apartment for a family of four. Compare light bulbs with energy-saving lights
	Alternatively: Calculate the lighting costs of our school. Compare your results with energy-saving lights
5	*Costs depending on various factors*
M	Graphs, linear functions, intersections
6	*Global benefit*
M	How much energy can potentially be conserved by EU citizens?
7 M/P	*Finalisation*: Debate about the pros and cons of banning the traditional light bulb

M mathematics, *P* physics

Table 14.3 Students' estimation of teaching practice in mathematics in Germany (1: disagree, 4: agree)

	Student-centred methods (in mathematics)	Discussion culture (in mathematics)	Application-oriented teaching (in mathematics)
Pretest	M=2.59	M=2.59	M=2.15
	SD=0.65	SD=0.80	SD=0.66
Posttest	M=2.78	M=2.63	M=2.35
	SD=0.55	SD=0.79	SD=0.62
Significance	Significant	Not significant	Significant

materials on their teaching. Interestingly, after implementing the COMPASS material, students perceive significant changes in teaching practices only in their mathematics classes. No changes in the science classes are visible (Fig. 14.4). This effect holds also for the other countries participating in COMPASS. In summary,

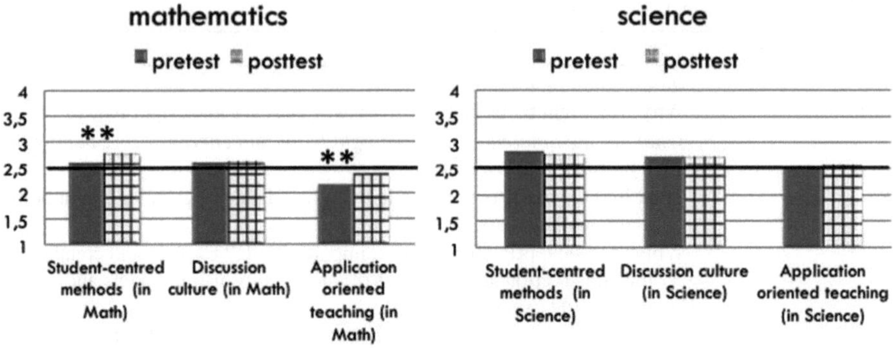

Fig. 14.4 Students' estimations on teaching practice (Germany)

we can observe that mathematics benefits most of all as a school subject from the COMPASS units. Having a higher starting level probably makes it more challenging to implement IBL-oriented teaching practices.

5 Conclusion

The baseline study provides a comprehensive survey on the variety of IBL approaches in mathematics and science teaching across Europe. It demonstrates the potential and the high expectations among teachers and delivers favourable evidence for the implementation of a more student-centred pedagogy. All over Europe there are teachers with initial IBL experience and who are keen to learn more about improving their pedagogical skills. There is a strong belief that IBL has the potential to overcome learning problems and to motivate students. With respect to the present status of practising IBL, it is important to react adequately to the existing high variability due to systemic restrictions and the different teaching and learning cultures.

However, in spite of the positive orientation on the teachers' part, results on implementing IBL are sobering. Inquiry-based education is not a well-defined idea. It is not even clear whether a teaching method, a learning style or something else is being described. It is a multifaceted concept subsuming many of the recent approaches concerned with improving the learning of science and mathematics, e.g. constructivist and/or learner-centred ideas. The implementation of inquiry-based education is not straightforward but strongly depends on the subject, the culture and the individual teacher. Mathematics and science teachers from the different science disciplines have distinct scripts for their daily practice, and therefore they start from different points with different needs that also depend on the specific subjects.

The actual status, the orientation towards inquiry-based education and the anticipated problems faced during implementation are strongly influenced by the cultural background and the subject-specific socialisation of the teachers. The latter is demonstrated by the difference of IBL approaches between mathematics and science subjects (in Germany). The evaluation within the COMPASS project demonstrates the potentials for improvement in mathematics, whereas in science classes the implementation of IBL-oriented material does not show the aspired effects. Probably, there is a saturation effect. The positive effects of moving to more IBL-oriented teaching practices are less pronounced when elements of IBL practices are already at a higher initial level.

As a result, it can be concluded that teaching practice can be changed towards more inquiry-based education through making appropriate material available. However, implementation is not straightforward. It has to take various subject-specific needs and boundary conditions of the teaching culture and systemic restrictions into account. A widely accepted and successful implementation of inquiry-based mathematics and science education is far from trivial. Although many good examples for inquiry-based learning in mathematics and science education have been put forward by researchers, teacher educators and experienced teachers, changing the prevailing deductive teaching style is a highly challenging issue. It is a great challenge for all teachers to keeping a proper balance between instruction and autonomous construction in the teaching and learning of mathematics and science. In order to manage this, teachers' professional competences are of crucial importance.

References

Anderson, R. D. (2002). Reforming science teaching: What research says about inquiry. *Journal of Science Teacher Education, 13*(1), 1–12.

Barrow, L. H. (2006). A brief history of inquiry: From Dewey to standards. *Journal of Science Teacher Education, 17*, 265–278.

Clausen, M. (2002). *Unterrichtsqualität: Eine frage der perspektive? [Quality of instruction: A matter of perspective?]*. Münster: Waxmann.

Colburn, A. (2000). An inquiry primer. *Science Scope, 23*, 42–44.

Colburn A. (2006). *What teacher educators need to know about inquiry-based instruction*. Paper presented at the Annual meeting of the Association for the Education of Teachers in Science, Akron, OH.

COMPASS. (2009). http://www.compass-project.eu. Accessed 06 Dec 2012.

Dewey, J. (1910). Science as subject-matter and as a method. *Science, 31*, 121–127.

Euler, M. (2011). WP9: Report about the survey on inquiry-based learning and teaching in the European partner countries. *PRIMAS: Promoting inquiry-based learning in mathematics and science education across Europe*.

Hattie, J. A. C. (2009). *Visible learning: A synthesis of over 800 meta-analyses relating to achievement*. Abingdon: Routledge.

Hayes, M. T. (2002). Elementary preservice teachers' struggles to define inquiry-based science teaching. *Journal of Science Teacher Education, 13*(2), 147–165.

Healey, M. (2005). Linking research and teaching: Exploring disciplinary spaces and the role of inquiry-based learning. In R. Barnett (Ed.), *Reshaping the university: New relationships between research, scholarship and teaching* (pp. 67–78). McGraw Hill: Open University Press.

Hmelo-Silver, C. E. (2004). Problem-based learning: What and how do students learn? *Educational Psychology Review, 16*(3), 235–266.

Kirschner, P. A., Sweller, J., & Clark, R. E. (2006). Why minimal guidance during instruction does not work: An analysis of the failure of constructivist, discovery, problem-based, experiential, and inquiry-based teaching. *Educational Psychologist, 41*(2), 75–86.

Linn, M. C., Davis, E. A., & Bell, P. (2004). *Internet environments for science education*. Mahwah: Taylor & Francis.

Lüdtke, O., Trautwein, U., Kunter, M., & Baumert, J. (2006). Reliability and agreement of student ratings of the classroom environment: A reanalysis of TIMSS data. *Learning Environments Research, 9*(3), 215–230.

Mayer, R. E. (2004). Should there be a three-strikes rule against pure discovery learning. *American Psychologist, 59*(1), 14–19.

OECD. (2009a). *Technical report- PISA 2006*. Paris: OECD Publishing.

OECD (Ed.). (2009b). *Creating effective teaching and learning environments: First results from TALIS*. Paris: Organization for Economic Cooperation & Development.

PRIMAS. (2010). http://www.primas-project.eu/en/index.do. Accessed 29 May 2012

Prince, M., & Felder, R. (2007). The many faces of inductive teaching and learning. *Journal of College Science Teaching, 36*, 14–20.

Rocard M., Csermely P., Jorde D., Lenzen D., Walberg-Henriksson H., Hemmo V. (2007) *Rocard report: "Science education now: A new pedagogy for the future of Europe"*. EU 22845, European Commission.

Staver, J. R., & Bay, M. (1987). Analysis of the project synthesis goal cluster orientation and inquiry emphasis of elementary science textbooks. *Journal of Research in Science Teaching, 24*, 629–643.

Stodolsky, S. S., & Grossman, P. L. (1995). The impact of subject matter on curricular activity: An analysis of five academic subjects. *American Educational Research Journal, 32*(2), 227–249.

Wagenschein, M. (1962). *Die pädagogische dimension der physik [the pedagogical dimension of physics]*. Braunschweig: Westermann.

Walker, M. D. (2007). *Teaching inquiry-based science – A guide for middle and high school teachers*. LaVergne: Lightning Source.

Walker, A., & Leary, H. (2009). A problem based learning meta analysis: Differences across problem types, implementation types, disciplines and assessment levels. *The Interdisciplinary Journal of Problem-Based Learning, 3*(1), 12–43.

Chapter 15
Measuring Chemistry Teachers' Content Knowledge: Is It Correlated to Pedagogical Content Knowledge?

Oliver Tepner and Sabrina Dollny

1 Introduction

While in many studies teachers' content knowledge and its relation to other teacher characteristics are not based on a direct measurement, a large-scale test instrument for quantifying chemistry teachers' content knowledge has been devised. The test development has been part of the project ProwiN (Professionswissen in den Naturwissenschaften – Professional Knowledge in Science) that explores content knowledge (CK), pedagogical content knowledge (PCK), and pedagogical knowledge (PK) of science teachers (Borowski et al. 2010; Jüttner and Neuhaus 2012). The chemistry test items are presented in a multiple-choice single-select format that was based on a theoretical model which considers different types of knowledge, different topics, and curricular classifications (Tepner et al. 2012). Besides evaluation of content knowledge, a scale for describing teachers' pedagogical content knowledge was used and background information was collected. This procedure allows for revealing aspects influencing teachers' content knowledge and information about the correlation between content knowledge and pedagogical content knowledge. Furthermore, presented results serve as a basis for discussion on teachers' content knowledge needed in school.

O. Tepner (✉)
Chemistry Education, University of Regensburg, Regensburg, Germany
e-mail: oliver.tepner@chemie.uni-regensburg.de

S. Dollny
Vestisches Gymnasium Kirchhellen, Bottrop, Germany
e-mail: sabrina.dollny@uni-due.de

C. Bruguière et al. (eds.), *Topics and Trends in Current Science Education: 9th ESERA Conference Selected Contributions*, Contributions from Science Education Research 1,
DOI 10.1007/978-94-007-7281-6_15, © Springer Science+Business Media Dordrecht 2014

2 Theoretical Background

Teachers' professional knowledge is widely believed to be one of the most important factors contributing to high-quality instruction (Abell 2007; Blömeke et al. 2009; Bransford and Darling-Hammond 2005; Kunter et al. 2009; Peterson et al. 1989): "Teachers with broad and deep disciplinary knowledge, including subject-specific knowledge, awareness of common alternative conceptions, and multiple levels of scientific models, can provide rich learning opportunities for their students" (Khourey-Bowers and Fenk 2009). Recent literature conceptualizes professional knowledge in three dimensions, mainly based on Shulman's ideas for teacher education (Baumert et al. 2010; Shulman 1986, 1987): Pedagogical knowledge, content knowledge, and pedagogical content knowledge. These are seen as common dimensions that should be taken into account when considering competencies a good teacher should possess (Baumert et al. 2010; Grossman 1990). This study on subject-specific knowledge deals with CK and PCK, only.

3 Content Knowledge

On a subject-related level, CK is conceptualized as knowledge of subject-specific facts and concepts (Cochran and Jones 1998). Referring to Krauss et al., CK is described as "teacher's understanding of the structures of his or her domain" (Krauss et al. 2008) and it is formed as a result of curriculum-related work, that is, the designing and structuring of subject-specific contents for teaching (Baumert et al. 2010). The understanding of the material should be deep enough to sophisticatedly expound certain contents that are to be mastered by the students (Krauss et al. 2008). It embodies much more than just factual, everyday knowledge based on academic fundamentals (Shulman 1987). Referring to Cochran and Jones (1998), CK is distinguished from subject matter knowledge. While subject matter knowledge (SMK) could be seen as composed of CK (facts and concepts of the subject matter), substantive knowledge (explanatory structures of the field), syntactic knowledge (processes by which new knowledge is generated), and beliefs about the subject matter, CK itself only refers to cognitive aspects (Cochran and Jones 1998). According to Shulman (1987), CK is conceptualized – besides, e.g., PCK and PK – as a dimension of teachers' professional knowledge. Both teachers' professional knowledge and professional beliefs, motivational orientations, and self-regulation skills influence teachers' competence and behavior (Baumert and Kunter 2006).

There are different approaches for conceptualizing teachers' CK. Generally, CK could be assumed to be domain specific even within a certain subject, so high knowledge in one topic does not imply necessarily high knowledge in another topic. Hill et al. (2004) showed this for the distinct dimensions "number concepts/operations" and "patterns/functions/algebra" in mathematics at elementary level. However, there are topics that are generally relevant for understanding further topics. For example, "Mechanics" is seen as such a fundamental topic in physics (Riese and Reinhold 2009a). Additionally, CK has been tested at different (school) levels. While Rowan et al. (1997) referred to knowledge at high school level, only, Ball and colleagues distinguished between "knowledge of mathematics that any well-educated adult should have and mathematical knowledge that is "specialized" to the work of teaching and that only teachers need know" (Ball et al. 2005). However, empirical data showed no clear distinction between both levels (Hill et al. 2004). In order to evaluate preservice physics teachers' knowledge, Riese and Reinhold (2008) have tested knowledge at school level, deepened knowledge, and knowledge at university level (Riese and Reinhold 2010). TEDS-M (Teacher Education and Development Study: Learning to Teach Mathematics) categorized mathematics content items into "novice (indicating mathematics content that is typically taught at the grades the future teacher will teach); intermediate (indicating content that is typically taught one or two grades beyond the highest grade the future teacher will teach); or advanced (indicating content that is typically taught 3 or more years beyond the highest grade the future teacher will teach)" (Tatto et al. 2008). Within the project COACTIV (Professional Competence of Teachers, Cognitively Activating Instruction, and the Development of Students' Mathematical Literacy; Baumert et al. 2010), mathematics teachers' knowledge is conceptualized on four different levels: (a) everyday knowledge that adults have after they have left school, (b) mastery of content included at the taught level, (c) profound comprehension of school mathematics, and (d) mathematical knowledge at university level (Baumert et al. 2010). However, COACTIV test items only referred to level (c): "a profound mathematical understanding of the curricular content to be taught" (Baumert et al. 2010). Summing up all different approaches, there is consensus that adequate preparation and instruction itself require CK that goes beyond a certain level taught in school (Blömeke et al. 2009; National Mathematics Advisory 2008).

4 Correlation Between Content Knowledge and Pedagogical Content Knowledge

In order to provide conceptual understanding for students, both CK and PCK are needed: "If teachers do not have in-depth knowledge of a topic themselves, it is clearly difficult for them to provide conceptual depth for their students – hence, the importance of PCK and the role of SMK in its development" (Rollnick et al. 2008). PCK can be described as a fusion of PK and CK. While PK is conceptualized as knowledge about broad principles and strategies of classroom management, organization, and learning strategies (Baumert et al. 2010), pedagogical content knowledge is used to transform a specific topic into teachable content. According to Shulman (1987), the conceptualization of PCK as an amalgam of PK and CK is useful for distinguishing between a content specialist and a teacher. "PCK [...] is constituted by what a teacher knows, what a teacher does, and the reasons for the teacher's actions" (Baxter and Lederman 1999). Although several conceptualizations of PCK have already been developed, there is no generally accepted definition of PCK (Kind 2009). However, two facets can be seen as key elements of PCK: knowledge of representations and strategies and knowledge of students' pre- and misconceptions (Blömeke et al. 2008; Grossman 1990; Krauss et al. 2011; Marks 1990; Park and Oliver 2008; Smith and Neale 1989; Tamir 1988; van Driel et al. 1998). While students' preconceptions include both correct and incorrect scientific conceptions, misconception can be seen as "any concept that differs from the commonly accepted scientific understanding of the term" (Nakhleh 1992). Preconceptions are relevant for students' understanding, because learning new information is based on it (Duit and Treagust 2003). Knowledge of representations and strategies could be specified in chemistry as dealing with models and dealing with experiments (Hofstein and Lunetta 2004). Especially, selection, preparing, and reworking of experiments seem to be relevant for learning processes (Hofstein and Lunetta 2004; Wahser 2008). Knowledge of the adequate use of models in chemistry classes is another important aspect of teachers' PCK (Oh and Oh 2011). For instance, teachers should know and teach the limitations of models used for instruction (Khourey-Bowers and Fenk 2009; Oh and Oh 2011). However, there are hints that concerning this topic, in particular, teachers' knowledge is rather limited (van Driel and Verloop 1999).

When exploring teachers' CK, it can be useful to take aspects of PCK into account, because an explicit connection between PCK and CK in a teacher's knowledge base is assumed. Large-scale studies show specific correlations between CK and PCK regarding different school forms and education phases (Krauss et al. 2008; Riese et al. 2009b). In this study, school forms were differentiated into intensified ("academic") and non-intensified ("nonacademic") general education. Intensified general education leads to the acquisition of a university entrance qualification and is part of the secondary education phase in school, whereas non-intensified general education (first education phase) ends when students are around 16 years old (Baumert et al. 2010; Bonsen et al. 2008). Mathematics teachers working at schools educating both first and second education phases in intensified general education show a higher correlation between CK and PCK than teachers in non-intensified schools (Krauss et al. 2008). However, there are hints that excellent expertise in CK may hinder recognizing "possible difficulties in understanding something that is very clear to themselves" (de Jong and van Driel 2004). Nevertheless, a profound understanding of CK can be assumed to be a precondition for developing PCK in order to teach the subject matter in an understandable way (Loucks-Horsley and Matsumoto 1999). It seems to be a "necessary, but far from sufficient, precondition for providing insightful instruction" (Baumert et al. 2010).

5 Rationale

The study is based on recent data gathered from mathematics teachers showing that teachers teaching 11–18-year-olds hold more secure subject-specific CK and demonstrate better quality PCK than those teaching 11–16-year-olds alone (Krauss et al. 2008). We were prompted to investigate if such findings hold true for chemistry teachers. In Germany, the locus for our study, a "non-intense," non-academic general education ends at 16. There are two school types, in which only this lower secondary age group is taught: basic general education (Hauptschule) and extensive general education (Realschule). In other secondary schools, a more intense, academic, general education is offered that ends at age 18 (intensified general education, Gymnasium) and at age 19 (comprehensive school, Gesamtschule) with the award of an entrance qualification for university. In general, CK can be assumed to be a necessary precondition for developing PCK (Baumert et al. 2010; Bonsen et al. 2008).

Our study therefore investigates:

1. What differences in chemistry CK are observed in teachers teaching at intensified and non-intensified levels?
2. To what extent does the quality of chemistry PCK correlate with teachers' chemistry CK?

6 Methods

One hundred sixty-six chemistry teachers working at non-intensified level (basic general education, extensive general education) and intensified level (intensified general education and comprehensive school) were asked to complete questionnaires probing their chemistry CK and PCK. Both were constructed in a paper/pencil form and required approximately 45 min to complete.

In order to compare and correlate tested dimensions, both tests were based on the same content and on a similar theoretical model (Fig. 15.1) (Jüttner and Neuhaus 2012; Tepner et al. 2012). The questionnaires were based on content themes centered on "structure of atoms and the periodic table," "chemical bonding," and "chemical reactions using acids and bases." The reason for choosing these topics was their curricular relevance both for Bavaria and North Rhine-Westphalia and for intensified- and non-intensified-level education. Because these topics are part of the curriculum of all middle schools, every chemistry teacher could reasonably be expected to hold CK and PCK about them.

The underpinning theoretical structure is provided by three axes (Tepner et al. 2012). Two, themes (listed above) and knowledge areas, are common to both tests. The knowledge axis comprises declarative, procedural, and conditional knowledge (Paris et al. 1983). While declarative knowledge mainly refers to facts, procedural knowledge takes operations and processes into account. Conditional knowledge is knowledge about reasons and conditions under which some aspects are relevant for instruction. The third axis differs between the two questionnaires. In the CK questionnaire, the third axis was curricular classifications, which permitted grading of questions by level of difficulty suitable for students aged 12–15 (grades 7–9), aged 16–18 (grades 10–12), and aged 18–21 (first years of undergraduate study). In the PCK questionnaire, the third axis was called "facets" (Park and Oliver 2008) and

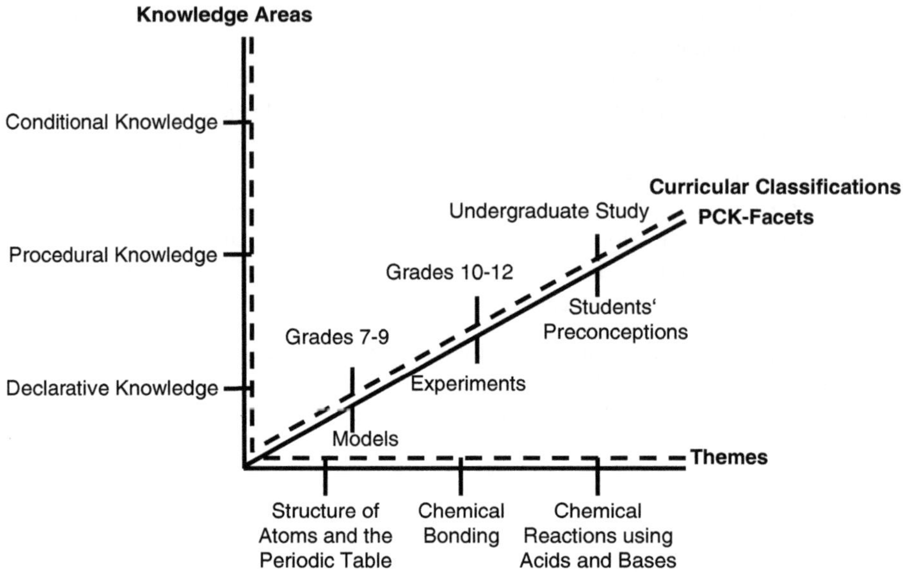

Fig. 15.1 Theoretical model adopted for item development in chemistry. CK is displayed as a *dashed line*, PCK as a *solid line*

comprised knowledge of models, experiments, and students' preconceptions, all of which are assumed to be relevant for effective instruction (van Driel et al. 1998).

In order to ensure content validity, both tests were developed in cooperation with teachers working in school for several years and lecturers in chemistry education at university. The CK test consists of 25 multiple-choice/single-select items with four answer alternatives per item (Walpuski et al. 2012). The PCK test includes 19 items comprising a description of a certain classroom situation and four alternatives how to react adequately. Each alternative has to be graded from 1 = "very good" to 6 = "unsatisfactory" (Walpuski et al. 2012). A reason for choosing this item format was to avoid featuring decisively right or wrong answer alternatives, as this is a PCK test. To analyze data, a further expert rating was required. Applying the same rating scale items, this expert rating was used as a reference to follow the teacher assessments. People involved in chemistry teacher education (e.g., professors and discipline leaders of chemistry education) served as experts. Referring to Thillmann (2008), results of the expert rating were analyzed using quasi-ranked pairs: each two answer alternatives were compared to each other. Instead of the exact numerical value, the relation between two alternatives (e.g., alternative "c" is better than "b") was relevant for scoring (Walpuski et al. 2012). So answer alternatives of each item were compared to each other in order to find relations that were rated identically by almost all experts ($\alpha = 0.91$). Only identically rated relations were considered for the analysis of teacher data.

After conducting a pilot study in 2010 involving 62 chemistry teachers, seven CK items and eight PCK relations have been removed, e.g., because item difficulty was not between 20 % and 80 %. In the end, 65 relations covering 19 items were used for die PCK test. While items had different scores, it was assured that only the aspects that agreed with the PCK experts were taken into account. The reliability of both tests is good (CK: $\alpha=0.84$, PCK: $\alpha=0.83$).

Because sample of pilot study ($n=62$) and sample of main study ($n=104$), conducted in 2010/2011, were comparable to each other, they have been combined for further analyses ($N=166$).

7 Results

The first research question deals with differences between teachers of intensified education and teachers of non-intensified education regarding their chemistry CK. Revealed CK differences are significant ($t(164)=-10.79$; $p<0.001$; $d=2.02$ (Fig. 15.2)).

Furthermore, CK has a substantial influence on developing PCK. While differences between teachers of intensified education and non-intensified education in terms of PCK are significant ($t(32.94)=-3.68$; $p=0.001$; $d=1.01$ (Fig. 15.3)), PCK differences decrease when CK is incorporated as a covariate.

If CK is taken into account, the ANOVA reveals a nonsignificant (n.s.) difference between intensified and non-intensified teachers' PCK ($F(1,163)=1.413$; $p=0.236$ (Fig. 15.4).

The second research question analyses correlations between CK and PCK. Analyzing the sample, a significant correlation of $r=0.36$ ($p<0.001$) between CK and PCK questionnaires can be found (Fig. 15.5): for example, teachers scoring highly on the CK questionnaire regarding a specific topic know much more about students' misconceptions in the same topic than those with poor-quality CK, who

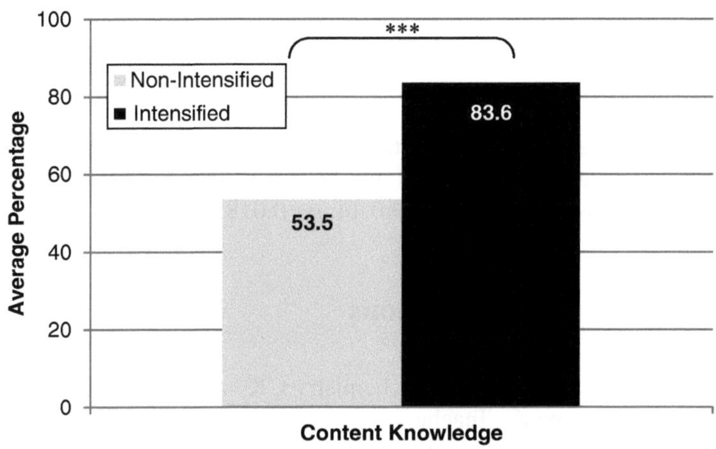

Fig. 15.2 Differences between teachers of different school levels regarding their CK

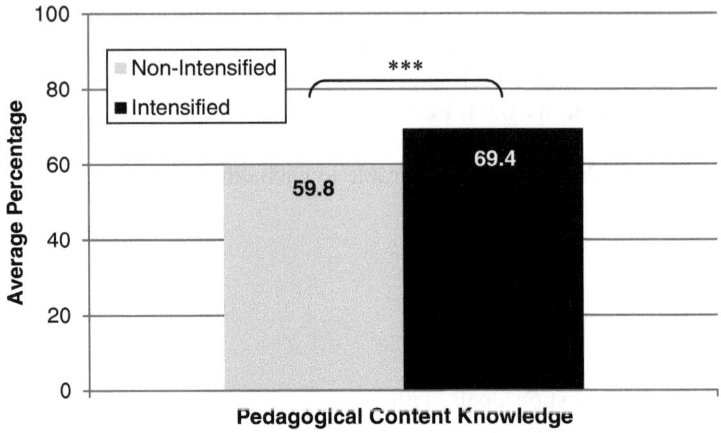

Fig. 15.3 Differences between teachers of different school levels regarding their PCK

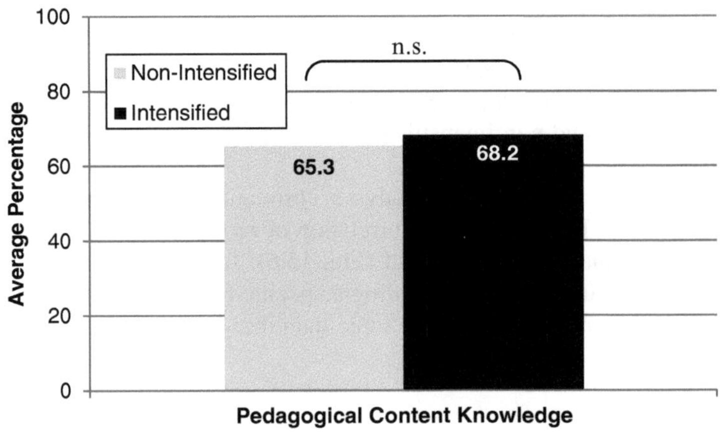

Fig. 15.4 Differences between teachers of different school levels regarding their PCK (covariate: CK)

scored badly in this aspect of the questionnaire. Nevertheless, the intensified education teachers' correlation between CK and PCK ($r=0.17$; $p=0.047$) is weaker than that of the non-intensified teachers ($r=0.44$; $p=0.018$).

8 Conclusions and Implications

Our evidence suggests that teachers' chemistry CK varies according to the type of school in which they work. Teachers not using all tested levels of CK in school everyday show limited chemistry CK. The development of CK is more related to school type than development of PCK. This might be a matter of learning opportunities in studies and a matter of requirements in daily work.

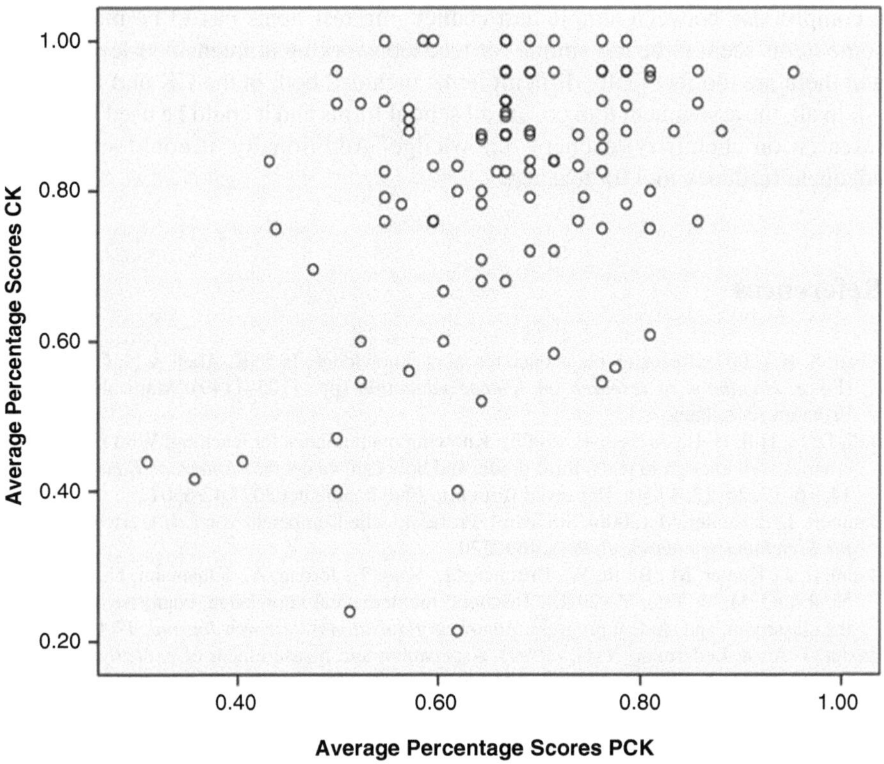

Fig. 15.5 Correlation between CK and PCK

Correlation between CK and PCK is sufficiently high to be interpreted as substantial, but low enough to see CK and PCK as different dimensions (Baumert et al. 2010). The moderate but significant correlation corroborates the increasingly accepted belief that possession of good CK is a precondition for developing high-quality PCK. Higher correlation between CK and PCK at non-intensified level might give a hint that a basic level of CK is a prerequisite for developing PCK, while a high CK does not affect a high PCK to the same extent. So CK might be a necessary but not a sufficient condition for developing PCK. Additionally, intensified education teachers seem to be specialists either in PCK or in CK.

As a consequence, focusing on school type, specific knowledge aspects should be reconsidered and CK immediately interlinked with PCK as it is needed in school. Discussion about this topic will contribute to recent international debate about the relative effectiveness of school- and university-based teacher education programs (Demir and Abell 2010; Department of Education 2010).

From a methodological point of view, this study reveals that developing and effectively using a large-scale test instrument for measuring chemistry teachers' CK and PCK is possible. Because CK and PCK tests should be adequate for measuring knowledge of teachers working both at intensified and non-intensified levels, a compromise between simple and challenging test items has to be made. Thus, some items seem to be too simple for teachers working at intensified-level schools and there are too few really difficult items included both in the CK and PCK tests. All in all, the assessment tool covers all school forms and it could be used for further research on chemistry teachers' knowledge. Additionally, it could serve as an adequate feedback tool for teachers.

References

Abell, S. K. (2007). Research on science teachers' knowledge. In S. K. Abell & N. G. Lederman (Eds.), *Handbook of research on science education* (pp. 1105–1149). Mahwah: Lawrence Erlbaum Associates.

Ball, D. L., Hill, H. H., & Bass, H. (2005). Knowing mathematics for teaching: Who knows mathematics well enough to teach third grade, and how can we decide? *American Educator, 29*(1), 14, 16–17, 20–22, 43–46. Retrieved from http://hdl.handle.net/2027.42/65072

Baumert, J., & Kunter, M. (2006). Stichwort: Professionelle Kompetenz von Lehrkräften. *Zeitschrift für Erziehungswissenschaft, 9*(4), 469–520.

Baumert, J., Kunter, M., Blum, W., Brunner, M., Voss, T., Jordan, A., Klusmann, U., Krauss, S., Neubrand, M., & Tsai, Y. (2010). Teachers' mathematical knowledge, cognitive activation in the classroom, and student progress. *American Educational Research Journal, 47*(1), 133–180.

Baxter, J. A., & Lederman, N. G. (1999). Assessment and measurement of pedagogical content knowledge. In J. Gess-Newsome & N. G. Lederman (Eds.), *Examining pedagogical content knowledge* (pp. 147–161). Dordrecht: Kluwer Academic.

Blömeke, S., Seeber, S., Lehmann, R., Kaiser, G., Schwarz, B., Felbrich, A., & Müller, C. (2008). Messung des fachbezogenen Wissens angehender Mathematiklehrkräfte. In S. Blömeke, G. Kaiser, & R. Lehmann (Eds.), *Professionelle Kompetenz angehender Lehrerinnen und Lehrer. Wissen, Überzeugungen und Lerngelegenheiten deutscher Mathematikstudierender und -referendare* (pp. 49–88). Münster: Waxmann.

Blömeke, S., Kaiser, G., Lehmann, R., König, J., Döhrmann, M., Buchholtz, C., & Hacke, S. (2009). TEDS-M: Messung von Lehrerkompetenzen im internationalen Vergleich. In O. Zlatkin-Troitschanskaia, K. Beck, D. Sembill, R. Nickolaus, & R. Mulder (Eds.), *Lehrprofessionalität. Bedingungen, Genese, Wirkungen und ihre Messung* (pp. 181–210). Weinheim: Beltz.

Bonsen, M., Bos, W., & Frey, K. A. (2008). Germany. In I. V. S. Mullis, M. O. Martin, J. F. Olson, D. R. Berger, D. Milne, & G. M. Stanco (Eds.), *TIMSS 2007 encyclopedia. A guide to mathematics and science education around the world. Volume I* (pp. 203–216). Chestnut Hill: TIMSS & PIRLS International Study Center Boston College.

Borowski, A., Neuhaus, B. J., Tepner, O., Wirth, J., Fischer, H. E., Leutner, D., Sandmann, A., & Sumfleth, E. (2010). Professionswissen von Lehrkräften in den Naturwissenschaften (ProwiN) – Kurzdarstellung des BMBF-Projekts. *Zeitschrift für Didaktik der Naturwissenschaften, 16*, 341–349.

Bransford, J. D., & Darling-Hammond, L. (2005). Introduction. In L. Darling-Hammond & J. D. Bransford (Eds.), *Preparing teachers for a changing world. What teachers should learn and be able to do* (pp. 1–39). San Francisco: Jossey-Bass.

Cochran, K. F., & Jones, L. L. (1998). The subject matter knowledge of preservice science teachers. In B. J. Fraser & K. G. Tobin (Eds.), *International handbook of science education. Part two* (pp. 707–718). London: Kluwer Academic.

de Jong, O., & van Driel, J. (2004). Exploring the development of student teachers' PCK of the multiple meanings of chemistry topics. *International Journal of Science and Mathematics Education, 2*(4), 477–491. doi:10.1007/s10763-004-4197-x.

Demir, A., & Abell, S. K. (2010). Views of inquiry: Mismatches between views of science education faculty and students of an alternative certification program. *Journal of Research in Science Teaching, 47*(6), 716–741.

Department of Education. (2010). *The importance of teaching. The schools white paper 2010.* London: The Stationery Office.

Duit, R., & Treagust, D. F. (2003). Conceptual change: A powerful framework for improving science teaching and learning. *International Journal of Science Education, 25*, 671–688.

Grossman, P. L. (1990). *The making of a teacher. Teacher knowledge and teacher education* (Professional development and practice series). New York: Teachers College Press.

Hill, H. C., Schilling, S. G., & Ball, D. L. (2004). Developing measures of teachers' mathematics knowledge for teaching. *Elementary School Journal, 105*(1), 11–30.

Hofstein, A., & Lunetta, V. N. (2004). The laboratory in science education: Foundations for the twenty-first century. *Science Education, 88*, 28–54.

Jüttner, M., & Neuhaus, B. J. (2012). Development of items for a pedagogical content knowledge test based on empirical analysis of pupils' errors. *International Journal of Science Education, 34*(7), 1125–1143.

Khourey-Bowers, C., & Fenk, C. (2009). Influence of constructivist professional development on chemistry content knowledge and scientific model development. *Journal of Science Teacher Education, 20*, 437–457.

Kind, V. (2009). Pedagogical content knowledge in science education: Perspectives and potential for progress. *Studies in Science Education, 45*(2), 169–204.

Krauss, S., Brunner, M., Kunter, M., Baumert, J., Blum, W., Neubrand, M., & Jordan, A. (2008). Pedagogical content knowledge and content knowledge of secondary mathematics teachers. *Journal of Educational Psychology, 100*(3), 716–725.

Krauss, S., Blum, W., Brunner, M., Neubrand, M., Baumert, J., Kunter, M., Besser, M., & Elsner, J. (2011). Konzeptualisierung und Testkonstruktion zum fachbezogenen Professionswissen von Mathematiklehrkräften. In M. Kunter, J. Baumert, W. Blum, U. Klusmann, S. Krauss, & M. Neubrand (Eds.), *Professionelle Kompetenz von Lehrkräften. Ergebnisse des Forschungsprogramms COACTIV* (pp. 135–161). Münster: Waxmann.

Kunter, M., Klusmann, U., & Baumert, J. (2009). Professionelle Kompetenz von Mathematiklehrkräften: Das COACTIV-Modell. In O. Zlatkin-Troitschanskaia, K. Beck, D. Sembill, R. Nickolaus, & R. Mulder (Eds.), *Lehrprofessionalität. Bedingungen, Genese, Wirkungen und ihreMessung* (pp. 153–165). Weinheim: Beltz.

Loucks-Horsley, S., & Matsumoto, C. (1999). Research on professional development for teachers of mathematics and science: The state of the scene. *School Science and Mathematics, 99*(5), 258–271.

Marks, R. (1990). Pedagogical content knowledge: From a mathematical case to a modified conception. *Journal of Teacher Education, 41*(3), 3–11.

Nakhleh, M. (1992). Why some students don't learn chemistry. Chemical misconceptions. *Journal of Chemical Education, 69*(3), 191–196.

National Mathematics Advisory Panel. (2008). Foundations for success: The final report of the National Mathematics Advisory Panel, Washington, DC.

Oh, P. S., & Oh, S. J. (2011). What teachers of science need to know about models: An overview. *International Journal of Science Education, 33*(8), 1109–1130. doi:10.1080/09500693.2010.502191.

Paris, S. G., Lipson, M. Y., & Wixson, K. K. (1983). Becoming a strategic reader. *Contemporary Educational Psychology, 8*, 293–316.

Park, S., & Oliver, S. J. (2008). Revisiting the conceptualisation of pedagogical content knowledge (PCK): PCK as a conceptual tool to understand teachers as professionals. *Research in Science Education, 38*(3), 261–284.

Peterson, P. L., Carpenter, T. P., & Fennema, E. (1989). Teachers' knowledge of students' knowledge in mathematics problem solving: Correlating and case analysis. *Journal of Educational Psychology, 81*(4), 558–569.

Riese, J., & Reinhold, P. (2008). Entwicklung und Validierung eines Instruments zur Messung professioneller Handlungskompetenz bei (angehenden) Physiklehrkräften. *Lehrerbildung auf dem Prüfstand, 1*(2), 625–640.

Riese, J., & Reinhold, P. (2009). Fachbezogene Kompetenzmessung und Kompetenzentwicklung bei Lehramtsstudierenden der Physik im Vergleich verschiedener Studiengänge. *Lehrerbildung auf demPrüfstand, 2*(1), 104–125.

Riese, J., & Reinhold, P. (2009b). Structure and development of physics student teachers' professional action competence. In: NARST (Ed.), *Grand challenges and great opportunities in science education*. Proceedings of the NARST 2009 Annual Meeting, Garden Grove.

Riese, J., & Reinhold, P. (2010). Empirische Erkenntnisse zur Struktur professioneller Handlungskompetenz von angehenden Physiklehrkräften. *Zeitschrift für Didaktik der Naturwissenschaften, 16*, 167–187.

Rollnick, M., Bennett, J., Rhemtula, M., Dharsey, N., & Ndlovu, T. (2008). The Place of subject matter knowledge in pedagogical content knowledge: A case study of South African teachers teaching the amount of substance and chemical equilibrium. *International Journal of Science Education, 30*(10), 1365–1387.

Rowan, B., Chiang, F., & Miller, R. J. (1997). Using research on employees' performance to study the effects of teachers on students' achievement. *Sociology of Education, 70*(4), 256–284.

Shulman, L. (1986). Those who understand: Knowledge growth in teaching. *Educational Researcher, 15*(2), 4–14.

Shulman, L. S. (1987). Knowledge and teaching of the new reform. *Harvard Educational Review, 57*, 1–22.

Smith, D. C., & Neale, D. C. (1989). The construction of subject matter knowledge in primary science teaching. *Teaching and Teacher Education, 5*(1), 1–20.

Tamir, P. (1988). Subject matter and related pedagogical knowledge in teacher education. *Teaching and Teacher Education, 4*(2), 99–110.

Tatto, M. T., Schwille, J., Senk, S., Ingvarson, L., Peck, R., & Rowley, G. (2008). *Teacher Education and Development Study in Mathematics (TEDS-M): Conceptual framework. Teacher Education and Development International Study Center*, College of Education, Michigan State University, East Lansing.

Tepner, O., Borowski, A., Dollny, S., Fischer, H. E., Jüttner, M., Kirschner, S., Leutner, D., Neuhaus, B. J., Sandmann, A., Sumfleth, E., Thillmann, H., & Wirth, J. (2012). Modell zur Entwicklung von Testitems zur Erfassung des Professionswissens von Lehrkräften in den Naturwissenschaften. *Zeitschrift für Didaktik der Naturwissenschaften, 18*, 7–28.

Thillmann, H. (2008). Selbstreguliertes lernen durch experimentieren: Von der Erfassung zur Förderung. Dissertation, Universität Duisburg-Essen.

van Driel, J., & Verloop, N. (1999). Teachers' knowledge of models an modeling in science. *International Journal of Science Education, 21*, 1141–1153.

van Driel, J., Verloop, N., & de Vos, W. (1998). Developing science teachers' pedagogical content knowledge. *Journal of Research in Science Teaching, 35*(6), 673–695.

Wahser, I. (2008). *Training von naturwissenschaftlichen Arbeitsweisen zur Unterstützung experimenteller Kleingruppenarbeit im Fach Chemie. Studien zum Physik- und Chemielernen, vol 72.* Berlin: Logos.

Walpuski, M., Tepner, O., Sumfleth, E., Dollny, S., Hostenbach, J., & Pollender, T. (2012). Multiple perspectives on students' scientific communication & reasoning in chemistry education: VISIONS 2011: Teaching. *ActaDidacticaNorge, 6*(1). Retrieved from http://adno.no/index.php/adno/article/view/206

Part IV
The Students: Multiple Perspectives

Chapter 16
Boys in Physics Lessons: Focus on Masculinity in an Analysis of Learning Opportunities

Josimeire M. Julio and Arnaldo M. Vaz

1 Introduction

In the last decades, government officials and educational researchers have stressed the importance of science education promoting the development of thinking skills, collaborative group work, investigative attitude and problem-solving, among other competencies (AAAS 1989; Bybee 2010; Davies 2004; Jenkins 2009; Millar and Osborne 1998). Once they have moved beyond the impartation of either information or processes, these new curricula are more difficult to implement. The refinement of the curricular aims demands a fairly elaborated pedagogy and sophisticated conceptualization of learning. The promotion of the above competencies requires teachers to change the way in which they interact with students. We dwell in an academic context in which scholars already take on board social interactions and other experiences lived by the students in the classroom (Barron 2003; Hofstein et al. 2005; Kanari and Millar 2004). Since 2004, we have conducted classroom observation combining longitudinal and in-depth techniques. Our studies take place in an authentic school located within a university. The experienced physics staff of that school has established partnerships with practising school teachers in other schools. The partnership has led to the formation of a research collective – the Inovar Group – whose purpose is to conduct scholarly investigation in science education which allows the design and implementation of research-based innovation in science teaching.

J.M. Julio (✉)
Departamento de Metodologia de Ensino, UFSCar – Universidade Federal de São Carlos, São Carlos, Brazil
e-mail: josimeire@ufscar.br

A.M. Vaz
Colégio Técnico, UFMG – Universidade Federal de Minas Gerais, Belo Horizonte, Brazil
e-mail: arnaldovaz@ufmg.br

C. Bruguière et al. (eds.), *Topics and Trends in Current Science Education: 9th ESERA Conference Selected Contributions*, Contributions from Science Education Research 1,
DOI 10.1007/978-94-007-7281-6_16, © Springer Science+Business Media Dordrecht 2014

In the educational context fostered by these teachers, students do develop physics thinking and competencies akin to those valued by government officials and researchers. We have established an eclectic programme of scholarly research to identify and characterize which factors jeopardize the attainment of more sophisticated curriculum targets. Interesting findings came from analyses of students' interactions within small groups devoted to open-ended investigations and certain problem-solving activities. These special settings allowed us to identify factors likely to impact the development of scientific thinking and modes of thought. Even when students are effectively engaged in special activities, their actions are constrained by their attitudes towards collaboration, their resistance patterns and power relations. Patterns of social interaction are not always explicit or conscious, since they result from cultural factors.

The naïve views that students usually hold about discovery in science are one of the starting points of the physics course in the school featuring in our study. The teachers there challenge students to solve problems and conduct investigations that involve engagement in collaborative group work. Hence, students realize the amount of groundwork and collective input that is involved in scientific research. To communicate findings from analysis of this physics programme, we have drawn on data from the very first activity of the course, as it is particularly helpful in characterizing several of the factors intervening in students' learning. The reference above to students' naïve ideas about discovery illustrates what systematic analysis of the initial activity can reveal: that such activity gives the students the chance to create opportunities for learning beyond concepts and formulae. Our analysis also showed what could be seen in many pedagogical strategies used in physics: a classroom activity should sometimes force students to become conscious of prejudices or habits of thought that impair their overall socio-affective-intellectual development.

Gender configurations are an important example of one such intervening factor. These configurations are unconscious processes that are perpetuated in social institutions and social practices (Connell 1995). Masculinity plays a central part in gender relations within our society and exerts a strong influence on power relations, resistance patterns and collaboration within learning groups (Julio and Vaz 2010). The present study focuses on the mutual influence between configurations of masculinity and elaborations of learning opportunities.

2 Underpinning Concepts

The development of scientific thinking is a curricular aim that can be achieved if learning situations involve students in collective action and collective thinking. In the study presented here, we investigate the influence that certain displays of masculinity had on students' initiatives and classroom interactions of this collective nature. We also focus on its reverse: the extent to which the opportunities that students create in a collaborative activity alter configurations of masculinity. We gave priority to aspects of these interactions that are related to the processes of construction and internalization of masculinities and that exert influence both in terms of the classroom dynamics and in the constitution of the social identities that are built up within the teaching/learning context of school physics. We did this by establishing a connection between those elements appealing to Connell's *social theory of gender* and to the concept of *learning opportunity* from interactional ethnography. The next two subsections refer to the original academic context of these key concepts and provide definitions operational to the present study. After, we interweave the two concepts in a separate section.

2.1 Masculinity

In the investigation reported here, we used *masculinity* as an analytical tool. Connell (1995) defines this concept by contrast and comparison. However, masculinity should not be associated simply with men. In Connell's own words, masculinity is essentially relational. It is 'simultaneously a place in gender relations, the practices through which men and women engage that place in gender and the effects of these practices in bodily experience, personality and culture' (Connell 1995).

Schippers (2007) summarizes this definition in three components. First, there is the social status an individual acquires due to their practices, notwithstanding their gender. Second, there are practices and characteristics understood to be 'masculine'. These vary according to the historical and cultural context, possibly relating to physical strength, virility and rationality, among others. Third, there is a generalization of the cultural and social effects of these practices when they are particularly embodied by men but also by women. There are, therefore, different configurations of masculinity, and these depend on the way in which personal experiences are embedded in a system of gender relations. Furthermore, configurations of masculinity are necessarily ephemeral and unstable, multiple, mutant and defined in different ways in different human groups.

Haenfler (2004) argues that, in western society, the hegemonic masculinity legitimizes and generally values competition, hierarchy, individualism, sexual prowess, body strength, rationality, emotional detachment, dominance and the courage to take risks. Analysing interactions within male groups, Connell was able to identify an alternation in the patterns of resistance, power relations and collaboration. Interestingly, although they alternate, Connell observed only four patterns of masculinity. He dubbed them: hegemony, subordination, complicity and marginalization (Connell 1995).

Hegemony refers to a cultural dynamics in which one group claims and sustains a leading position in social life. *Hegemonic masculinity* 'guarantees (or is taken to guarantee) the dominant position of men and the subordination of women'. *Subordinate masculinity* relates to the lower and stigmatized position of women and male groups outside the circle legitimized by hegemonic masculinity. Groups that have symbolic similarity with the feminine are also associated with lower positions, being addressed by an abusive vocabulary – nerd, mama's boy, coward, effeminate, among others.

Complicit masculinity guarantees men in general an opportunity to enjoy privileges gradually acquired by patriarchy. Complicity also enables connection with the current hegemonic project, even if hegemonic rules are not strictly followed. Hegemony, subordination and complicity are relations that are internal to the gender order. However, gender interplays with other social structures, such as class and race. That creates further relationships between masculinities (Connell 1995). *Marginalization* is the term Connell uses 'to refer to relations between the masculinities of dominant and subordinated classes or ethnic groups' (Connell 1995).

According to Connell, masculinities and femininities are not only ideas in the head, or personal identities. They are also extended in the world, merged in organized social relations, institutions, professions, the labour market, etc. From this point of view, social institutions are subject to 'genderfication' in case one gender is massively represented there. That is the situation of areas of science and technology that bear a gendered cultural character that 'masculinizes' them (Connell 1995).

2.2 Learning Opportunity

A concept like Connell's masculinity prompts the reader to think about self and not self. Besides raising issues of identity and alterity, that concept stimulates reflection about interactions. We are interested in interactions that take place in the classroom. More precisely, we want to know which classroom interactions give students opportunity to appreciate the value of scientific reasoning and learn how scientists' thoughts and intellectual skills can be useful. We adopted the idea of *learning opportunity* to refer to the initiatives of individual students to redefine the task or re-elaborate scientific thinking as a result of their interaction with peers, the teacher and further elements present in the particular classroom context.

Interactional ethnographers coined the expression *learning opportunity* to refer to situations in which students construe and reify meaning through verbal and non-verbal social discourse in classrooms (Bloome and Bailey 1992; Rex et al. 2006). We define learning opportunity as each and all initiatives in which one or more students undergo change, possibly in an array of domains. This array might include meaning-making, interpersonal relationships with colleagues, rational perception or affective connection with the object of knowledge or with the teacher. It also does not matter whether the change was planned or not. What is important is students' agency, the fact that the student assumed responsibility and took initiative, not necessarily consciously.

The same activity can lead different students to experience different learning opportunities. *Learning opportunity* represents a different experience for each individual. That is because each person has their own social identity (namely, class, gender, and ethnicity), learning potential and relationship with school. From the point of view of interactional ethnography, knowledge, practices and actions are located outside people's thinking. Each of these phenomena manifests itself through the interaction that takes place among these individuals when they share the same social environment (Jordan and Henderson 1995). We also assume that students experience the development of scientific thought and thinking while they create learning opportunities. As a last assumption, when teaching happens to stimulate students to experience learning opportunities, it offers them the chance to grasp the general structure of the discipline that is being taught (Bruner 1960).

Having defined masculinity and learning opportunity, now comes the point where we interweave these ideas. The next section brings an important argument for doing that. At least in physics lessons, the dynamics of creating learning opportunities apparently does an important job in respect to masculinity. Although configurations of masculinity are deeply ingrained in the cultural milieu students are immersed, a learning opportunity might at times shake unconscious cultural tenets they hold. Once certain cultural cornerstones are shaken, students are freed from biases and habits of thought that had been impairing their development as human beings, let alone as physics students.

2.3 Doubly Favourable Condition

At the outset, we mentioned the competencies that students would be expected to develop. We will now argue that physics lessons might represent a doubly favourable condition to investigate students' development in that direction. The investigation made into this development is based on the concept of *masculinity* and on the concept of *learning opportunity*. The previous sections provide operational definitions for them in reference to their theoretical frameworks. Here is where we state our perspective about the interaction between masculinity and students' learning opportunity, thus paving the way to describe the methodological design of this study.

Masculinity emerges in situations involving interactions; the same applies to learning opportunities. Therefore, these are relational concepts. We are committed to studying the interaction between one and the other. In the classroom, the relation between these concepts might give us clues about the obstacles to students' development in terms of the competencies mentioned earlier. This assumption results from our systematic observation of physics lessons (Julio and Vaz 2010; Vaz and Julio 2005). We consider such lessons to be a singular context. Physics lessons can be an opportunity for students to go through personal transformation, besides learning about physics beyond concepts and formulae. Those lessons can also be an opportunity to reduce the obstacles to students' full engagement on vital experiences and to achieve an overall cognitive, personal and collective development.

Authors have drawn attention to the cultural patterns that become ingrained in science (Easlea 1983). Somewhere in the history of science, the rationality, objectivity, determinism, mathematical rigour, analytical capability and harnessing of nature observed in physics became prominent due to the masculine domination within this science. As a consequence, intuition, collaboration, imagination and creativity were subliminally lowered in importance among scientists and the general public. From the moment when students enter the classroom to have a physics lesson (no matter whether they are male or female students), their configurations of masculinity will impact their learning opportunities. Depending on the values at play, this impact will be negative. On the other hand, if that physics lesson gives emphasis to the second set of traits of science, collaboration included, it creates learning opportunities that will inevitably influence those students' configurations of masculinity.

The relationship between masculinities that is established in the classroom results first from conviviality among individuals, each one carrying naturalized models of physics and of school. Those individuals also have a self-image regarding their personal characteristics and learning potential. During their interactions, students negotiate courses of action, make meaning about topics of the syllabus and form a social identity. In the course of such events, therefore, the interaction of intervening masculinities leads to specific practices. The masculinities at play have mutual influence on each other, according to the situation.

One usually expects classroom activities to change students' conceptual understanding and knowledge repertoire. However, it is not always conceived that a situation can play a significant role in changing students' personal traits and values. Once it is seen that configurations of masculinity can have a negative impact on students' learning, it is desirable to investigate the extent to which classroom activities interferes with students' configurations of masculinity. One way to do this is to observe how students create learning opportunities in the course of a series of activities. Once students hold different configurations of masculinity, they certainly create different learning opportunities. Besides, the same student might create different learning opportunities in different activities, due to the differences between those activities in regard to what they represent in terms of gender roles in that context.

3 Methods

The university committee on ethics in research involving human beings approved the broad plan laid out here and in the next section for data collection, handling procedures and analysis. The audio and video recordings made in the classroom include male and female students. The recordings captured only those who voluntarily agreed and whose parents approved the use and diffusion of these data for research. Not all observation was aided by recording equipment. Over the course of 1 year, one of us – Ms. Josimeire Julio – did systematic observation of all the physics lessons in two Grade 10 classrooms. The longitudinal observations were essential for the conduct of ethnographic in-depth microanalysis of the audio and video material. Mr. Arnaldo Vaz regularly teaches in the school and was in charge of the two classrooms.

As in descriptive research, the familiarity gained during the year helped to highlight the dominant characteristics of individual and group work during the video- and audio-recorded lessons. The descriptions allowed further analyses of episodes. We focused on (a) the relation between students within a group, (b) the relation between the group and the activity, (c) the relation that the group established with physics and (d) the relation between the group and other context elements, such as the teacher and other groups. In the present work, we tried to identify (i) the relations that boys commonly established in learning situations and (ii) the factors that interfered in their actions – either to facilitate or to inhibit those actions.

The identification of the latter aspects was possible because of the dovetailing of masculinity and learning opportunity concepts. For the microanalysis, we chose a group in view of their characteristics and the generalizability of its analysis. Each of the three boys in the group acted according to one of the different masculinity patterns conceived by Connell (1995).

3.1 Diegesis

Some information about the educational context is necessary to appreciate the statements that we make about the relations established by the students and our interpretation of them. Herein, we describe the school and classroom scenario, the individual characteristics of the students at play and the particular activity in which the actions that we chose to describe take place. We have restricted the report of our fieldwork to this activity because of the representativeness of the interactions observed there, in terms of their quality and diversity.

We collected data on entrance classes of a technical high school located within a public university. The classes are designated as 'Grade 10' so as to facilitate international comparisons. The students had three physics lessons a week (100 min on 1 day, 50 min on another). Half the class had another 100 min of physics laboratory once every 2 weeks. A 3-year-long recursive physics curriculum was in its third year of adoption. Rather than being propaedeutic, the curriculum aimed at the development of scientific thinking and modes of thought. Class and laboratory involve group work. Three to four students are put together according to various criteria, for instance, no more than one repeater and no more than one former student from the middle school located within the university.

The introductory activity for the physics course[1] is six lessons long. It is only after a sequence of semi-structured investigations, measurements, data plotting and analyses of graphs that students will learn that the activity is about stellar evolution. At its outset, however, the teacher says nothing about stars. With students sat in groups of three to four, the teacher simply gives them the first task: to analyse a set of 18 photographs and decide whether or not there is a phenomenon worth

[1] A former upper secondary public school teacher and physics lecturer from Universidade de São Paulo – Dr. Norberto C. Ferreira – designed that role-playing activity in the 1970s in the light of the concerns that he had with activities that referred to scientific procedures in the 1960s science teaching projects. Early in 2013, there is as yet no published material about the activity. The intention is to revert that in the near future.

investigating. Students are requested to act like in a game: to pretend their group is a team of scientists working with data not yet analysed. As the other teams work in other countries, conversation with the colleagues sat in the other groups is supposed to be nonexistent. The 'plenary sessions' are an exception. As the narrative below illustrates, plenary sessions are pivotal in adjusting the direction of the lesson, restating the task, establishing guiding principles for argumentation, negotiating rules for group work, etc.

The first task begins a series of semi-structured investigations for students to engage. It forces them to consider what is relevant to observe. Although it does not require recollection of information, it demands the ability to establish adequate references. Spontaneously, students put events in chronological order. Their attention to spatial reference, in contrast, is not automatic. Usually students detect a displacement in the photographs that in fact is just apparent. To see it that way, an adequate frame of reference is required. At that point in the lesson, sometimes a polarization emerges in the class. For example, some students choose the borders of photographs as spatial reference and disagree with those who argue that the sensible thing to do would be to adopt something represented in the pictures as point of reference. Every time that kind of tension emerges, the teacher ought to intervene for teams should work in near synchronicity on each new semi-structured investigation. Many plenary sessions serve a reflective purpose that is key in the dynamics of shaking students' unconscious cultural tenets that help them to create learning opportunities.

From such an intricate teaching unit, we drew events where that dynamics most clearly challenge students in terms of two distinct sets of values and beliefs: one about knowledge, the other about learning; to be more specific, scientific knowledge, on the one hand, and collective and collaborative learning, on the other. These subtle tenets of theirs could only be noticed because a systematic observation was conducted along the year. But it was the video and audio data that allowed us to pin them down. Although mandatory, the personal and in-depth portrait of the students could not possibly give readers elements to perceive the role such tenets play in the intertwinement of configurations of masculinity and learning opportunities. Even a typical ethnographical description could not give readers such elements. The mutual influence between configurations of masculinity and elaborations of learning opportunities can only emerge when an analysis such as the one we report herein is conducted.

Three male students form the group where we identified the best story line to produce here in order to communicate our findings. Here come their profiles.

Nicolau is a repeater who experiences difficulties and resistances towards school. He has low performance in physics and low involvement in classroom tasks. He acknowledges physics as a highly prestigious discipline, lying beyond his own range of possibilities. He shows characteristics of protest masculinity, a subtype of subordination.

Roger is eager to express to colleagues that he is capable of mastery and finds it easy to assimilate physics knowledge and concepts. That results from physics being displayed as a highly prestigious intellectual activity characteristic of people of exceptional ability. He tries to demonstrate autonomy in physics lessons but diverts colleagues' attention in class. He strives to keep the status of 'brilliant student' among his peers and some teachers. He shows characteristics of western hegemonic masculinity.

Tales shows affinity with the intellectual elaboration of ideas as well as their categorization and refinement. He is disciplined, focused and engaged in school tasks. He is sympathetic towards physics as a rational activity guided by systematic procedures. He shows the characteristics of men of reason, a particular kind of complicity masculinity.

The dynamics of this group illustrates the relational nature of both configurations of masculinity and learning opportunities. Our analysis of their interplay follows.

3.2 Description of Interactions in the Lessons

The actions took place in a basic classroom with desks arranged in clusters. Some students opted not to be videoed. During the lesson, classroom discussion and group work predominated. The teacher usually opened lessons with a 10-min talk and made summaries at other times.

4 First Lesson

Mr. Vaz, the teacher, announces that after handing an envelope to each group, the first challenge will be proposed. In the envelope there is a sequence of slides, deliberately out of order. The task is to find out whether or not any phenomenon worth an investigation is shown in the pictures. Initially puzzled, all the students eventually demonstrate curiosity. The teacher avoids giving any information to a single group. Occasionally, he encourages groups to exchange views among themselves.

Tales, Roger and Nicolau are more dispersive than others; nevertheless they commit to the activity. Nicolau suggests that they begin working; they notice the dates on the pictures' labels and organize them into chronological order. Meanwhile, Roger becomes restless: he shakes his legs, slaps the table and tries to distract Tales and Nicolau and calls colleagues from other groups. Tales and Nicolau compare the pictures. Nicolau investigates whether a star appears in one of the pictures. Tales shows him that the stars change place and helps him to locate each one of them. Nicolau asks Roger to take part in the activity. Roger suggests that they should stop doing the activity to 'have a nap'. Meanwhile, Roger grabs one of the pictures that the two colleagues were analysing. Nicolau grows impatient with the colleague's attitude, tries to deter him, asks for the photo to be returned and rebukes him. Tales and Nicolau go on comparing the pictures and trying to locate some stars that disappear in one picture and appear in another. Roger interrupts them and says ironically that the stars have not disappeared, they have just got out of the picture's visible area due to the Earth's movement. He frequently interrupts the two colleagues and criticizes the way that they communicate their observations. Nicolau reacts to Roger's irony and derisive remarks are made. Tales notes that the stars turn in a single direction. Nicolau looks at the pictures, trying to evaluate that assertion. Both try to ignore Roger's interventions, eager to discover one phenomenon that is worth investigating.

After a few moments, Roger leaves the jests aside, says that it is all about a 'description of the rotation in relation to the constellations'. Tales and Nicolau try to understand how that movement occurs. Roger affirms that the movement observed is a consequence of the Earth's movement and argues that this would be the reason for some stars no longer appearing in the pictures. Tales tries to investigate the direction of the movement of the stars from this observation, with the help of Nicolau. Roger begins kidding around again: he slaps the table and jokes with a colleague in another group. Nicolau reacts again, making derisive remarks. Tales suggests that they should measure the distances between the stars to check if they move among themselves. Nicolau agrees. Roger suggests that they sleep. Tales asserts that the stars rotate around a point and do not move between each other. Roger disagrees, corrects the colleague and says that it is in fact the point that is rotating around the stars. He demands that his colleagues take note of this answer. Tales does not agree that this is a phenomenon. Nicolau suggests Roger should write down his idea in the notebook. Roger refuses and asks them to call the teacher to communicate the discovery. He calls the teacher himself while announcing that he already has a notion of what it is. Tales disagrees. He does not consider that this is something worth investigating and suggests that Roger wait a little longer. Roger continues to call the teacher. Tales protests. Before Roger can be heard, the teacher calls the groups for collective discussion, which he calls a 'plenary session'.

The group work was interrupted after 15 min. At the beginning of the plenary session, this is compared to a scientific conference. The teacher asks that students do not say straightaway what phenomenon they believe to be worth investigating. He wants groups first to report 'what they did in their search for the phenomenon'. Rules on taking turns to speak are established, used, evaluated and re-established. Finally, he suggests that the groups use the information shared in plenary session when working on the second challenge.

Still in plenary session, the teacher moves on to the second part of the question 'what phenomenon is worth studying?' The teacher organizes the discussion. On the one hand, he asks statements to be classified either as attempts to explain or as observations. The teacher asks groups to get back to work, with the guidance that this should take into account a spatial reference, a temporal reference and the fact that the stars do not move between each other.

The group comprising Tales, Roger and Nicolau did not speak in the plenary session at any time. They listened to the discussion and hardly made any comment to each other. In the final 5 min of the lesson, Tales and Nicolau return to see the photos, but this time they are more distracted. Roger comments on other lessons that they will have later and Nicolau gets distracted paying attention to Roger. Tales looks at the photos for a few minutes but then also gets distracted.

The teacher ends the lesson by announcing that the investigation will continue in the next lesson.

5 Second and Third Lessons

In the following two lessons, before returning the slides to the students, the teacher summarizes what was discussed in the previous lesson's plenary session.

A student reveals that she has discovered that a star is changing size. The teacher tries to put all the groups in context by asking her to reveal how she identified the star that changes. He then summarizes her guidelines in order that the groups can identify other stars that change size. A new competition begins. The groups must now identify all the stars that change size.

Within moments, a student calls the teacher and shows another variable star. Tales has finished sorting out the photographs and begins to examine them against the light. Roger takes two and examines them in the same way. Nicolau is a little distracted and just watches the colleague. About 4 min later, Roger throws the photographs onto the table and says that he has the explanation for the phenomenon; Nicolau pays attention to his colleague. For Roger, the phenomenon is due to the translational movement, as the stars get smaller. Nicolau agrees and Roger formulates his theory, saying that the evidence is that big stars almost do not shrink and that small stars decrease only slightly in size. Nicolau asks his colleague to tell the teacher, but he says he will not talk.

Roger calls colleagues from other groups and advises them to give up because his group has discovered the answer. Nicolau insists that Roger should speak to the teacher and hears one more refusal. Nicolau calls the teacher and says that they have discovered the answer. Roger says that all the 'little' stars change size. Before he can continue, the teacher asks: Which [stars]?

Nicolau says that this task is more difficult. The teacher in turn determines that they should first make clear which stars change size and leaves. Roger complains:

Roger: 'Aaah!' (immediately dropping on the table the picture that he was holding)

Nicolau suggests that they take a pencil and mark the stars, in another group. While Tales is still studying the photographs, Roger and Nicolau become distracted. (...)

When all the students have identified the stars that vary, the teacher speaks about positioning systems and instructs the students to use the quadrille ruling grid to find the stars through ordered pairs (x; y). The teacher signals the start of a new race: the groups are to write on the blackboard the coordinates of the stars that they have discovered and name them with the group number.

Once again, the groups work very similarly. Even students who initially refused to participate go to the blackboard to see the possibility of recording the coordinates of a star that has not been discovered by other groups. Roger is one of the newly engaged.

At the end, the teacher and students in plenary session analyse the coordinates on the board. He discusses mistakes and rules about ordered pairs. However, the new challenge is to answer the question 'how do the stars change size?' Discussions begin in the groups. Soon after, the teacher begins distributing scales to measure the 'size' of the stars. The groups must now fill in a table, recording the size of each star for each week. The lesson ends; the task is to be completed the following week.

6 Fourth Lesson

The teacher begins the fourth lesson with a recap of what was discussed in the previous lessons. He explains that they are studying the changing size of the stars and will understand how the stars change through a description of the changes.

Nicolau tells Tales and Roger to hurry up and take measurements and complete the table so that their work can be graded. Roger takes note of the stars' coordinates. His colleagues begin to complete the table with the measurements. While Nicolau and Tales fill in the table, Roger becomes distracted again. Nicolau insists that he should help in the work and do something. He does not answer. After a while, Roger gets up and walks around, visiting the other groups. He engages with his peers. When he comes back and sits down, he tells the partners that they are ahead of the other groups [i.e. *their table with data about the stars is more complete than those of others*]. *That thrills Nicolau and Tales. Colleagues from other groups begin to query them about the stars' coordinates. They are the second group to finish the table and begin calling the teacher repeatedly. When the teacher goes to the group, they boast that their colleagues have asked for help.*

The teacher draws the lesson to a close while some groups are still completing the table. He draws attention to the fact that in two groups, two people did not participate in the tasks, which is something that undermines the functioning of the group. He says that in such cases, one member of the group took on all the tasks, which also affects the functioning of the group. He warns that when a group does not complete the task, that might well be a malfunction of the group. For the next lesson, the next day, students should bring graphs tracing how the stars change in size. The groups that did not complete the tables should copy the measurements obtained by another group. It should be noted that at the end of the learning sequence, in the fifth and sixth lessons, only the group of Nicolau, Tales and Roger had not drawn the graphs and contributed to the discussion in the classroom.

7 Analysis

The hegemonic masculinity appeared around the relationships that occurred within the group analysed from the first lesson. Roger was the pivot of the power relations that were established within the group; there was a tension between his conduct and the engagement that Tales and Nicolau had in the activity. While Roger wanted to take command of the group and subordinate colleagues to his will, Tales and Nicolau wanted to do research and discover the phenomenon before the other groups. The research and discovery of the phenomenon consisted of learning opportunities that the two colleagues wanted to experience.

Roger proved to be bothered by the immediate involvement of Tales and Nicolau in the investigation of the phenomenon and tried to sabotage the activity, disqualifying it as homework and diverting the attention of colleagues. Roger defied the rules of the classroom and tried to impose his control and authority over his peers. Nicolau found it difficult to concentrate on the investigation in the face of provocations from Roger. He responded with some aggression to Roger's appeals (in some way he responded when provoked and was controlled by Roger), but at the same time he tried to convince him to engage in research.

It is apparent that Roger was trying to sabotage the activity and that at the same time he was seeking to show his colleagues that he could easily win the challenge proposed by the teacher. When he formulated explanations for the phenomenon, he did not take into account what their colleagues said. He refused to share learning opportunities with them. He decided that his explanation was sufficient to identify the phenomenon without paying attention to Tales's observations. He refused to take notes and wanted to communicate his discovery directly to the teacher.

The colours of the hegemonic masculinity pattern were defined around the relationship that Roger established with Nicolau and Tales. It denoted hierarchical position, individualism and dominating rationality. They did not dispute the control of the group. On the one hand, Roger imposed his command, challenging the ideas of colleagues without giving them the opportunity to argue the case with him. He examined the photographs and presented his interpretation without discussing with them and did not allow them to move on. On the other hand, Nicolau and Tales gave him the command and did not contest his position.

In our interpretation, the authority was granted to Roger on account of his intellectual prestige, as evidenced by the dominance that he showed in relation to scientific knowledge. It was a rational strategy of the group to stay ahead of the other groups. This is evidenced by the fact that the explanation elaborated by Roger in analysing the phenomenon was accepted with enthusiasm by Nicolau, since they could gain an advantage over the other groups. This reinforced Roger's arrogant and individualistic attitude, whereby he chose to disregard the comments made by Nicolau and Tales, inhibiting them with explanations that took into account only his own point of view and assumed prior knowledge.

Roger's relationship with his colleagues showed two patterns of masculinity. The first was the hegemonic masculinity of rationality and knowledge that led him to assist the group only in matters that required 'intellectual work', while refusing to take measurements or notes. However, he did not respect the opinions of his colleagues and their explicit desire to engage in the activity sharing learning opportunities, as other groups were doing. That compromised their performance in the investigation and left the group at a disadvantage compared to others.

The second model of masculinity is close to what Connell (1995) calls 'protest masculinity'. In this group, it was manifestly linked to the 'complicity' between Roger and Nicolau. Protest masculinity is one way to gain prestige among peers, outline differences and have pleasure in confronting rules and challenging authority. Nicolau had no other means to achieve academic success in that school. That made him likely to resort to protest masculinity as a means of obtaining prestige among colleagues. At various times, he did want to engage in the activity. But the influence of Roger demoted him. There was complicity between the two about the jokes and jests in class. It is noticeable that Roger had a tendency to challenge the rules in the classroom and deter his group and other groups from focusing on the task.

Even when he appeared to be curious or intrigued by the phenomenon, Roger tried to give the impression that the tasks were not a challenge to him.

Tales sought alternatives and committed to the task, so that they remained at the level of the other groups. He tried to take advantage of Roger's rationality and of Nicolau's aid to perform the tasks, avoiding confrontation with them. That was his strategy to stay on task, in the running for prizes in the competitions, while creating new learning opportunities.

The plenary sessions enabled the socialization of all the groups' research strategies and findings. Those sessions placed the groups on the same mark, before they started the next challenge, which was always within everyone's reach. The groups gradually got themselves organized in a more collaborative mode, sharing tasks and confronting evidence following the teacher's interventions. Both when he intervened at group level and in plenary session, the teacher acted to manage the tensions that resulted from the participation of some boys. He inhibited actions and relations of dominance in favour of interactions guided by cooperation and dialogism, based on respect for the ideas of colleagues. This dialogism is present in situations where the interlocutors take into consideration the voice of one another through the 'interanimation of ideas'. These interactions did not occur in events in which hegemonic masculinity was predominant, so it was necessary for the teacher to value alternative configurations of masculinities, opposed to that model.

8 Conclusion

It is almost a truism that some obstacles to students' development are associated to naturalized cultural traces of the time and the society in which those students are immersed. In physics lessons, naturalized ideas about the science of physics might do that. What we have come to realize with the work we report here is more controversial. According to our analysis of data from a determined school context, there is good reason to assert that configurations of masculinity and learning opportunities might be key elements of educational processes. As far as the present work is concerned, those two relational concepts were crucial in the analysis of the obstacles to student development, particularly the development of thinking skills, collaborative group work, investigative attitude and problem-solving, among other sophisticated competencies science education is expected to develop.

We had recourse to interactional analysis to problematize patterns of masculinity and their influence on the grasping of learning opportunities. We used the example of a group of boys in a physics class as the basis for our study. Learning opportunities involved the experience of the discovery process, of the search for evidence that would prove or rule out students' tentative hypotheses, of the discipline to carry out measurements, of the negotiation of expectations and of the organization of a work plan.

In this paper, we gave special attention to the fact that Roger, a boy with a strong tendency to individualism and a great need to show his presumed superiority, brought about protest masculinity in Nicolau, a boy who had been captivated by the activity. Although Nicolau was facing difficulties in school, he was engrossed in creating learning opportunities until Roger's hegemonic masculinity fully entered their interaction. Equally importantly, we have also noticed that the tensions between different forms of masculinity were attenuated when those involved were engaged by the activities in the classroom and could envision possibilities for growth and success in relation to their colleagues or to other masculinities.

When resorting to challenging situations in the classroom, such as semi-structured investigation activities, it is necessary to proceed with some care. Important dimensions in the education of boys are put at risk when in a discipline such as physics, they are guided by Western hegemonic masculinity patterns, which notably legitimize and value competition, an ultrarational objectivity, individualism, emotional distance and domination.

The three boys portrayed formed a special group in the sense that it allowed us to produce the present research report and to communicate our findings. But systematic observation of the other groups leads us to the same conclusion. That is, the tensions among different manifestation of masculinity were regulated when those involved got engrossed in the activity and foresaw chances of growth and success in respect to colleagues or to other masculinities. Therefore, there are indications that exciting tasks or activities that place individuals at the same level can give them all an opportunity for growth even in prestigious disciplines. According to our analysis of the whole groups set, the best learning opportunities occurred in more collaborative interactions; groups realized this little by little in the plenary sessions and through the teacher's interventions. Those plenary sessions and interventions thus ended up being occasions for learning opportunities which led to configurations of masculinity that emphasize alliance relations among group members.

Acknowledgements The authors wish to thank the Brazilian 'Conselho Nacional de Desenvolvimento Científico e Tecnológico – CNPq', for financial support (grant numbers: 309215/2003-0; 481496/2008-4); Marialice Emboava, Tarciso Borges, Oto Borges, Larissa de Camargo, for discussions and generic support; the students, their parents and further people, for their involvement and formal consent to take part in the study; the staff of Colégio Técnico (UFMG), for the general support; and the editors of this volume for the thorough analysis of our manuscript, thoughtful and pertinent suggestions.

References

AAAS (American Association for the Advancement of Science). (1989). *Science for all Americans: A project 2061 report on literacy goals in science, mathematics and technology*. Washington, DC: AAAS.

Barron, B. (2003). When smart groups fail. *The Journal of the Learning Sciences, 12*, 307–359.

Bloome, D., & Bailey, F. M. (1992). Studying language and literacy through events, particularity and intertextuality. In R. Beach et al. (Eds.), *Multidisciplinary perspectives on literacy research*. Urbana: National Council of Teachers.

Bruner, J. (1960). *The process of education*. Cambridge: Harvard University Press.

Bybee, R. (2010). A new challenge for science education leaders: Developing 21st century workforce skills. In J. Rhoton (Ed.), *Science education leadership: Best practices for a new century*. Arlington: NSTA Press.

Connell, R. W. (1995). *Masculinities: Knowledge, power and social change*. Berkeley: University of California Press.
Davies, I. (2004). Science and citizenship education. *International Journal of Science Education, 26*(14), 1751–1763. doi:10.1080/0950069042000230785.
Easlea, B. (1983). *Fathering the unthinkable: Masculinity, scientists and the nuclear arms race*. London: Pluto Press.
Haenfler, R. (2004). Manhood in contradiction: The two faces of a straight edge. *Men and Masculinities, 7*(1), 77–99. doi:10.1177/1097184X03257522.
Hofstein, A., Navon, O., Kipnis, M., & Mamlok-Naaman, R. (2005). Developing students' ability to ask more and better questions resulting from inquiry-type chemistry laboratories. *Journal of Research in Science Teaching, 42*(7), 791–806.
Jenkins, E. W. (2009). Reforming school science education: A commentary on selected reports and policy documents. *Studies in Science Education, 45*(1), 65–92. doi:10.1080/03057260802681813.
Jordan, B., & Henderson, A. (1995). Interaction analysis: Foundations and practice. *The Journal of the Learning Sciences, 4*(1), 39–103.
Julio, J. M., & Vaz, A. M. (2010). Latent masculinity in school physics investigation activities: Microanalysis of small groups and whole class interactions. In G. Çakmakci & M. F. Tasar (Eds.), *Contemporary science education research: Scientific literacy and social aspects of science* (pp. 333–337). Ankara: Pegem Akademi. 5.
Kanari, Z., & Millar, R. (2004). Reasoning from data: How students collect and interpret data in science investigations. *Journal of Research in Science Teaching, 41*(7), 748–769.
Millar, R., & Osborne, J. (1998). *Beyond 2000: Science education for the future*. London: King's College London School of Education.
Rex, L. A., Steadman, S. C., & Graciano, M. K. (2006). Researching the complexity of classroom interaction. In J. L. Green et al. (Eds.), *Handbook of complementary methods in education research (hardcover)* (3rd ed.). Mahwah: Lawrence Erlbaum Associates.
Schippers, M. (2007). Recovering the feminine other: Masculinity, femininity and gender hegemony. *Theory and Society, 36*(1), 85–102.
Vaz, A. & Julio, J. M. (2005). School physics investigation activities: A psychoanalytical analysis of small groups. In *Proceedings ESERA 2005 contributions of research to enhancing student interest in learning science* (vol. 5, pp. 1377–1379). Barcelona.

Chapter 17
Which Effective Competencies Do Students Use in PISA Assessment of Scientific Literacy?

Florence Le Hebel, Pascale Montpied, and Andrée Tiberghien

1 Introduction

The Program for International Student Assessment (PISA – http://www.pisa.oecd.org) was launched by the Organization for Economic Cooperation and Development (OECD) in 1997 to assess to what degree students near the end of compulsory education have acquired some knowledge and skills that are essential for full participation in society. In all PISA cycles, the domains of reading, mathematics, and scientific literacy are covered not merely in terms of mastery of the school curriculum but in terms of the knowledge and skills needed in adult life. Four assessments have so far been carried out (in 2000, 2003, 2006, and 2009). Administered every 3 years, PISA alternates emphasis on reading, mathematics, and scientific literacy. In 2006, the focus was on scientific literacy and was administered in 57 countries to approximately 500,000 students (OECD 2007).

PISA Science 2006 situated its definition of scientific literacy and its science assessment questions within a framework that used the following components: scientific contexts of the situations on which the questions are based (i.e. life situations involving science and technology), the scientific competencies that are assessed (i.e. identifying scientific issues, explaining phenomena scientifically, and using scientific evidence), the domains of scientific knowledge involved in the questions (i.e. students' understanding of scientific concepts as well as their understanding of the

F. Le Hebel
Université Lyon 1, UMR ICAR, CNRS, Université Lyon 2, ENS-Lyon, Lyon, France
e-mail: Florence.Le-Hebel@ens-lyon.fr

P. Montpied
UMR ICAR, CNRS, Université Lyon 2, ENS-Lyon, Lyon, France
e-mail: pascale.montpied@ens-lyon.fr

A. Tiberghien (✉)
UMR ICAR, CNRS, University of Lyon 2, ENS-Lyon, Lyon Cedex, France
e-mail: andree.tiberghien@univ-lyon2.fr

C. Bruguière et al. (eds.), *Topics and Trends in Current Science Education: 9th ESERA Conference Selected Contributions*, Contributions from Science Education Research 1,
DOI 10.1007/978-94-007-7281-6_17, © Springer Science+Business Media Dordrecht 2014

nature of science), and student attitudes toward science (i.e., interest in science, support for scientific inquiry, and responsibility toward resources and environments) (OECD 2007; Bybee et al. 2009).

PISA data provide information on the school systems of participating countries and can exert a profound impact on science education policies. Additionally, PISA results lead to the development of research projects about secondary analysis of PISA results (Olsen and Lie 2006). We present very succinctly the studies following Pisa results.

Some studies focus on the effects of PISA results on education policymaking in different countries. In a study included in the European project *Knowledge & Policy* (Mons and Pons 2009), the authors show that consequently to low PISA results, educational reforms have been implemented in some countries, whereas in others, the same low score did not impact on education policymaking. For instance, PISA results in Germany and Switzerland, showing that the education systems were apparently less efficient and equitable than the average of OECD countries, lead to strong reactions, media coverage, and political, pedagogical, and institutional debates (Moreau et al. 2006; Prenzel and Zimmer 2006). Similarly, Milford et al. (2010) present the reaction to PISA within the media of both North and South America. In their study, Dolin and Krogh (2010) discuss the relevance of PISA as a catalyst for the educational actions taken in Denmark, arguing that their analysis shows crucial differences between the PISA science assessment framework and test system and the Danish educational goals in science.

PISA results lead also to the development of research projects about the meaning of the content of the complex PISA data resulting from both sophisticated methods and the competency contents in different domains (mathematics, reading, and science literacy) in the PISA framework (Olsen and Lie 2006). For instance, several studies question the meaning of literacy as defined in the PISA framework and, consequently, the interpretation of literacy measurement (Rocher 2003; Bain 2003; Goldstein 2004; Vrignaud 2006). Secondary and complementary analyses of PISA 2000 reading literacy tests in France argue that the required competencies to answer PISA literacy questions are in fact much more complex and heterogeneous than those that the PISA designers claim to test (Bautier et al. 2006; Rochex 2006). The authors reconsider the concept that it is possible to assess a limited number of competencies with rather few questions as PISA proposes.

Several studies reanalyzed PISA students' scores and obtained complementary data. Fensham (2009) reanalyzed students' responses using the contextual sets of items as the unit of analysis.[1] He investigates if we can observe score differences in favor of girls or boys according to the context of questions. His results suggest that gender/context effects may be present, but these are probably less than the effects of gender on scores based on competencies. Nentwig et al. (2009) propose a new categorization of PISA Science 2006 items by the degree to which they required students to extract and apply information from the contexts provided. Their analysis shows that, in the majority of countries, the students perform equally well with

[1] The PISA main assessment consists of a series of units; each unit has an introduction with a text and possibly photos, drawings, and diagrams presenting a situation followed by a series of questions called items (an example of an introduction and one item is given in (Fig. 17.1).

both kinds of items (with a high or low level of contextualization), that is, they demonstrate both abilities – the recall of knowledge and the extraction of information from the test-unit stimulus – to the same extent, in other terms the ability to decontextualize and recontextualize information.

Some researchers develop secondary analysis about measurements of students' interest in science (Olsen and Lie 2011; Drechel et al. 2011; Bybee and McCrae 2011), particularly from data collected in the associated PISA students' questionnaires on background and attitudes to the assessment questionnaire.

Other research studies discuss scientific literacy as defined in the PISA framework. For instance, Ratcliff and Millar (2009) are very critical of teaching science using contexts such as those developed for PISA Science 2006.

Some studies call into question the adequacy between the definition of scientific literacy in the PISA framework and the scientific literacy effectively assessed in items. For instance, Lau (2009) shows that many questions supposed to test knowledge about science actually assess knowledge of science.

Our present work also focuses on this adequacy, but we do not a priori question the definition of scientific literacy in the PISA framework and the defined competencies. Our aim is to investigate to what extent the competencies that each item is supposed to test correspond to the competencies involved when French students construct their answer.

As scientific literacy was the major domain in 2006, our study is based on PISA Science 2006 data. In France, PISA Science 2006 indicated average scores overall, and a particularly high proportion (compared to the OECD average) of students in difficulties, at level 1, meaning that "students have such limited knowledge that they are only able to apply that knowledge in a very small number of familiar situations," or below level 1, meaning that "students are not able to use scientific knowledge to understand and do the easiest PISA tasks." The levels are defined on a scale from 6 to below 1, a posteriori by PISA, that is, only after correction and scoring in all OECD countries (30 in 2006). We keep those levels in our study.

Moreover, PISA Science 2006 in France shows striking results concerning differences between the scores for the three main competencies, "identifying scientific issues," "using scientific evidence," and "explaining phenomena scientifically." The latter especially presents a particularly low score in France compared to the other two, whereas scores for the three major competencies are close in most of the other countries.

The aims of this project are to investigate the links between:

- Students' performances in PISA science, that is, their scores
- Students' understanding of the context of the unit, proposed in the unit stimulus or items
- Competencies involved when the students effectively construct the answer to PISA items

This introduction presents the situation on which the items are based.

Our study is based on the observation of students in a situation in which they answer PISA questions and on the analysis of their actions. This analysis should lead us to the competencies involved and their possible diversity.

2 Theoretical Framework

In order to reconstruct the competencies involved when students answer PISA items, our theoretical framework is linked to the mental and behavioral processes when solving problems or answering questions. When students construct their answer to PISA science items, they combine the processes of reading and of solving a scientific task. Indeed, in our study, we chose a framework based on a theoretical model, assuming that we cannot completely separate the reading and the solving tasks (Bergqvist and Osterhölm 2010). These authors analyze connections between the process of reading and the process of solving a mathematical task focusing on mental representation based on Kintsch's theory (1998) and the mathematics education perspective focusing on facets of behavioral aspects of task solving based on Lithner's theory (2008). The Kintsch's theory (1998) describes the mental representation of texts by distinguishing between three different levels or components: the surface component referring to the encoding of the words and the phrases themselves; the textbase representing the meaning of the text, which is the semantic structure of the text; and the last level corresponding to the associative processes with the reader's prior knowledge. The other part of the theoretical model is based on theory developed by Lithner (2008), who identifies two types of strategies: imitative reasoning (memorized reasoning and algorithmic reasoning) based on surface properties of the task and creative reasoning based on the scientific content of the task.

This model has a structure that takes into account the mutual constraints between aspects of mental representation and aspects of behavior. It includes a cyclic component that allows for the behavioral component to affect the mental representation and not only the other way around. In this model, the first reading of the text (or a part of it) creates a mental representation that can lead the student either to the end of the task by giving a final answer with the activation of prior knowledge or stopping the task or to carry out a strategy in a wider sense like rereading the text, choosing, recalling, constructing, discovering, or guessing (Lithner 2008). A main idea is that any strategy implementation will affect the mental representation since what the solver does becomes part of the comprehension of the task. There is thus a cyclic process.

The study of the students' mental and behavioral processes when solving tasks should allow us to answer the following research questions:

- Which competencies defined by PISA 2006 are involved when the students answer PISA science items?
- Which differences can we observe between high- and low-achieving students while solving PISA science items?

The first question leads us to compare the effective competencies to those that the questions are supposed to assess. To reconstruct the students' processes, it is necessary to collect data in situations of answering the items, and therefore we chose a case study methodology as presented below.[2]

[2] The French Ministry of Education (DEPP – Evaluation, Forecasting, and Performance Department) in agreement with the OECD has allowed us to use the PISA 2006 results in France and in other OECD countries.

3 Methodology

In order to study the students' behavioral and mental processes and to compare the students' competencies when answering the questions and to analyze the differences between students, we proceed in two steps, first an a priori analysis of PISA units and second an analysis of the data collected when students construct their answers.

The a priori analysis of the PISA units consists in:

- Analyzing each unit according to several criteria, in order to select a set of units and items to test
- Characterizing possible mental representations that would be a priori stimulated by the PISA unit in their leading text and the associated items

To select relevant units for studying French students' processes, we base our a priori analysis of all the units on the following criteria:

- The diversity of scientific knowledge required for the item (knowledge of science or knowledge about science included in the school curriculum or not, daily knowledge, etc.)
- The different content areas tested in the item (Physical Systems, Living Systems, Earth and Space Systems, Technology Systems, Scientific Inquiry, etc.)
- The usefulness or not of the introduction of a unit and/or item, the question format (multiple choice, open, etc.), and the competency evaluated by PISA for the item
- The scores obtained for each item in France compared to OECD average scores

The second step consists of an analysis of answering processes from students' oral and written data when they construct their answer and/or during the following interview. Cognitive processes are characterized from students' observations, mainly their verbal behaviors and gestural behaviors associated with the reading of the text. We observed 21 students of 15 years of age (the age for the PISA evaluation) and of differing academic levels and identified their actions that allow us to draw conclusions about:

- The diversity of the cognitive processes mobilized while they try to solve the problems
- The knowledge which they use or lack to answer the item
- The effective competencies involved when students effectively answer PISA items
- The answering strategies when they construct their answers

We collected data on two types of situations. First, nine pairs of students required to answer the items together were videotaped. In parallel, three students answered the PISA questionnaire individually. Among these 21 students, we interviewed the three students answering individually and six pairs, immediately after the written test.

The two methods are complementary. The disadvantage of the first method is that the test conditions differ from those in the PISA test. However, through student discussions, it makes some thinking processes more explicit. The second method respects the conditions of PISA testing. However, the interview implies that we collected only a delayed recall process, which might be missing some elements of what the student really did. During the following interview, we asked the students to make their thinking process explicit about how they answered the questions. We called that the "explicitation interview" (Vermesch and Maurel 1997.[3]

We collected audiotaped and/or videotaped data from nine pairs of students answering the items and nine interviews (three with the students answering individually); thus 12 written questionnaires were collected. First, we videotaped seven grade 10 students of 15 years of age at a suburban upper secondary school, displaying social diversity. The seven students are high achievers, according to their science teacher. Then, we collected data from 14 grade 9 students of 15 years of age at middle secondary schools (as in France students can repeat a class). These 14 students came from two schools situated in a disadvantaged neighborhood, and 13 of them are low-achieving students. In both cases, the students working in each pair had the same academic level according to their teacher. Sometimes the science teacher suggested the choice of pairs, but most of time, the students formed the pairs themselves.

All 21 students answered the same 30 items from 12 PISA units. Students 1–7 and student 21 are high achievers, whereas students 8–20 are low achievers. The data were independently analyzed by three researchers, who conferred to obtain a common analysis in case of different interpretations.

4 Data Analysis

We present our way of analysis in the case of an item of the released unit "Grand Canyon" (Fig. 17.1) to show how we proceeded. We present only a few items among those used, because all the others are confidential.

In order to evaluate scientific literacy, PISA categorized three main competencies, each including three sub-competencies as described below:

Identifying Scientific Issues (ISI)

- Recognizing questions that it is possible to investigate scientifically (ISI 1)
- Identifying keywords to search for scientific information (ISI 2)
- Recognizing the key features of a scientific investigation (ISI 3)

Explaining Phenomena Scientifically (EPS)

- Applying knowledge of science in a given situation (EPS 1)
- Describing or interpreting phenomena scientifically and predicting change (EPS 2)
- Identifying appropriate descriptions, explanations, and predictions (EPS 3)

[3] We used Transana software to analyze the videotapes (http://www.transana.org).

The Grand Canyon

The Grand Canyon is located in a desert in the USA. It is a very large and deep canyon containing many layers of rock. Sometime in the past, movements in the Earth's crust lifted these layers up. The Grand Canyon is now 1.6 km deep in parts. The Colorado River runs through the bottom of the canyon.

See the picture below of the Grand Canyon taken from its south rim. Several different layers of rock can be seen in the walls of the canyon.

Question S426Q03: The temperature in the Grand Canyon ranges from below 0°C to over 40°C. Although it is a desert area, cracks in the rocks sometimes contain water. How do these temperature changes and the water in the rock cracks help to speed up the breakdown of rocks?

A. Freezing water dissolves warm rocks
B. Water cements rocks together
C. Ice smoothes the surface of rocks
D. Freezing water expands in the rock cracks

Fig. 17.1 Extract of a released unit ("Grand Canyon") from PISA Science 2006

Using Scientific Evidence (USE)

- Interpreting scientific evidence and drawing conclusions (USE 1)
- Identifying assumptions/evidence/reasons (USE 2)
- Reasoning societal implications of sciences and technologies (USE 3)

First, in order to understand more precisely the students' actions, according to our theoretical framework, we analyzed cognitive processes from mental representations and strategies constructed by each student or pair of students from situations proposed in items and answering strategies. In the second part, we focused on the competencies they effectively developed while solving the items, in order to compare with the competency that PISA claims to test in the item.

The item example that we chose to show how we analyze is a part of the released unit "Grand Canyon" and is coded S426Q03. The right answer is D. The mean score obtained in OECD countries for this item is 67.60 % and is much lower in France (49.61 %). The competency that PISA claims to evaluate in this item is "applying knowledge of science in a given situation" (EPS 1).

According to our a priori analysis, the mental representations stimulated when reading the introduction are that the Grand Canyon is in the desert, is a deep canyon, and has many layers of rocks that have different names (from the photo key). We do not develop this analysis to the extent that the associated question (S426Q03) presented here can be answered with a minimal representation of the situation. On the other hand, we develop the representations of this question. In Table 17.1, we present a part of the possible representations. All the representations R1 to R10 are correct, but only R1 to R6 are necessary to answer the item, even if they are not really made explicit.

Table 17.1 Students' answers, competencies, and mental representations for the item "Grand Canyon" S426Q03

	An	Comp	R1	R2	R3	R4	R5	R6	R7	R8	R9	R10
St1/St2	D	EPS1	Y	Y	Y	Y	Y	Y	Y			
St3/St4	D	EPS2 (EPS1)	Y	Y	Y	Y	Y	a	Y	Y		
St5/St6	D	EPS1	Y							Y		
St7	D	EPS1	Y	Y	Y	Y	Y	Y	Y			
St8/St9	A D	EPS2			Y	No	b	No	Y	Y	Y St9/no St8	Y
St10/St11												
St12/St13	A D	No	Y	Y	No for St12 c	Y for St13	Y	d	Part	No St 12		No e
St14/St15	C	No		Y	No		Y St14 no St15	f	g	h	h	h
NSt16/St17 -	D	No		Y	Y		Y St17	No	No			
St18/19	D	EPS2	Y St 19	Y	Y	Y	No	Y				
St20	D	EPS2	Y	Y	Y	Y	Y		Y			
St21	D	EPS1/ EPS2	Y	Y	Y	Y	Y	Y	Y			

An: answer
Comp.: competency
Y: representation involved in students' construction of answer
No: representation not constructed by the student
Nothing in the cell: no information from data
*R*1: the temperature in the Grand Canyon ranges from 0 °C to 40 °C
*R*2: the transformation of water to ice when the temperature in the Grand Canyon ranges from over 40 °C to below 0 °C and the reverse
*R*3: cracks sometimes contain water
*R*4: dilatation as a volume increases makes the rock breakdown
*R*5: the link between the changes of water temperatures and the breakdown of the rocks
*R*6: ice has a bigger volume than liquid water
*R*7: cracking increase in rocks speeds up the breakdown of rocks
*R*8: representation of dissolution
*R*9: representation of cementation of rocks by water
*R*10: representation that ice smoothes the surface of rocks
a: Correct representation from the beginning for St3 student; for St4, the correct representation of the particular motion of gas/liquid and solid in conflict with the exception of water, then agreement with St3 proposal (ice volume increasing comparing to liquid water)
b: Representation of the link between water and breakdown of the rock, but they do not take temperature changes into account
c: St12 thinks that if water freezes, it cannot move in the cracks
d: Lack of knowledge of the difference of volume between water and ice for St12
e: No understanding of the phenomena "ice smoothes the surface of rocks"
f: The representation of the situation does not call for this knowledge
g: Built a representation giving opposite sense to the word "breakdown"
h: Built her representation according to her goal of the item

Table 17.2 Representations involved in students answering that are not a priori involved with answer item S426Q03

St3/St4	Link between temperature and molecular agitation
	Representation of an everyday situation (bottle of water in freezer)
St8/St9	Link between porosity of different rocks and breakdown of those rocks
St14/St15	They built a new item goal from association of cementation with cracks – the new goal is to understand how it increases sealing of cracks
St16/St17	Representation of volume change due to freezing and thawing for a small stone
St21	Representation of an everyday situation (bottle of water in freezer)

Table 17.1 shows the multiple-choice answers, the competencies, and the constructed representations. It appears that out of the 11 answers (one pair did not answer), 8 are correct. For two answers, two proposals were selected of which one is correct, and finally one pair chose a wrong proposal. However, even if there is not a large variety of answers, and almost all students proceeded by elimination, there appear to be different representations and strategies. In Table 17.1, we distinguish the case where the students are in contradiction with the representation (indicated by "No" in the cell) and the case where we have no information from our data (indicated by an empty cell).

In Table 17.2, we mention other unexpected representations that were involved in the students' construction of answers that we did not envisage in our a priori analysis of the representations stimulated when reading the introduction of the unit and the item.

Concerning the students' representations and strategies, we illustrate our analysis with examples.

The first example shows how some low-achieving students constructed their answer. Students 12 and 13 chose two options, A and D.

After reading the question aloud, the two students take the propositions one by one and proceed by elimination:

St12	Wait, what's the question? How do these temperature changes and the water in rock cracks help to speed up the breakdown of rocks?
St13	Speed up the breakdown of the rock?
St12	Freezing water dissolves warm rock, yeah, because you know the changes, you saw as for temperature changes...
	(The students ticked answer A.)
St13	Ice smoothes the surface of rocks.
	[St13 asks the researcher the meaning of "smoothes." The researcher explains the word]
St12	Ah ok, so no.
St13	No, so freezing water expands in the rock cracks.
	(The students go back to read the text of the question.)
St12	Freezing water dissolves warm rocks, I would say A and D.
	(The students ticked answer D.)

This discussion and the following interview with the researcher (R) allow us to understand more precisely how the two students proceeded to choose answers A and D, as shown below:

R	So you ticked answers A […] and D […]. How did you arrive at your answer?
St12	A, that's er… temperature changes.
R	So A, that's temperature changes, which do what?
St12	Bah which, bah which do some, which, some, I don't know how to say it, that is written, but I can't remember it… that's it, which breaks down the rock.
	(*St 12 looks for the sentence in the text.*)
R	Which breaks down the rock, ok, and how did you choose answer A?
St13	Bah with that (*St13 shows the text*), with the introduction there. (*St13 shows the sentence "How do these temperature changes and the water in rock cracks help to speed up the breakdown of rocks?"*)

The answering strategy of this pair of students is that they select propositions for which they can associate words picked from the text of the question with the words from the options. They associate the formulation "temperature change" with the word "freezing" that we can find in options A and D. Moreover, they make the correspondence between words like "rock cracks" that we can find in the text of the proposition as well as in the question. They eliminate propositions B and C, because the words "cements" and "smoothes" cannot be associated with the breakdown.

During the following interview with the researcher (R), they specify that they recalled previous academic knowledge to choose answer A.

St12	And with our knowledge.
R	Your knowledge?
St12	Because we know that the temperature changes, it does that.
St13	It dissolves rocks.
St12	Yeah, we learned that last year, I think.

During this exchange, St12 claims to use previously known everyday knowledge, but this element of knowledge establishing a causality between temperature change and dissolution of the rock is incorrect.

R	So you also ticked answer D.
St13	Water expands in the rock cracks.
R	So how did you choose D?
St12	I can't remember any more, er, freezing water expands in the rock cracks …ah it's wrong! Because if it freezes, it doesn't move in rock cracks, it's wrong.
R	So you don't agree with what you thought anymore?
St12	No, I don't agree.
St13	But there's a sentence saying that rock cracks sometimes contain water.
R	Yes.
St12	Yes but here, it says freezing water expands.
St13	Ah no.

St12	If water freezes, it cannot move into the cracks.
St13	No, but actually we don't agree with each other because I was helped by the sentence "cracks in the rocks sometimes contain water." Actually, I use it to say that water can expand rock cracks.

This extract shows that St13 constructed a representation of cracks sometimes containing water, whereas St12 did not and no longer agrees with answer D (Table 17.1); that can be reason enough to explain her wrong choice. For this example, we could not identify any PISA competency involved in the solving process.

Conversely, a pair of high achievers chose answer D directly. We report below the exchange between Student 1 (St1) and Student 2 (St2).

St1	I would say answer D.
St2	Freezing water expands in the rock cracks.
St1	OK, that's right.

It took them 20 s to answer. In the following interview, St1 mentions that they already knew that "when water freezes, it dilates." They apply the appropriate knowledge in a context. This supposes that they constructed a mental representation of the situation where water is inside the rocks and changes state with temperature change. Thus, they use the concept of water changing state by a process of contextualization-decontextualization which leads them to directly apply the element of knowledge "when water freezes, it dilates." The competency they effectively use when they construct their answer is "applying knowledge of science in a given situation" (EPS1).

For the same item, the pair of students 8 and 9, who are both low achievers, ticked answers A and D (time to answer: 2 min 40 s). First, they focus on the breakdown of the rock. They check if the breakdown is possible for each one of these options. They do not construct a stable representation of the situation; they start to focus on breakdown, then on expansion, and finally on the type of rocks. One of the students recalls academic knowledge, triggered by the picture, and tries to link porosity and breakdown. These focuses lead them to eliminate options B and C and construct the representations "link between water and breakdown of the rock" (but they do not take temperature changes into account) and "link between porosity of different rocks and breakdown of these rocks" (Table 17.2). It never brings any clarification to help them choose between answers A and D. For this pair of students, our interpretation is that the effective competency they used while answering this question is "describing or interpreting phenomena scientifically and predicting change" (EPS 2), which differs from the competency that PISA proposes to evaluate with this proposition (Table 17.1).

Nevertheless, for this example of a question, although the sub-competency effectively developed by some students and the one that PISA claims to assess can differ, they both still remain within the main competency "explaining phenomena scientifically" (EPS).

5 Results

We propose first to present our results on the competencies effectively involved by students when they construct their answer to PISA items and second to focus on the representations and strategies that students develop while solving PISA items.

5.1 Competencies Effectively Involved When Students Solve PISA Items

In Table 17.3, we reported on part of our results consisting of 30 selected items for 21 students (eight pairs and three individual students). In this table, we respectively reported for each item the competency that PISA claims to assess, the mean score in OECD countries, the score obtained in France, and the competencies which, according to our analysis, the students effectively develop while constructing their answer to the item. Students 1–7 and student 21 are high achievers, whereas students 8–20 are low achievers.

The results show that the number of right answers is much higher for the high achievers (87 out of 115 possible answers) than for the low achievers (48 out of 161).

As we can observe, effective competencies developed by students when solving the item and those that PISA claims to assess can either be the same or differ for the same item. Our data show that the percentage of right answers involving the same competency is much higher for the high achievers (70 %) than for the low achievers (60 %). Moreover, the percentage of right answers involving different competencies is around 30 % for the high achievers and 42 % for the low achievers. These results should be taken carefully because of the high level of no answering by the low achievers. However, the discrepancy between the percentage of right answers for which the effective competencies and those that PISA claims to test are the same is much less for the high achievers. This may indicate that PISA is more appropriate to test high- or possibly middle-level achievers.

When the competencies differ, they can both still belong to the same main competency, as we observed with the example of the released item "Grand Canyon" S426Q03. On the contrary, for some other items, effective sub-competencies involved in the students' answers and the competency that PISA is supposed to assess do not even belong to the same category of competencies. For instance, for question S304Q03a (Table 17.3), the competency that PISA claims to test is "reasoning societal implications of sciences and technologies" (USE 3) included in the category of competencies "using scientific evidence" (USE), whereas we find in our analysis that the competency which students can effectively involve while constructing their answer is "applying knowledge of science in a given situation" (EPS 1), included in "explaining phenomena scientifically" (EPS).

Table 17.3 Competencies effectively developed by students while solving PISA items compared to competencies that PISA claims to test for selected items. The *gray-colored cells* highlight questions for which effective competencies when solving the items differ from competencies that PISA is supposed to test. *X*, the competency developed by the students cannot be identified; *Nans*, item not analyzed by the students; *, the students obtain full credit but no PISA competency was involved; (*competency in brackets*), the students do not have the credit, although they develop this competency; *competency n= not appropriate competency*, the students use a competency which is not appropriate to give the right answer. *Italics* indicate the wrong answers

Item Nb	PISA	OECD	France	Competencies effectively developed by students while solving PISA items											
				High Achievers						Low Achievers					
				1&2	3&4	5&6	7	21	8&9	10&11	12&13	14 &15	16&17	18&19	20
S268Q01	**ISI 3**	72.48	69.74	ISI 3	ISI 3	ISI 3	*X*	ISI 3	ISI 3	Nans	Nans	*X*	*X*	*	ISI 3
S268Q06	**EPS 1**	55.20	52.60	*(EPS 1)*	EPS 1	X	EPS 1	EPS 1	*X*	Nans	Nans	*X*	*X*	*X*	*X*
S304Q01	**USE 1**	43.63	44.86	EPS2-1	EPS2-1	EPS2-1	*EPS2 n*	*EPS2n*	*X*	EPS2-1	*X*	*EPS2 n*	*X*	*X*	*X*
S304Q02	**EPS 1**	62.07	63.74	EPS 1	EPS 1	EPS 1	EPS 1	EPS 1	EPS 1	EPS 1	*X*	*X*	EPS 1	*X*	EPS 1
S304Q03a	**USE 3**	39.00	47.85	EPS 1	EPS 1	EPS 1	EPS 1	*EPS 1 USE 3*	EPS 1	EPS1	*X*	*X*	EPS1	*X*	X
S304Q03b	**USE 2**	50.68	47.96	EPS2-1	EPS1-2	EPS1-2	EPS1-2	EPS2-1	*X*	*X*	*X*	EPS1-2	EPS1-2	*X*	*X*
S408Q01	**EPS 1**	62.95	47.76	EPS 1	EPS 1	X	*X*	EPS1-2	*X*	*X*	*(EPS1)*	EPS 1	EPS 1	*X*	*X*
S408Q03	**EPS 1**	30.49	29.26	*X*	*	EPS 1	*(EPS 1)*	EPS 1	*X*	USE 2	*X*	X	USE 2	*X*	*X*
S426Q03	**EPS 1**	67.60	49.61	EPS 1	EPS2-1	EPS 1	EPS 1	EPS1-2	EPS2	Nans	X	X	X	EPS 2	EPS 2
S428Q01	**USE 1**	61.68	56.87	USE 1	USE 1	USE 1	USE 1	USE 1	USE 1	Nans	Nans	USE 1	USE 1	*X*	*X*
S428Q03	**USE 2**	71.35	55.21	USE 1	USE 1	USE 1	USE 1	USE 1	USE 2	Nans	Nans	USE 1	USE 1	*USEn)*	*USE1)*
S476Q01	**EPS 1**	70.73	66.60	EPS 1	EPS 1	EPS 1	*X*	*(EPS1)*	*X*	Nans	EPS 1	*X*	*(EPS 1)*	*X*	*X*
S476Q02	**EPS 1**	70.87	64.32	EPS 1	EPS 1	EPS 1	*(EPS 1)*	X	X	Nans	EPS 1	EPS 1	EPS 1	*X*	*X*
S476Q03	**EPS 1**	60.11	55.94	EPS1	EPS 1	X	*X*	EPS 1	*X*	Nans	*X*	*X*	EPS 1	*X*	*X*
S485Q02	**EPS 1**	57.68	42.70	X	EPS 1	X	X	EPS 1	Nans	Nans	Nans	*X*	EPS 1	*X*	*X*
S495Q01	**USE 2**	42.13	45.40	USE 1	*(USE 1)*	*(USE 1)*	USE 1	*USE1)*	*X*	*X*	*X*	X	USE 1	X	*(USE)*
S495Q02	**USE 2**	57.62	54.18	USE 1	USE 1	USE 1	USE 1	USE 1	USE 1	USE 1	*X*	*X*	USE 1	X	*(USE)*
S495Q03	**USE 1**	38.56	39.33	USE 1	USE 1	USE 1	USE 1	USE 1	*X*	*X*	Nans	Nans	*X*	*X*	*X*
S498Q02	**ISI 3**	46.91	48.49	ISI3/EPS3 USE2	ISI 3/ EPS 1	ISI 3	*(ISI 3)*	*(ISI 3)*	*X*	Nans	*X*	ISI 3	*X*	*X*	*(ISI 3)*
S498Q03	**ISI 3**	42.61	48.20	ISI 3	ISI 3	*(ISI3)*	ISI 3	ISI 3	ISI 3	Nans	*(ISI 3)*	*(ISI 3)*	ISI 3	*X*	ISI 3
S498Q04	**USE 1**	59.93.	74.90	USE 1	USE 1 /USE 2	USE 1 /USE 2	USE 1 /USE 2	USE 1	USE 1	Nans	Nans	USE 1	ISI 3	EPS 1	USE 1
S508Q02	**ISI 3**	60.95	72.20	ISI 3	ISI 3	ISI 3	*X*	ISI 3	*X*	Nans	USE n	*USE n*	*	*	*X*
S508Q03	**ISI 3**	73.55	58.61	ISI 3	ISI 3	ISI 3	*X*	ISI 3	*X*	Nans	*(ISI 3)*	ISI 3	ISI 3	*X*	ISI 3

This item requires two very common pieces of knowledge for all students. Nevertheless, the mean score obtained in OECD countries for this item is 39.0 % and is 47.8 % in France.

Our analysis shows that low-achieving students, even if they have the required knowledge to answer the question, are not able to implement that knowledge in the context of the PISA item. They are able to build neither a stable representation of the aim of the item nor a representation of the technical structures described in the item. On the other hand, high achievers do not show any difficulty in mobilizing their knowledge in this new context. But they do not have to think about the societal implication of the technical structures in order to find the right answer, only to apply common scientific knowledge.

As we can observe in Table 17.3, students in fact most often use the "applying knowledge of science in a given situation" competency included in the major competency "explaining phenomena scientifically" to give the right answer, even if it was not the competency that the item is supposed to test according to PISA.

Furthermore, no scientific competency is involved rather often in the low achievers' solving process.

5.2 Representations and Answering Strategies Developed by Students While Solving PISA Items

Our results show that low achievers (Students 8–20), contrary to high achievers, have difficulties in constructing coherent, global, and stable representations of the unit goals and the item goals. Consequently, we observed certain specific types of behavior:

- They focus on one word or group of words or a single component of the question or of the leading text, generally one easy to understand, and build a new question goal around it that can be unstable, and thus evolves during the discussions.
- They initiate a reflection on a fuzzy representation which fades easily while they discuss the answer and replace it immediately with another unstable representation and so on.
- They do not build any representation of the situation, system, or item goal and answer at random or build the sentence with a grammatical routine integrating words and formulations picked out from the item's leading text and/or core question.

This analysis leads us to define answering strategies involving these representations and to observe how much they differ between the low and high achievers. We categorize different solving strategies:

- Direct with or without argumentation
- Elimination with or without argumentation in the case of multiple-choice question items

- Returning to the leading text or core question to find information or data
- Searching for some wording consistent with the item words in the leading text of the unit or the core question
- Graph or drawing interpretation
- At random

We observe that some items appear to be more suitable for a large diversity of answering strategies generally related to the question format or the complexity. For instance, for an open question, the answer requires an argumentation, so we more often find "direct with argumentation" answering strategies, particularly for high-achieving students who understand the question's goal better and are able to argue. On the other hand, the "elimination" answering strategy is most often used in the case of a multiple-choice question by all students.

In addition, we can immediately observe that, most of the time, high achievers and low achievers do not use the same diversity of answering strategies for a given item. High achievers most often use direct and elimination strategies, with argumentation when it is required.

When argumentation is required for an open item, high achievers are able to go further in the discussion about the item, the consequence of an experiment, for instance, which is not required by the question. The high achievers show a wide range of skills allowing them to diversify their analysis, solving strategies, and vocabulary to discuss and answer the items and which lead most of the time to the right answer. In contrast, the low achievers are frequently limited to only a few strategies, as presented above for most of the questions in relation to their unstable representations. This analysis clearly shows that the answering strategies used by the students when they construct their answer are linked to the question format, the vocabulary that it requires, and consequently to the representations that students build from the leading text and core questions of the items.

We did not find any typical difficulties in the high achievers, whereas we identified systematically some difficulties in the low achievers, even when they gave the right answer. They often show a lack of knowledge and a lack of motivation and concentration linked to an unfamiliar and effortful task.

6 Concluding Remarks

Our results show that for the same item, the competencies effectively developed by the students while solving items can differ from the competency that PISA claims to test. In particular, we observe that the students most often use the competency "applying knowledge of science in a given situation" included in the major competency "explaining phenomena scientifically." Consequently, we have to think more precisely about the real meaning of the "explaining phenomena scientifically," when the student applies an element of knowledge which is not necessarily an explanation process. It would thus lead to a different interpretation of PISA scores in France, particularly the low score that French students obtained in the major competency "explaining phenomena scientifically." These findings are consistent with previous studies (e.g., Lau 2009), showing that different interpretations can be made regarding the actual meaning of PISA's measurement of scientific literacy.

In addition, we observe that the low achievers have difficulty in obtaining coherent, global, and stable representations of the unit goals and the item goals and consequently use different answering strategies compared to the high achievers. We systematically identified certain difficulties in the low achievers, even when they gave the right answer. As the competencies may appear far upstream from the final competency evaluated by PISA, the PISA items on low achievers may not give relevant indications of what is effectively acquired by these students. The competencies should be refined to evaluate the scientific literacy competencies, in particular for low-achieving students.

Furthermore, our results could be of interest in the case of a reflection on science evaluations carried out by teachers in class. Indeed, we can question the consistency between the competencies that teachers a priori believe that they are assessing in an evaluation and the competencies effectively involved when students answer questions. More precisely, we can question whether the proportion of the restitution of knowledge in an evaluation is not greater than what teachers really want to assess, in the light of our results showing the frequency of the competency "applying knowledge of science in a given situation," even if the questions were not supposed a priori to test that competency.

Acknowledgments This project is supported by the European Project S-Team and the French Ministry of Education (DEPP – Evaluation, Forecasting and Performance Department). We are grateful to our colleagues from the DEPP and to the schools, students, and teachers who made it possible to collect the data.

References

Bain, D. (2003). PISA et la lecture: un point de vue de didacticien. *Revue Suisse des Sciences de l'Education, 25*, 59–78.

Bautier, E., Crinon, J., Rayon, P., & Rochex, J.-Y. (2006). Performances en littéracie, mode de faire et univers mobilisés par les élèves; analyses secondaires de l'enquête PISA 2000. *Revue Française de Pédagogie, 157*, 85–101.

Bergqvist, E., & Österholm, M. (2010). A theoretical model of the connection between the process of reading and the process of solving mathematical tasks. In C. Bergsten, E. Jablonka, & T. Wedege (Eds.), *Mathematics and mathematics education: cultural and social dimensions*. Proceedings of the MADIF 7: The seventh mathematics education research seminar, Stockholm 2010. SMDF Linköping, Sweden.

Bybee, R., & McCrae, B. (2011). Scientific literacy and student attitudes: Perspectives from PISA Science 2006. *International Journal of Science Education, 33*, 7–26.

Bybee, R., McCrae, B., & Laurie, B. (2009). PISA 2006: An assessment of scientific literacy. *Journal of Research in Science Teaching, 46*, 865–883.

Dolin, J., & Krogh, L. B. (2010). The relevance and consequences of PISA science in a Danish context. *International Journal of Science and Mathematics Education, 8*(3), 565–592.

Drechel, B., Carstensen, C., & Prenzel, M. (2011). The role of content and context in PISA interest scales: A study of the embedded interest items in the PISA Science 2006 assessment. *International Journal of Science Education, 33*, 73–95.
Fensham, P. (2009). Real world contexts in PISA science: Implications for context-based science education. *Journal of Research in Science Teaching, 46*, 884–896.
Goldstein, H. (2004). International comparisons of student attainment: Some issues arising from the PISA study. *Assessment in Education, 11*, 319–330.
Kintsch, W. (1998). *Comprehension: A paradigm for cognition*. Cambridge: Cambridge University Press.
Lau, K. C. (2009). A critical examination of PISA's assessment on scientific literacy. *International Journal of Science and Mathematics Education, 7*, 1061–1088.
Lithner, J. (2008). A research framework for creative and imitative reasoning. *Educational Studies in Mathematics, 67*, 255–276.
Milford, T., Ross, S. P., & Anderson, J. O. (2010). An opportunity to better understand schooling: The growing presence of PISA in the Americas. *International Journal of Science and Mathematics Education, 8*(3), 453–473.
Mons, N., & Pons, X. (2009). *PISA dans l'élaboration des politiques éducatives en France: instrument de connaissances ou outil de communication? Rapport pour le projet européen 6ème PCRD, Knowledge & Policy* (Delevirable 12) Retreived on October 2013 from http://knowandpol.eu/IMG/pdf/o31.pisa.france.fr.pdf.
Moreau, J., Niddeger, C., & Soussi, A. (2006). Définition des compétences, choix méthodologiques et retombées sur la politique scolaire en Suisse. *Revue Française de Pédagogie, 157*, 43–53.
Nentwig, P., Roennebeck, S., Schoeps, K., Rumman, S., & Carstensen, C. (2009). Performance and levels of contextualization in a selection of OECD countries in PISA 2006. *Journal of Research in Science Teaching, 46*, 897–908.
OECD. (2007). *PISA 2006 science competencies for tomorrow's world* (Vol. 1). Paris: OECD.
Olsen, R., & Lie, S. (2006). Les évaluations internationales et la recherche en éducation : principaux objectifs et perspectives. *Revue Française de Pédagogie, 157*, 11–26.
Olsen, R., & Lie, S. (2011). Profiles of students' interest in science issues around the world: Analysis of data from PISA 2006. *International Journal of Science Education, 33*, 97–120.
Prenzel, M., & Zimmer, K. (2006). Etudes complémentaires de PISA 2003 en Allemagne : principaux résultats et enseignements. *Revue Française de Pédagogie, 157*, 55–70.
Ratcliffe, M., & Millar, R. (2009). Teaching for understanding of science in context: Evidence from the pilot trials of the twenty-first century science courses. *Journal of Research in Science Teaching, 46*, 945–959.
Rocher, T. (2003). La méthodologie des évaluations internationales de compétences. *Psychologie et psychométrie, 24*, 117–146.
Rochex, J.-Y. (2006). Social, methodological and theoretical issues regarding assessment. Lessons from a secondary analysis of PISA 2000 literacy tests. *Review of Research in Education, 30*, 163–212.
Vermesch, P., & Maurel, M. (1997). *Pratique de l'entretien d'explicitation*. Paris: Editions ES.
Vrignaud, P. (2006). La mesure de la litéracie dans PISA: la methodologie est la réponse, mais quelle était la question? *Revue Française de Pédagogie, 157*, 27–41.

Chapter 18
Development of Understanding in Chemistry

Hannah Sevian, Vicente Talanquer, Astrid M.W. Bulte, Angelica Stacy, and Jennifer Claesgens

1 Introduction

Considerable research in science education in Europe and the United States (USA) in recent years has been focused on studying how students' understandings of core ideas in science develop over time. This work goes by many names, including "learning progressions" and "learning trajectories" in the USA and "teaching sequences" and "teaching experiments" in Europe. In this chapter, we mostly consider work associated with the "learning progressions" perspective. Learning progressions (LPs) describe successively more sophisticated ways of thinking about a topic (Corcoran et al. 2009; NRC 2007) and are based on research about how people learn as well as on the critical analysis of the structure of the associated disciplinary

H. Sevian (✉)
Department of Chemistry, University of Massachusetts-Boston,
Boston, MA 02125, USA
e-mail: Hannah.Sevian@umb.edu

V. Talanquer
Department of Chemistry and Biochemistry, University of Arizona,
Tucson, AZ 85721, USA
e-mail: vicente@email.arizona.edu

A.M.W. Bulte
Freudenthal Institute for Science and Mathematics Education, Utrecht University,
Princetonplein 5, NL 3584 CC, Utrecht, The Netherlands
e-mail: A.M.W.Bulte@uu.nl

A. Stacy
Department of Chemistry, University of California, Berkeley, CA 94720-1460, USA
e-mail: astacy@berkeley.edu

J. Claesgens
Center for Science Teaching and Learning, Northern Arizona University,
Flagstaff, AZ 86011, USA
e-mail: jennifer.claesgens@nau.edu

C. Bruguière et al. (eds.), *Topics and Trends in Current Science Education: 9th ESERA Conference Selected Contributions*, Contributions from Science Education Research 1,
DOI 10.1007/978-94-007-7281-6_18,

knowledge. To date, educational researchers have developed LPs in science for diverse topics such as atomic-molecular structure (Smith et al. 2006), properties of matter (Smith et al. 1985), carbon cycling (Mohan et al. 2009), and force and motion (Alonzo and Steedle 2009). Much current thought on LPs is collected in a recent volume (Alonzo and Gotwals 2012). However, there is still ample debate on issues such as what constitutes progress in a given area (Foster and Wiser 2012), how more sophisticated ways of thinking are characterized (Mohan and Plummer 2012), and whether progress can adequately be described as a series of successive levels of understanding (Sikorski and Hammer 2010). There are also discussions about how to apply findings from LP research to the development of standards (Foster and Wiser 2012), to curriculum development (Wiser et al. 2012), and to assessment design (Alonzo and Gotwals 2012).

The promise of LPs lies in the potential to guide the coordination of teaching practices, instructional resources, and assessment tools with students' cognitive, metacognitive, and sociocultural resources so that learning builds coherently. However, much work needs to be done to fulfill such promise. Thus, in this chapter we highlight critical issues in the development of LPs that can actually serve as effective curriculum models and assessment frameworks in the teaching of chemistry across educational levels and in diverse contexts. In particular, we underscore the need for better understanding (a) how students' reasoning evolves with training in the discipline, (b) what assessment frameworks can better uncover actual progression in understanding central concepts and ideas, and (c) what instructional sequences are likely to foster development of more sophisticated ways of thinking about core topics. To frame these discussions, in the following section we analyze critical aspects of recent research in the area of LPs in chemistry.

2 Learning Progressions

A recent paper by Duschl et al. (2011) provides an analytical review of LP research in science education, with connections to learning trajectories work in mathematics education, across work occurring in the USA and Europe over the past decades. This comprehensive review focused on how LPs are being created and how they are being validated and described. In particular, the authors isolate four major aspects in which existing LPs tend to vary.

First, Duschl et al. find that LPs tend to focus either on scientific knowledge without integrating science practices or on science practices without integrating domain knowledge. Where there is integration between concepts and practices, there is variation as to how LPs are constructed. Some LPs treat concepts and practices separately and then merge the two; a second set stresses science content over practices, while a third strand embeds or situates science practices into domain-specific contexts.

Second, Duschl et al. find that how lower and upper levels or stages of an LP (also called the lower and upper anchors) are defined tend to vary. The idea of lower and upper levels is based upon research on the differences of expert and novices. While some LPs provide explicit definitions of a lower anchor (novice), others do so more implicitly. For example, lower anchors may be expressed as descriptions of students' intuitive accounts of familiar events. Upper anchors of LPs are defined more clearly in most LP work, corresponding to descriptions of scientific knowledge and practices that students are expected to master.

A third variation in LPs relates to how intermediate levels of understanding are studied, described, and related to instruction. Some authors, for example, describe intermediate levels as a linear sequence of steps somewhat disconnected from instruction (Alonzo and Steedle 2009). In other cases, intermediate levels are described as "stepping stones" in students' learning which represent productive ways of thinking that may support important reconceptualizations with proper instruction (Smith et al. 2010).

Finally, the fourth variation identified by Duschl et al. refers to the explicit or implicit model of conceptual change associated with the LP. In particular, the authors describe two classes of conceptual change models, which they identify as the misconception-based "fix it" view and the "work with it" view. These two classes correspond to two types of LPs described by the same authors: validation LPs, which tend to view learning as a linear and rather predetermined path toward canonical forms of scientific understanding, and evolutionary LPs, which conceptualize learning as developing productive networks of conceptual knowledge. Lower and intermediate levels of progression in validation LPs often seek to elicit and confront students' misconceptions, while evolutionary LPs seek to identify productive ideas or ways of reasoning that can be used to bolster meaning making.

In addition to the characterization by Duschl et al. (2011), other attempts to describe differences among approaches to developing LPs exist. Furtak (2009) observed two types of LPs: Type 1 are sequences of correct ideas organized in a logical order deriving from consultation with experts and/or standards documents, and Type 2 are maps of student ideas bounded by naïve pre-instructional ideas about the natural world on the lower end and by scientifically accepted explanations on the upper end. Wiser et al. (2013) described two views of LPs in terms of their "knowledge paths" based on how they use empirical data on students' ideas and what students can do and by the relationships between the knowledge paths and curriculum. The first view is based on cross-sectional studies of students of assessment data, without establishing a relationship to curricula. For example, Liu and Lesniak (2005) analyzed responses to TIMSS items that assess students' understandings of different aspects of matter – structure, conservation, and change – to identify waves of understanding in students as they progress from grade 3 to 12. The second view includes relating students' beliefs longitudinally to the curriculum that students are experiencing and seeks to uncover paths of learning as series of conceptual changes bringing students' structures of knowledge more in line with scientific theories. The primary difference between these two views, then, is the question of whether LPs and curricula are distinguishable.

Our view is that LPs cannot exist outside of the conditions of student learning. However, we recognize that LP research is complex and demands comprehensive attention over long periods of time. Thus, there are multiple entry points to the development of an LP, as exemplified by existing investigations on LPs for different topics in chemistry such as atomic-molecular structure (Smith et al. 2006), properties of matter (Smith et al. 1985), the concept of substance (Johnson and Tymms 2011), and the nature of matter (Stevens et al. 2010). In this chapter, we seek to contribute to the knowledge base in the field by describing three particular approaches to the development of LPs in chemistry that emphasize different aspects of learning and teaching. The first of these approaches highlights the need for better characterizing how implicit cognitive constraints may guide and limit student reasoning at different learning stages. The second approach stresses the importance of developing a coherent content framework to track progression in conceptual understanding. The third approach challenges the dominant content-focus view in LP research by concentrating attention on instruction that motivates and engages students in purposeful activity in relevant contexts. All three approaches can lead to LPs that describe pathways of how student understanding develops over time and the conditions that optimize students' progression through these pathways. All three approaches require assessment of student understanding as part of iterative cycles of validation of the LP, moving from hypothesis of the LP, to assessment of student understanding, to interpretation of student data that revises the LP.

3 Mapping Cognitive Constraints

The development of LPs could be facilitated if we had a more extensive and thorough understanding of how students' ideas and ways of reasoning are likely to evolve with training in a discipline. Talanquer (2006, 2009) has proposed that student reasoning in chemistry seems to be guided by implicit assumptions about the nature of chemical substances and processes. These assumptions act as cognitive constraints that guide and support but also limit student reasoning. Specifically, cognitive constraints help students make decisions about what behaviors are possible or not and about what variables are most relevant in determining behavior. These constraining ideas also support the development and application of dynamic mental models of systems of interest.

A variety of researchers have identified diverse implicit cognitive elements that seem to guide, but also constrain, students' reasoning in different domains. They have referred to them in different ways, such as implicit presuppositions (Vosniadou 1994), ontological beliefs (Chi 2008), and phenomenological primitives (diSessa 1993). However, there is considerable debate on the extent to which these types of cognitive elements form coherent integrated knowledge systems or more fragmented collections of cognitive resources. It is likely that their level of integration may vary depending on the nature of the knowledge domain and the prior knowledge and experiences of each individual. The nature of constraining cognitive elements can be expected to change over time with development and learning; some of these constraints may lose or gain strength depending on existing knowledge and perceived salient cues and goals of a task. From this perspective, defining progress in understanding, or LPs, may be facilitated by first mapping the landscape of cognitive constraints that most commonly guide student reasoning when engaged in learning a given topic.

3.1 Goals and Methodology

In recent years, Talanquer (2006, 2008, 2009) and colleagues (Maeyer and Talanquer 2010) have carried out research guided by the following overarching questions:

- What central assumptions and reasoning strategies constrain students' ideas and reasoning about chemical entities at various learning stages?
- What does this set of cognitive constraints reveal about characteristic LPs for core chemical ideas and ways of thinking?

The search for answers to these questions has been pursued using different research strategies that include analysis of prior research studies in the field, paying particular attention to longitudinal and cohort studies that explore students' ideas at various grade levels, open questionnaires, and individual interviews. These latter research projects have involved college students enrolled in first- and second-year chemistry courses. In general, data analysis has used iterative constant comparison methods in which common assumptions and reasoning strategies are identified within each question or interview task. The analytical process seeks to identify (a) types of agents invoked in building explanation or making predictions (e.g., active, passive), (b) types of properties noted (e.g., compositional cues, explicit structural factors, implicit molecular properties), (c) explanatory mechanisms indicated (e.g., centralized causal, teleological), and (d) conditions judged to be relevant in explaining properties and behavior (e.g., external vs. internal factors, single vs. multiple variables). These different elements are used to build hypotheses about core implicit assumptions and reasoning strategies underlying students' explanations and predictions.

3.2 Illustrative Findings

To illustrate results generated by the studies described above, Table 18.1 summarizes some of the core implicit assumptions derived from the analysis of students' alternative conceptions about the structure of matter (Talanquer 2009). The assumptions are arranged along different dimensions (e.g., properties, structure) to indicate the possibility of semi-independent evolution with learning or development. Although the representation is linear, it does not imply that learning follows a linear path, that all individuals move sequentially through every stage, or that old assumptions fully replace new ones. The representation implies, for example, that naïve chemistry students can be expected to think of a piece of matter as a "continuous" medium that can be divided into smaller pieces that have the same properties as the original part ("inheritance" assumption). With training in the discipline, many students begin to assume that a substance is made of a collection of particles ("corpuscularity") embedded in some sort of material medium ("embedding" assumption), but many of them still consider that these particles have the same properties as the macroscopic sample.

Table 18.1 Progression of a subset of implicit assumptions that seem to constrain student reasoning about the structure of matter at different learning stages (Talanquer 2009)

Dimension	Progression			
	Naïve	*Novice*		*Expert*
Structure	Continuity ——	Granularity ——		Corpuscularity
		Embedding ——		Vacuum
Properties	Inheritance ——			Emergence
Dynamics	Static ——	Causal-Dynamic ——	Contingent-Dynamic ——	Intrínsic-Dynamic

In general, the results of these types of investigations suggest that students' ideas about chemical entities are constrained by sets of implicit assumptions that evolve with learning by addition, coalescence, differentiation, and reorganization of the core elements. Although many implicit assumptions seem to be interrelated, some of them lose or gain strength independently from one another (with age and experience some assumptions may be activated more or less frequently). Overlapping or competing assumptions about the properties of chemical substances and processes are able to coexist at any given time, particularly at intermediate learning stages. The activation of certain cognitive constraints seems to be highly dependent on judgments of similarity among systems or tasks, cognitive availability, and framing of the task based on salient cues and perceived goals.

3.3 Application

The analysis of students' assumptions and reasoning strategies about chemical entities and processes indicates that it is possible to identify a number of cognitive constraints that seem to guide student thinking in different areas and learning stages. Constraint maps, such as that illustrated in Table 18.1, can then be used to design assessment instruments to diagnose and place students along the different dimensions in the progression and also to revise the framing of the LP. As an example, Stains et al. (2011) have applied this cognitive framework to the development and validation of a survey, the Structure and Motion of Matter (SAMM) survey, designed to assess students' understanding of diffusion. Data collected from 485 students from grade 8 (age 13) to upper-level undergraduate (fourth year of university) indicate that an approach based in the identification of implicit assumptions is fruitful in characterizing progression in understanding along three relevant progress variables: (1) structure of solute and solvent substances in a gas solution, (2) origin of the motion of gaseous solute particles, and (3) nature of particle trajectories (Sevian and Stains 2013). As students apply underlying assumptions to the phenomenon of diffusion in a gas, the first progress variable appears to depend on assumptions in the structure dimension (see Table 18.1), the second upon assumptions in the dynamics dimension, and the third upon combinations of assumptions in both dimensions. However, these studies also underscore the complexity of tracking the evolution of students' ideas across many grade levels, in diverse school contexts, and with different curricula. For example, grade 8 students learning through a half-year curriculum that required them to reason using more sophisticated assumptions about structure, but not about dynamics, consistently demonstrated thinking patterns in the first and third progress variables that were more advanced than those expressed by students at grades 9–12 and university levels in which the curriculum did not explicitly require reasoning using sophisticated assumptions about structure.

4 Assessing Conceptual Progression

The development of LPs also requires the creation of coherent content frameworks to track progression in conceptual understanding. Such was the goal of the ChemQuery assessment system developed to measure and describe how students learn chemistry around the big ideas in the discipline (Claesgens et al. 2009). This project led to the development of the "Perspectives of Chemists" framework, which is built on the theoretical idea that school chemistry is largely based on three core conceptions: *matter* (matter is composed of atoms arranged in various ways), *change* (change is associated with rearrangements of atoms), and *energy* (energy is associated with changes that occur). In terms of measurement, these conceptions are considered progress variables that help characterize how far students progress in their conceptual understanding of a topic. The framework then can be used to (i) measure students' understandings in reliable and valid ways, (ii) explicitly identify relationships between the explanatory models that facilitate student understanding in chemistry and discrete standards that instructors must teach, (iii) make the goals of instruction clear enough to facilitate students' participation in regulating their own understanding, and (iv) yield information helpful to understanding how pacing, sequence, and structure of learning activities might improve student-learning outcomes.

4.1 Goals, Methodology, and Framework

The Perspectives of Chemists framework focuses on describing and mapping student conceptual understanding in chemistry. The goals are to describe what chemistry students actually learn at different educational stages and to characterize what successful learning looks like. The approach to assessment and measurement is comprised of various steps and methods. The process begins with qualitative analysis of student work through classroom observations, cognitive task analysis, and phenomenography, to elicit patterns in student response data. Scoring and quantitative data are then used to reveal additional complexity in student learning. These areas of complexity are further explored using qualitative research methods such as interviewing, verbal protocol analysis, and continued classroom observations. The approach to measurement is based on a partial credit item response that generates validity and reliability evidence, as well as estimates of how precise a student score is likely to be (Claesgens et al. 2009).

The Perspectives of Chemists framework emerged from many iterative rounds of qualitative and quantitative data collection, analysis, and interpretation of findings by groups of experts. During this process, patterns in student responses were identified, and answers were grouped to reflect similarities in thinking approaches and strategies. Construction of performance levels for different progress variables followed a generalizable pattern somewhat similar to concepts associated with the SOLO taxonomy (Biggs and Collis 1982). This taxonomy allocates student responses on assessment tasks to a hierarchy of stages. In the case of the Perspectives of Chemists framework, it included five levels:

1. *Prestructural*: Student answer is an irrelevant response to the assessment task.
2. *Unistructural*: Student response focuses on a single aspect of the information available.
3. *Multistructural*: Student response uses multiple aspects of the information available.
4. *Relational*: Student takes the information available and relates it to aspects of external information in one or more other structures, schemas, or scripts.
5. *Extended abstract*: Student response draws on and relates structures to additional information and concepts.

The ChemQuery assessment system developed as part of the research project consists of detailed descriptions of the progress variables and a scale of progression in understanding across each variable, illustrated in Table 18.2 for the "matter" variable. The system includes over 20 open-ended items associated with each progress variable, scoring rubrics, and item exemplars (Scalise et al. 2006). This system has been used to map student performance in chemistry across high school and university levels. Data were collected from 418 high school students (ages 14–17) after 1 year of chemistry instruction and from 116 university students (ages 18–20) after they completed college-level introductory chemistry. In the following subsection, results of applying this approach to the development of progression along the "matter" variable are summarized.

4.2 Summary of Results

The application of the Perspectives of Chemists framework revealed that most high school students in the sample were moving from a "notions" level as described in Table 18.2 to beginning to describe and explain properties of matter at a particulate

Table 18.2 Assessment framework: perspectives of chemists on "matter"

Levels (low to high)	Essential questions and big ideas	Description of level	Examples
1. Notions	*What do you know about matter?* Matter has mass and takes up space. It can be classified according to how it occupies space	Students articulate their ideas about *matter* and use prior experiences, observations, logical reasoning, and knowledge to provide evidence for their ideas. The focus is largely on *macroscopic* descriptions of *matter*	Students describe and explain materials or activity based on observable properties
2. Recognition	*How do chemists describe matter?* Matter is categorized and described by various types of subatomic particles, atoms, ions, and molecules	Students begin to explore language used by chemists to describe matter. They relate *atomic structure* and *motion* to *composition* and *phase*. Ways of thinking about matter are limited to relating one idea to another at a simplistic level of understanding	Students represent matter through arrangements of atoms as discrete particles
3. Formulation	*How can we think about interactions between atoms?* Composition, structure, and properties of matter are related to how electrons are distributed among atoms	Students are developing more coherent understanding that matter is made of *particles* and the *arrangements* of *particles* relate to *properties of matter*. Student reasoning is limited to causal instead of explanatory mechanisms	Students recognize that matter has characteristic properties due to the arrangement of atoms into molecules and compounds
4. Construction	*How can we understand composition, structure, properties, and amounts?* Structure and properties of matter are explained by varying strengths of interactions between particles and by particle motion	Students reason using normative *models* of chemistry and use these models to *explain* and *analyze* the *phase*, *composition*, and *properties of matter*. They use appropriate chemistry models in explanations and understand assumptions used to construct the models	Students explain molecular behavior and properties in terms of stability and energies involved in intra- and intermolecular bonding
5. Generation	*What new experiments can we design to gain a deeper understanding of matter?* Bonding models are used as foundation for the generation of new knowledge	Students are becoming experts as they gain proficiency in generating new understanding of *complex systems* through the development of new instruments and new experiments	Students design experiments to explore the structure-property relationships in macromolecular systems

level (Claesgens et al. 2009). For example, many students were starting to relate numbers of electrons, protons, and neutrons to atomic properties (e.g., identity, mass) and arrangements and motions of atoms to phase behavior. In general, high school students could articulate their ideas of matter, using prior experiences and logical reasoning to justify their thinking, but much of the evidence they provided was out of scope, off-topic, or distant from normative models of chemistry. Many students seemed to answer questions and solve problem based on hybrid mental models that merged learned chemistry concepts with intuitive understandings about chemical systems.

The results of the field study indicated that after 1 year of high school and 1 year of college chemistry, most university students scored in the range of recognition of basic models of matter (upper region of Level 2 and lower region of Level 3 in Table 18.2). Only a small fraction of students demonstrated sound conceptual understanding of multi-relational interactions in chemical systems and ability to generate accurate causal mechanisms. Many college students tended to overgeneralize the application of concepts and ideas as they engaged in problem-solving. These results indicate that many students at this level should not be expected to effectively build models relating physical and chemical properties to molecular structure (Formulation level in Table 18.2).

4.3 Conclusions and Implications

The Perspectives of Chemists framework is an illustration of how a generalizable conceptual construct calibrated with item response modeling can be used to characterize progress in the understanding of core ideas in chemistry. The described approach can be used to reliably measure student learning conceived not simply as a matter of acquiring more knowledge and skills, but as progress toward higher levels of competence and knowledge integration. The results suggest that it takes substantial time for students to achieve conceptual understanding of chemistry. However, many students seem to be able to significantly improve their thinking given time and opportunity. This progress seems to require extensive opportunities to explore, use, and recreate the language and models of chemistry. Understanding how students transition between different levels along each progress variable may help us identify more effective instructional strategies that support student learning by taking advantage of the somewhat incorrect, but often productive ways of reasoning that many students develop.

5 Engaging in Purposeful Activity

LPs are frequently conceived as descriptions of progression of conceptual understanding toward specified big ideas (upper anchors) over extended periods of time. Many studies in this area seem then to be constrained by the following assumptions:

- LPs are to be considered through a lens of development of conceptual understanding.

- Teaching influences the development of conceptual understanding by carefully planned confrontation with a certain sequence of pre-established concepts.

However, it could be argued that there may be alternative guiding premises for the elaboration of LPs. In particular, one could assume that:

- LPs should be considered through a lens of development *in activity*.
- Teaching influences the development of conceptual understanding by carefully planned confrontation with a certain sequence of *types of activity*.

In this section we elaborate on this alternative conceptualization to the development of LPs.

5.1 Activity as a Basis for Learning

Conventional LPs tend to pay close attention to the sequencing of concepts and ideas from more simple to more complex. For example within the area of "matter and materials," a suggested LP may start by focusing on understanding what objects are and then proceed to substances, to elements, and then introduce modeling using the particulate model of matter. Both the content-focused tradition, which pays close attention to the conceptual "architecture" of expert understanding, as well as the cognition-focused tradition, which attends to the cognitive "architecture" of the mind of learners, tend to prioritize the learning of content. Instructional planning under these educational paradigms focuses thus on the analysis of whether component X of the targeted content area should be taught before or after component Y. However, this approach to the development of LPs does not create opportunities for learning to be regulated by the students' own motives. In this regard, context-based approaches to chemistry education (Gilbert 2006) may provide insights into how to open spaces for student self-regulation of their own learning progress. A key element in context-based approaches to teaching and learning is the engagement of the learner in purposeful activity. Luntley (2008) describes this as follows:

> We need a notion of a kind of purposeful activity with respect to things that … captures the idea that [a] subject is putting her life in order in acting with respect to X and yet lacks concepts for discriminating X. If both conditions are met [purposeful activity with respect to X and the lack of concept X], there will then be scope for explaining the conceptual development of acquiring a concept for X out of this more basis purposeful activity. (Luntley 2008, p. 7)

This way of describing conceptual development stresses the idea that it is the motives, the affective components, the purposefulness, and the usefulness of an activity (as behavioral environment) that drive the progression of learning.

5.2 LPs as a Sequence of Purposeful Activities

According to activity theory, activity is a cultural-historical phenomenon in which human beings master their world by purposefully changing natural and social reality (Vygotsky 1978). In order to observe progression in learning over a certain time span, there needs to be a motive for this progression. Involvement and meaningful participation in purposeful activity is a critical condition for learning (Luntley 2008). From this perspective, one can argue that the planning of learning tasks should be situated within authentic social activity (Bulte et al. 2006; Gilbert 2006, Gilbert et al. 2011). Consequently, LPs should require student involvement in different types of activities with increasing levels of complexity, mimicking authentic human practices in a way that acknowledges what lies within the students' zone of proximal development.

If an activity as focal event can successfully serve as a context for learning, we should identify authentic practices within each discipline that can be adapted for educational purposes. Then, we need to sequence such activities in such a way that progression in learning can take place over time with a clear purpose and direction. Given the focus of this chapter, let us consider how one could accomplish such task in the case of teaching and learning chemistry, from primary education until the level of tertiary education.

In our complex societies, chemists engage in different types of activities, including production of food and goods, evaluation of the quality of such products, and conceptual design of new substances and processes. These can be characterized by a set of actions in diverse behavioral environments. Engaging in the activities requires developing understanding about the composition and behavior of chemical substances and processes. How could students be engaged in these types of activities to help them develop more complex or sophisticated ways of understanding? For example, one could propose the following progression:

A. *Production*: At a first level, one could work with students guided by the overarching idea that in our society we deal with all kinds of consumer products: foods and goods. Within this scenario, a variety of essential questions could be posed: What are these products good for? What are they made of? Looking for answers to such questions opens a path for constructing central ideas about *objects and materials*.
B. *Evaluation of quality*: In a next stage, the following questions may be posed: What is the quality of these products? What does it take to evaluate whether a product is good for use or consumption? Answers to these questions could be explored in different relevant contexts such as evaluating the quality of products for personal hygiene. While engaged in these activities, discussions could focus on identifying the main components of a product or on quantifying their amounts. Central ideas about chemical composition in terms of *substances and mixtures* and *elements and compounds* could come into focus for students to make sense of practical activity at this level.
C. *Conceptual design*: The evaluation of the quality of chemical products could naturally lead students to wonder about issues related to the synthesis of new substances and the production of materials. Questions such as "How is this product made?" and "How do we prepare an alternative product that better satisfies our quality criteria?" could help engage students in a next set of activities

involving the design of desired products. For example, the cleaning agents for a shampoo or a washing powder. At this stage, concepts and ideas related to the *atomic-molecular theory of matter* come into focus as powerful tools for explanation, prediction, and decision making.

D. *Research/inquiry*: The proposed sequence of activities would gradually lead to analysis of difficulties that chemists encounter to accomplish their goals: What if we cannot synthesize what we want? What if we need to explore new types of chemical substances or alternative synthetic routes? These questions could be used to motivate students to become involved in research projects that demand the development of *models* and engagement in *scientific argumentation*.

The sequences and examples of activities described above exemplify how to enact LPs through a lens of development in activity and how to facilitate the development of conceptual understanding by carefully planned confrontation with a certain sequence of types of activities. While we do not report here on results of using this approach to elucidate an LP, other studies that approach LP research through designing purposeful activities that deliberately promote a particular sequence of reconceptualizations have done so (e.g., Wiser et al. 2013).

6 Final Comments

The research endeavors described above highlight different critical aspects in the research and development of LPs. First, there is a call to more thoroughly understand implicit cognitive elements that may constrain student reasoning at different stages in development and training in a domain. The construction of detailed constraint maps like those described in Sect. 3 has proven to be useful in the design of assessment instruments that effectively diagnose students' places along a given progression. Second, there is an invitation to more clearly define and characterize what it means to advance in conceptual understanding of big ideas in a discipline. These conceptual frameworks are needed to generate coherent assessment systems that can be used to map student performance across different educational levels. Finally, there is a call to review and expand current conceptualizations of LPs to recognize the central role that engagement in purposeful activity plays in the development of meaningful understandings. From this perspective, the separation that is frequently made between learning and instruction in current LP research needs to be challenged.

Although very different in their perspectives, the three approaches to LP work described in this chapter suggest a similar progression in the understanding of core ideas about the structure of matter. This progression is visualized in Table 18.1 (Sect. 3), in the assessment framework presented in Table 18.2 (Sect. 4), and in the sequence of types of activities described in Sect. 5. In all these examples, student understanding moves from stages in which explanations and predictions are based on perceptive cues and macroscopic conceptualizations of matter, to levels in which macroscopic and particulate ideas are either merged or selectively used depending on the context, to stages in which properties, chemical entities, and processes are explained using atomic-molecular models of matter.

However, the three types of studies summarized in this chapter also elicit tensions to be navigated. The development of LPs demands the identification and definition of variables along which progress of learning can be characterized. These progress variables must be measurable. The three measurable progress variables described in Sect. 3 – structure, origin of motion, and particle trajectories – may or may not overlap well with the *matter* variable of Sect. 4. Additionally, how student understanding is intended to be measured in Sect. 3 differs from how it is measured in Sect. 4. To measure student understanding according to the LP described in Sect. 3, measurements must be designed to identify the implicit assumptions that guide student reasoning (e.g., Stains et al. 2011). However, in Sect. 4, assessment instruments are framed in terms of what students can do with their chemical knowledge (i.e., performance assessments).

The curriculum approach described in Sect. 5 provides students with purposeful practices through which they can reconceptualize their understanding. However, the opportunities for reconceptualization should ideally derive from the results of formative assessment employed by the teacher, as facilitator of student learning. Formative assessment derived from the LP study should provide the teacher with a means for creating learning opportunities that challenge students at the proper zone of proximal development. Thus, the development of teaching tools and strategies depends on the cognitive or conceptual framework that is used and on how student understanding is actually measured.

These different tensions raise a larger challenge: a need for better integration of different perspectives in the development of LPs. Attempts have been made to propose hypothetical LPs that merge findings from various LPs covering overlapping science content and age ranges (e.g., Rogat 2011). However, such a merger overlooks inconsistencies in theoretical assumptions about how cognitive constraints evolve, how students represent understanding, and how activity motivates students and fosters understanding and ability to use knowledge productively.

Notwithstanding the challenges involved, further understanding of students' development along LPs should inform the coordinated development of curricula and assessments that scaffold student learning in chemistry. Only then will LPs be more likely to achieve the promise of guiding the coordination of teaching, instructional resources, and assessment tools with students' cognitive and metacognitive resources so that learning builds coherently.

Acknowledgments The authors wish to acknowledge funding sources that provided support for their research: HS acknowledges US NSF award 0412390, AMWB acknowledges the Department of Chemistry of Utrecht University, and AS acknowledges US NSF award 0125651. Part of this work was conducted while one of the authors (HS) was under employment of the US National Science Foundation (NSF). Any opinions, findings, conclusions, or recommendations expressed in this material are those of the authors and/or of those whose work is drawn upon and do not necessarily reflect the views of any of the funding sources. The authors acknowledge their coauthors who significantly contributed to the work summarized in this chapter: Kathleen Scalise, John Gilbert, and Albert Pilot.

References

Alonzo, A. C., & Gotwals, A. W. (2012). *Learning progressions in science: Current challenges and future directions*. Rotterdam: Sense Publishers.

Alonzo, A., & Steedle, J. T. (2009). Developing and assessing a force and motion learning progression. *Science Education, 93*(3), 389–421.

Biggs, J. B., & Collis, K. F. (1982). *Evaluating the quality of learning: The solo taxonomy*. New York: Academic.

Bulte, A. M. W., Westbroek, H. B., de Jong, O., & Pilot, A. (2006). A research approach to designing chemistry education using authentic practices as contexts. *International Journal of Science Education, 28*, 1063–1086.

Chi, M. T. H. (2008). Three kinds of conceptual change: Belief revision, mental model transformation, and ontological shift. In S. Vosniadou (Ed.), *International handbook of research on conceptual change* (pp. 61–82). New York: Routledge.

Claesgens, J., Scalise, K., Wilson, M., & Stacy, A. (2009). Mapping student understanding in chemistry: The perspectives of chemists. *Science Education, 93*(1), 56–85.

Corcoran, T., Mosher, F. A., Rogat, A. (2009). Learning progressions in science: An evidence-based approach to reform. Consortium for policy research in education report #RR-63, Philadelphia.

diSessa, A. A. (1993). Toward an epistemology of physics. *Cognition and Instruction, 10*(2&3), 105–225.

Duschl, R., Maeng, S., & Sezen, A. (2011). Learning progressions and teaching sequences: A review and analysis. *Studies in Science Education, 47*(2), 123–182.

Foster, J., & Wiser, M. (2012). The potential of learning progression research to inform the design of state science standards. In A. C. Alonzo & A. W. Gotwals (Eds.), *Learning progressions in science: Current challenges and future directions* (pp. 435–459). Rotterdam: Sense Publishers.

Furtak, E. M. (2009). Toward learning progressions as teacher development tools. Paper presented at the learning progressions in science, Iowa City. http://www.education.msu.edu/projects/leaps/proceedings/Furtak.pdf. Accessed 4 May 2012.

Gilbert, J. K. (2006). On the nature of "context" in chemical education. *International Journal of Science Education, 28*(9), 957–976.

Gilbert, J. K., Bulte, A. M. W., & Pilot, A. (2011). Concept development and transfer in context-based science education. *International Journal of Science Education, 33*(6), 817–837.

Johnson, P., & Tymms, P. (2011). The emergence of a learning progression in Middle School Chemistry. *Journal of Research in Science Teaching, 48*(8), 849–877.

Liu, X., & Lesniak, K. M. (2005). Students' progression of understanding the matter concept from elementary to high school. *Science Education, 89*(3), 433–450.

Luntley, M. (2008). Conceptual development and the paradox of learning. *Journal of Philology Education, 42*(1), 1–14.

Maeyer, J., & Talanquer, V. (2010). The role of intuitive heuristics in students' thinking: Ranking chemical substances. *Science Education, 94*, 963–984.

Mohan, L., & Plummer, J. (2012). Exploring challenges to defining learning progressions. In A. C. Alonzo & A. W. Gotwals (Eds.), *Learning progressions in science: Current challenges and future directions* (pp. 139–147). Rotterdam: Sense Publishers.

Mohan, L., Chen, J., & Anderson, C. W. (2009). Developing a multi-year learning progression for carbon cycling in socio-ecological systems. *Journal of Research in Science Teaching, 46*(6), 675–698.

National Research Council (NRC). (2007). *Taking science to school: Learning and teaching science in grades K-8*. Washington, DC: National Academies Press.

Rogat, A. (2011). Developing learning progressions in support of the new science standards: a RAPID workshop series. http://www.cpre.org/developing-learning-progressions-support-new-science-standards-rapid-workshop-series-0. Accessed 19 Sept 2012.

Scalise, K., Claesgens, J., Wilson, M., Stacy, A. (2006). ChemQuery: An assessment system for mapping student progress in learning general chemistry. Paper presented at the NSF conference for assessment of student achievement, Washington, DC.

Sevian, H., & Stains, M. (2013). Implicit assumptions and progress variables in a learning progression about structure and motion of matter. In G. Tsaparlis & H. Sevian (Eds.), *Concepts of matter in science education* (pp. 69–94). Dordrecht: Springer.

Sikorski, T. R., Hammer, D. (2010). A critique of how learning progressions research conceptualizes sophistication and progress. In K. Gomez, L. Lyons, J. Radinsky (Eds.) Learning in the disciplines: Proceedings of the 9th international conference of the learning sciences, vol. 1, pp. 1032–1039. International Society of the Learning Sciences: Chicago.

Smith, C., Carey, S., & Wiser, M. (1985). On differentiation: A case study of the development of the concepts of size, weight, and density. *Cognition, 21*, 177–237.

Smith, C. L., Wiser, M., Anderson, C. W., & Krajcik, J. (2006). Implications of research on children's learning for standards and assessment: A proposed learning progression for matter and atomic-molecular theory. *Measurement, 4*(1&2), 1–98.

Smith, C. L., Wiser, M., Carraher, D. W. (2010). Using a comparative, longitudinal study with upper elementary school students to test some assumptions of a learning progression for matter. Paper presented at the annual meeting of the National Association for Research on Science Teaching, Philadelphia.

Stains, M. N., Escriu-Suñé, M., Molina, M., & Sevian, H. (2011). Assessing secondary and college students' understanding of the particulate nature of matter: Development and validation of the Structure And Motion of Matter (SAMM) survey. *Journal of Chemical Education, 88*(10), 1359–1365.

Stevens, S. Y., Delgado, C., & Krajcik, J. S. (2010). Developing a hypothetical multi-dimensional learning progression for the nature of matter. *Journal of Research in Science Teaching, 47*(6), 687–715.

Talanquer, V. (2006). Common sense chemistry: A model for understanding students' alternative conceptions. *Journal of Chemical Education, 83*(5), 811–816.

Talanquer, V. (2008). Students' predictions about the sensory properties of chemical compounds: Additive versus emergent frameworks. *Science Education, 92*(1), 96–114.

Talanquer, V. (2009). On cognitive constraints and learning progressions: The case of structure of matter. *International Journal of Science Education, 31*(15), 2123–2136.

Vosniadou, S. (1994). Capturing and modeling the process of conceptual change. *Learning and Instruction, 4*(1), 45–69.

Vygotsky, L. S. (1978). *Mind in society. The development of higher psychological processes.* Cambridge: Harvard University Press.

Wiser, M., Smith, C. L., & Doubler, S. (2012). Learning progressions as tools for curriculum development: Lessons from the inquiry project. In A. C. Alonzo & A. W. Gotwals (Eds.), *Learning progressions in science: Current challenges and future directions* (pp. 359–403). Rotterdam: Sense Publishers.

Wiser, M., Frazier, K., & Fox, V. (2013). At the beginning was amount of material: A learning progression for matter for early elementary grades. In G. Tsaparlis & H. Sevian (Eds.), *Concepts of matter in science education* (pp. 95–122). Dordrecht: Springer.

Chapter 19
Learning Affordances: Understanding Visitors' Learning in Science Museum Environment

Hyeonjeong Shin, Eun Ji Park, and Chan-Jong Kim

1 Introduction

A recent report from the US Board of Science Education says that there is abundant evidence that informal science education settings, such as everyday family environment, museums, and community-based learning programs, are important contributors to people's knowledge and interest in science (Bell et al. 2009). Among informal settings, learning in science museums is characterized by visitors' interaction with exhibits as well as with other people. Physical environment of science museums, exhibits, has been the focus of research because many researchers believe that the types and characteristics of the exhibits can make differences in visitors' learning. Multisidedness, accessibility, multimode and relevancy (Borun and Dritsas 1997), and open-endedness and technological novelty (Sandifer 2003) are examples of the characteristics of exhibits investigated. Such studies have contributed to the understandings of the relationship between visitors' learning and physical environments in museums. However, to understand visitors' informal science learning processes in science museums, we need to focus on more fundamental aspects of exhibits for motivating and triggering visitors' learning.

We adopt the perspectives of learning in science museums as meaning-making processes (Mortimer and Scott 2003) using resources and capabilities of the visitors and resources provided by the exhibits and through social interactions with other people in science museums. Museum exhibits can be regarded as mediating artifacts designed for learning and entertaining visitors. Artifacts and visitors' perception about them influence their behavior. As exhibits are mediating artifacts, they provide affordances for acting and learning to the visitors. We would like to suggest the concept of "learning affordance" and elaborate the concept by investigating the visitors' interaction with exhibits in a science museum.

H. Shin (✉) • E.J. Park • C.-J. Kim
Seoul National University, Seoul, South Korea
e-mail: masic12@gmail.com; starryblueej@gmail.com; chajokim@gmail.com

C. Bruguière et al. (eds.), *Topics and Trends in Current Science Education: 9th ESERA Conference Selected Contributions*, Contributions from Science Education Research 1,
DOI 10.1007/978-94-007-7281-6_19,

Fig. 19.1 Categorization of affordances based on the perceptible information (Adapted from Gaver 1991)

Perceptual information		Affordance: Yes	Affordance: No
	Yes	Perceptible affordance	False Affordance
	No	Hidden affordance	Correct Rejection

1.1 Affordances and Learning Affordances

The concept of "affordance" was formulated by Gibson (1979/1986) and popularized by Norman (1988). Gibson (1979/1986) first defined affordances as all "action possibilities" latent in the environment, objectively measurable and independent of the individual's ability to recognize them, but always in relation to the actor and dependent on their capabilities.

Norman (1988) appropriated Gibson's affordances in the context of human-machine interaction to refer to those action possibilities that are readily perceivable by an actor. It makes the concept dependent not only on the physical capabilities of an actor but also the actor's goals, plans, values, beliefs, and past experiences. In general, affordances are thought to be different from abstract physical properties, because they are emergent properties of an object in relation between an organism and an object (Hammond 2010). Meanwhile, an affordance exists relative to the action capabilities of a particular actor (McGrenere and Ho 2000). The affordances of environment are perceived not by all agents but by a specific agent within specific space and time (McGrenere and Ho 2000). For instance, the affordance of a doorknob is perceived to an adult who can turn it, but not to a child who cannot reach it or a dog which cannot turn it. Since perception of an affordance is related to an agent's intention, the affordance cannot be perceived without the agent's specific goals and abilities (Young et al. 2000). For example, a turning round metal piece on the door can be perceived by who intends to open the door. Even though there is a myriad of information about affordance, only the agent who has a specific intention can pick up the information. In this sense, the concept of affordances is similar to the concept of "umwelt" (Uexkull 1992). The umwelt of an organism is a section carved out of the environment with perceptual cues and reacting.

By distinguishing affordances from perceptual information about them, Gaver (1991) offered further insight about affordances. Gaver proposed three types of affordances: "perceptible affordance," "false affordance," and "hidden affordance" (Fig. 19.1). If there is explicit information about the affordance in the objects, then it is perceptible affordance. Without perceptible information, affordance is hidden to agents. If there is perceptible information even though there is no affordance, it is false affordance. Also, he introduced other affordances for complex actions such as "nested affordances" which are grouped in space and "sequential affordances" which are revealed over time.

As objects, scientific exhibits have their own affordances. Moreover, based on the concepts of Gibson, Gaver, and Norman, the quality of a scientific exhibit depends on whether or how visitors perceive the affordances of an exhibit. However, affordances of the scientific exhibits have rarely been studied extensively. Rowe (2002) regarded affordance as "the potential" of a museum object, and it may afford certain practices and constrains others. For Allen (2004), affordance refers to "immediate apprehendability," which reduces cognitive overload and frees visitors to focus on those aspects of the environment that are worthy of their attention. She said that the immediate apprehendability can promote both intrinsic motivation and science learning, and one way to achieve this immediate apprehendability and reduce cognitive overload in exhibits is through "user-centered design" (also called "end-user" or "natural" design). In these studies, affordances are assumed as certain characteristics of exhibits to be perceived immediately by the visitors and to promote certain practices for the visitors. However, the nature of the affordances for learning has rarely been investigated.

Exhibits in science museums are known to encourage visitors' learning science (Afonso and Gilbert 2007). Learning occurs in the science museums because visitors perceive the affordances for learning in the exhibits. Learning affordances were conceptualized by the researchers as the intersection of affordances of exhibits and visitors' goals and abilities. Learning affordances are affordances directly related to learning. They are also decided by the reciprocal processes between the affordances and the goals and ability of the visitors. In these senses, learning affordances are part of general affordances.

1.2 Research Question

The purposes of the study are to investigate the possibilities and usefulness of the concept of learning affordance. From this goal, we set up the research question as below: Which characteristics of learning affordance occur during visit to the science museum?

2 Research Design

The research site of this study was Gwacheon National Museum in Korea, especially in the Basic Science Hall. This museum is very popular among Korean students and parents because of its new and interactive exhibits in various halls. Among them, Basic Science Hall has 56 exhibits in the areas of mathematics, physics, chemistry, biology, and Earth science. Data collection was conducted for 6 months from June to November 2010. The total number of the participants was 82. Participants were

Analysis Elements of exhibit	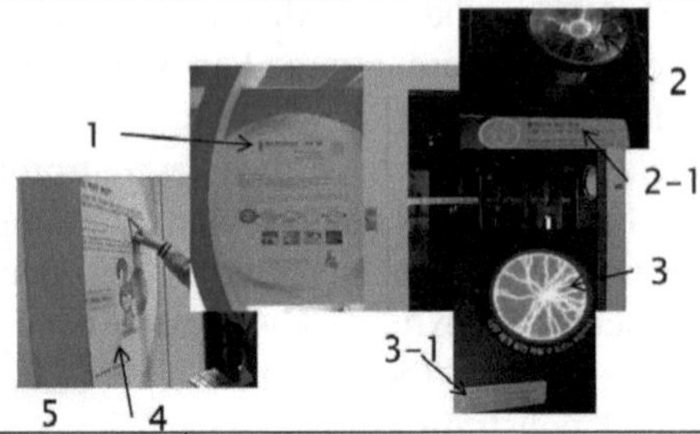

Physical affordance → (scenario, usability of interface)	Contents affordance (usefulness)
1(Exp. Panel): read	Four phases of elements, definition of plasma
2-1, 3-1(Ope. Label): read → 2, 3(touch)	observation Plasma discharge
4(Exp. Panel): read	The reason why lights gathered finger tips
5(movie panel (3'35")): watch	What is plasma, usage of plasma.

Fig. 19.2 Affordance analysis of World of Plasma Exhibit. Each element not only has physical affordances which bring about visitor's behavior, such as read, touch, and watch, but also has content affordances being used during interaction

chosen from 7th- to 11th-grade students (from 12 to 16 years old) visiting the museum, because the exhibits of the hall are optimized for 7th grade and above.

The data gathering process was divided into three phases: analyzing exhibits, observing participants visiting exhibits, and analyzing the data collected. In the first phase, the physical segments and contents of each exhibit were analyzed according to Webb (2005) and McGrene and Ho (2000) (Fig. 19.2). Especially, the physical segments were divided and given numbers by the analysis of microcontext (Botelho and Morais 2006). Moreover, the basic behaviors which can be led to from the physical segments, such as "read," "touch," and "watch," were identified. The contents of exhibits as well as the physical segments were analyzed and described because they introduced what to learn to visitors. Related information on exhibits was collected from the website of the science museum. Exhibit manuals of the hall were also collected and used for the analysis. Participants' interviews, writings, and worksheets completed during their visit were also collected after their visits of the hall. The information was compared with the result of analysis of the physical segments and contents of each exhibit.

In the second phase, eight pairs of participants were investigated. They were observed and video recorded during their visits and interviewed right after their visits to the hall. Two participants formed a dyad and asked to visit and talk to each other about exhibits. The purposes of forming dyads were twofold: Situate participants in a typical situation of science museum visit and let them show meaning-making process to the researchers through conversation (Coffee 2007; Rahm 2004; Rowe 2002; Zimmerman et al. 2010). It was also expected that they could understand and interact with exhibits better with others than alone (Draper 1985; Rennie and

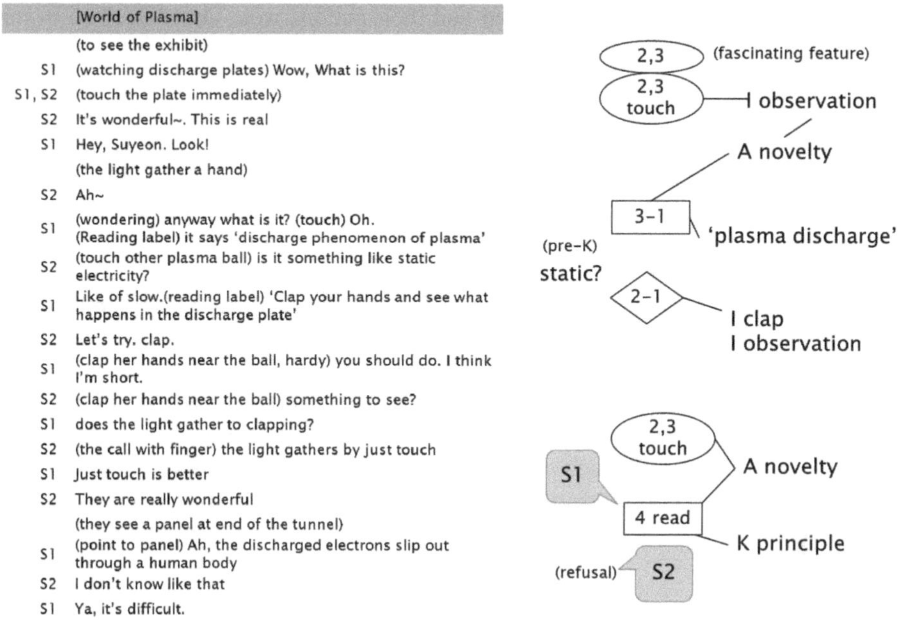

Fig. 19.3 An example of analysis of one group's interaction on World of Plasma Exhibit. *Numbers* in the diagram are elements of exhibit (refer to Fig. 19.2). *Lines* mean learning affordances brought about learning (Index: *pre-K* previous knowledge, *I* inquiry learning, *A* affective learning, *K* knowledge or cognitive learning)

McClafferty 2002). All participants visited exhibits freely with average visiting time 33′25″. A researcher followed them keeping a distance, observing and keeping field notes. Two video recorders and an extra voice recorder were also used for every pair of participants to minimize lost data in open spaces like science museums. After their visits, each of them was asked to choose three exhibits which were most impressive and interviewed in front of the exhibit chosen by them. During the interview, questions such as the reasons why they chose them, how they worked, and what they did and why they did it while visiting the exhibits were asked.

In the third phase, mostly data analyses were conducted. Participants' discourses, writings, and interviews were fragmented into units based on interaction scene between an exhibit and participants. Interaction scene is participants' talking, writing, or discourse about an exhibit. Each unit was analyzed according to meaning-making processes and learning outcomes. Meaning-making processes in science museums were combinations of acting (reading, watching, touching, pressing, turning, clapping, etc.) and talking (asking questions, answering, reading aloud, asking to do something, etc.). Five types of learning outcomes in science museums were categorized based on previous studies such as generic learning outcomes (Hooper-Greenhill 2007) and domains of informal science learning (Bell et al. 2009), and they were used for the analysis: knowledge, inquiry, affective aspects, value, and implementing. An example of analysis about World of Plasma Exhibit is shown in Fig. 19.3. Discourses of the two participants (S1, S2) were transcribed and their actions were also described in the parenthesis on the left side of Fig. 19.3. Their interactions and learning outcomes were summarized on the right side of Fig. 19.3. The numbers represent physical elements of the exhibit (Fig. 19.2). Participants' interactions with physical elements are represented by the shape of the boxes (acting, ellipse; explanatory labels of science contents, box; explanatory labels of how to operate, diamond). Learning outcomes are represented by signs (I, inquiry learning; A, affective learning; K, knowledge). By looking at the right side of Fig. 19.3, we can understand how participants interacted with the exhibit and what they learned from the exhibit.

All data were analyzed and interpreted qualitatively. Researchers' coding and interpretation are important elements in this research. To secure validity and reliability, we try to use triangulation by using and comparing data from multiple sources such as participants' writings, activity sheets, video recordings, and interviews. We also tried to find and describe episodes related to the concept of learning affordances as much as possible to understand the possibilities and usefulness of the concept. Research team also has had continuous discussions about the data and interpretation from the beginning to check and improve the validity of the research.

3 Results: Elaborating the Concept of Learning Affordances

The purpose of this study is developing and elaborating the concept of learning affordances in science museum learning context. The results of the study are described along the development of the idea. The concept of learning affordance was devised and evolved successively along the process of the study.

3.1 First Stage: Defining Learning Affordance

Elements of exhibits are thought to have certain affordances because it is assumed that visitors have action abilities. Although general meaning of affordance focuses on direct and eidetic perception, perceptions of an exhibit's elements are not always connected to learning. For instance, "seeing" a panel of a certain exhibit means that the "seeing" action is caused by recognizing affordance of the exhibit, but whether the visitor recognizes the information in the element meaningfully or not is a different matter. Learning affordance is conceptualized as a special kind of affordance which brings about visitors' learning. It would be codetermined by visitors' intentions and abilities and characteristics of exhibits. The initial model of learning affordance is the intersection of affordances in exhibits and learner's ability (Fig. 19.4). Topic, contents, and methods of students' meaning-making activities are influenced by learning affordances in exhibits, and, at the same time, whether those affordances are functioned as "learning affordances" is decided by learners as shown in Fig. 19.4.

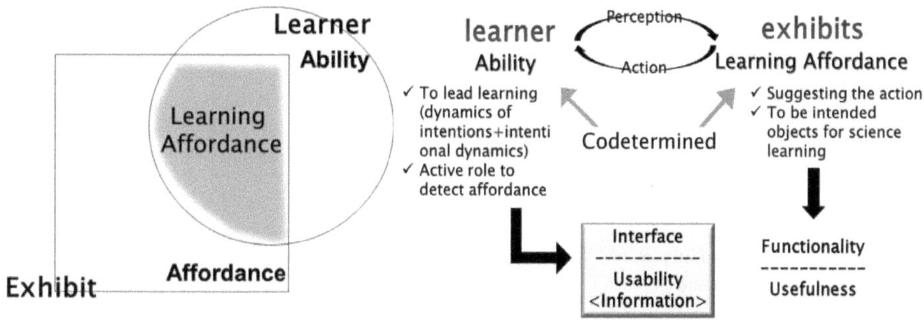

Fig. 19.4 Learning affordance is defined as the intersection of affordance and visitors' ability

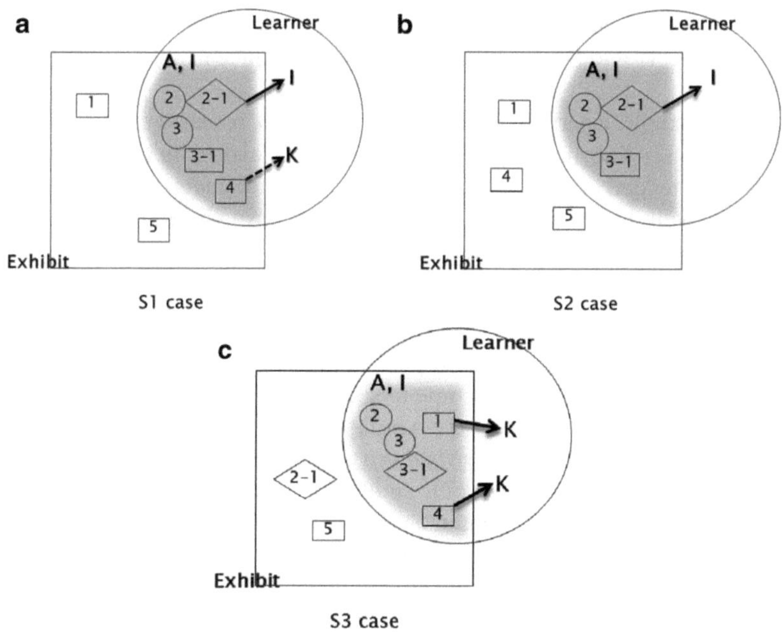

Fig. 19.5 An element of exhibit sometimes cannot offer learning affordances. In this World of Plasma Exhibit, Panel 4 provided learning for S3 and insufficient learning (*broken line*) for S1, while not for S2. The *numbers* in the diagram represent the elements of the exhibit. *I* inquiry learning, *A* affective learning, *K* knowledge, cognitive learning

The dynamic, reciprocal relationships between visitors and exhibits in shaping learning affordance were found in this study (Fig. 19.5). For example, during a couple's (S1 and S2) visit to World of Plasma Exhibit, S1 and S2 interacted with the exhibit and experienced how plasma works. This ended up with learning mostly in affective and inquiry domains. However, around the end of their visit to

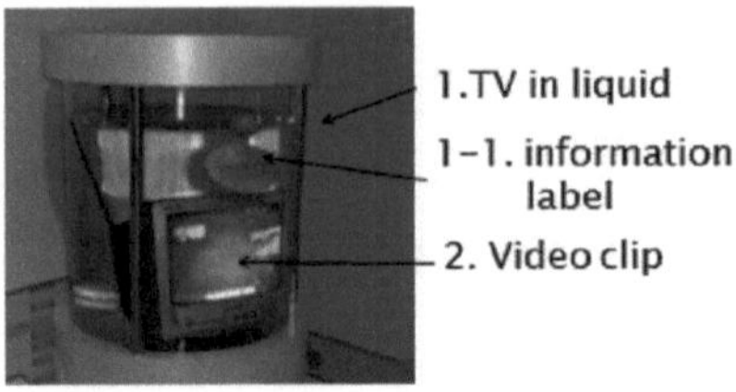

Learning Affordance	
Physical affordance	Contents affordance
1. TV in liquid	Silicon oil as insulation material
1-1. Information label	About silicon oil
2. Video on TV	Art with chemical material

Fig. 19.6 Affordance analysis of Chemical Art Exhibit. No explicit information in video contents made it hidden affordance

this exhibit, they faced Panel 4. Panel 4 has basic information about plasma. Panel 4 functioned as learning affordance to visitor S1 which leads to cognitive learning although it was not sufficient. However, it was not recognized as learning affordance to visitor S2, and no learning happened to her. The differences in ability between visitor S1 and S2 contributed to their difference in shaping their learning affordances.

For another example, a participant in another group, S3, revealed affordance different from S1 and S2. She had high level of interests and achievement in science. Moreover, she tended to read panels carefully and could understand the explanation in them very well. For that reason, even though she also visited the Plasma Exhibit for the first time like S1 or S2, she could recognize Panel 1, could read it, and had more cognitive learning (Fig. 19.5). During interview, she answered that she learned the new concept, plasma, from the exhibit.

Researcher: Have you known about plasma before?
S3: No. This is my first time to see it.
Researcher: How ;do you know the word, plasma?
S3: ***I read that when I visited this exhibit.*** That was something separated. Like this. An atom? Or an electron? Anyway, they were separated. I saw this movement on TV before, but, when I touched it, they followed my hand. From that, I knew for the first time that the electrons were escaping and came to my finger. It was interesting.

3.2 Second Stage: Hidden Affordance

Sometimes, learning affordance was not formed because of the lack of perceptual information in the exhibit. Learning affordance depends on perceived information related to the topic. Without relevant information, affordance could be hidden. For instance, an exhibit titled Chemical Art is fairly popular in "Basic Science Hall" (Fig. 19.6). Visitors are astonished and attracted by this exhibit because in this exhibit, TV seems to be working in the water. The label of the exhibit explains about the liquid (silicon oil). However, as the title shows, the major goal of this exhibit is introducing chemical ingredient of dye used by artists. Visitors can only realize the

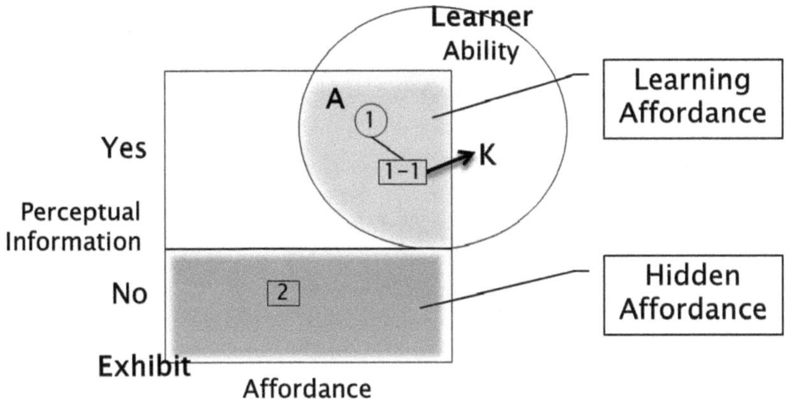

Fig. 19.7 The result of analysis of a small group's interaction with Chemical Art Exhibit. Hidden affordance is identified as affordance without perceptual information. The *numbers* in the diagram represent the elements of the exhibit. *I* inquiry learning, *A* affective learning, *K* knowledge or cognitive learning

goal of this exhibit when they watch the video contents. Therefore, few participants recognized this goal because there was no explicit information to introduce this aspect in the exhibit. By this case, we can differentiate affordance into two areas, affordance and hidden affordance. Without perceptual information about relevant aspects of an exhibit, visitors hardly shape learning affordance even though they enjoyed the exhibit (Fig. 19.7).

3.3 *Third Stage: False Affordance and Correct Rejection*

It is expected that affordance would usually allow visitors to perceive properly and act accordingly, hence enhance learning in science museums. However, sometimes certain affordance would make visitors confused. Principle of Particle Accelerator Exhibit shows an example. It uses a metal ball and coils to simulate particle accelerator. The metal ball represents a charged particle and coils represent drift tubes of particle accelerator. If a visitor pushes start button, small metal ball would start along the circular rail and goes gradually faster. After 1 min, the metal ball stops automatically and Lamp 6 is turned on. When the light of Lamp 6 is turned on, the exhibit wouldn't work. Visitors have to wait for 2 min for next operation. Unfortunately, the shape of the Lamp 6 is exactly the same with start button nearby (Fig. 19.8). Besides, the Lamp 6 looks even more attractive when it is turned on. When it was turned on, participants usually pushed the Lamp 6 and start button in turn repeatedly without knowing that it was in the waiting mode and said, "It doesn't work! What is it?" In this example, some elements, such as Lamp 6, offered false affordance. Visitors perceived it as a button because of its shape and tried to push it

Fig. 19.8 Part of Principle of Particle Accelerator Exhibit. Though these two elements look the same, the *left* one is button 5 with name tag "start," and the *right* one is Lamp 6 without any tag

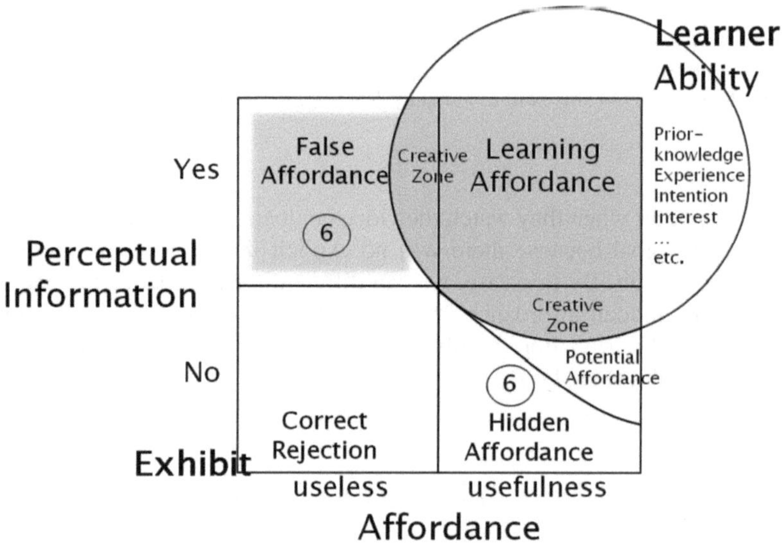

Fig. 19.9 Affordance is differentiated further by its usefulness. As a result, four types of affordance are defined: learning affordance, hidden affordance, false affordance, and correct rejection

to operate the model. There is no information about the role of Lamp 6 or the meaning of light on Lamp 6 in the exhibit. Although Lamp 6 plays its role as an indicator, its shape offers perceptive information like button which arouses useless push action to the visitors.

This case exemplifies the need to differentiate affordance into useless as well as useful. With this differentiation, affordance can be divided into four main categories: learning affordance, false affordance, hidden affordance, and correct rejection (Fig. 19.9). When an exhibit or its elements have useful perceptual information, it has learning affordance. However, if it or its elements have useless perceptual information, then it has false affordance. When an exhibit or its elements have useful unperceptual information, then it has hidden affordance. When an exhibit or its

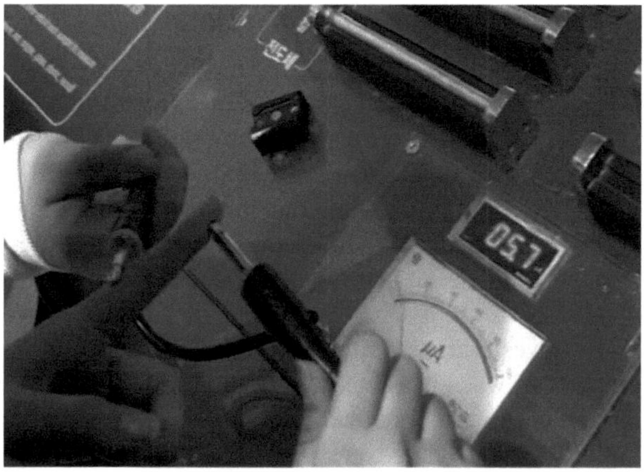

Fig. 19.10 Visitors' interaction beyond the affordance at "Free Electron and Metallic Bond" Exhibit

elements have useless unperceptual information, then it is correct rejection. False affordance usually disturbs or cuts learning experience. When participants visited Principle of Particle Accelerator Exhibit, they tried in vain or gave up and left the exhibit because they could not interact with the exhibit.

3.4 Fourth Stage: Creative Zone

Visitors interact with and learn from exhibits in various ways. Many times, their interactions are well aligned to the intentions and expectations of the exhibit designers or educators. However, sometimes, they interact with exhibits in unexpected ways and built their own meanings integrated with their own interests and experience. Visitors' unexpected interactions usually occur in hidden or false affordance areas in Fig. 19.9. We call the learning affordance in hidden or false affordance as creative zone.

For example, Free Electron and Metallic Bond Exhibit consists of three kinds of metal rods, three kinds of nonmetal rods, two electrode sticks, and an ammeter. Visitors are supposed to touch both sides of a rod with the sticks, to see ammeter whether its needle is moving or not, and if it moves, how much. Most participants interacted with the exhibit in anticipated ways and succeeded to recognize conductors and nonconductors. However, a small group went further by touching their fingers with the electrode sticks because they wanted to check if electric currents pass through their bodies as they learned in schools. They found the needle of ammeter was moving, and they were very excited. They performed additional interaction even though there was no explicit information about testing human body in this exhibit. This case illustrates that hidden affordance can serve as learning affordance if visitors have abilities and motives even though there is no explicit information in the exhibit (Fig. 19.10).

4 Discussion and Implications

To understand visitors' learning in informal science context, the concept of learning affordance was theorized and elaborated by investigating visitors in Gwacheon National Science Museum in Korea. Learning affordance was conceptualized as a special kind of affordance which brings about learning to the visitors. It would be codetermined by visitors' intentions and abilities and affordances of exhibits. The initial model of learning affordance was the intersection of exhibit's affordance and learner's ability.

By investigating visitors' learning in science museums, the concept of learning affordance was elaborated. First, learning affordance was confirmed as codetermination of affordances of exhibits and visitors' intentions and abilities. Second, learning affordances usually require explicit perceptual information about affordances. Third, without explicit perceptual information about affordances, the exhibit has hidden affordances. With hidden affordances, learning affordances are not shaped usually and visitors have difficulties to interact with the exhibits. Fourth, perceptible but useless affordance is categorized as false affordance. With false affordance, learning affordances can hardly be formed. False affordances even can disturb or stop learning processes. Fifth, sometimes, visitors with high ability or interests can develop creative zones with hidden or false affordances and arrive at learning outcomes unexpected or integrated with personal interests and experience.

Learning affordances have dynamic and reciprocal features because of the interplay between the exhibits and visitors. With this concept of learning affordance, we can understand the nature of the exhibit's educational potential and interaction processes between the visitors and exhibits. By analyzing the discourses of visitors, their behavior, conversation, and learning can be understood. The learning affordance diagram (Fig. 19.9) seems to be a powerful tool to represent and understand the visitors' learning process visually.

Learning affordances also provide valuable information about the appropriateness of the exhibits depending on the abilities of the visitors. The area of learning affordances and creative zones in Fig. 19.9 widens while false affordances and hidden affordances decrease as a visitor's ability increases. However, special attention and provision are needed to low-ability visitors with limited learning affordances to the exhibits. More explicit and useful perceptual information should be provided to them. In addition, if visitors' creative zones of affordances in exhibits are found, this can also be considered in development or improvement of the exhibits for low-ability visitors.

The concept of "learning affordance" may help us to understand interaction between visitors and exhibits. From the ecological perspective, we can discuss various aspects of science learning during free-choice visits in science museums. Therefore, the concept of affordance can lead a holistic view about the nature of the exhibit's educational potential to inspire thinking and behavior related to science learning.

In addition, the concept of learning affordance can be applicable to other learning environments or resources such as school classrooms or learning materials including printed and audiovisual media. Also, this concept can be used to understand science teachers' talking and explanations. Moreover, this concept can also contribute to online learning and human-computer interaction (HCI) fields in developing and revising learning programs and understanding learning processes. Further research and continuous efforts are needed to elaborate and expand the theoretical framework of learning affordances.

References

Afonso, A. S., & Gilbert, J. K. (2007). Educational value of different types of exhibits in an interactive science and technology center. *Science Education, 91*(6), 967–987. doi:10.1002/sce.20220.

Allen, S. (2004). Designs for learning: Studying science museum exhibits that do more than entertain. *Science Education, 88*(S1), S17–S33. doi:10.1002/sce.20016.

Bell, P., Lewenstein, B., Shouse, A. W., & Feder, M. A. (2009). *Learning science in informal environments: People, places, and pursuits*. Washington, DC: National Academies Press.

Borun, M., & Dritsas, J. (1997). Developing family-friendly exhibits. *Curator: The Museum Journal, 40*(3), 178–196. doi:10.1111/j.2151-6952.1997.tb01302.x.

Botelho, A., & Morais, A. M. (2006). Students-exhibits interaction at a science center. *Journal of Research in Science Teaching, 43*(10), 987–1018. doi:10.1002/tea.20135.

Coffee, K. (2007). Audience research and the museum experience as social practice. *Museum Management and Curatorship, 22*(4), 377–389. doi:10.1080/09647770701757732.

Draper, L. (1985). *Friendship and the museum experience: The interrelationship of social ties and learning*. Unpublished doctoral dissertation, University of California, Berkeley.

Gaver, W. W. (1991). *Technology affordances*. Paper presented at the SIGCHI conference on human factors in computing systems: Reaching through technology, New Orleans, doi: 10.1145/108844.108856.

Gibson, J. J. (1986). *The ecological approach to visual perception*. Hillsdale: Lawrence Erlbaum (Original work published 1979).

Hammond, M. (2010). What is an affordance and can it help us understand the use of ICT in education? *Education and Information Technologies, 15*(3), 205–217. doi:10.1007/s10639-009-9106-z.

Hooper-Greenhill, E. (2007). *Museums and education: Purpose, pedagogy, performance*. London: Routledge.

McGrenere, J., & Ho, W. (2000). *Affordances: Clarifying and evolving a concept*. Paper presented at the Graphics Interface 2000, Montreal.

Mortime, R. E., & Scott, P. (2003). *Meaning making in secondary science classrooms*. New York: McGraw-Hill.

Norman, D. A. (1988). *The psychology of everyday things*. New York: Basic Books.

Rahm, J. (2004). Multiple modes of meaning-making in a science center. *Science Education, 88*(2), 223–247. doi:10.1002/sce.10117.

Rennie, L. J., & McClafferty, T. P. (2002). Objects and learning: Understanding young children's interaction with science exhibits. In S. G. Paris (Ed.), *Perspectives on object-centered learning in museums* (pp. 174–194). Mahwah: Lawrence Erlbaum.

Rowe, S. (2002). The role of objects in active, distributed meaning-making. In S. G. Paris (Ed.), *Perspectives on object-centered learning in museums* (pp. 17–32). Mahwah: Lawrence Erlbaum.

Sandifer, C. (2003). Technological novelty and open-endedness: Two characteristics of interactive exhibits that contribute to the holding of visitor attention in a science museum. *Journal of Research in Science Teaching, 40*(2), 121–137. doi:10.1002/tea.10068.

Von Uexkull, J. (1992). A stroll through the worlds of animals and men: A picture book of invisible worlds. *Semiotica, 89*(4), 319–391.

Webb, M. E. (2005). Affordances of ICT in science learning: implications for an integrated pedagogy. *International Journal of Science Education, 27*(6), 705–735.

Young, M. F., Barab, S. A., & Garrett, S. (2000). Agent as detector: An ecological psychology perspective on learning by perceiving-acting systems. In D. H. Jonassen & S. M. Land (Eds.), *Theoretical foundations of learning environments* (pp. 147–172). Mahwah: Lawrence Erlbaum.

Zimmerman, H. T., Reeve, S., & Bell, P. (2010). Family sense-making practices in science center conversations. *Science Education, 94*(3), 478–505. doi:10.1002/sce.20374.

Chapter 20
Modelling and Assessing Experimental Competencies in Physics

Heike Theyßen, Horst Schecker, Christoph Gut, Martin Hopf, Jochen Kuhn, Peter Labudde, Andreas Müller, Nico Schreiber, and Patrik Vogt

1 Introduction

Experimental work is regarded as an essential part of physics education. Science curricula from primary school to university request that students learn to carry out experiments (e.g., England (QCA 2004 (revised)), Canada (NRC 1996) and Germany (KMK 2004)). In contrast to its importance, experimental work of students is often criticised as ineffective. Most findings from science education research do not support the hypothesis that students working with experimental materials grasp science more easily (Singer et al. 2006). In a review of research, Hofstein and Lunetta (2004) demand more support for teachers to overcome the limitations of labwork in teaching science. Section 3 of this chapter presents a problem-based guided enquiry approach and describes its effects.

H. Theyßen (✉) • N. Schreiber
University of Duisburg-Essen, Essen, Germany
e-mail: heike.theyssen@uni-due.de; nico.schreiber@tu-dortmund.de

H. Schecker
University of Bremen, Bremen, Germany
e-mail: schecker@uni-bremen.de

C. Gut
Zurich University of Teacher Education, Zurich, Switzerland
e-mail: christoph.gut@fhnw.ch

M. Hopf
University of Vienna, Vienna, Austria
e-mail: martin.hopf@univie.ac.at

J. Kuhn
University of Kaiserslautern, Kaiserslautern, Germany
e-mail: kuhn@uni-landau.de

C. Bruguière et al. (eds.), *Topics and Trends in Current Science Education: 9th ESERA Conference Selected Contributions*, Contributions from Science Education Research 1,
DOI 10.1007/978-94-007-7281-6_20, © Springer Science+Business Media Dordrecht 2014

Experimenting as an important domain of scientific literacy needs to be structured by explicit models. This chapter introduces two approaches for modelling the phases of experimenting and the development of experimental competencies (Sect. 4). The focus of this chapter lies on assessment procedures for students' performances in experimenting. Section 5 shows product-based as well as process-oriented analyses. Data from empirical studies serve to evaluate chances and limitations of the assessment tools. The chapter starts with a brief overview of the theoretical background.

2 Theoretical Background

2.1 Modelling Experimental Competencies

Although the significance of experimental competence is non-controversial, the construct needs to be clarified as a basis for teaching and assessment. As a starting point, normative models have to be developed, which are then to be validated empirically. Most approaches to experimental competencies, like those presented by Hammann (2004) and Walpuski (2006), distinguish three phases of the experimental process: The first phase comprises the "preparation", e.g. to generate questions for an experimental investigation and to plan an adequate experiment. The second phase contains the performance of the experiment and the third phase the conclusions, e.g. to answer the initial question with regard to the experimental results.

Most models of experimental competencies do not get more detailed concerning the second phase, the actual performance of the experiment. This may account for the fact that most of the tests used to validate these models neither measure a student's ability to actually perform a real experiment nor do they differentiate between components of this ability. Klahr and Dunbar (1988), e.g., used very simple computer-based experiments in which the test persons only had to choose the right constellation of experimental parameters and the system displayed the results that the corresponding experiment would have yielded. Others (e.g. Möller et al. 2007; Hammann 2004) used written tests only.

Section 4 of this chapter presents and compares two models for structuring the components of experimental competence in more detail.

P. Labudde
University of Applied Sciences North-West Switzerland, Basel, Switzerland
e-mail: peter.labudde@fhnw.ch

A. Müller
University of Geneva, Geneva, Switzerland
e-mail: Andreas.Mueller@unige.ch

P. Vogt
University of Education Schwäbisch Gmünd, Schwäbisch Gmünd, Germany
e-mail: patrik.vogt@ph-gmuend.de

2.2 *Assessing Experimental Competencies*

Usually experimental competencies are assessed by written tests (e.g. Henke and Sumfleth 2006; Grube et al. 2007) or hands-on tests (e.g. Harmon et al. 1997; Shavelson et al. 1999; Chen and Klahr 1999). In some studies, computer-based simulations are used as assessment tools (e.g. Shavelson et al. 1999; Klahr and Dunbar 1988). The choice of tools depends on the underlying model of experimental competencies and the focus of the investigation. Studies focusing on *process* aspects typically use tasks with hands-on experiments and a process-oriented analysis of students' actions (e.g. Walpuski 2006). Since the method is very resource consuming, a product-based analysis is often preferred to evaluate hands-on tests (e.g. Shavelson et al. 1999; Harmon et al. 1997). There is a need to further develop and evaluate assessment procedures. Section 5 focuses on this topic.

3 A Problem-Based Guided Enquiry Approach

There has been a long discussion about effects of labwork in science education with a lot of research and very heterogeneous results. One reason for this heterogeneity is that labwork is a very broad term for all sorts of activities. For the purpose of measuring experimental competencies, this is a real problem, which will be discussed in the following.

Hopf (2011) studied the effects of labwork on conceptual understanding, interests and motivation. One of the most cited criticisms of traditional "cookbook"-style labwork is that students are not fully engaged in thinking about their labwork while doing it. In the literature one can find many ideas how to improve labwork learning environments to change this unwanted student behaviour. Hopf evaluated one of these approaches in depth (for more details see Hopf 2011). Adapting an idea from van Heuvelen (1995) several typical labwork activities from introductory optics and electricity were reformulated into an "ill-defined" problem. For example, instead of finding the right additional resistance for a voltmeter, the problem stated was to measure an unknown voltage with an antique instrument without damaging it. The problems were constructed in a way that trial-and-error-strategies could not successfully be applied for the solution. Interesting topics were chosen for contextualisation.

To assess the effects of this problem-based learning environment, a comparative empirical study in 9th- and 10th-grade classrooms was planned and carried out. 410 students from 17 classes participated in this study. In the experimental group, students worked for ten 45-min lessons with the new materials. In the control group cookbook-style labs were used. In all the groups, knowledge tests and a questionnaire including adapted and translated subscales of SLEI (Fraser et al. 1995) as well as a subscale of CLES on personal relevance (Taylor et al. 1997) to measure students' perception of the learning environments were used as pre- and posttests. To compare students' activities during labwork, a small sample of students from the control and from the experimental group were videotaped. These videos were analysed in 10-s loops with an adapted version of the CBAV-category system (Niedderer et al. 1998).

The results are very heterogeneous: On the one hand the measurement of students' activities with video analysis showed advantages of the problem-based labwork. Students spent significantly more time on task and talked much more about physics and of the relationship between their experimental data and the physics theory than students from the control group. On the other hand, practically no effects could be found on students' conceptual understanding or on students' attitudes. The scales were of limited use for measuring effects in the learning environments studied: Only one scale of the SLEI ("student cohesiveness") could be replicated. Thus, in terms of learning *processes*, the problem-based labwork activities were successful, but not in terms of *outcomes*.

A possible interpretation of these results is that measuring effects of labwork is very context dependent. While standard instruments, e.g., for labroom perception (like the SLEI) work very well in standard laboratory classrooms, they do not work in non-standard learning environments as the one discussed above. It can be assumed that this could also be a problem for the measurement of experimental competencies. Assessing the effects of students' activities in the lab is difficult – the more so as a lot of different activities are subsumed under the term "labwork".

4 Modelling Experimental Competencies

The following Sects. 4 and 5 focus on experimental competencies as important instrumental and epistemological components of scientific literacy. The overall research question joining these parts is: How can experimental competencies be described and assessed based on normative models? Section 4 presents and compares two approaches for modelling experimental competencies. These models form the theoretical background of the assessment studies described in Sects. 5.2 and 5.3.

4.1 Modelling the Performance of Experiments

Competencies related to the actual performance of experiments are difficult to operationalise and to measure. Nevertheless, the performance of experiments should be modelled in detail in order to support adequate teaching and assessment. If performance aspects are neglected, students may learn theoretically about scientific reasoning, but they will not be enabled to carry out and learn from own experimental investigations.

Within the German project eXkomp (diagnostics of experimental competencies), a model of experimental competence (Fig. 20.1) was developed that accentuates performance components (Schreiber et al. 2009, 2012). Like most proposals to

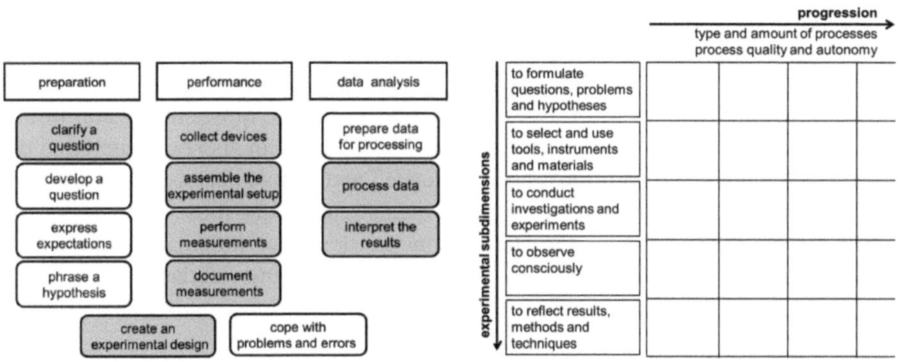

Fig. 20.1 Models of experimental competencies (*left*: according to Schreiber, Theyßen and Schecker (2012), for the *grey* components see Sect. 5.2; *right*: according to Gut and Labudde (2010))

structure experimental competencies (see Sect. 2.1), this description distinguishes between three phases: preparation, performance and analysis. Each phase consists of several components. The model focuses on the middle phase: assembling devices and setting up the experiment, measuring and documenting data. The model does not suggest a linear order of steps to be followed in the experimental process. The components can be passed in an iterative and a not necessarily complete sequence. The normative model was validated with an expert rating (teachers and teacher educators; cf. Schreiber 2012). The results confirm the comprehensiveness of the model and the high relevance of all its components for the description of labwork in science teaching.

4.2 Modelling the Progression of Experimental Competence

While the eXkomp model was designed as a starting point for empirical research, the aim of the Swiss project HarmoS (harmonisation of compulsory school) was to formulate national science education standards. For that purpose a competence model had to be developed. The HarmoS model includes three dimensions: skills, domains of contents and achievement levels (Labudde et al. 2012; Ramseier et al. 2011). The model distinguishes eight skills addressing cognitive and social competencies as well as skills for practical laboratory work. For each skill, a priori achievement levels were formulated, ranging consecutively from the 2nd to the 9th grade (Labudde et al. 2009). In correspondence to the achievement levels, based on the evaluation of students' performances in large-scale tests, standards were defined by the political authorities (Labudde 2007; EDK 2011).

The HarmoS model distinguishes five sub-dimensions that refer directly to experimental competence as given on the right part of Fig. 20.1. For each sub-dimension standards were formulated. The progression of the standards depends on the type and the amount of processes that are addressed by a task as well as on the quality and the autonomy with which a task has to be accomplished.

The five sub-dimensions of the HarmoS model describe quite similar abilities as the components of the eXkomp model (Fig. 20.1). They can also be assigned to the three main phases "preparation", "performance" and "data analysis". The eXkomp model is more differentiated in its components, especially those concerning the performance phase, whereas the HarmoS model describes the *progression* within each sub-dimension. The latter point is important for setting standards.

5 Assessment Tools for Experimental Competencies

With regard to the research question given in the introduction of Sect. 4 ("How can experimental competencies be described and assessed based on normative models?"), this part is dedicated to model-based assessment. In the following we present a process-based and two product-based approaches to the assessment of experimental competencies. Issues of validity are discussed.

5.1 Rubrics as Tools for Measuring Experimental Competence

Rubrics are standardised tools for performance measurement of multidimensional, composite competencies. They explicitly state the relevant criteria (or assessment components) and the levels of attainment for each of them. The term "rubric" goes back to the red colour usually used for correction (cf. Latin *rubrica*, red ochre or chalk). Rubrics take the form of two-dimensional matrices, with lines (usually) given by criteria, and columns by level. Both for their explicitness of the assessment components and attainment levels and for their clear 2×2 arrangement, rubrics are widely used and discussed as practical and reliable way of assessment (Jonsson and Svingby 2007).

The aim of this study was to develop a viable and reliable rubric for experimental competencies.

5.1.1 Design

As for experimental competencies, several rubrics have been proposed in the literature, which partially overlap (e.g. Nadji et al. 2003; RPAEG 2008 for a practitioner and a research point of view, respectively). In order to arrive at a comprehensive, well-validated instrument, Vogt, Müller and Kuhn proceeded in two steps:

1. Conceptual analysis and synthesis of existing rubrics.
2. Psychometric analysis and characterisation of synthesis rubrics of step 1.

For Step 1 the existing rubrics were considered as a background of expert statements about important aspects of experimental competence. In order to subsume this expert knowledge, a qualitative analysis was undertaken, where all the criteria were entered into a common text-base and treated as follows:

- *Splitting*: combined criteria (such as "response summarizes data, draws conclusions from these results, and evaluates them relative to the problem", Nadji et al. 2003, under heading "conclusions") were first split into separate statements.
- *Matching and Formulation*: On the basis of a one-by-one comparison, it was then decided, whether a given criterion statement matched another one from a different source; for a group of matching statements, the clearest formulation was then sought (sometimes by paraphrasing, if none of the existing formulations seemed adequate).
- *Subsuming*: For clarity and usability of the instrument, appropriate groupings of the assessment components are necessary. We decided to group them according to the different stages of the experimentation process (see below). Other groupings may be possible, but this is one which is both appropriate for the practical scoring process and conceptually sound, as it conveys the basic conceptual and chronological structure of experimentation.

Step 1 led to a first version of a synthesis rubric of experimental competence (ShREC) with 50 assessment criteria in five subgroups ("preparation", 5; "design", 3; "procedure", 13; "analysis", 14; "presentation", 15). The procedure described above still involves a subjective element on the side of the researcher (as length and formulation of an instrument always do). In order to address questions like whether formulations are understandable or whether criteria are missing or superfluous, a subsequent quantitative study within a larger expert community was undertaken. This is Step 2 and described below.

5.1.2 Methods: Analysis of Rubric Validity and Reliability

Within a first round of expert rating, ten raters (physics teachers and university lecturers) assessed whether the items of the instrument represent a criterion, which is either "essential", "useful but not essential", or "not useful" for measuring experimental competence (content validity). To determine furthermore such items which could be dropped without impairing content validity, the content validity ratio (CVR; Lawshe 1975) for each item was calculated. Moreover, experts were asked to add items they considered as pertinent for measuring experimental competencies.

After adapting the instrument to the rating feedback (by deleting and adding items as described above) a second round of the expert-rating process had to be conducted to determine the content validity of the modified instrument. Items with $CVR < 0.62$ were deleted again. Then, the mean of the remaining items was calculated, obtaining the content validity index (CVI), which represents a measure of the content validity of the instrument in total (Lawshe 1975).

Moreover, convergent validity was tested as cross-validity with traditional lab report assessment, based on the following sample: From the labwork courses taught by two of the authors, there is a continuously growing sample of lab reports by physics teacher students for which traditional assessment reports exist, i.e. plain text comments together with a final grading mark. In the 2010 summer term, 122 such reports and their evaluations were collected and analysed; no kind of selection on this sample was carried out. For the traditional and ShREC assessment, the measure of cross-validity according to standard regression-correlation techniques (Diehl and Staufenbiel 2002) was determined, yielding a measure of concurrent validity of rubrics and traditional labwork assessment.

The last step was a reliability analysis of the synthesis rubric for experimental competence under two perspectives. First, reliability in the sense of internal consistency was calculated (Cronbach's α). Second, inter-rater reliability was assessed: ten raters (physics teachers and university lecturers; the same rater sample as mentioned above) evaluated the same three lab reports (a bad one, an average one and an excellent one).

5.1.3 Results and Discussion

In a pilot study on validation of ShREC according to above methods and based on the sample above, the following results were obtained:

The reliability of the instrument as a whole was satisfactory (Cronbach's $\alpha > 0.70$), and the reliabilities of the item subgroups were smaller but still acceptable (Cronbach's $\alpha > 0.64$). The cross-validation between the instrument and traditional lab report evaluation reveals reasonable convergent validity (correlation = 0.75). The subgroups based on the practical and chronological structure of labwork did not express themselves in a corresponding factor analytic structure. Still, they are justified for both practical and conceptual reasons (see above).

Based on the work so far, the experimental competence rubric investigated here shows satisfying reliability and validity (the latter, in particular, being based on a systematic synthesis of expert knowledge and corroborated by psychometric analysis). Henceforth, it could be considered as a step towards meeting the triple requirement of conceptual analysis and adequacy, of practical feasibility and of psychometric validation.

5.2 *Process-Oriented Assessment by Hands-On and Mouse-On Tests*

While rubrics are useful for rating lab protocols as the final *products* of experimenting, a *process*-oriented analysis of labwork activities has to take students' actions in the lab into account. Process-studies usually refer to video recordings of students' hands-on activities. Video analysis consumes much more resources than a product-based approach. Schreiber, Theyßen and Schecker present a method that may allow for a process-oriented assessment with reduced effort (for details see Schreiber et al. 2012). The idea is to replace assessment procedures looking at hands-on experimenting by mouse-on experiments, i.e. workbenches that simulate real experimental environments. Log files are used for a time-efficient analysis of the experimental process. Thus, two research questions arise:

1. Are mouse-on tests based on computer-simulation workbenches a suitable substitute for hands-on tests?
2. Are ratings of process-qualities based on log-file analysis as reliable as ratings based on the qualitative analysis of screen recordings?

5.2.1 Design, Assessment Tools and Data Collection

The research questions were answered in a comparative study within a larger project on the diagnostics of experimental competence (cf. Schreiber et al. 2012). Two similar physics tasks in the domain of electric circuits were implemented both as hands-on experiments and as mouse-on experiments. Based on a pretest, two groups of students (grades 10-12, upper secondary schools) with about 35 students each were put together according to selected cognitive parameters like content knowledge and general and spatial intelligence. After a training session, in which the students were introduced into the experimental devices and the simulation workbench, they performed the actual tests.

One of the tasks was "Here are three bulbs. Find the one with the highest power at 6 V". The second task was to find out which of three given wire materials has the lowest specific resistance. Each student worked on one task in the hands-on setting and the other in a simulation environment ("mouse-on"). The order of the assessment settings was varied systematically. The simulation provided the same experimental apparatus as the hands-on experiment and allowed for all the relevant manipulations of the devices (selecting and assembling devices, measuring, even the destruction of equipment in case of incorrect use). A pre-structured worksheet prompted the students to clarify the question (task), create and conduct a suitable experiment, document their measurements and interpret the result. The students had 30 min to plan, perform and analyse each experiment. Within this period, the timing and the course of action were up to the students; no intermediate results had to be presented.

Figure 20.2 shows a screenshot of the simulation workbench (right) and a photograph of the corresponding hands-on experiment (left). In the hands-on test, students' actions were videotaped. Their paper and pencil worksheets were collected. In the "mouse-on" experiment, students' actions were documented by screen recordings. The students filled in an online worksheet. Detailed log files captured each manipulation of the virtual experiment and each entry to the online worksheet.

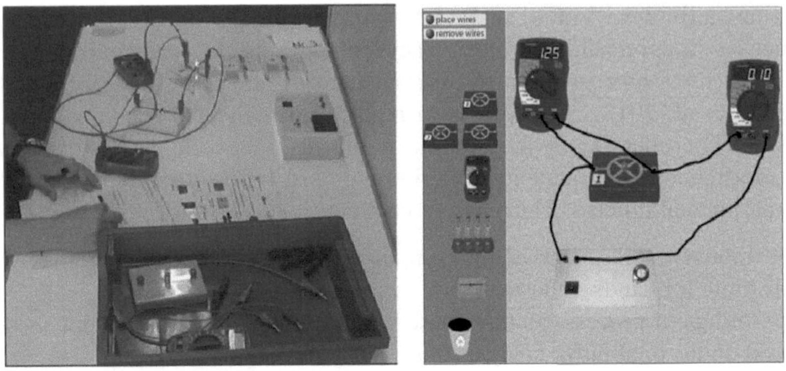

Fig. 20.2 The hands-on environment (*left*) and a screenshot of the mouse-on environment (*right*)

5.2.2 Methods: Process-Oriented Data Analysis

For a process-oriented data analysis, a set of categories was developed for coding students' experimental actions and their sequence. The categories represent the eight model components covered in the assessments (highlighted in Fig. 20.1). They can be coded with a high inter-coder reliability (Cohen's kappa = 0.84). For instance, when a student connects electrical devices on the table or on the screen, the category "assemble the experimental setup" is attributed. Selecting devices from the pool and placing them on the table (or on the screen) is coded as "collect devices".

A subsequent step of analysis takes into account the *quality* of actions, e.g. whether the experimental setup is correct or incorrect. The *substructure* of actions within a model component is also evaluated, e.g. whether the setup improves during several trials. This detailed *sequence analysis* yields a score on a five-stage ordinal scale for students' actions within each model component.

Sequence analysis was applied to the videos of the hands-on experiments as well as to the screen recordings of the mouse-on experiments and even to the log data of the mouse-on experiments. The inter-coder reliability is high (e.g. Cohen's kappa = 0.91 for the setup phase). The validity was confirmed by comparison with a high inferent expert rating of students' experimental competencies (Dickmann et al. 2012).

5.2.3 Results

Research question 2 refers to the reliability of log-file analyses for coding students' experimental actions in a simulation environment. Our data show that algorithms can provide dependable action diagrams. Human coders then only have to make the qualitative coding decisions according to the correctness (e.g. of the setup). This semiautomatic approach significantly simplifies the analysis and leads to reliable final scores (Cohen's kappa > 0.77).

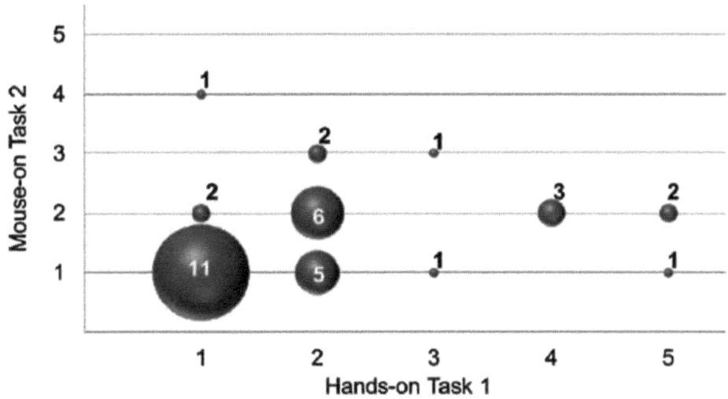

Fig. 20.3 Bubble diagram showing the scores students achieved in the hands-on task and in the mouse-on task (Kendall-tau-b = 0.337*)

While research question 2 can be answered positively, this is not the case for research question 1.

The sequence analysis yields two scores for a student's experimental competence: one score from the hands-on test and one from the mouse-on test. Although from the physics point of view the tasks are very similar as far as experimental actions are concerned, the correlations between the achievement scores are much lower than expected (Kendall-Tau-b < 0.4). There are dramatic changes of individual students' scores between tasks (cf. Fig. 20.3).

5.2.4 Discussion

At first glance, it seems that mouse-on tests are no valid substitutes for hands-on tests with regard to the diagnostics of experimental competence. Figure 20.3 shows a ground effect (both tasks seem to be too difficult/there are too many low achievers) as well as the instability of students' achievements. Several students reach high scores for the one task and only low scores for the other task. On the other hand, a qualitative comparison of the hands-on and mouse-on tests across all test persons shows very similar patterns of successful experimenting and very similar mistakes in both environments, e.g. students use a battery in addition to the power supply or choose wrong settings for the multimeters in both settings. Obviously, the challenges and difficulties that lead to success or failure rather result from the physics content than from the test format. Although the two experimental tasks require very similar physics content knowledge as well as similar experimental actions, students could not apply their knowledge consistently in both experimental situations. Subsequent analyses (cf. Schreiber 2012) show that the results of hands-on and mouse-on tests do correlate, when one compares the distributions of the achievements in the hands-on and the mouse-on assessment instead of comparing individual students' scores. Thus, the "mouse-on" technique appears to be promising for large-scale assessments.

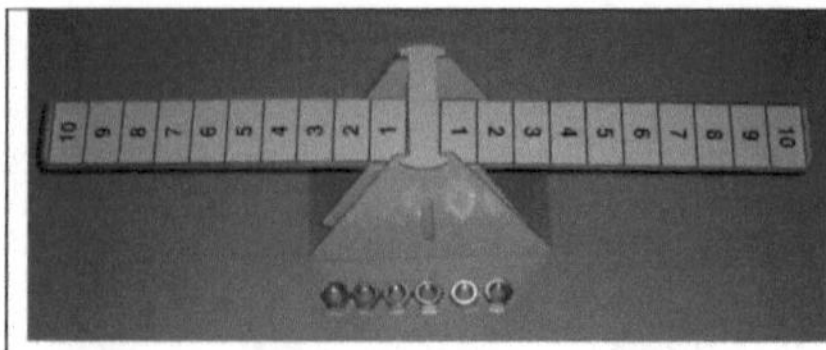

Hypothesis: A balance with symmetrical loads is always in equilibrium.
Carry out two experiments to test this hypothesis. Use the drawing to describe your experiments.

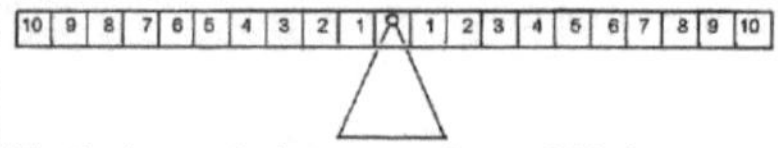

What is the result of your experiments? Mark:
□ the hypothesis is true
□ the hypothesis is false

Fig. 20.4 The balance experiment (*left*) and an excerpt from the paper and pencil worksheet (*right*; reduced reproduction)

5.3 *Item Difficulties of Large-Scale Hands-On Tests*

Students' performances in an experimental competence test are influenced by more aspects than only the demands of the experimental task or the students' content-related subject matter knowledge. Item characteristics like the format of the problem presentation and the answer format can be expected to affect item difficulties. The aim of Gut's study (2012) was to find out to what extent the HarmoS experimental test is sensitive to item features that are directly relevant to the experimental competencies compared to other features like format aspects.

5.3.1 Design, Assessment Tools and Data Collection

As part of the project HarmoS (Labudde et al. 2012), a hands-on test was carried out with 738 students of the 6th and the 9th grade in the German-speaking part of Switzerland in spring 2008. Fifteen experimental units, similar to the units of the TIMSS performance assessment in Harmon et al. (1997), covering different content domains within physical or living systems were used. The units contain 95 items that relate to one of the five sub-dimensions of laboratory work mentioned in Sect. 4.2.

The aim of the test was to provide evidence about the experimental competencies of Swiss students and to formulate basic standards for labwork on the base of these test results. Therefore, the tasks had to ensure an authentic assessment of a wide range of meaningful classroom activities (see Messick 1994). Furthermore, the results of the test were used to validate the HarmoS competence model (see above; Gut and Labudde 2010; Labudde et al. 2009). In order to make the test sensitive for low achievers, the tasks were developed according to Solano-Flores and Shavelson (1997) by repeated pilot tests with small groups of students. The task sheets were simplified after each repetition. The simplifications addressed item features closely related to experimental competencies as well as general item features (e.g. language, answer format). In the main test, each student worked on two experimental tasks during a session of 60 min.

Figure 20.4 shows an example of a hands-on experiment with an excerpt from the corresponding paper and pencil worksheet. In this task, the students had to

Table 20.1 Item features used for the analysis of item difficulty (for the grey features see below)

Features irrelevant to exp. competencies		Features relevant to exp. competencies	
"To grasp the task"	"To give the answer"	"To solve the problem"	"To code the solution"
Language	Gap formats	Problem	Correctness
Text length	Empty space	Processes	Heuristics
Sentence structure	Empty lines	Task type	Theory
Text coherence	Figures	Subject/context	Evidence
	Multiple choice		Logic
	...		Technique/practical
Content input	Fill formats	Solution	Quality
Textual inputs	Describing	Structuredness	Precision of measurement
Figural inputs	Terming	Openness	Precision of observation
	Marking		
	Drawing		
Description	Description		Completeness
Problem description	Task description		Gaps
			Specifications

"verify" or "falsify" a given hypothesis about the equilibrium of a lever by performing two experiments with a mathematical balance and six weights.

5.3.2 Methods: Analysis of Item Difficulty

Explorative analyses of the labwork tasks were done post hoc (Gut 2012). For clarifying the content validity of the test, the correspondence of item features and item difficulties was studied.

For the analysis, a conceptual framework was developed, distinguishing four dimensions of item difficulties with corresponding sets of item features. These dimensions relate to the four stages of performing and evaluating a task: "to grasp the task", "to solve the problem", "to give the answer" and "to code the solution", comprising about three dozens of item features (cf. Table 20.1). The two stages "to grasp the task" and "to give the answer" correspond to demands such as reading and writing difficulties that are not relevant to the experimental competencies. The stage "to solve the problem" relates to the complexity of the experimental task, whereas in the stage "to code the solution" the quality of the experimental processes is rated. Item features corresponding to both stages are relevant to the experimental competencies. The rather large number of analysed features was necessary to cope with the heterogeneity of the tasks with respect to the experimental problems, the task description and the coding systems.

The analysis was performed in three steps. First, the difficulties of the 95 items were calculated within a one-dimensional Rasch analysis by the program ConQuest 2.0. Second, the variables corresponding to the item features were dichotomised, mainly by a median split. At last, a multiple regression was calculated with the item difficulty as the dependent variable and the item features as the predictor variables. According to the method described by Prenzel et al. (2002) the item variables were reduced by iterative regression calculations.

5.3.3 Results

As result of the data analyses, 12 item features, highlighted in Table 20.1, can be identified that explain significantly 44 % of the variance of the item difficulties (Gut 2012). The analysis shows that the HarmoS hands-on test is sensitive to demands relevant to experimental competencies: For instance, the coding of the adequacy of practical-technical handling and the precision of measurement make items more difficult. But, as shown in Table 20.1 with the significant item features highlighted, the test is not sensitive to the crucial working step "to solve the problem". Therefore, the analysis of item difficulty post hoc does not help to explain competence progression. On the other hand, the test is also sensitive to demands not relevant to practical experimenting. The majority of significant item features belongs to one of the two working steps "to grasp the task" and "to give the answer". We assume that the test scores are influenced by the reading and writing competencies of the students to a rather large extent.

5.3.4 Discussion

From a psychometric point of view, the test analysis is tenuous to a certain extent. In relation to the 12 significant predictor variables, the number of 95 explained items is rather small. Furthermore, some results cannot be explained plausibly, e.g. that the use of the gap format "empty space" makes items easier. Both results may be a consequence of the difficulty to model heterogeneous experimental tasks with respect to the types of problem and the scoring systems. On the other hand, the heterogeneity of the test is necessary when authentic and meaningful classroom activities are to be assessed.

6 Conclusion

The studies reported in this chapter emphasise that the assessment of experimental competencies as well as of the effects of labwork still pose challenges to science education research. The study presented in Sect. 3 applied elaborate empirical research methods for comparing a problem-based guided enquiry approach with a conventional "cookbook"-style labwork setting. The conclusion is that the effects of non-standard laboratory learning environments cannot be measured with standard instruments. The validation of rubrics (Sect. 5.1) yields that these instruments are useful tools for the assessment of experimental competencies in science education studies with respect to content validity and reliability. However, their potentials and limitations have to be considered adequately. Rubrics can be used to analyse the *documentations* and *products* of labwork (summative data). The analysis of *processes* in the lab, i.e. the quality of students' *actions*, needs formative data and further methods. In the eXkomp study (Sects. 4.1 and 5.2) students worked on non-trivial tasks compatible with the syllabus of physics instruction. Using hands-on and mouse-on tests, the researchers found that individual students often worked on two consecutive similar experimental tasks in rather different ways. The very particular physics content of the experimental task seemed to play an important role for the individual student. This leads to the thesis, that the more advanced the physics topic of an experimental task is, the more challenging becomes the reliable measurement of experimental competencies distinct from content knowledge. A similar conclusion is drawn from the evaluation of the HarmoS test (Sects. 4.2 and 5.3): The more authentic an experimental test is, in the sense of real and meaningful classroom activities, the more challenging becomes the valid measurement of experimental competence. The analysis of the relationships between item features and item difficulties in a large-scale hands-on test yielded that students' performances were very sensitive to the formulation of the task and the prescribed answer format. The findings of the projects eXkomp and HarmoS correspond with other research reports showing a volatile picture of students' achievements in studies on experimental competence (e.g. Shavelson et al. 1999; Rosenquist et al. 2000). The question is to which extend this instability is due to the assessment methods or to inherent performance variations on the side of the students. Thus further investigations are necessary with a broader variety of experimental tasks within a specific content area and across different content areas.

In summary, the results of the studies presented in this chapter emphasise that the various objectives of experimenting in physics education and the complexity of the construct "experimental competencies" itself demand the development of highly adapted assessment tools. Summative and formative data should be included in the analyses of students' performances. The tools have to be validated carefully. Researchers should cooperate closely to exchange instruments and to share experiences.

References

Chen, Z., & Klahr, D. (1999). All other things being equal: Acquisition and transfer of the control of variables strategy. *Child Development, 70*(5), 1098–1120.

Dickmann, M., Schreiber, N., & Theyßen, H. (2012). Vergleich prozessorientierter Auswertungsverfahren für Experimentaltests. In S. Bernhold (Ed.), *Konzepte fachdidaktischer Strukturierung für den Unterricht* (pp. 449–451). Berlin: Lit-Verlag.

Diehl, J. M., & Staufenbiel, T. (2002). *Statistik mit SPSS – Version 10 + 11*. Eschborn: Verlag Dietmar Klotz.

EDK, Schweizerische Konferenz der kantonalen Erziehungsdirektoren. (2011). *Grundkompetenzen für die Naturwissenschaften. Nationale Bildungsstandards*. Bern: EDK. http://edudoc.ch/record/96787/files/grundkomp_nawi_d.pdf. Accessed 30 Apr 2012.

Fraser, B. J., Giddings, G. J., & McRobbie, C. J. (1995). Evolution and validation of a personal form of an instrument for assessing science laboratory classroom environments. *Journal of Research in Science Teaching, 32*, 399–422.

Grube, C., Möller, A., & Mayer, J. (2007). Dimensionen eines Kompetenzstrukturmodells zum Experimentieren. In H. Bayrhuber et al. (Eds.), *Ausbildung und Professionalisierung von Lehrkräften, internationale Tagung der Fachgruppe Biologiedidaktik im Verband Biologie, Biowissenschaften, Biomedizin* (pp. 31–34). Kassel: Universität Kassel.

Gut, C. (2012). Modellierung und Messung experimenteller Kompetenz: Analyse eines large-scale Experimentiertests. Ph.D. thesis. Berlin: Logos.

Gut, C., & Labudde, P. (2010). Assessment of students' practical performance in science: The Swiss HarmoS project. In G. Çakmaki & M. Taşar (Eds.), Contemporary science education research: Learning and assessment. ESERA proceedings 2009 (pp. 295–298). Istanbul: Pegem Akademi.

Hammann, M. (2004). Kompetenzentwicklungsmodelle: Merkmale und ihre Bedeutung – dargestellt anhand von Kompetenzen beim Experimentieren. *MNU, 57*, 196–203.

Harmon, M., Smith, T. A., Martin, M. O., Kelly, D. L., Beaton, A. E., Mullis, I. V. S., Gonzalez, E. J., & Orpwood, G. (1997). *Performance assessment in IEA's third international mathematics and science study*. Chestnut Hill: TIMSS International Study Center, Boston College.

Henke, C., & Sumfleth, E. (2006). Leistungsmessung in der Oberstufe mit chemischen Experimentalaufgaben. In A. Pitton (Ed.), *Gesellschaft für Didaktik der Chemie und Physik: Lehren und lernen mit neuen Medien* (pp. 340–342). Berlin: Lit-Verlag.

Hofstein, A., & Lunetta, V. N. (2004). The laboratory in science education: Foundations for the twenty-first century. *Science Education, 88*(1), 28–54.

Hopf, M. (2011). Measuring effects of non conventional labwork. In C. Bruguière, A. Tiberghien, & P. Clément (Eds.), *E-book proceedings of the ESERA 2011 conference*, Lyon. http://lsg.ucy.ac.cy/esera/e_book/base/ebook/strand10/ebook-esera2011_HOPF-10.pdf. Accessed 1 May 2012.

Jonsson, A., & Svingby, G. (2007). The use of scoring rubrics: Reliability, validity and educational consequences. *Educational Research Review, 2*, 130–144.

Klahr, D., & Dunbar, K. (1988). Dual space search during scientific reasoning. *Cognitive Science, 12*, 1–48.

KMK. (2004). *Bildungsstandards im Fach Physik für den Mittleren Schulabschluss*. München: Luchterhand.

Labudde, P. (2007). How to develop, implement and assess standards in science education, 12 challenges from a Swiss perspective. In D. Waddington, P. Nentwig, & S. Schanze (Eds.), *Making it comparable: Standards in science education* (pp. 277–301). Münster: Waxmann.

Labudde, P., Nidegger, C., Adamina, M., & Gingins, F. (2012). The development, validation, and implementation of standards in science education: Chances and difficulties in the Swiss project HarmoS. In S. Bernholt, K. Neumann, & P. Nentwig (Eds.), *Making it tangible: Learning outcomes in science education* (pp. 235–259). Münster/New York/München/Berlin: Waxmann.

Labudde, P., Metzger, S., & Gut, C. (2009). Bildungsstandards: Validierung des Kompetenzmodells. In D. Höttecke (Ed.), *Chemie- und Physikdidaktik für die Lehramtsausbildung, GDCP Jahrestagung 2008* (pp. 307–317). Berlin: LIT Verlag.

Lawshe, C. H. (1975). A quantitative approach to content validity. *Personnel Psychology, 28*, 563–575.

Messick, S. (1994). The interplay of evidence and consequences in the validation of performance assessment. *Educational Researcher, 23*(2), 13–23.

Möller, A., Grube, C., & Mayer, J. (2007). Kompetenzniveaus der Erkenntnisgewinnung bei Schülerinnen & Schülern der Sekundarstufe I. In H. Bayrhuber et al. (Eds.), *Ausbildung und Professionalisierung von Lehrkräften, internationale Tagung der Fachgruppe Biologiedidaktik im Verband Biologie, Biowissenschaften & Biomedizin* (pp. 55–58). Kassel: Universität Kassel.

National Research Council (NRC) (1996). *National science education standards*. Washington, DC: National Academy Press.

Nadji, T., Lach, M., & Blanton, P. (2003). Assessment strategies for laboratory reports. *The Science Teacher, 4*(1), 56. ftp://ftp.aip.org/epaps/phys_teach/E-PHTEAH-41-022301/Rubric.doc 2-12-2010. Accessed 7 May 2012.

Niedderer, H., Tiberghien, A., Buty, C., Haller, K., Hucke, L., Sander, F., Fischer, H. E., Schecker, H., Aufschnaiter, S., & Welzel, M. (1998). Category based analysis of videotapes from labwork (CBAV) – Method and results from four case-studies. http://www.idn.uni-bremen.de/pubs/Niedderer/1998-WP9.pdf. Accessed 8 August 2013.

Prenzel, M., Häussler, P., Rost, J., & Senkbeil, M. (2002). Der PISA-Naturwissenschaftstest: Lassen sich die Aufgabenschwierigkeiten vorhersagen? *Unterrichtswissenschaft, 30*(1), 120–135.

QCA (Qualifications and Curriculum Authority) (Eds.) (2004). *Science – The national curriculum for England* (Revised 2004). London: Department for Education and Skills.

Ramseier, E., Labudde, P., & Adamina, M. (2011). Validierung des Kompetenzmodells HarmoS Naturwissenschaften: Fazit und Defizite. *Zeitschrift für Didaktik der Naturwissenschaften, 17*, 17–33.

Rosenquist, A., Shavelson, R. J., & Ruiz-Primo, M. A. (2000). *On the "exchangeability" of hands-on and computer-simulated science performance assessment.* Technical report, Center for the Study of Evaluation; National Center for Research on evaluation, Standards, and Student Testing. Graduate School of Education and Information Studies; University of California, Los Angeles.

RPAEG. (2008). *Scientific ability rubrics (Rutgers physics and astronomy education group).* http://paer.rutgers.edu/ScientificAbilities/Rubrics/default.aspx. Accessed 7 May 2012.

Schreiber, N. (2012). Diagnostik experimenteller Kompetenz. Validierung technologiegestützter Testverfahren im Rahmen eines Kompetenzstrukturmodells. Berlin: Logos.

Schreiber, N., Theyßen, H., & Schecker, H. (2009). Experimentelle Kompetenz messen?! *Physik und Didaktik in Schule und Hochschule, 8*(3), 92–101.

Schreiber, N., Theyßen, H., & Schecker, H. (2012). Experimental competencies in science: A comparison of assessment tools. In C. Bruguière, A. Tiberghien, & P. Clément (Eds.), E-book proceedings of the ESERA 2011 conference, Lyon. http://lsg.ucy.ac.cy/esera/e_book/base/ebook/strand10/ebook-esera2011_SCHREIBER-10.pdf. Accessed 1 May 2012.

Shavelson, R. J., Ruiz-Primo, M. A., & Wiley, E. W. (1999). Note on sources of sampling variability in science performance assessments. *Journal of Educational Measurement, 36*(1), 61–71.

Singer, S. R., Hilton, M. L., & Schweingruber, H. A. (2006). *America's lab report: Investigations in high school science*. Washington: National Academies Press.

Solano-Flores, G., & Shavelson, R. J. (1997). Development of performance assessments in science: Conceptual, practical and logistical issues. *Educational Measurement: Issues and Practice, 16*(3), 16–25.

Taylor, P. C., Fraser, B. J., & Fisher, D. L. (1997). Monitoring constructivist classroom learning environments / advances in research on educational learning environments. *International Journal of Educational Research, 27*(4), 293–302.

Van Heuvelen, A. (1995). Experiment problems for mechanics. *The Physics Teacher, 33*(3), 176–180.

Walpuski, M. (2006). *Optimierung von Kleingruppenarbeit durch Strukturierungshilfen und Feedback.* Berlin: Logos.

Chapter 21
Understanding Students' Conceptions of Electromagnetic Induction: A Semiotic Analysis

Jennifer Yeo

1 Introduction

Students' conceptual learning in science can be perceived in two ways – as an acquisition of the knowledge structures of science or as a participation in the scientific practices of science (Scott et al. 2007). The acquisition framework perceives science learning to be acquiring some knowledge, stored in the heads of students, and refined to form richer cognitive structures. While studies taking a cognitive lens have generated much knowledge about the misconceptions students have in science, and the internal conceptual structures they might possess after instruction, the extent in which these externalizations are representative of what goes on in their head is still uncertain. Instead of focusing on what is in the brain, which may still remain a black box, a participation approach provides an alternative perspective to conceptualizing science learning. It takes the view that science learning is a process of learning to be engaged in the theory-building practices that characterize the science community. This includes not only appropriating the inquiry processes of theory building and the use of physical tools that scientists use to construct models and theories but also the conceptual tools that scientists use to think and reason about the world's phenomena. These conceptual tools include signs and symbols (e.g., linguistic tools, pictorial representations, gestures, mathematical symbols) that scientists use to represent often abstract and invisible entities and processes inferred from the phenomenon and to communicate the interpretation of the members of the science community.

Electromagnetic induction (EMI) is one example in science that makes extensive use of semiotic tools such as visual imagery, linguistic representations, gestures,

J. Yeo (✉)
Nanyang Technological University, Singapore, Singapore
e-mail: jennifer.yeo@nie.edu.sg

C. Bruguière et al. (eds.), *Topics and Trends in Current Science Education: 9th ESERA Conference Selected Contributions*, Contributions from Science Education Research 1,
DOI 10.1007/978-94-007-7281-6_21, © Springer Science+Business Media Dordrecht 2014

and mathematical symbols to theorize the production of current when a conductor experiences a change in magnetic field (Rieber 1995). Nersessian (1985) found that Michael Faraday himself made use of "lines" of force as a pictorial representation of a magnetic field to think about the actions taking place that led to the production of induced current. From a situative perspective, students need to appropriate these visual tools in thinking and exploring the phenomenon of EMI in order to make sense of that phenomenon and interact with one another in the science community. Understanding EMI is therefore analogous to being able to understand the conventions of these culturally consistent semiotic tools and to make use of and transform them in the reconstruction of EMI phenomena. Thus while studies taking a cognitive perspective might generate a list of misconceptions related to EMI, it does not inform what it takes to think and communicate like a scientist. A participation approach, on the other hand, could inform one's competencies in using culturally specific tools in this enculturation process of becoming like a scientist. In other words, students' use of culturally consistent tools in making sense of EMI should thus be one key factor when considering students' conceptualization of the topic.

Studies in science education on students' visualization capability have typically made use of instruments taken or adapted from educational psychology for testing students' spatial operations, such as the Purdue visualization and rotation test (PVRT) (Bodner and Guay 1997), mental rotation tests (MRT) (Vandenberg and Kuse 1978), or the Wechsler adult intelligence scale for children (Wechsler 1974), to measure students' capacity in spatial perception, spatial visualization, mental rotation, spatial relations, and spatial orientation (Sorby 1999). While these studies may inform about students' visual-spatial learning capacity, the decontextualized test items from any scientific content may not provide much useful information to help us understand how students' ability to recognize culturally specific visual tools use them to communicate information or apply them to solve problems could contribute to or hinder their conceptual understanding of EMI. Instead, an alternative perspective, consistent with the sign-making practices of science, could perhaps be borrowed from the field of semiotics, in particular social semiotics.

Social semiotics is a branch of study that perceives signs and symbols to be imbued with a meaning by community/context which in turn shapes the context in which these representations are used (Halliday and Hasan 1985). Adopted by Lemke (1990) as a theoretical lens to highlight the importance of the language of science in science learning, he showed how students' conceptions could be analyzed in terms of the thematic patterns of language use and other forms of meaningful human action. Lemke (1998) later extended the framework to analyze how multimodal representations are combined in scientific genres. The goal of this paper is thus to draw upon the works of Lemke (1998) and others in multimodality (e.g., Kress et al. 2001) to propose the use of a multimodal framework to analyze students' visualization of EMI and to understand students' conception of EMI. The research question that I ask is, "What meanings of EMI were made by students using semiotic tools when constructing an explanation of an EMI phenomenon?"

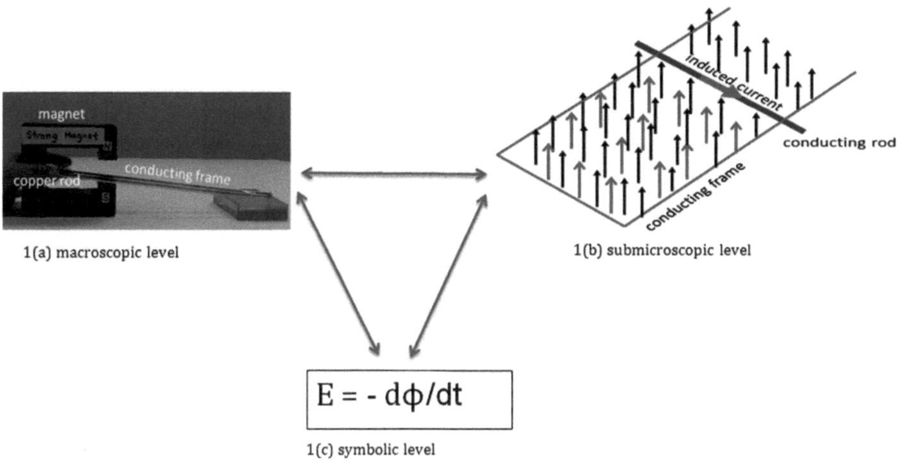

Fig. 21.1 An EMI phenomenon reconstructed at different levels of representation

2 Theoretical Framework

From the perspective of social semiotics, science is a discourse about the materiality of the natural world, construed with a system of interdependent representations that the science community has institutionalized (Lemke 1998). This idea can be illustrated with a phenomenon of EMI – a copper rod slowing down when it moves between the poles of a magnet (refer to Fig. 21.1a). This phenomenon can be reconstructed with different forms of signs and symbols at different representational levels – macroscopic, submicroscopic, and symbolic (Gilbert 2007).

At the macroscopic level, iconic figures are used to represent the physical objects (magnet, copper rod, and conducting frame – refer to Fig. 21.1a), construing the spatial relationships among the visible objects, thus constructing the phenomenon at its most naturalistic form. Textual representations could be used to further construe consequential relations that may not be afforded by iconic representations. At the submicroscopic level, arrows are drawn as indexical signs (refer to Fig. 21.1b) to construe meanings of properties of the inferred entities (strength and direction of magnetic fields) abstracted from the phenomenon and the interaction processes between them. This same phenomenon can also be represented with mathematical symbols (Fig. 21.1c) that construe the quantitative relationships among the inferred entities, thus distancing it further from its naturalistic form. Each of these representation levels is a reconstruction of the same phenomenon, albeit at different levels of abstraction and of varying forms of relations. In other words, making sense of an EMI phenomenon is essentially a sign-making practice that entails the use of representation in construing meaning. The signs and symbols thus form a system of semiotic tools which, when institutionalized, become the materials tools for scientists to think about the interaction of the (visible or inferred) entities as well as serving as communication tools to support social discourse among scientists about EMI. It is with semiotic tools that meanings are constructed, constituting the development of the scientific field. In a nutshell, science is essentially a sign-making enterprise, described to be specific, abstract, and convention based (Halliday and Martin 1993; Kress et al. 2001; Schleppegrell 2004). Students learning science will need to appropriate this sign-making process, which entails understanding the culturally specific meanings imbued in these signs and symbols, selecting appropriate semiotic tools to construct meanings, and transforming them to ascribe them new meanings in a given situation. In other words, science learning is essentially an active process of sign making, which goes beyond simple encoding of representations. Instead, it is "an active shaping and reshaping of resources that one has available, in the wish to make representations match intentions as closely as possible" (Kress et al. 2001, p. 2).

The practice of using signs in sense making is influenced by one's cultural background. The choice of semiotic resources, the organization of the tools, and their transformation to indicate relationships and processes of the perceived context are reflections of one's cultural ways of thinking. Taking a mechanically powered torch as an example of an EMI phenomenon, the general public may understand it as an action (shaking of the torch) producing a consequence (bulb lights up), construed with mainly iconic figures; on the other hand, someone schooled in physics would probably visualize it using diagrammatic arrows as indexical signs to think about the causal relations arising from the interaction of magnetic fields and resulting in a current flow. Accordingly, the kinds of meanings made are thus a reflection of one's way of thinking about a phenomenon (Lemke 1990) and, hence, one's conception of a phenomenon. The use of representations and meaning construed reflect one's competency in making sense of a phenomenon in a scientifically consistent pattern. Kress et al. (2001) refer to this competency as "scientificness." In other words, studying students' use of signs and symbols in their interpretation of the world's phenomena would provide insights into the extent to which students have appropriated the culturally consistent semiotic tools in making sense of the world, and hence their conception of the world.

Derived from the framework of social semiotics, semiotic tools can construe different forms of meanings, which Halliday and Matthiessen (2004) refer to as metafunctions. According to Lemke (1998), the selection, usage, and combination of these resources can provide information about the events and relations one makes (*presentational* meaning), the attitude one has towards the content and others (orientational meaning), and the degree of coherence of the text (*organizational* meaning). The *presentational meaning* "defines the sense in which we speak about something, construct a theme or topic, make predictions and argument" (Lemke 1998, p. 93). Drawing a parallel with a similar concept used in linguistics (ideational meaning), our deployment of semiotic resources (including linguistics, gestures, diagrams) can be used to specify processes or relationships, semiotic participants, and circumstances. For example, in Fig. 21.1c, the mathematical equation construes identification relations among the three entities of the phenomenon, electromotive force (E), magnetic flux (Φ), and time (t). Semiotic resources can thus present what is supposed to be "there" to be happening, or relations. Besides presenting the state of affairs, semiotic resources can also construe an orientational stance toward that state of affairs at the same time, which Lemke (1998) refers to as *orientational meaning*. Drawing closely with interpersonal meaning in linguistics, semiotic resources can construe a relative positioning of the producer and "text"

Table 21.1 Description of three metafunctions of semiotic resources

Presentational	Orientational	Organizational
Identifies the kinds of relations construed among semiotic resources in terms of events and relationships among the entities in the phenomenon (e.g., Φ = BA indicates identifying relations among magnetic field strength (B), area (A), and magnetic flux (Φ))	Identifies the level of abstractions by studying the kinds of semiotic resources used in meaning-making (e.g., the expression, Φ = BA, indicates the use of mathematical symbols, which in turn indicates the level of abstraction made out of the phenomenon)	Identifies the rules and conventions used, by studying how semiotic resources are organized (e.g., the expression, Φ = BA, makes use of mathematical conventions as a rule in its organization)

(semiotic production), indicated by one's choice of representations in the whole social space of possible discourses and viewpoints on the phenomenon. For example, in Fig. 21.1c, the construction of the phenomenon at the symbolic level against the other possible viewpoints and depictions (macro or submicro) is an indication of the level of abstraction at which the producer is orientating the readers/audiences. While the presentational and orientational meanings demonstrate the functions of semiotic resources in the construction of content and interaction (between text and producer, between producer and audience), a producer of a "text" has the choice of how the semiotic resources are to be organized, such as the parts to foreground or background, connections to create between parts of a message, and so on. Referred to as organizational meaning by Lemke (1998), it indicates the compositional function of semiotic tools to organize a "text" into elements and regions and to link disjointed parts into a coherent whole. For example, the symbols in the equation E = -d Φ/dt are organized in accordance with the conventions of the mathematical concept of differentiation. The left positioning of E (electromotive force) highlights the entity as the theme in its descriptive relationship between the two other entities, magnetic flux (Φ), and time (t). The mathematical genre thus unites the different symbolic representations into a whole in the system of scientific constructs.

In consideration of the relationship between semiotic tools and the meaning construed, the analysis of the three metafunctions of semiotic resources would potentially provide insights into the conceptions that one forms from a phenomenon – the events visualized to be happening in the phenomenon (presentational meaning), the level of abstraction one makes out of the phenomenon (orientational meaning), and the kinds of rules and conventions used in meaning construction (organizational meaning). Table 21.1 summarizes how and what each metafunction can inform about students' meaning-making of a phenomenon with semiotic tools.

3 Method

In this study, a case study approach is used to answer the research question, "What meanings of EMI were made by students using semiotic tools when constructing an explanation of an EMI phenomenon?" Two students, Peter and

Problem:	
A copper disc spins freely between the poles of an unconnected electromagnet as shown in the figure on the right. Describe and explain what will happen to the speed of the rotation of the disc when a direct current is switched on in the electromagnet.	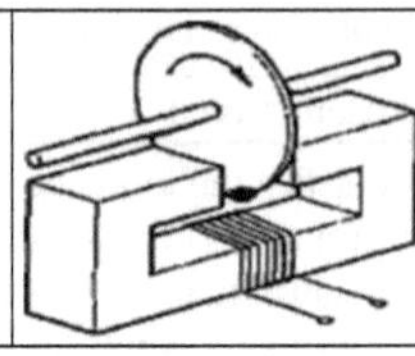

Fig. 21.2 An EMI phenomenon

Nancy, were chosen from one high school physics class. They had completed 6 h of instruction on EMI, which includes demonstration of EMI experiments and visually rich graphics to explain the concept of EMI at the macro (physical objects), micro (effects on electrons), and symbolic (formulas) levels of representation. After the instruction, students were chosen at random for a think-aloud protocol interview to understand how they make use of representations in meaning construction of the EMI phenomenon. Peter and Nancy were the two students interviewed. They were chosen for the case study as their contrasting explanation of the same phenomenon would be useful to illustrate the use of the derived multimodal framework as a lens to students' conceptualization of EMI. For the interview, Peter and Nancy were presented with an EMI phenomenon, as shown in Fig. 21.2, which the students had not come across during instruction, but aligned with the intended learning outcomes of the visualization-based instruction the students had received.

This problem was also chosen for the variety of levels of visualization at which the phenomenon could be thought about: macro, submicro, and symbolic. The correct prediction for the given problem would be that the speed of rotation of the disc would slow down. Given the macroscopic level of visualization of the phenomenon, the prediction can be derived in two possible ways: using Lenz's law or the action of force on a moving charged particle in a magnetic field.

1. Using Lenz's law, the copper disc could be said to slow down because the induced current produced in the copper disc is such as to oppose the change in the magnetic flux experienced by that disc when interacting with the magnetic field. In this case, induced current moving in a circular path in the copper disc would be formed in a direction that is opposite to the polarity that it is facing. Such an explanation involved an interpretation of a law, inscribed symbolically (linguistics tools) in the context of a macroscopic phenomenon.
2. A second explanation makes use of the notion of a force acting on a moving charged particle in a magnetic field, as emphasized in the instructional package as a link between the earlier topic the students had learned about and the idea of EMI. Taking the view that the electrons in the copper disc would be moving along with it as it rotates, the moving electrons could be imagined to be moving through the magnetic field of the electromagnet. The electrons will thus experience a force, whose direction can be determined by Fleming's left-hand rule, thus explaining the production of an induced current. As this induced current is

essentially moving electrons, its presence in the magnetic field produces another force that eventually slows down the copper disc. In this case, the macroscopic phenomenon was reasoned at the submicroscopic level as the invisible electrons were reasoned about.

During the interview, the students were encouraged to draw, write, say, or use objects as representations in their explanation. The interview was video-recorded to capture the use of semiotics (verbal, textual, and other modes of representations) during problem-solving and transcribed to capture the actional, visual, and linguistic resources used to make meaning of the phenomenon.

As meanings are made at the intersection of the different modes of representations, the meaning of each mode of representation was first analyzed. This was done by examining each mode of representation in terms of identifying the types of representations used (orientational), the event's happening as construed with the representations used (presentational), and the organization or structure of bringing the different semiotic forms together to form the explanation (organizational). The next stage involved viewing and comparing across the meaning functions of each mode and examining how meanings were made jointly by co-occurring modes of representation.

4 Students' Meaning-Making of the Electromagnetic Induction Phenomenon

The two students predicted correctly that the copper disc would slow down when a current flowed in the solenoid. However, the meanings they constructed, the level of abstraction visualized, and the conventions of organization were different. Excerpts 21.1 and 21.2 show the explanations given by the two students, Peter and Nancy, respectively.

Using drawings and gestures as iconic representations of specific and observable objects and processes of the phenomenon (refer to Excerpt 21.1), Peter identified "a particular point" on the disc (by pointing out a specific physical point on copper disc in Excerpt 21.1(i)) to be in the "same situation when the magnet is trying to enter a coil" (as he used his fingers to curl like a coil of wires in Excerpt 21.1(ii)). Drawing the parallel with a magnet pulling out of a coil, he concluded that the copper disc "will ... induce another phenomenon" (Excerpt 21.1(iii)). In Peter's reconstruction of the phenomenon, he drew associative relations as he perceived the given phenomenon to be similar to another he had encountered (i.e., "same situation when the magnet is trying to enter a coil" and "same as pulling the magnet out of the coil"). By using hand gestures as representations of the physical objects of the other phenomenon, he demonstrated their similarities in terms of their spatial arrangement. The iconic representations of the gestures and diagram used provide evidence of the naturalistic orientation (macroscopic level of representation) of Peter's explanation.

1(i)	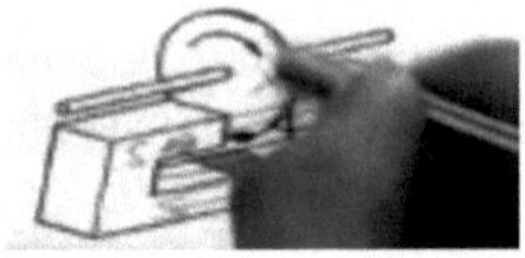	*(Symbols of S and N were already written on the magnet to indicate its polarity)* … this is a particular *point (drawing an x on the copper disc)* before it was subjected to these magnetic field lines …
1(ii)		Same situation when the magnet is trying to enter a coil *(curling fingers like a coil of wires)*… Before you try to enter … this *copper disc* … will try to oppose the change …
1(iii)	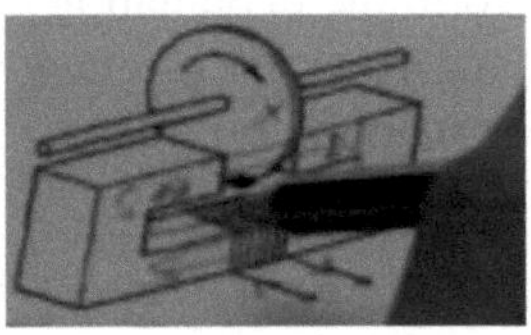	… it's the same as pulling the magnet out of the coil … will like induce another phenomenon … attract *this point (gesturing a point on copper disc beyond the magnet with pencil)* back to the magnetic field.

Excerpt 21.1 Peter's meaning-making of the EMI phenomenon

In Nancy's reconstruction of the same phenomenon, she visualized the existence of invisible electrons, which she represented with an enlarged dot on the drawing. Using her fingers orientated to indicate the spatial direction of each abstract entity of the phenomenon (current, force, and magnetic field) in Excerpt 21.2(ii) and (iii), she drew consequential relations about the direction of current flow (e.g., "there will be a force going downwards") and consequently causal relations (i.e., "then there will be force going out of the page … so it will slow down") as she deduced the subsequent motion of the electron resulting from its motion in a magnetic field using Fleming's left-hand rule. Here, drawings, hand gestures, and language have mediated Nancy's construction of the invisible events happening to equally microscopic objects. A scientifically consistent semiotic tool (i.e., Fleming's left-hand rule) was used to unite the invisible entities in terms of their spatial relationship. Table 21.2 summarizes the types of meanings made by Peter and Nancy.

5 Discussion and Conclusion

Learning is a transformative action of sign making (Kress et al. 2001). The case studies showed the two students selecting, using, adapting, and transforming signs that they had come across in the classroom or otherwise to construct meanings from

2(i)

I think about the *particle* inside this rotating disc … the *electrons* … into the page (drawing a dot, circle and arrow on the right of the given diagram) …

2(ii)

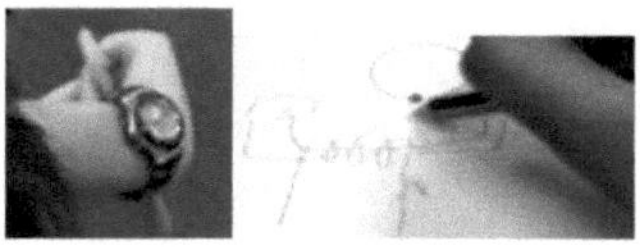

so there is a *current* going out of the page *(right picture: with her forefinger, center finger and thumb aligned at 90° to one another, she pointed her centre finger towards her, and the forefinger to her left, with her thumb pointing downwards)*… so there will be *force going downwards (left picture: pointing to a point at the bottom of the circle and drawing a downward arrow from the point).*

2(iii)

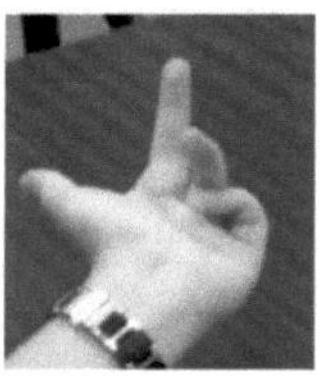

Current is induced by the *movement of electrons*. … it (the disc) slows down … because of *current flowing upwards (with her forefinger, centre finger and thumb aligned at 90° to one another, she pointed her forefinger away from her and thumb to her left, with her centre finger pointing upwards)* …

2(iv)

then there will be *force going out of the page* … so it *(pointing to the dot on the circle)* will slow down.

Excerpt 21.2 Nancy's meaning-making of the EMI phenomenon

Table 21.2 Summary of the meanings made by Peter and Nancy

	Peter's reconstruction of the phenomenon	Nancy's reconstruction of the phenomenon
Presentational meaning	1. Identification of physical object in the phenomenon (i.e., "a particular point") 2. Associative relations (i.e., "same situation when …") 3. Consequential relation (i.e., "will oppose the change")	1. Identification of abstract entities (e.g., "electrons," "current," "force") 2. Spatial relationships (e.g., "into the page," "going out of the page," "going downwards") 3. Conditional relation (e.g., "then there will be force going out of the page") 4. Causal relation (i.e., "then there will be force going out of the page … so it will slow down")
Orientational meaning	1. Phenomenon visualized at the macroscopic level with gestures as iconic representations of physical objects and spatial arrangement of objects and verbal text to indicate naming and consequential relations 2. Assumed common ground with interviewer ("same situation when the magnet is trying to enter a coil")	1. Phenomenon visualized at the submicroscopic level, with mainly gestures representing the spatial arrangement of abstract entities inferred from the physical phenomenon 2. Articulating thoughts ("I think") to a third party
Organizational meaning	Spatial arrangement of the iconic figures, reflective of the naturalistic setup	Using an EMI-consistent rule (Fleming's rule) to organize the different entities (representations), thus affording the potential to identify the subsequent motion of the copper rod

the given problem. Comparing the two reconstructions of the same phenomenon, we can see that Nancy's meaning-making involved imageries consistent to those of EMI, which were absent in Peter's. Nancy's imageries include identification of invisible and abstract entities such as electrons, current, and force, using Fleming's rule as a conditional convention to connect the different entities so that consequential relations can be made about the movement of the copper rod. These imageries were abstractions from the naturalistic phenomenon, which characterized the abstract ideas of EMI (Chabay and Sherwood 2006; Planinic 2006; Saglam and Millar 2006). On the other hand, Peter's reconstruction was based on the naturalistic setup, with few signs of abstraction. Focusing on the macro details of the phenomenon, Peter's diagrams and gestures were representative of physical objects/parts in the experimental setup of the problem or of a similar phenomenon he had come across (macro level). Such meaning-making differs from the technical, abstract, and causal relations characterizing scientific concepts. There were few EMI-consistent relations or rules used to explain the subsequent motion of the copper rod.

Comparing the two students in the case study, Nancy's ability to transit fluently from the frame of naturalistic realism (macroscopic level of representation) to that of scientific realism (submicroscopic level of representation) is evident of the higher degree of scientificness she had achieved in the conception of EMI. Gilbert (2007) describes this fluency in visualization involving the ability to acquire, monitor, integrate, and extend learning from representations as metavisualization. Peter's visualization, on the other hand, was mainly rooted in the macroscopic level as he made associations with another phenomenon that resembled the given problem. Such association with specific context made his explanations nongeneralizable. Generalizability is one key characteristic of scientific texts (Schleppegrell 2004). In terms of the scientificness, Peter did not seem to have shown convincing evidence that he was able to select appropriate tools and rules to reconstruct the phenomenon in a scientific manner. While it may be possible that Peter's explanation could have arisen from his assumption that the interviewer shared the same common ground as him (as the interviewer was present in the EMI lessons he attended), nonetheless, it was indicative of his sense of scientificness in such a context. Similar explanations should perhaps be further questioned by the teacher to find out if the student could shift from a concrete level of visualization to a more abstract level with the appropriate semiotic tools.

In conclusion, this paper aims to propose an alternative lens to students' conception of EMI. The social semiotic multimodal framework focuses on the study of the kinds of semiotic resources used by the students by drawing inferences to the kinds of meaning made, the levels of representations, and coherence of their sense making, to provide evidence about students' conceptual understanding of EMI. The analysis framework proposed in this paper helps to illuminate the differences in meaning-making between two students, albeit predicting the outcome of the phenomenon correctly. Nancy's fluency in transiting between macro and micro levels of representations using different kinds of semiotic resources as she considered the effect on a moving electron in a copper disc (micro) as the copper disc rotated (macro) was more consistent with the kinds of culturally specific tools used to think about EMI phenomena. It also reflects her higher degree of metavisualization and scientificness. Peter's visualization, on the other hand, was mainly that of associative relations as he focused on macro representations (a point on a copper disc indicated by pictorial symbols) in his visualization and making associations with equally concrete experiences he had come across. Such a meaning-making pattern tends to be less generalizable to other situations and may be an indication of possible difficulties with the use of representations consistent with the field of study. The difference in the levels and types of representations used by the two students during problem-solving – Nancy shifting between macro and micro levels using culturally specific representations, Peter relying mainly on macro objects and events – could perhaps point to another source of difficulty students might have with EMI, thus extending the comprehensive list of conceptual difficulties students have with already established content-specific concepts. This study also shows that the use of social semiotics as a theoretical lens to understand students' conception of EMI could also inform the degree of enculturation into the practices of science.

References

Bodner, G. M., & Guay, R. B. (1997). The purdue visualization of rotation test. *The Chemical Educator, 2*(4), 1–17.

Chabay, R., & Sherwood, B. (2006). Restructuring the introductory electricity and magnetism course. *American Journal of Physics, 74*(4), 329–336.

Gilbert, J. K. (2007). Visualization: A metacognitive skill in science and science education. In J. K. Gilbert (Ed.), *Visualization in science education* (pp. 9–28). Dordrecht: Springer.

Halliday, M. A. K. (1993). Some grammatical problems in scientific English. In M. A. K. Halliday & J. R. Martin (Eds.), *Writing science: Literacy and discursive power* (pp. 69–85). London: The Falmer Press.

Halliday, M. A. K., & Hasan, R. (1985). *Language, context, and text: Aspects of language in a social-semiotic perspective*. Victoria: Deakin University Press.

Halliday, M. A. K., & Matthiessen, C. M. I. M. (2004). *An introduction to functional grammar* (3rd ed.). London: Arnold.

Kress, G., Jewitt, C., Ogborn, J., & Tsatsarelis, C. (2001). *Multimodal teaching and learning: The rhetorics of the science classroom*. London: Continuum.

Lemke, J. (1990). *Talking science*. London: Ablex.

Lemke, J. (1998). Multiplying meaning: Visual and verbal semiotics in scientific text. In J. R. Martin & R. Veel (Eds.), *Reading science* (pp. 87–113). London: Routledge.

Nersessian, N. J. (1985). Faraday's field concept. In D. Gooding & F. A. J. L. James (Eds.), *Faraday rediscovered* (pp. 175–187). New York: Stockton.

Planinic, M. (2006). Assessment of difficulties of some conceptual areas from electricity and magnetism using the conceptual survey of electricity and magnetism. *American Journal of Physics, 74*(12), 1143–1148.

Rieber, L. P. (1995). A historical review of visualization in human cognition. *Educational Technology Research and Development, 43*(1), 45–56.

Saglam, M., & Millar, R. (2006). Upper high school students' understanding of electromagnetism. *International Journal of Science Education, 28*(5), 543–566.

Schleppegrell, M. (2004). *The language of schooling*. Mahwah: Lawrence Erlbaum.

Scott, P., Asoko, H., & Leach, J. (2007). Student conceptions and conceptual learning in science. In S. K. Abell & N. G. Lederman (Eds.), *Handbook of research in science education*. Mahwah: Lawrence Erlbaum.

Sorby, A. S. (1999). Developing 3-D spatial visualization skills. *Engineering Design Graphics Journal, 63*, 21–32.

Vandenberg, S. G., & Kuse, A. R. (1978). Mental rotations, a group test of three dimensional spatial visualization. *Perceptual and Motor Skills, 47*, 599–604.

Wechsler, D. (1974). *Manual for the Wechsler intelligence scale for children –revised*. New York: Psychological Corporation.

Part V
Relationships Between Teaching and Learning

Chapter 22
Analyzing Classroom Activities: Theoretical and Methodological Considerations

Gregory J. Kelly

1 Introduction

This chapter is organized into three main sections. First, I provide a rationale for discourse analytic studies in science education and identify the theoretical commitments supporting the research methodology. Second, I provide examples of the research approach drawing from two empirical studies. In each case, I identify how the theoretical orientation of the studies informed methodological decisions. Finally, I summarize the themes of the chapter and provide suggestions for future research in science education.

2 Rationale for the Study of Classroom Discourse

The current situation facing the global community presents challenges and opportunities for research in science education. First, many countries are becoming increasingly multicultural and multilingual societies. While cultural diversity provides many opportunities for developing mutual understanding and awareness across cultural differences, to realize these opportunities, educational systems need to adapt to the range of diversity in their students. Second, our planet faces a threatening ecological crisis. This provides an opportunity to advance science and develop thinking about science and technology that can contribute to solving such problems. Third, there is

---

Invited plenary session at the bi-annual conference of the European Science Education Research Association
Lyon, France
September 2011

G.J. Kelly (✉)
Penn State University, University Park, PA, USA
e-mail: gkelly@psu.edu

C. Bruguière et al. (eds.), *Topics and Trends in Current Science Education: 9th ESERA Conference Selected Contributions*, Contributions from Science Education Research 1,
DOI 10.1007/978-94-007-7281-6_22,

acute interest in many parts of the world in developing a stronger science and engineering workforce. While this is often narrowly construed with the goal of economic development, the possibility of developing a new vision of science for the twenty-first century may emerge from the current attention on science and technology education. Fourth, in many parts of the world, there is an intensified focus on testing of students. Perhaps related to the increases in testing, there is a marked decrease student interest in science in many countries. Thus, the current situation provides reasons for action and reform in science education.

In this chapter, I argue that to address the constraints facing science education and to capitalize on the opportunities made available, the field needs to develop a robust ecology of educational research. Current research directions too often focus on experimental, statistical studies that, when taken as the primary methodological standard, can lead to narrowing of research approaches. Large-scale studies (such as TIMMS, PISA, ROSE) do offer valuable insights, but only of a certain kind. For example, the Relevance of Science Education (ROSE) (see http://www.uv.uio.no/ils/english/research/projects/rose/) study provided valuable information about worldwide students' attitudes and interests in pursuing science and technology fields. In addition to such large-scale views of education, the field also needs studies of classroom interaction that can provide the bases for understanding the social accomplishment of everyday life in education settings.

3 Methodological Themes for Analyzing Classroom Activities

This chapter is focused primarily on the methodological processes of understanding classroom interaction. Through theoretical considerations and the two illustrative studies, five methodological themes are identified for analyzing classroom activities. First, *discourse processes are situated in social practice* and need to be examined in the relevant context of use. In the examples provided, I show how to look over time and levels of analysis. The emphasis on situating discourse aims to explicate how discourse processes must be understood in a broader context of use and how discourse processes both shape social practices and are embedded in such practices. Second, I explore *different ways of representing social action and practices* by "zooming in" to understand instances of action and "zooming out" to view patterns of activity. In this way, the reader, or audience for the research, is able to understand the evidentiary basis for the claims put forth. The reader is put in the position to make informed decisions about the nature of the research and what can be inferred by the conclusions put forth. Third, I offer examples of *how to make analyses systematic*. While the complexity of classroom life suggests the need for researchers to be creative and adaptive to events as constructed in situ, some degree of systematicity develops credibility and renders the analytic procedures transparent for other researchers. Fourth, the ethnographically informed discourse analysis perspective I present needs to identify how *pedagogical change in specific educational settings can be informed through research*.

While descriptive research has merits in its own right regarding understanding the forms of life constructed in sociocultural situations, it can also serve to inform pedagogical practices. Finally, it is important for research to develop reflexivity and *recognize the contingency of our own language* (Rorty 1989). Research takes a textual form in the context of presentation, and, in doing so, commits to certain vocabularies and ways of conceptualizing phenomena. By recognizing the contingency of such choices, our research can be more reflexive and sensitive to the limitations of any one given framework for observation, description, and analysis.

4 Theoretical Framework

Three intellectual fields contribute to the theoretical framework supporting my research program: science studies (such as sociology, anthropology, rhetoric of science), interactional sociolinguistics, and activity theory. Each of these fields offers different insights into how to understand and characterize what counts as science in different educational contexts.

Science studies encompass a variety of fields that take science as an institution as the subject of study. In science education, the most prevalent and dominant of these fields have been philosophy and history of science. Among other areas, these fields have contributed to understandings about the theory change and the development of models in science. These fields have generally provided a normative, disciplinary view of science (Kelly 2008). Studies of epistemology have generally focused on personal views of knowledge and how such views contribute to learning science (Kelly et al. 2012). While philosophy has focused on normative dimensions of disciplinary knowledge and psychology has focused on personal views of knowledge, sociology and anthropology have contributed by investigating empirically the actual practice of scientific communities. These studies of science-in-the-making identify the contextual, contingent nature of the construction of scientific knowledge (Latour 1987). These empirical studies of scientific practice offer much to science education. While such studies can be misconstrued as providing a view of "real" or "authentic" science, a preferred interpretation suggests that the empirical study of the everyday actions of professional science can expand the range of discourse and social processes that count as science and offer a methodological stance toward the study of everyday action. Following Wittgenstein's (1958) advice – "Don't think, look!" – such studies consider epistemological questions by examining how knowledge is accomplished through interaction (Kelly 2011).

Sociolinguistics is a "field of inquiry which investigates the language usage of particular human groups" (Gumperz 1982, p. 9). To study language usage, specific discourse processes are examined in contexts of use, and thus studies are tied to ethnographic description. Sociolinguistics investigates ways that everyday life is accomplished through discourse processes. To research the accomplishment of everyday life, a number of substantive assumptions define the orientation (Kelly and Green 1998): As members of a group affiliate over time, they create through social interaction particular ways of talking, thinking, acting, and interacting. These ways of acting come to define cultural practices, become resources for members, and evolve as members internalize the common practices, and also transform them through externalization. The cultural practices that constitute membership in a community are created interactionally through discourse processes. Local group members are also members of other groups and thus bring frames of reference to each interaction, including experiences, beliefs, values, knowledge, and practices (e.g., ways of knowing, doing, interpreting, and so forth) that may match or clash with local ones.

Interactional sociolinguistics begins with an initial period of ethnographic research that seeks to understand insights into local communicative ecologies, discover recurrent communicative patterns, and identity how local actors define problems (Gumperz 2001). These analyses become the basis for selecting sequentially bounded units or events, denoted by co-occurring shifts in content, prosody, or other stylistic markers, which are represented by transcripts. Thus, ethnographic description provides a basis for selection decisions and theoretical sampling in large data sets (Gee and Green 1998; Kelly and Chen 1999). By drawing from the ethnographic descriptions, the detailed discourse analysis is informed by the broader contexts of use of the discourse processes in question.

Science studies and sociolinguistics take a thoroughly sociocultural view of human learning and experience. This focus on sociocultural practices coalesces well with activity theory, which, in addition to analyzing social activity, places importance on the mediated nature of learning and the prominent role of language in culture. A focus on sociocultural practices suggests new units of analyses for research of the study of learning.

Activity theory provides a candidate for the unit of analysis of such a system: "object-oriented, collective, and culturally mediated human activity, or activity system" (Engestrom and Miettinen 1999, p. 9). Such a focus situates learning not in the individual mind of the learners but rather across multiple actors and mediating artifacts, each with affordances constructed by the roles and relationships of the actors, the norms and expectations of the relevant social groups, and the history of sociocultural practices and knowledge. This view of learning seeks to understand how historical and cultural dimensions of society influence learning contexts and need to be taken into consideration for the study of human learning. Thus, the theory suggests an examination of learning across timescales from the micro-moments of interaction, to mesolevel, to ontogenetic, to the sociohistorical (Wortham 2003).

5 Application of Framework to Studies of Science Education

Consistent with the sociolinguistic view of language use, an activity theory view of learning recognizes the embedded nature of interaction in sociohistorical contexts. To discuss how these three dimensions of the theoretical framework contribute to the development of the research methodology, the themes of this chapter are illustrated in two specific studies.

5.1 Study 1: Framing Evidence Use in Oceanography

The first study described here was conducted in a writing-intensive, university oceanography course for nonscience majors (Kelly et al. 2000). A central goal of the course was to develop scientific literacy among the students. This was conceptualized as developing the ability to use and critique scientific evidence regarding socio-scientific issues. The study was conducted from an ethnographic perspective, and a variety of data types were collected including videotapes of classroom interaction, artifacts from the course, student products, and ethnographic interviews.

These data were analyzed in the following manner. The videotapes were the basis for the construction of event maps to document the range and types of activities, the sequence of topics, time spent on each activity, as well as the communicative setting associated with the activity. Video analysis was completed for the entire corpus of data, including each of the lectures and small group sessions under the direction of graduate student teaching assistants. Through a review of the video, ethnographic field notes, and classroom artifacts, the prominence of writing disciplinary knowledge was identified as an important cultural practice. This led to a sampling of the video data records dedicated to discussion and activity around writing. The focused video analysis at the level of phase and sequence was completed for all 27 sessions identified as connected to writing. An event map was created for each of these events such as the one shown in Fig. 22.1. These event maps then served as basis for selection decisions regarding more focused detailed discourse analysis based on transcripts of talk and action (Green et al. 2003; Kelly and Crawford 1997). The data representation involved examined the most highly specific, micro-moments of interaction, *message units* (considering the spoken word with contextualization cues, e.g., pitch, stress, intonation, pause structures, physical orientation, proxemic distance, and eye gaze), building toward the construct of the events through *interaction units* (turns, spoken word with contextualization), *sequence units* (cohesive, thematically tied interactions, with ideational content redundantly marked by contextual cues), and *phase units* (set of sequence units, built by concerted and coordinated action among participants, and reflecting a common content focus of the group). In the event map shown in Fig. 22.1, the sequence and phase units are depicted; the research note shows the flow of activity including verbatim quotes from the episode. In this event, the course professor is describing the importance and purposes of writing in science. In doing so, he drew from his personal experience as a scientist and provided insight into the relationship of the knowledge and communication in scientific practice. In this and other instances, the course instructor, his teaching assistants, and the texts and technologies of the oceanography course provided an epistemic framing of the disciplinary knowledge. This led to more focused observations.

Through the process of data collection, initial analysis, and focused analysis of event maps, the study's research questions evolved from general questions to more specific questions about details of the educational context: What is happening here? How is technology being used to support science learning? How is writing being

line #	phase units	time	sequence units	*research notes & comments*/ "transcribed talk"
101 102 103 104 105	Goal of midterm paper: Do what scientists do	00:00:05	Reference to lab notebook on how to write a technical paper	*Begins with :* "There's a lot of possible ways you could go on this midterm paper." *Professor describes his balancing act between too much information and too little information; enough information to let students know professor is expecting, but not to o much so that it becomes "fill in the blanks ."*
106 107	↓	00:00:52	Struggle of scientists	*Scientists go through a progress of struggle:* "trying to make sense out of what seems like chaos a lot of times."
108 109 110 111 112 113 114		00:01:10	Personal experience of not understanding what the data means	"So NSF has given us ah 80 grand or something. We've gone out, done field experiments, measured earthquakes for a couple of months, and we look at it and 'oh boy' and you start looking at it. you start, 'well let's try this.' you start looking at it and plotting things up and trying to make some consistency out of it. Then looking at this." *Professor says for students not to feel alone. or dumb if don't understand right away.*
115 116 117		00:01:54	Actions students should take	*Professor mentions need to turn over and struggle with concepts; talk with others; discuss in sections; in the process of "doing science" as scientists do.*
118 119		00:02:20	Earthquake data	*Professor indicates that students will be analyzing current, real data.*
120 121 122 123 124 125 126 127 128 129 130 131		00:02:51	Identifying and explaining a problem	"There's another part of science that you need to know about is you can't explain everything." "Part of doing science is to figure out kinda (winnow) out what can I explain and what can't I explain. What's an interesting problem or what's an interesting thing to explain or what uh you know when when a scientist does research, one of the talents you have or has to be developed is knowing first off what's an interesting problem to study. Second off, whether studying it, and what data you can take can actually solve that problem or get some new information on that problem. So there's a lot of problems in the world uh that that are unsolvable you know, sort of classic problems you might not want to tackle."
	Onset on new phase of activity focused on the waves.			Professor walks back to podium saying, "OK uh waves."

Fig. 22.1 Event map of oceanography professor's talk about science writing (From Kelly et al. 2000), labeled as "what scientists do" in Fig. 22.2

used to support the learning of evidence use? How is writing supported by the framing of epistemic practices in oceanography? Through this process, the research team was able to examine specific events, situate them in ongoing coordinated actions, and make decisions about the patterns of activity. Examples included the description of the rationale and importance of writing in science by the course professor, a detailed analysis of the semiotics of a "good observation" in a class discussion, and various examples of how written materials and writing heuristics framed the knowledge of technical writing in oceanography. Following the sociolinguistic framing of the study, the research team culled the episodes ($n=27$) in the course around writing and knowledge and created a set of interactions for further, more detailed, discourse analysis (Fig. 22.2). While the large data set and numerous transcript examples pointed to ways that writing and knowledge were framed in the course, the use of domain analysis allowed the research team to detect patterns across instances and develop systematicity in the process. Domain analysis is a way to understand and organize the discourse processes, actions, and artifacts of the participants of a social group. Domain analysis includes a cover term (or category) that includes a number of folk terms (in this case direct quotes from spoken and written text).

Two sets of domains were identified. One set concerned practices constitutive of writing in science: ways to write in science, ways to distinguish technical writing

Date	(#) Analytic episode	Duration (min:sec)	Short description
			Week 1: Negotiating entry - - no videotaping
			Week 2: Scientific writing and communication
10/12	(1) Students working in groups	2:22	Karen (TA_1) explains how to do a peer group evaluation. Uses observation process of scientific method as analogy.
10/12	(2) Qualitative vs. quantitative observations in science writing	5:29	Karen explains beach assignment as observation exercise: qualitative data contrasted with "numbers."
10/13	(3) Observation & writing	2:49	Karen contrasts "English paper" with science writing. Introduces idea of genres of written discourse. Starts with observations: suggests qualitative vs. quantitative differences, importance of audience.
			Week 3: Maps and profiles
10/17	(4)Observation/Interpretation & writing	2:21	Earl (TA_2) starts with student's writing, looks at observation/interpretation distinction.
10/17	(5) Doing science -- Geology	1:09	Earl uses real data of ship track lines, makes explicit "what oceanographers do", i.e.. they try to find features.
10/20	(6) Reading and writing lab notebook	2:31	Karen stresses need to make outline of the midterm paper before writing, as is recommended in the lab manual.
10/20	(7) Citation & credit: using other's work	1:34	Karen discusses sources for the paper. Credit must be given for data source.
			Week 4: Exploring the deep
10/23	**(8) Technical writing: "What scientists do"**	**3:57**	**Bill (Prof.) explains importance of writing. Uses examples from personal history as scientist, scientific problems, data, and doing science.**
10/25	(9) Writing, using data, role of audience	4:04	Bill explains writing process, importance of revision and consideration of audience.
10/25	(10) Power of written language	1:01	Bill makes reference to adage concerning writing and power: "pen is mightier than the sword."
10/27	(11) How to make an observation & interpretation	0:50	Karen suggests a cause and effect relationship for the observation/interpretation dichotomy. One is needed for the other; "use observation to make interpretation."
10/27	(12) Map as observation	2:45	Karen brings up the crucial question of how to find data. Purpose of using "tools" is to find evidence for plate tectonics. Data then become part of the process for making this argument. The necessity of "proof" is invoked here.
10/27	(13) How to display data in the paper	3:34	Karen discusses layout of "technical paper." Tries to make explicit how data should be organized to make argument.
10/27	(14) Mechanics of writing science	5:32	Karen discusses the "details" of the technical paper: illustrations, references, citation form, footnotes.
10/27	(15) More mechanics of writing science	2:52	Karen addresses students' software and documentation questions, how to reference map points and use coordinates.

Fig. 22.2 Selection of events for detailed discourse analysis from first 4 weeks of class (From Kelly et al. 2000). Details for episode 8 are provided in Fig. 22.1

from other types of writing, reasons for writing as a scientific practice, and ways to distinguish observation and interpretation in science. This first set of domains focused on the ways to accomplish writing for the purposes of engaging in the internal practices of scientific practices – that is, those practices concerned with establishing knowledge within the field. For example, one such domain concerned the "characteristics of doing the work science." The discussion of the importance of choosing "an interesting problem to study" (line 126, Fig. 22.1) was grouped in this domain. A second set focused on contextual and societal dimensions of writing as a scientific practice: kinds of scientific practices identified by the social mediators; characteristics of doing the work of scientists; kinds of scientific norms identified by the social mediators; kinds of social, political, and economic ramifications of science; and attributes of socially responsible use of science/scientific knowledge. This second set of domains focuses on how science is part of society, draws from societal resources, but also enters into debates about socioscientific issues. For example, one of the illustrative examples used by the professor concerning the various audiences of science writing concerned the importance of persuading agencies to fund research (sequence 9, Fig. 22.2).

Across the domains, two thematic stances toward scientific writing emerged from the ethnographic analysis. Writing in science was presented as a practice that (a) was shaped by a community's procedures, practices, and norms and (b) required an understanding of the reasons, uses, and limitations of written knowledge specific to the discipline. Thus, writing can be seen as a particular type of epistemic practice within science. Each of the three components of the theoretical framework evinces how such practices were constructed in this educational setting. Science studies provided a stance toward science that sought not only to identify the products of social negotiation but also to examine the science-in-the-making (Latour 1987). Sociolinguistics provided a set of theoretical and methodological tools to consider how such practices were constructed through discourse processes. Although not part of the original study, activity theory is useful in interpreting the ways that sociohistorical artifacts enter into and are appropriated in the everyday life of science students.

The thematic stances concerning disciplinary writing raised questions about the student experience and learning in the course: What are the consequences for student learning given the epistemological framing of the disciplinary practices of oceanography? How do students engage with disciplinary knowledge and practice through writing? These questions led to a series of studies of argumentation analysis examining the epistemic and rhetorical features of student writing, including the epistemic level of students' claims, the use of lexical cohesion, and ways that theoretical and empirical terms function to tie theory to inscriptions (e.g., Kelly et al. 2010, 2008; Takao and Kelly 2003).

Across the studies of oceanography and argumentation, a number of specific pedagogical changes have been implemented based on research results. These changes include the development of heuristics for writing evidence (including ways of thinking about the nature and practice of using data to establish claims through warranted argument), changing the nature of genre and constraints from data sets and rhetoric task (from using inscriptions regarding plate tectonics to studies of global climate change, to ocean fisheries policy), and gradual development of peer-review processes, from informal to calibrated peer-review scoring (Prothero and Kelly 2008).

The study of the epistemic framing of oceanography as a discipline represents an example of how individual and social group learning can be contextualized over time. From a sociocultural framework, learning by individual students occurs within a community; furthermore, we can consider the learning of the collective itself as the community develops the ability to function differently and more competently. The four timescales developed earlier are mapped out in Fig. 22.3 (micro-moments of interaction, to mesolevel, to ontogenetic, to the sociohistorical) illustrating how the learning of the individual within the social group and the learning of the social group can be considered across time. Figure 22.3 shows the

Time frames for activity :	interactional - >	< -- mesolevel -- >	< -- ontogenetic -- >	< -- sociohistorical -- >
Learners within social group	**Discourse & actions:** e.g., discussion of meaning of scientific writing	**Communication of conventionalized practices:** e.g., sequences of events framing the epistemological framework of discipline		
Collective learning of social group		**Appropriation of discourse features:** e.g., recognizing conventionalized ways of presenting data		**Historical development of scientific writing:** e.g., changing mores of genre conventions

Fig. 22.3 Contextualization of learning oceanography across timescales (Modified from Kelly 2008)

mapping of the learning dimensions over time. While each cell was not examined in this study, a number of dimensions were considered. The figure shows how timescale and social practice can be contextualized in any given ethnographic study. For the learners within a social group, there was analysis of the discourse and actions at the micro-interactional level. For example, the numerous discussions of meaning regarding scientific writing were articulated by the participants. Such discussions were the focus of the discourse analysis at the message unit level (see Kelly et al 2000). At the mesolevel, there was communication of conventionalized practices – patterns of discourse over the course of the 10-week educational experience that identified for the students ways to take up scientific practices. These ways were sequences of events framing the epistemological framework of the discipline. Mesolevel analysis can be seen in sequencing of events, such as presented in Fig. 22.1 and summarized in the domain analyses. At the level of social group, at the mesolevel, there was appropriation of conventionalized discourse features by the collective. For example, through discussion, writing, and peer review, the members came to recognizing conventionalized ways of presenting data (synthesized in domain analysis). At the sociohistorical timescale, the collective was influenced by the inherited genre conventions with specific mores around uses of citation, evidence, and reference to data. These conventions are evoked in the micro-moments of interaction and can be identified through developments of genres within disciplines over time (Bazerman 1988). The contextualization of learning over time shows how activity theory and ethnography are mutually informing frameworks for the study of educational practices.

5.2 *Study 2: Learning to Reflect on One's Own Teaching*

The second study described here was conducted in a university course for prospective (preservice) teachers (Sezen et al. 2009). The course was designed to introduce prospective teachers to issues around student learning and conceptual change, lesson design and assessment, the nature of science, and reflective practice. One part of the course involved pairs of prospective teachers ($n = 23$) providing instruction to visiting groups of four to five middle school students (ages 11–12 years). The purpose for the teachers was to learn from teaching. The purpose for the visiting students was to learn about new, supplemental science concepts. The academic task for the preservice teachers was to develop analytic processes of reflection through viewing and commenting on video of their own teaching. The preservice teachers recorded the teaching episodes (about 20–25 min each), created a digital video file, and were instructed to use iMovie to create a second audio track for reflective commentary on a video of their teaching. The exact prompt for the preservice teachers was the following: "For a non-interrupted, 5 min segment, create a voice-over in which you explain what you see and hear happening in your lesson."

The study is based on a view of professional vision first introduced by Goodwin (1994) as "socially organized ways of seeing and understanding events that are answerable to the distinctive interests of a particular social group" (p. 606). The participants in the study were new to teaching, and the analysis of this, their first formal reflection of teaching, offered the possibility of establishing a baseline for their ways of viewing and observing their own teaching practice. The study's goal was to identify aspects of teachers' professional vision for interpreting classroom interaction, as part of the continuum on learning to teach. This led to the central research question: What do new teachers notice and attend to in their observations of their own teaching? The study addressed this question by drawing from sociolinguistics and an application of cultural historical activity theory (CHAT) for analysis of the teachers' reflections. Activity theory is well suited for the analysis of the event due to its focus on the multiple discursive and social aspects of the complex activity including the multiple actors, mediating artifacts (tools, technologies, social languages), roles (constructed, established, and positioned), norms and expectations, and history of sociocultural practices and knowledge associated with learning to teach through reflection (Kelly 2008).

Much like the first study presented, the data collection included videotapes of classroom interaction, audiotape of the students' voice-over reflections, and collection of artifacts, including the students' lesson plans and written critiques and reflections

Phase	#	Time stamp	Sequence	Voice -Over
Engaging students; Introducing polarizing sheets	1	1:59	Teachers introduce and demonstrate polarizing sheets. Adam holds two sheets up and rotates the back one, which causes the light going through the sheets to be blocked.	Adam's voice-over commentary: 01:58 - 06:58
	2	2:46	Teachers show students a "polarizing tube" and allow them to manipulate it. Yolanda asks for the students' observations.	
	3	3:49	Teachers ask the students for guesses about what is happening with the polarizing tube and sheets. Students wonder if there is something inside the tube that's causing it to block the light. One student guesses that the polarizing sheets have the same behaviors as transition lenses.	
Providing explanation & Evaluating students	4	4:50	Teachers ask students what they think the nature of light is (particle, fluid, gas), and then the teachers explain the nature of light as a wave and that it travels. Teachers make an analogy to water waves. Adam draws an arrow representation of light ("up and down and left and right") and explain how polarizing sheets block "up and down" or "left and right" parts of light.	
	5	6:32	Teachers ask students why the second polarizing sheet causes all light to be blocked. One student answers that turning a second polarizing sheet so its orientation is different than the first one will block all the light. Adam then draws the arrow representation of what the student describes.	
	6	21:28	The two student pairs reconvene as one class to discuss what a polarizing sheet would look like in order to block all the light.	

Fig. 22.4 Selection of event map of Adam's reflective voice-over showing the phases and sequences of action used for the 5-min reflection

of their lessons. Following the sociolinguistic framework, our research group constructed event maps to document the range and types of activities, the sequence of topics, time spent on each activity, and overall organizational structure of events. These served as a basis for further, more detailed analysis, based on transcripts of talk and action for each of the teaching events and corresponding reflections. This allowed the research team to examine ways that events are interactionally accomplished and how, through observation of the discourse events and reflections, the preservice teachers came to see their own teaching practice in this instance. A representative example of an event map is shown in Fig. 22.4.

For this study, I present three types of representations of discourse events. The first is a transcript of the two preservice teachers setting up ideas regarding their lesson on the polarization of light with the visiting middle school students. The transcript shows the speaker (the two teachers are Yolanda and Adam), line numbers, spoken discourse separated by message unit, and nonverbal actions.

Speaker	Line	Talk (message units /)	Actions
Yolanda	1.1.1	Well/that's what we're/gonna start doing	
	1.1.2	So/we have these two/sheets, they look like sunglasses/and these are called/polarizing sheets	Yolanda and Adam each pick up and unwrap a polarizing sheet
	1.1.3	Now/you see/like/there's the light/and you can see through/the sheet/like that	With one polarizing sheet in her hand, Yolanda gestures to the ceiling, then holds the polarizing sheet straight out in front of her and looks through it
	1.1.4	So/Adam has one/as well	Adam holds a sheet straight out in front of him. Yolanda points to Adam
Adam	1.1.5	Okay?	Adam and Yolanda speak at the same time
Yolanda	1.1.6	And/if you/	
	1.1.7	if you see right now/like/you can/like/it blocks it/some light	Yolanda holds her left palm out next to the polarizing sheet in her right hand
	1.1.8	But/when you put them together/it blocks it	Yolanda holds her polarizing sheet behind Adam's. Adam takes hold of Yolanda's sheet
Adam	1.1.9	It looks about the same/when I put em/this way	Adam turns Yolanda's sheet and aligns it with his own
Yolanda	1.1.10	Mm-hmm	

The entire teaching episode was transcribed in message units, and sequences and phases were identified in the process. Figure 22.5 shows how the transcript, sequences, and phases can be represented in Studiocode (a commercial video analysis tool) and can be represented and linked to the digital video record (not shown here). Studiocode provides a number of affordances, as seen in the representation in Fig. 22.5. In this case, the digital video, transcript window, and coding of phases, sequences, and interactions allowed for analysis of instances of talk and action and understanding of such actions in the overall flow of social interaction in time.

The next part of the analysis focused on the students' reflective discourse (the second audio track) regarding their teaching practice. This is the second of the two activity systems (see Fig. 22.6): First, the teaching event itself is an activity system, and second, the reflective commentary represents a second activity system. The reflective system is mapped out following Engestrom's (1999) view of activity, showing the subject, object, mediating artifacts, the sociocultural practices or rules, community, and the division of labor, leading to an outcome. This mapping of the activity system makes visible the interactions relevant to a learning context and the possible tensions across dimensions of the system.

Our analysis was organized in two steps. First, we conducted a line-by-line analysis of voice-over reflection identifying the key referents in the preservice teachers' observation of their teaching events. This was designed to consider what these early teachers notice about their teaching. Second, we conducted an additional analysis of voice-over reflection identifying aspects of activity system observed in the preservice teachers' observation of their teaching. This allowed us to consider ways that the

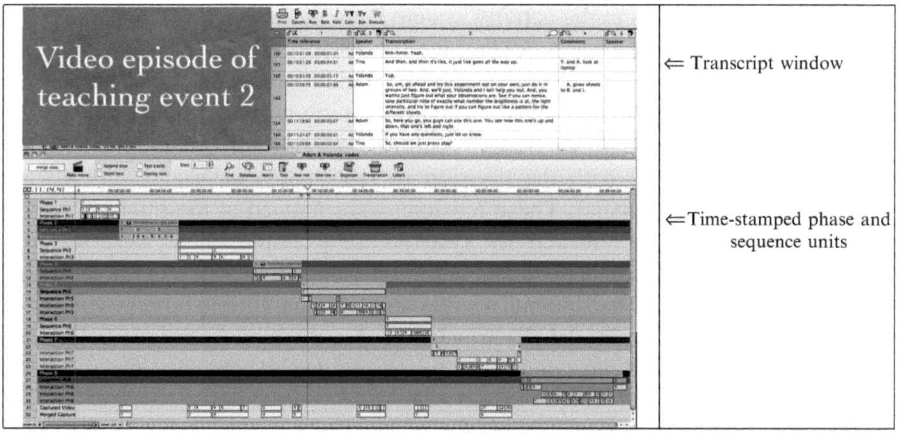

Fig. 22.5 Video episode from analysis in Studiocode showing space for digital video, transcript, and phase and sequence units marked with time stamps. Message units do not appear in transcript window. The phrases are divided into message units in word processor at a later time

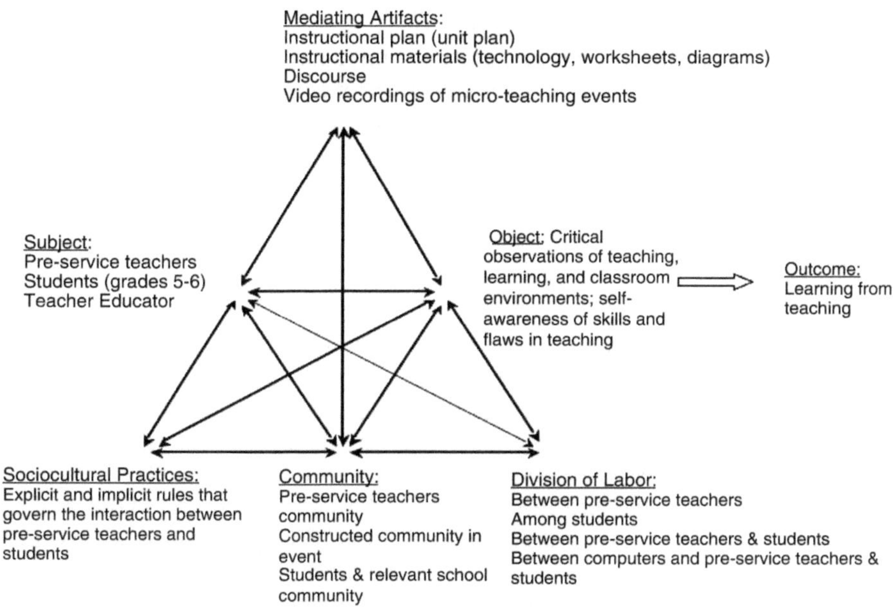

Fig. 22.6 Activity system of teachers' reflections on the microteaching event

components of an activity system are invoked in the teachers' reflective discourse. Through the analysis, we were able to consider the tensions across the activity system and what this may mean for teacher education. Figure 22.6 depicts the activity system of the reflective event, which represents commentary on the teaching activity system.

A number of tensions in the activity system were noticed by the teachers. When referring to the *learning from teaching*, the preservice teachers ($n=23$) identified the following: lack of their ability to assess the middle school students' knowledge (through discourse, in the moment), difficulty helping middle school students use mediating artifacts (graphs), challenges with role as educator (maintaining focus), and intricacies involved with the division of labor (physical stance not always aligned with cognitive goals). These tensions were noticed at the intersections of different components of the activity system. For example, there were tensions between the subjects (students and teaching) with mediating artifacts for achieving their collective objects. The students needed to integrate multiple representations of light waves, gestures, and models for wave motion in a complex semiotic field. Such sensemaking posed problems communicating meaning and achieving the collective goals of the subjects.

The study raised a number of important questions for the research program. Moving forward the research agenda needs to include evidence about how expert teachers learn from teaching. Questions include the following: What sort of professional vision do they develop and employ? Can such vision be useful for teacher education? This may lead to the development of experiences and scaffolds to foster new ways of viewing teaching practice for preservice teachers. Furthermore, while activity theory was used for analysis in this study, the commitment to understanding the sociocultural nature of teaching entailed by this approach may have pedagogical implications. We will continue to explore the use of activity theory in pedagogy of learning to teach. Viewing teaching from an activity theory point of view may foster a more social and interactive view of teaching.

6 Conclusion

Through these two studies of classroom life, I showed how a sociolinguistic perspective, informed by educational ethnography, provides an important contribution to the overall ecology of educational research in science education. By asking "what counts as science, evidence, teaching, and reflection," these studies of the interactional accomplishment of educational events render the familiar strange. The study and representation of everyday life allow educational events to be rendered observable, witnessable, and subject to discussion and reflection. Through this process, research can demystify scientific and teaching practices. Since this chapter was primarily focused on the theory and methods of educational research, I sought to illustrate five methodological themes: discourse processes as situated in social practice, social action and practices rendered witnessable through representation, importance of making analyses systematic, development of pedagogical implications from research, and recognition of the contingency of our own language.

Science education faces a number of important challenges. First, we face a global ecological crisis. While such a crisis cannot be addressed by science education alone, studies of how students learn to use evidence, address socioscientific issues, and develop an understanding of the consequences of their actions can inform our educational institutions about pedagogies that develop such competencies. Second, developing interest in science and engineering and providing access for all students, regardless of ethnicity and language background, is a central concern for many societies. Access to scientific practices occurs through social interaction and discourse. Studies of classroom discourse can identify ways to improve access to science for all students and exemplars for developing student interest in science and engineering.

References

Bazerman, C. (1988). *Shaping written knowledge: The genre and activity of the experimental article in science*. Madison: University of Wisconsin Press.

Engestrom, Y. (1999). Activity theory and individual and social transformation. In Y. Engestrom, R. Miettinen, & R.-L. Punamaki (Eds.), *Perspectives on activity theory* (pp. 19–38). Cambridge: Cambridge University Press.

Engestrom, Y., & Miettinen, R. (1999). Introduction. In Y. Engestrom, R. Miettinen, & R.-L. Punamaki (Eds.), *Perspectives on activity theory* (pp. 1–16). Cambridge: Cambridge University Press.

Gee, J. P., & Green, J. L. (1998). Discourse analysis, learning, and social practice: A methodological study. *Review of Research in Education, 23*, 119–169.

Goodwin, C. (1994). Professional vision. *American Anthropologist, 96*(3), 606–633.

Green, J. L., Dixon, C. N., & Zaharlick, A. (2003). Ethnography as a logic of inquiry. In J. Flood, D. Lapp, J. R. Squire, & J. M. Jensen (Eds.), *Handbook of research on teaching the English language arts* (2nd ed., pp. 201–224). Mahwah: Lawrence Erlbaum.

Gumperz, J. J. (1982). *Discourse strategies*. Cambridge: Cambridge University Press.

Gumperz, J. J. (2001). Interactional sociolinguistics: A personal perspective. In D. Schiffrin, D. Tannen, & H. E. Hamilton (Eds.), *Handbook of discourse analysis*. Malden: Blackwell.

Kelly, G. J. (2008). Inquiry, activity, and epistemic practice. In R. Duschl & R. Grandy (Eds.), *Teaching scientific inquiry: Recommendations for research and implementation*. Rotterdam: Sense Publishers (pp. 99–117; 288–291).

Kelly, G. J. (2011). Scientific literacy, discourse, and epistemic practices. In C. Linder, L. Östman, D. A. Roberts, P. Wickman, G. Erikson, & A. McKinnon (Eds.), *Exploring the landscape of scientific literacy* (pp. 61–73). New York: Routledge.

Kelly, G. J., & Chen, C. (1999). The sound of music: Constructing science as sociocultural practices through oral and written discourse. *Journal of Research in Science Teaching, 36*, 883–915.

Kelly, G. J., & Crawford, T. (1997). An ethnographic investigation of the discourse processes of school science. *Science Education, 81*(5), 533–559.

Kelly, G. J., & Green, J. (1998). The social nature of knowing: Toward a sociocultural perspective on conceptual change and knowledge construction. In B. Guzzetti & C. Hynd (Eds.), *Perspectives on conceptual change: Multiple ways to understand knowing and learning in a complex world* (pp. 145–181). Mahwah: Lawrence Erlbaum.

Kelly, G. J., Chen, C., & Prothero, W. (2000). The epistemological framing of a discipline: Writing science in university oceanography. *Journal of Research in Science Teaching, 37*, 691–718.

Kelly, G. J., Regev, J., & Prothero, W. A. (2008). Analysis of lines of reasoning in written argumentation. In S. Erduran & M. P. Jimenez-Aleixandre (Eds.), *Argumentation in science education: Recent developments and future directions* (pp. 137–157). New York: Springer.

Kelly, G. J., Bazerman, C., Skukauskaite, A., & Prothero, W. (2010). Rhetorical features of student science writing in introductory university oceanography. In C. Bazerman, R. Krut, K. Lunsford, S. McLeod, S. Null, P. Rogers, & A. Stansell (Eds.), *Traditions of writing research* (pp. 265–282). New York: Routledge.

Kelly, G. J., McDonald, S., & Wickman, P. O. (2012). Science learning and epistemology. In K. Tobin, B. Fraser, & C. McRobbie (Eds.), *Second international handbook of science education* (pp. 281–291). Dordrecht: Springer.

Latour, B. (1987). *Science in action: How to follow scientists and engineers through society*. Cambridge: Harvard University Press.

Prothero, W., & Kelly, G. J. (2008). Earth data, science writing, and peer review in a large general education oceanography class. *Journal of Geoscience Education, 56*(1), 61–72.

Rorty, R. (1989). *Contingency, irony, and solidarity*. New York: Cambridge University Press.

Sezen, A., Tran, M.-D. T., McDonald, S., & Kelly, G. J. (2009). *Preservice science teachers' reflections upon their micro-teaching experience: An activity theory perspective*. Paper presented at the annual meeting of the National Association for Research in Science Teaching, Garden Grove.

Takao, A. Y., & Kelly, G. J. (2003). Assessment of evidence in university students' scientific writing. *Science Education, 12*, 341–363.

Wittgenstein, L. (1958). *Philosophical investigations* (3rd ed.) (trans: Anscombe, G. E. M.). New York: Macmillan.

Wortham, S. (2003). Curriculum as a resource for the development of social identity. *Sociology of Education, 76*, 229–247.

Chapter 23
The Impact of a Context-Led Curriculum on Different Students' Experiences of School Science

Indira Banner and Jim Ryder

1 Introduction

This chapter looks at students' experiences of a science curriculum with a greater emphasis on context-based science and investigates how different groups of students talk about their experiences of science in the classroom and the potential use and enjoyment of such a curriculum to their future lives. In England, in 2006, the curriculum reform was nationally imposed and mandatory in most state schools. It describes the science that must be taught to students at different stages of their school career (in this case aged 14–16 years). However, different interpretations of the curriculum are produced by different companies who design science courses and examinations, with schools free to choose between these available courses. This chapter is derived from a larger study looking at the enactment and impact of this science education reform (EISER).[1]

Context-based science learning has its roots in theories of situated learning which emphasise the importance of acquiring knowledge set within real contexts rather than as an abstract cognitive process set in a narrowly academic context (Mandl and Kopp 2005). The intended outcomes are that students gain a broad understanding of science that they recognise as being related to their own lives using methods of teaching that encourage students to be active participants (op. cit.). This locating of science teaching in examples of real scientific and/or technological products and processes is similar to STS (science-technology-society) education. Context-based science education can also contain teaching about socio-scientific issues (SSI) and the nature of science (NoS), for example, encouraging students to

[1] http://www.education.leeds.ac.uk/research/projects/enactment-and-impact-of-science-education-reform-eiser

I. Banner (✉) • J. Ryder
Centre for Studies in Science and Mathematics Education, School of Education, University of Leeds, Leeds LS2 9JT, UK
e-mail: i.banner@education.leeds.ac.uk; j.ryder@education.leeds.ac.uk

C. Bruguière et al. (eds.), *Topics and Trends in Current Science Education: 9th ESERA Conference Selected Contributions*, Contributions from Science Education Research 1,
DOI 10.1007/978-94-007-7281-6_23,

think about the place of science in society and how it can (and cannot) provide answers to a range of problems. The term student 'experience' is an intentionally broad term encompassing more than interest or enjoyment of school science and including aspects of what purportedly goes on in science lessons and how students respond to and engage with these experiences. Students' personal feelings about school science and their identities will also form a part of this experience (Aikenhead 1996; Carlone 2004).

2 Context-Based Science

It has been shown that students learning about relevant, often local, issues with scientific as well as social factors show increased attainment and positive attitudes towards school science (Lee and Erdogan 2007; Kanter and Konstantopoulos 2010) and that this can be an engaging route for teaching traditional science content (Sadler et al. 2007). There is also work showing that science topics set within a social context are important as a vehicle for encouraging teaching students in areas such as argumentation (Albe 2008), reflective judgement (Zeidler et al. 2009) and high-quality decision-making (Grace 2009) that are perceived to be important in the development of critical thinking and skills which may help students in their lives as adult citizens.

Most of this research involves intervention studies, like those reviewed by Bennett et al. (2007), with volunteer teachers where the course has been designed collaboratively with the teacher (Albe 2008) or where teachers have undergone specific professional development surrounding the use of, for example, SSI in the classroom (Lee and Erdogan 2007). The extent of the teacher's content and pedagogical knowledge is recognised as being very important in outcomes of more relevant (e.g. project based) science curriculum (Kanter and Konstantopoulos 2010). Students need to be taught to engage with SSI through learning about the nature of science and how to argue (Sadler 2004).

The change in the science curriculum in 2006 was a radical one for school science and teachers in England. The courses contain a noticeable shift towards context-based science teaching and the inclusion of socio-scientific issues (SSI) and the nature of science (NoS). It is known that many teachers have a naive understanding of SSI and NoS (Abd-El-Khalick and Lederman 2000) and a range of views towards the use of SSI in teaching in the science classroom (Sadler et al. 2006). Some teachers therefore may not promote the necessary use of aspects of NoS or argumentation in the SSI discussions to make them worthwhile learning activities (Walker and Zeidler 2007). It has been reported that teachers' motivation to teach SSI comes from personal experiences or values and not from reform (Lee and Witz 2009). Thus, it follows that the quality of teaching SSI may also be independent of the reform. In addition, our research has shown that in the 2006 curriculum reform in England, many teachers were unhappy about the swing away from canonical science, others did not have any particular training to teach the new courses and others reported not being aware of any important differences between the new and old curricula (Ryder and Banner 2013). Therefore this research brings a new perspective to students' experiences of a context-based curriculum where teachers who have not chosen to teach this emphasis nor had opportunities to develop their own understanding of these areas of science and associated teaching approaches.

3 Experiences of Different Types of Students

Students are not a homogenous group. Thus it is possible, indeed likely, that not all students will be interested in and enjoy science in a social context. Much of the research cited above looks at groups of students without differentiation. Our research aims to address some of these issues by looking at experiences of different types of student. The idea of 'real-life' science came out of interviews with the students and refers to, in their terms, what they think they will use in their lives, usually in the future, both in and out of work.

Here we examine research showing how different types of students respond to context-based teaching approaches. Research in Singapore by Caleon and Subramaniam (2008) shows that 'average ability' students are less likely to enjoy science, want a career in science and appreciate the social implications of science compared to their 'above average' and 'gifted' peers. However, Zacharia and Barton (2004) and Barton and Tan (2010) report that when urban and black students (found to have 'exceptionally negative attitudes towards school science') were taught science through a relevant project-based approach, termed 'science in action', students acted as expert scientists, and their attitudes towards their learning experience was positive. 'Disengaged learners' become motivated when topics are of interest to them and tasks are useful and of value (Daniels and Arapostathis 2005). There is thought to be a disparity between many students' personal identity and their perceived identity of science careers and scientists that prevents them from pursuing course and careers in science (Schreiner and Sjoberg 2007). Thus, if courses are made relevant and useful, it is likely that students, who otherwise would not be interested in school science, will have a greater affinity with the topic and better, more meaningful experiences in the science classroom.

Students considered to be gifted and talented can also be engaged by less prescriptive material which may allow freedom for more discussion, for example, about SSI, in the classroom, such as is in the new curriculum in England. However, Kortland (2005) found that context-led physics teaching had to be modified for preuniversity students, compared to lower-attaining students, introducing systematic concept-based units, in order to provide 'more sophisticated and prominent' concepts to allow better application in a variety of contexts. This suggests that context-based teaching is not always the best way to teach science concepts to all students. Kind (2007) notes that context-based courses can provide the challenge and interest to engage talented learners but warns that teachers should be trained and confident in teaching the different approaches necessary to provide the intellectual stretch and other benefits.

4 The Research Context

In the state school system in England, this context-based curriculum with a greater emphasis on the nature of science and socio-scientific issues became compulsory in September 2006. Leading up to this there had been academic debate (e.g. Millar 1996; Millar and Osborne 1998) and published government documents (e.g. Roberts 2002; HM Treasury, DfES et al. 2004) describing the need for a curriculum which allowed flexibility to match the needs of all students, not just those who would be scientists. There were multiple aims ascribed to this curriculum reform with various stakeholders involved its development (Ryder and Banner 2010). The new national curriculum was introduced before the final evaluation of the pilot. However, early insights reported positive feedback from teachers about the course and students' reactions to it (Millar 2006).

As shown, research into context-led curricula provides evidence of the impact of a context-led school science curriculum. A curriculum which is relevant to students' everyday lives and will be useful to them as adult members of society in knowing about the world around them and for making decisions is supported widely (Driver et al. 1996; Jenkins 1999). Other authors write similarly about how this kind of curriculum will improve students' attitudes to school science and maybe encourage students to take science further as a result of such a curriculum (Millar and Osborne 1998). What is new in the study reported here is the nature of the reform itself. A nationwide, compulsory, top-down implementation is likely to have very different implications as compared to the introduction of small-scale and local interventions. By engaging students from a range of schools where teachers have differing opinions about the new curriculum (Ryder and Banner 2013), this study aims to present a broad perspective in contrast to small-scale intervention and pilot studies often associated with this kind of curriculum development. It gives arguably a more realistic view of the impact of teaching a context-based science curriculum. It is also important to consider the experiences of different groups of students and how the context-led curriculum impacts on them. Our research questions therefore are:

How do students experience a national curriculum with an emphasis on a context-based approach?

How do different groups of students talk about learning about 'real-life' science in the science classroom?

5 Methods

The research was carried out in 19 schools around England. Heads of science in these schools were asked to select two groups of six children aged 14–15 years old studying one of the recently reformed science qualifications. Although the final decision was up to the teachers, it was requested that students represented the gender balance in the school and the range of attainment and, if possible, attitudes in the class. We also asked for students who would be willing to contribute their points of view. Students were interviewed in groups of six in their penultimate and final years of compulsory schooling.

The first set of discussions was based on open interview questions to gather a range of students' ideas and experiences of their school science. When necessary to guide the talk, we used broad terms like 'usefulness' of school science to encourage students to think about aspects of what they were learning in a different way. Students were also asked about any plans they may have for after compulsory schooling. All students also completed short questionnaires immediately prior to and after the interview. This was to help students clarify their ideas before and after discussions that could act as a comparison to verbal responses. It also gave us quantitative data with which could be used to characterise the groups. Speaking to the students broadly and openly about aspects of their experiences enabled us to draw on students' own language in their descriptions of school science and science lessons allowing us, for example, to conceptualise 'real-life' science in the language of the students and not the researcher.

In the second year of the study, students' comments and use of language regarding school science were used to ask more specific questions about topics and teaching/ learning activities experienced in science lessons. Students did not refer specifically to SSI, so SSI were framed in activities like class discussions and looking at newspaper reports with typical examples of discussed topics taken from school textbooks, e.g. the safety of mobile phones, where to site open quarries and the safety of vaccines. In addition to 'real-life science', students were also asked about practical work, class discussion, independent work, teacher presentation with questions and how data can be used and what it means. Short descriptions of these headings were also on cards, and these were talked through with the students to ensure a consensus of understanding before that part of the interview. Again students completed short questionnaires immediately prior to and after the interview in the presence of the researcher.

6 Data Collection and Analysis

The group interviews were audio-recorded and transcribed. Discussions generally lasted between 30 and 60 min. Transcripts were analysed and coding was supported by MaxQDA software. Codes were derived from a subset of the data responsively in year 1, by two researchers independently, and a coding scheme agreed partly based on a framework describing relevance and interest (see below). The second set of interviews were coded in much the same way except that the specific classroom activities discussed in year 2 were created as codes prior to analysis, and other code categories were added responsively on analysis of the data.

Whilst being aware that all students have different abilities, interests and backgrounds, there was a need to categorise students to look at how different student groups experienced 'real-life' science. Thus, students were categorised depending on their 'choice' of courses at age 14–16 years. The curriculum change in 2006 increased the flexibility of routes through science education for 14–16-year-olds. Most of our case study school sample of students took a 'Core' science course in the first year of our research which has a greater emphasis on context-based science learning than subsequent courses. Our research distinguishes between the interviews with students taking the more academic pathway, resulting in qualifications in the three separate science subjects (biology, chemistry and physics) from 15 schools (known henceforth as *Triple* students) and students taking the less academic pathways. The less academic pathways encompassed students taking one single science qualification (the 'Core') over 2 years and/or taking 'Applied Science' courses, designed to have a more vocational element within them and often a larger proportion of centre-moderated coursework as part of the examination (termed here *Applied* students).

We can establish some general characteristics of these two student groups using national data bases held by government and a national survey that was sent out as a part of the project to be completed by all final year students in 19 schools. The databases are the National Pupil Database (NPD) and the Pupil Level Annual School Census (PLASC) which contain information about all pupils in state schools in England, their public examination results and personal characteristics like gender, ethnicity and measures of socio-economic status. In addition specific data were taken from the interviewed students in written questionnaires. There is another group of students, not considered here, who take a science course that awards two qualifications in science (made up of biology, chemistry and physics). These students form 46 % of the student cohort nationally.

On a national level students in England completing 'Applied' courses had a much lower average prior attainment than students completing the Triple science courses. The data also show that students on one of the 'Applied' courses scored about 30 % lower in national examinations than their peers taking Triple science (Homer et al. 2011). Students studying the 'Applied' courses were also much more likely than average to come from low-income households, whereas those studying Triple science were much less likely than average to come from low-income households (op. cit.). In data collected in our survey, with 580 Triple students and 280 students taking 'Applied' courses, nearly 50 % of Triple students agreed or strongly agreed with the statement that they would like a job that involves science compared to fewer than 15 % of 'Applied' students. Results from written questionnaires completed at the time of the second interview supported these characterisations.

Models of different types of relevance (Van Aalsvoort 2004) and interest (Lavonen and Laaksonen 2009) were used to help analyse the data. These models were important in the construct of experience, as students often referred to these ideas in their discussions, and students' concepts of relevance and interest changed with different personal contexts. Van Aalsvoort (op. cit.) identified different meanings of relevance by looking at its use in different studies which showed that 'relevance' was viewed in a range of ways. The term was used to mean: personal relevance

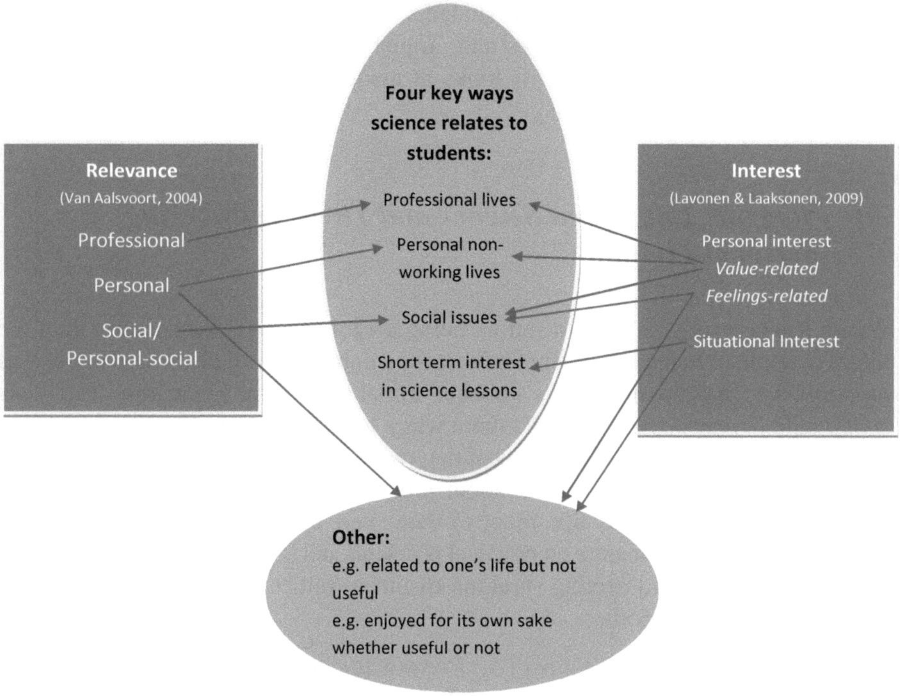

Fig. 23.1 The overlap between two published models of relevance and interest

to students' lives, professional relevance to future jobs and social relevance to show the importance of science in society. A report discussing the outcome of the PISA results in Finland (Lavonen and Laaksonen 2009) identified different types of interest and the bases for them, such as a 'value-related' interest, e.g. in things related to a student's future career and 'situational' interest, generated from what is going on in a student's surroundings, e.g. a lesson.

There is evidently overlap between these models which form a framework to link the ideas of relevance and interest (see Fig. 23.1). Using this framework the transcripts were analysed to identify which statements related to the relevance and interest of students' experiences in four key ways in which students spoke about school science: in relation to personal lives that was not work and in relation to their professional lives, to social issues and to short-term interest in science lessons.

7 Findings

The aims of this research were to look at different types of student in the context of a curriculum emphasising 'science for all' and socio-scientific issues. This chapter then focuses on the two areas considered to be closely related to the nature of this contextualised course, looking at the students' ideas about school science in relation to their nonworking personal lives and, arguably, a subset of this, where school science relates to social issues. It looks at evidence in each of these areas that answers both research questions.

7.1 Students Talk About Science and Their Nonworking Personal Lives

Where science was not recognised as being useful for jobs and careers, then the initial reaction to being asked whether science might be useful elsewhere in life was *probably not*. In many interviews students' initial ideas were vague, they said *some stuff* would *pop up somewhere*, and there was a feeling, perhaps almost a hope, that there *must* be something useful in what they were being taught.

When students started to think about more specific examples of how science might be useful in everyday life, many of them were naïve. Students thinking about everyday lives tended to centre on practical problems and issues, for example, cooking, health and first aid and driving (braking distance) rather than decision-making or other issues related to SSIs.

Girl 1: *Probably in driving a car (...) Because* [*the teacher*] *was* [*saying*] *how long it takes for you to stop with different masses of people in the car and -*
Girl 2: *Yeah, but if you're driving but you stop, you're not going to quickly calculate this is the mass of people in car, this is the distance.*
Girl 3: *So you crash.*
(Laughter).
Girl 1: *It's not hard just to work out in your head before you leave.*
Girl 2: *You wouldn't need to.*
Girl 1: *No, you wouldn't need to but you could.*
Girl 2: *Yeah you could.*
Girl 1: *You wouldn't need to but you could and they say six seconds stopping distance but...*
Girl 3: *Do you really think the millions of drivers in the whole world really know about the braking distance* [*like the teacher said*]?
(School 15, Triple, first year)

This suggests that students did not have a clear idea about how science might be useful in everyday life. They found it hard to see how things were relevant to them (even the example of learning about the safety of mobile phones). The students in this group all felt, with the exception of health-related knowledge in biology, that science was not useful. This was perhaps even more evident where teachers and students spoke about the importance of 'real' canonical science compared to things like discussing SSIs (see Sect. 3.2).

Both Triple and 'Applied' students felt that it was *good to know stuff* – some again for more practical reasons like helping their own children with homework in years to come or just talking with their parents. There was some, albeit limited, recognition that just being aware of and understanding science was a good thing in itself – the kind of cultural argument for learning science explored by Millar (1996).

Yes, it's just nice to like have knowledge really. If you don't know much it's kind of boring and if you're with your mates and talking about something you don't really know what's going on it's a bit like – you just feel a bit left out, but if you are – yes, I think science in a way is just gives you something to talk about. And is just interesting really (School 18, Triple, first year).

Differences from groups emerged more in what students wanted from their science education. As mentioned earlier, students taking the Triple science course were much more likely to have intentions to take science at post-16, and their comments regarding course content, particularly in the second year of interviews, reflected the importance to them of learning *real science* rather than *social science* and things that were *common sense*. Students following the 'Applied' courses, however, being much less likely to want to take science any further, frequently described that they were being taught science for future scientists and would *like* to learn about *real-life* science – implying that this was not what they were experiencing now as illustrated in the quote below:

Girl 1: *I just think science is a waste of time. I think you should choose if you want to be a scientist.*
IB: *So you don't think* [*school science*]*'s got any real life applications?*
Girl 1: *It actually doesn't.*
(School 15, 'Applied', second year)

Students from the 'Applied' groups in particular thought that more real-life science in their lessons would be better. It was understandably difficult for students to be able to describe what this was – not knowing what adult life held and the naivety of students was apparent here as well, with most students thinking of First Aid as an important science subject to be taught.

It was in the second year of interviews, when most students were taking a slightly different set of modules than the previous year that students recognised 'real-life' science, or indeed 'real science' (canonical science, traditional school science), as being more or less present in their science lessons. For some students, particularly those taking the course leading to two qualifications in science (not discussed in detail here), it was not until the second year of study that they realised the first year modules *had* actually been relevant and might be useful to them. They had not recognised this at the time. This highlights the methodological advantage of longitudinal interviewing.

7.2 Students Talk About Science and Social Issues

As mentioned previously, how school science relates to social issues could be considered a subset of school science in relation to students' personal non-working lives. The sorts of SSI raised in the curriculum seem to be perceived by many students as distant from their current or future lives. Students taking Triple science were more likely to recognise the importance of learning about issues like organic farming and global warming in particular. Students spoke about 'ethics' similarly – using 'ethics' interchangeably with SSI – often recognising the importance of studying 'ethics' but feeling that this was not science or an enjoyable aspect of science.

I think it's important to understand the ethics and to be aware of them, but I don't think it should be the main point of the subject because it's just.. [...] (it) should be all science (School 13, Triple, second year).

Students taking 'Applied' courses were less likely than Triple students to mention examples of SSI and related topics when asked about real-life science. This suggests that they had less exposure to this teaching or did not perceive it in the same way. However, in some groups of these students had obviously experienced discussions and other teaching about fair trade and organic food, religion and science and global warming and, with some prompting to recall these topics, seemed to feel that they learned something and that it was useful to know about them.

One common example of SSI in the new science curriculum is that of the safety of mobile phones. It was clear that students do not always connect with what adults think that they will:

IB: *But is it good learning about things like mobile phones?*
Girl 1: *No, not to me.*
IB: *Not to you. What about you? Have you got a mobile phone?*
Boy 1: *Yeah.*
IB: *So is it good to learn about it?*
Boy 1: *No.*
IB: *No?*
Boy 1: *You just use it.*
(School 2, Triple, first year)

Noteworthy points from these findings are that the majority of the students did not talk about learning from or about the processes involved in SSI – argumentation, reflection and decision-making, for example. When decision-making was mentioned, it was in relation to voting when the students were older, but mostly it was the content of the topics that was considered to be important, e.g. building mobile (cell) phone masts. It was also rare for these students to talk about their learning in relation to the nature of science (NoS), e.g. the quality of evidence and the fact that science cannot answer everything (which are explicit in the National Curriculum). Of course this does not mean that such learning was not happening; students referred to ideas about NoS more frequently in schools where we were confident that they had been taught these ideas either explicitly or in an integrated scheme (from teacher interviews or observed lessons). Nevertheless this was still rare and usually mentioned only after some prompting by the interviewer.

Boy 1: *It was like when you had say which ... there's ones there when you've got six different people with their opinions and you've got say which one's got the most clearest opinion or something like that, it's just like easy marks because it's just dead obvious.*

IB: *Do you think that sort of thing though will help you when you're older? Sort of training for that, thinking about opinions, listening to people or do you think ...?*

Boy 2: *I think we would have probably learnt it already by that age* (...) *just growing up really.*

(School 3, second year)

This quotation shows how students appeared to think that the skills used in these situations were common sense – which might indicate how well they had been taught or might be a true reflection of the extent of their skill (we did not test this). However, in most cases the ideas behind and the importance of these processes did not seem to be apparent to students or an important consequence of learning science.

8 Discussion

Here we return to the key questions of the study – How do students experience this national curriculum with an emphasis on a context-based approach, and how do different groups of students talk about learning about 'real-life' science in the science classroom?

It is not surprising that students' discussions about their experiences of school science vary considerably and that these reported experiences are heavily dependent on what happens in the classroom. Teachers' experiences of this reformed curriculum are examined elsewhere (Ryder and Banner 2013) showing that the many different contexts in which teachers live and work legitimately influence practice, resulting in a variety of content emphases, teaching and classroom management styles. There is much that is evident from students' discussions about how what teachers say and do has a direct impact on the experience of science in the classroom. However, the new curriculum does appear to influence the way some students talk about their experiences in science lessons. It was particularly striking that two groups of students following a particular applied science course in the second year of interviews were noticeably more engaged with the applied work than the 'Core' modules the previous year, giving the reason that it was much more relevant and useful to their lives. What was also evident was that most groups of Triple students preferred the more academic modules to the 'Core' modules. Thus curriculum, under the influence of the teacher, does have an important influence on the experiences of students in these groups.

However, students still struggled to see the relevance of many topics (e.g. the periodic table) or to describe what a useful science curriculum might look like.

Some students also reported that science is facts and that discussion is for religious education and English lessons. It is evident that a context-led curriculum written by adults does not necessarily seem 'real life' to students. Students also mostly talked about the subject matter of SSI and not some of the key ideas like the processes of and evidence in science, and it seems likely that the important ideas behind SSI are not being taught explicitly. Furthermore, it may be that SSI are not taught through interactive methods. Much of the literature which espouses SSI teaching does so in conjunction with use of argumentation and other group activity in lessons and/or integrating science learning and teaching into local issues. Data collected as part of both this project and related research, from interviews with teachers and some lesson observation suggests that this approach is not widespread, certainly in our case study schools (Morris 2012). The new approach may be having some impact however; students stated that much of what they were learning was *common sense*, and some said things like *of course you can't trust everything you read in the paper* although we do not know that similar students would not have said this before the new curriculum.

Differences between groups of students are difficult to analyse because of confounding variables such as different teachers, classroom behaviour and management; it seemed, for example, that students taking Applied science courses have less exposure to teaching of SSI, for example, than their Triple science peers. The different levels of and types of motivation that students have for learning science also need to be considered. It is also important to recognise the differences between individual students within these groups that have not been explored here. Furthermore, it is possible that students may misrepresent their true feelings in a group interview situation. However, this work goes some way to show that although heavily moderated by teachers, different students have different needs and desires for different curricula both in terms of content and teaching methods. The other factors to be considered such as the students' chosen career paths are discussed more fully elsewhere and obviously play an important role in shaping students' identities especially their science identities (Archer et al. 2010). Experiences, expectations and future plans mean that what is suitable for one student may not be suitable for another. The fact that many of the sample of Triple students did not see SSI as 'real science' despite seeming to understand the importance of learning about these issues perhaps reflects their perception of their future science needs, completing post-compulsory science qualifications with appropriate grades and going on to university courses, with a focus on traditional science content. This is not to suggest that this outlook is right or indeed that these students should not learn about SSI and the processes of science. However, the inclusion of this type of real-life science is not necessarily seen as motivating and relevant by all students. The responses of students taking Applied science and related courses were perhaps influenced by the same issues – expectations, experiences and future plans – very few of which included science. What were designed as worthy vocational routes through science are frequently used as courses for low-attaining students. It is important to think about the relevance to these students of a work-based science course and the types of activities therein: a science course based on everyday life for these students, who certainly felt that more real-life science was what they wanted, seems to be more appropriate.

8.1 Implications

Many students freely admit they do not know what their future personal futures hold, and students from lower socio-economic backgrounds are less likely to take high-status science courses. Therefore not taking high-status science courses may seem like a sensible decision at age 14 but could have far-reaching consequences. It is important that assumptions and decisions, especially by adults, about certain students should not risk limiting students' opportunities in the future. The current situation with one 'Core' course for all students and a range of follow-on courses designed to meet a range of differing needs of students may be the most suitable option.

However, students in the 'Applied' group may also find it hard to go on to study science at post-16, if they wanted to. This is because this less academic route through compulsory science education effectively prevents or discourages them from going on to take traditional high-status courses and because few post-16 providers have suitable applied or other post-16 courses available. We need to make sure that students who take the 'Applied' courses at age 16 have routes of progression through into post-compulsory science, giving diversity without hierarchy.

Many students still cannot understand the point of learning science in school. Although it seems paradoxical, some students need more defined guidance as to what is relevant to them and why – understandably for a future life somewhat distant and itself irrelevant to many of them at age 15. The science curriculum features contexts that adult curriculum designers consider will be 'relevant' to young people. However, our discussions with students show, first, that different students have varying notions of relevance in the context of the science lessons and, second, that these meanings of relevance can differ significantly from those held by adults.

Our data shows, for example, that not all students are engaged by socio-scientific issues or that they all enjoy discussion and real-life scenarios; there is a population of students, maybe influenced by their teachers, who enjoy learning canonical science. There also appear to be groups of students who are not exposed to the interactive teaching methods associated with teaching SSI. There is a need for both guidance and flexibility in the provision and teaching of science courses so that the student voice is heard and acted upon and at the same time students are provided with a broad and balanced curriculum to maximise opportunities for students both in everyday life and for their careers.

Acknowledgements The work reported here is one outcome of a 3 year study funded by the Economic and Social Research Council and the Gatsby Charitable Foundation, grant number RES-179-25-0004. Thanks are extended to the teachers and students who participated in this research.

References

Abd-El-Khalick, F., & Lederman, N. G. (2000). Improving science teachers' conceptions of nature of science: A critical review of the literature. *International Journal of Science Education, 22*(7), 665–701.

Aikenhead, G. S. (1996). Science education: Border crossing into the subculture of science. *Studies in Science Education, 27*(1), 1–52.

Albe, V. (2008). When scientific knowledge, daily life experience, epistemological and social considerations intersect: Students' argumentation in group discussions on a socio-scientific issue. *Research in Science Education, 38*(1), 67–90.

Archer, L., DeWitt, J., et al. (2010). Doing science versus being a scientist: Examining 10/11-year-old schoolchildren's constructions of science through the lens of identity. *Science Education, 94*(4), 617–639.

Barton, A. C., & Tan, E. (2010). *We Be Burnin'!* Agency, identity, and science learning. *The Journal of the Learning Sciences, 19*((2), 187–229.

Bennett, J., Lubben, F., et al. (2007). Bringing science to life: A synthesis of the research evidence on the effects of context-based and STS approaches to science teaching. *Science Education, 91*(3), 347–370.

Caleon, I. S., & Subramaniam, R. (2008). Attitudes towards science of intellectually gifted and mainstream upper primary students in Singapore. *Journal of Research in Science Teaching, 45*(8), 940–954.

Carlone, H. B. (2004). The cultural production of science in reform-based physics: Girls' access, participation, and resistance. *Journal of Research in Science Teaching, 41*, 392–414.

Daniels, E., & Arapostathis, M. (2005). What do they really want?: Student voices and motivation research. *Urban Education, 40*(1), 34–59.

Driver, R., Leach, J., et al. (1996). *Young people's images of science*. Buckingham: Open University Press.

Grace, M. (2009). Developing high quality decision making discussions about biological conservation in a normal classroom setting. *International Journal of Science Education, 31*(4), 551–570.

HM Treasury, DfES, et al. (2004). *Science and innovation investment framework: 2004–2014*. London: HM Treasury.

Homer, M., Ryder, J., et al. (2011). The use of national data sets to baseline science education reform: Exploring value-added approaches. *International Journal of Research & Method in Education, 34*(3), 309–325.

Jenkins, E. (1999). School science, citizenship and the public understanding of science. *International Journal of Science Education, 21*, 703–710.

Kanter, D. E., & Konstantopoulos S. (2010). The impact of a project-based science curriculum on minority student achievement, attitudes, and careers: The effects of teacher content and pedagogical content knowledge and inquiry-based practices. *Science Education, 94*, 855–887.

Kind, V. (2007). Context-based science: A 'gift-horse' for the talented? In K. Taber (Ed.), *Science education for gifted learners*. London: Routledge. xv, 240 p.

Kortland, K. (2005). Physics in personal, social and scientific contexts: A retrospective view on the Dutch physics curriculum development project PLON. In P. Nentwig & D. J. Waddington (Eds.), *Making it relevànt : Context based learning of science*. Munchen/New York: Waxmann. 359 p.

Lavonen, J., & Laaksonen, S. (2009). Context of teaching and learning school science in Finland: Reflections on PISA 2006 results. *Journal of Research in Science Teaching, 46*(8), 922–944.

Lee, M. K., & Erdogan, I. (2007). The effect of science–technology–society teaching on students' attitudes toward science and certain aspects of creativity. *International Journal of Science Education, 29*(11), 1315–1327.

Lee, H., & Witz, K. G. (2009). Science teachers' inspiration for teaching socio-scientific issues: Disconnection with reform efforts. *International Journal of Science Education, 31*(7), 931–960.
Mandl, H., & Kopp, B. (2005). Situated learning: Theories and models. In P. Nentwig & D. J. Waddington (Eds.), *Making it relevant: Context based learning of science* (pp. 15–34). Munchen: Waxmann.
Millar, R. (1996). Towards a science curriculum for public understanding. *School Science Review, 77*(280), 7–18.
Millar, R. (2006). Twenty first century science: Insights from the design and implementation of a scientific literacy approach in school science. *International Journal of Science Education, 28*(13), 1499–1521.
Millar, R., & Osborne, J. (Eds.). (1998). *Beyond 2000: Science education for the future*. London: King's College, School of Education.
Morris, H. E. (2012). *Girls' responses to the teaching of socioscientific issues*. Leeds: University of Leeds. School of Education, PhD.
Roberts, G. (2002). *SET for success. The supply of people with science, technology, engineering and mathematics skills*. London: H. Treasury.
Ryder, J., & Banner, I. (2010). Multiple aims in the development of a major reform of the national curriculum for science in England. *International Journal of Science Education, 33*(5), 709–725.
Ryder, J., & Banner I. (2013). School teachers' experiences of science curriculum reform. *International Journal of Science Education, 35*(3), 490–514.
Sadler, T. D. (2004). Informal reasoning regarding socioscientific issues: A critical review of research. *Journal of Research in Science Teaching, 41*(5), 513–536.
Sadler, T. D., Amirshokoohi, A., et al. (2006). Socioscience and ethics in science classrooms: Teacher perspectives and strategies. *Journal of Research in Science Teaching, 43*(4), 353–376.
Sadler, T. D., Barab, S. A., et al. (2007). What do students gain by engaging in socioscientific inquiry? *Research in Science Education, 37*(4), 371–391.
Schreiner, C., & Sjoberg, S. (2007). Science education and youth's identity construction – two incompatible projects? In D. Corrigan, J. Dillon, & R. Gunstone (Eds.), *The re-emergence of values in the science curriculum*. Rotterdam: Sense Publishers.
Van Aalsvoort, J. (2004). Logical positivism as a tool to analyse the problem of chemistry's lack of relevance in secondary school chemical education. *International Journal of Science Education, 26*(9), 1151–1168.
Walker, K. A., & Zeidler, D. L. (2007). Promoting discourse about socioscientific issues through scaffolded inquiry. *International Journal of Science Education, 29*(11), 1387–1410.
Zacharia, Z., & Calabrese Barton, A. (2004). Urban middle-school students' attitudes toward a defined science. *Science Education, 88*(2), 197–222.
Zeidler, D. L., Sadler, T. D., et al. (2009). Advancing reflective judgment through socioscientific issues. *Journal of Research in Science Teaching, 46*(1), 74–101.

[illegible]

Chapter 24
Students' Experienced Coherence Between Chemistry and Biology in Context-Based Secondary Science Education

Hilde J. Boer, Gjalt T. Prins, Martin J. Goedhart, and Kerst Th. Boersma

1 Introduction

Creating coherence between the content of science subjects has been a primary aim of certain reforms in science education and is often proposed in policy documents in various countries (Osborne and Dillon 2008; Schmidt et al. 2005; Osborne and Collins 2001). One of the problems that emerges from literature on coherence is the lack of consensus on its definition and conceptualization (Schmidt et al. 2005; Johnson and Ratcliff 2004).

In this research, the emphasis will be on coherence between the secondary school subjects chemistry and biology. In current educational practice, chemistry and biology are often treated as completely independent and unrelated subjects (Bardeen and Lederman 1998). This is incongruent with current scientific practice, in which the emphasis is shifting towards more multidisciplinary and interdisciplinary approaches (Bulte et al. 2006). However, chemical concepts like acid, pH, chemical reactions, and chemical bonds are frequently used in biological texts. Chemistry is considered to be a prerequisite for biology, and students are expected to grasp the interrelatedness of chemistry and biology by themselves.

H.J. Boer (✉) • G.T. Prins • K.Th. Boersma
Freudenthal Institute for Science and Mathematics Education (FIsme),
University of Utrecht, Utrecht, The Netherlands
e-mail: H.J.Boer@uu.nl; G.T.Prins@uu.nl; K.T.Boersma@uu.nl

M.J. Goedhart
Faculty of Mathematics and Natural Sciences, University of Groningen, Groningen,
The Netherlands
e-mail: m.j.goedhart@rug.nl

C. Bruguière et al. (eds.), *Topics and Trends in Current Science Education: 9th ESERA Conference Selected Contributions*, Contributions from Science Education Research 1,
DOI 10.1007/978-94-007-7281-6_24, © Springer Science+Business Media Dordrecht 2014

This study strives to contribute to the development of a learning and teaching (LT) strategy that emphasizes coherence between chemistry and biology from a students' perspective for upper secondary, preuniversity students (grade 11, age 16–17). The aim of this research is to enable students to experience coherence between chemistry and biology in a context-based curriculum unit. Based on activity theory, a context is defined as a social practice: the culturally defined activity or activities performed by participants with common goals, motives, and use of a common set of tools. A social practice has also been denoted as an authentic practice (Boersma et al. 2005; Westbroek 2005). In this study, an authentic practice covering the domains of chemistry and biology will be used as a source of inspiration for the design and development of a curriculum unit. We will report on the results of the first research cycle.

The central question addressed is: to what extent do students experience and express coherence between chemical and biological concepts by attending an authentic practice-based curriculum unit embodying the domains of chemistry and biology?

2 Conceptual Framework

2.1 *Coherence*

The lack of consensus on an operational definition of coherence affects educational research on coherence. Rudduck et al. (1994) describe in their research on coherence and students' learning experiences that it was hard to measure coherence. The absence of an operational definition may be caused by the observation that coherence in educational research can relate to different levels of the curriculum. Another reason is that it can be perceived from different perspectives, for example, from a students' perspective and from an organizational perspective. From an organizational perspective, the emphasis may be laid on the structure of a whole curriculum or on curriculum components, whereas coherence from a students' perspective involves, for example, coherence in and between contents of subjects and coherence in learning activities (Geraedts et al. 2006; Venville et al. 2002).

Coherence between the subjects biology and chemistry often remains implicit from a students' perspective. For students to be able to experience relations between the subjects, to actively construct links between those subjects, and to express multidisciplinary and interdisciplinary perspectives when engaging in learning activities, the coherence needs to be made explicit during the learning and teaching activities (Boersma et al. 2007).

This calls for a learning and teaching strategy that emphasizes coherence by placing students in the position to actively construct relationships between chemical and biological concepts themselves. This "conceptual structure" needs to be expressed and visualized (Novak and Cañas 2007).

2.2 Authentic Practices

Cultural-historical theory emphasizes that people learn by participating in culturally defined activities performed in a social practice (Engeström 1987). It is presumed that learning is facilitated by focusing on the way in which knowledge (e.g., concepts and strategies) is embedded in contexts (Van Aalsvoort 2004). The situated learning perspective claims that the meaning of a concept (knowledge) is dynamic when used in various contexts (Lave 1988).

The context-based approach derives from cultural-historical theory and situated learning theory and recognizes that knowledge should be learned by exploration or simulation of the context in which the knowledge is used (Boersma et al. 2005; Gilbert 2006). The social practice provides for heuristic guidelines for the design of a curriculum unit consisting of a sequence of learning and teaching activities. The context-based approach can arouse general interest and offers an opportunity for an orientation on a social practice. Furthermore, it can provide a steering question, a "need to know," and the confidence that the desired outcomes are attained (Boersma and Waarlo 2009). The social practices found in society are also denoted as *authentic* practices (Bulte et al. 2006).

In a social or authentic practice, a group of people works on real-world problems and societal issues in a "community." These people are connected by three characteristic features: having common motives and goals, working according to a similar type of characteristic procedure leading to an outcome, and displaying apparent necessary knowledge about the issue that they are working on (Bulte et al. 2006; Van Aalsvoort 2004; Engeström 1987).

2.3 Coherence and Authentic Practices

Consisting of activities in which coherence naturally occurs in the procedures and the knowledge about the issue worked on, authentic practices can be transformed in an educational context in which this coherence is maintained. When chemical and biological knowledge and procedures are applied in an authentic practice, it can provide a learning and teaching context that emphasizes this specific coherence for students.

Not all authentic practices covering the domains of chemistry and biology are suitable to use as context in secondary science education. A suitable authentic practice must meet the following criteria: the issue can interest and motivate students; purposes and motives are recognizable for students; the characteristic procedures of the practice are in line with procedures for general application; students are able to deal with the complexity of an issue; students must be familiar with the issue, where familiarity depends on the amount of preknowledge and routine skills versus the amount of new information presented in the practice; procedures as laboratory work must be feasible in the school environment; and the interdisciplinary character concerning both biology and chemistry is present and recognizable for students (Prins et al. 2008).

In this study, we selected an authentic practice which combines a systems thinking approach with the macro-micro thinking approach. Connecting macroscopic phenomena with microscopic entities or micro-macro thinking is an underlying basic structure in chemistry and in chemistry education (Bulte et al. 2006). In biology, thinking in terms of levels of biological organization as a form of systems thinking is an underlying basic structure (Knippels et al. 2005). Structures and processes at various levels of organization are frequently explained by interactions at the molecular level (Verhoeff et al. 2008). In multidisciplinary and interdisciplinary authentic practices, in which knowledge and procedures originating from the discipline chemistry as well as the discipline biology are applied, these underlying basic structures can be recognized. The embodied interrelatedness between underlying basic structures provides for a structure to conceptualize and operationalize the specific coherence between chemistry and biology.

2.4 Transposition of Authentic Practices

In an authentic practice, the participants do not question the relevance of the skills and issue knowledge involved, since they have clear motives to use these skills and the issue knowledge in their actions. In the module in which an instructional version of the authentic practice functions as context, students are asked to work towards certain goals. By searching for a path to reach the goals, students are required to develop a need for specific knowledge, skills, and motives to proceed in a certain way to solve the problems that they encounter and experience their actions as meaningful and coherent, for every action is necessary in order to reach the goal. In searching for this path, students perform, within the constraints of the classroom, similar actions or procedures as performed by the actual participant in the authentic practice. Due to their complex nature, authentic practices need to be transformed and restructured into educational contexts; this process can be referred to as didactic transposition (Ogborn 2005). Distinctions should be made between experts and students. Experts and students differ in knowledge, experiences, attitude and way of thinking (e.g., on a scientific level), and in motivation (Prins et al. 2008; Engeström 1987). In this study, a problem-posing approach (Klaassen 1995) was applied to structure the LT activities in the transposed authentic practice. A problem-posing approach emphasizes the structuring and formulation of teaching activities in a bottom-up approach, i.e., from the point of view of understanding, coherence, and purpose for students. The learning process is driven by developing content-specific motives, including a need to know and confidence that this need will be met by learning the next activity (Lijnse and Klaassen 2004; Lijnse 2005; Boersma and Waarlo 2009). In a problem-posing approach, a problem is developed in consultation with students. In this way students obtain ownership of the problem, in contrast to a problem-solving approach in which students focus on solving a given problem.

3 Methodology

In this research, an educational design research approach was adopted (Van den Akker et al. 2006; Lijnse 1995). Design research will be considered as research into and through the design of a learning and teaching strategy (LT strategy). In the exploration of a research cycle, development and research activities alternate.

The design activities will be informed by theoretical and empirical evidence from literature review, explorative studies, and practical knowledge. By testing the hypothetical LT strategy consisting of designed and sequenced learning and teaching activities (LT activities), the LT strategy can be adjusted and refined and will result in a contribution to domain-specific theory on achieving coherence between chemistry and biology from a students' perspective (Boersma and Waarlo 2009).

In this study, the design of the LT strategy started with a literature review and a practical orientation which comprises an orientation on the forms of coherence present in (a) currently used school textbooks for chemistry and biology and (b) modules of the new Dutch integrated school subject NLT (Nature, Life, and Science) and an orientation on authentic practices that provide a framework for the sequence of LT activities. From this, a route for achieving coherence between the subjects chemistry and biology is elaborated in a preliminary LT strategy. This LT strategy informs the construction of the scenario and the sequence of LT activities, also referred to as the learning and teaching sequence (LT sequence). The scenario describes the overall learning goals of the LT strategy, the contributions of the respective learning and teaching activities to these goals, and the learning process that the students will undergo while engaging in the planned LT activities. In this way, the subject and pedagogical knowledge, expectations, and theoretical perspective are made explicit in detail (Lijnse 1995). In this paper, the main focus lies on that part of the scenario which includes the conceptualization of coherence between chemistry and biology in the authentic practice (Fig. 24.1) and derived propositions between chemical and biological concepts and entities (Table 24.1). The propositions were prepared by the first and second author and further discussed by the research team. Here, propositions are considered as linking phrases between pairs of concepts (Cañas et al. 2003). The selection, interpretation, and analysis of data are guided by the question to what extent the attained learning outcomes of the learning and teaching process (LT process) are in accordance with the intended outcomes, in terms of formulated propositions. Along with empirical data from the actual enactment of the LT activities combined with a further analysis of the theory on achieving coherence between chemistry and biology from a students' perspective, the scenario will guide the redesign of the LT strategy.

The results provide characteristics of an LT strategy for upper secondary science education based on an authentic practice that enables the students to experience coherence between the subjects chemistry and biology in terms of propositions between chemical and biological concepts organized in a conceptual structure.

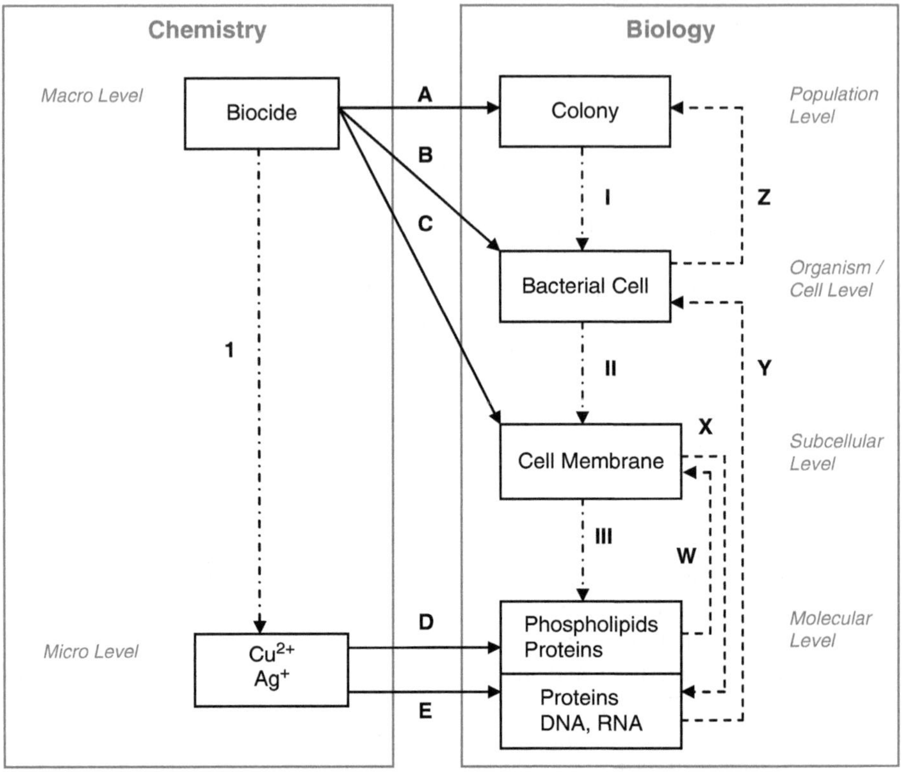

Fig. 24.1 Schematic representation of the conceptualization of coherence between chemistry and biology in the authentic practice of "*Legionella* contamination of the tap water system"

3.1 Design of the LT Strategy

The selected authentic practice informing the LT strategy is "*Legionella* contamination of the tap water system" or more precise "the danger and eradication of *Legionella pneumophila* by means of copper-silver ionization." The contamination of tap water by *Legionella pneumophila* forms a major problem in the complex tap water systems of modern buildings like schools and hospitals. The eradication by means of biocide copper-silver ionization offers an effective but potentially risky solution due to the use of heavy metals. One of the key features of the authentic practice is the presence of both biological and chemical subsystems that interact with each other. The knowledge and processes applied in the authentic practice originate from the key-conceptual area of biochemistry.

Figure 24.1 is a schematic representation of the conceptualization of coherence between chemistry and biology. Figure 24.1 depicts the disciplines chemistry and biology, the micro and macro levels, several levels of organization, and the chemical

Table 24.1 Propositions as presented in Fig. 24.1

Label	Proposition
1	1_1: Electrolysis of the metallic solid copper produces copper ions
	1_2: Electrolysis of the metallic solid silver produces silver ions
	1_3: The metallic solids copper and silver represent the macro level; copper ions and silver ions represent the micro level
A	A_1: Biocide acts on colony
	A_2: As a consequence, the total number of colonies decreases
	A_3: This takes place at population level
B	B_1: Biocide acts on bacterium
	B_2: As a consequence, the bacterium dies
	B_3: This takes place at organism and cell level
C	C_1: Biocide acts on cell membrane
	C_2: As a consequence the structure of the membrane changes, and the membrane becomes more permeable
	C_3: This takes place at subcellular level
D	D_1: The copper ions interact with phospholipids and proteins
	D_2: As a consequence, the molecules of the phospholipids and proteins change
	D_3: This takes place at molecular level
E	E_1: The silver ions interact with proteins, DNA, and RNA
	E_2: As a consequence, the molecules of proteins, DNA, and RNA become dysfunctional
	E_3: This takes place at molecular level
I	I_1: Zoom in from population level to organism and cell level
	I_2: A colony consists of multiple bacteria
II	II_1: Zoom in from organism and cell level to subcellular level
	II_2: A bacterium consists of subcellular components
III	III_1: Zoom in from subcellular components to molecules
	III_2: Subcellular components consist of molecules
W	Due to the change in the molecules of the phospholipids and proteins, the structure of the cell membrane changes and becomes more permeable for ions
X	As a consequence of the increased permeability of the cell membrane, silver ions pass the membrane and interact with proteins, DNA, RNA
Y	Due to the dysfunctional proteins, DNA, and RNA, the bacterium or the cell dies
Z	Due to the decrease in the number of bacteria or cells, there is a decrease in the number of colonies

and biological entities that were derived from the authentic practice. The concepts in the small boxes represent the subsystems in terms of chemical or biological entities, and the arrows represent the propositions between the concepts. Each arrow was labeled, and relations between the concepts were described as apparent and identified in the authentic practice. The propositions that are depicted in Fig. 24.1 are described in Table 24.1.

The design of the LT sequence, consisting of LT activities and learning and teaching materials (LT materials), was based on the conceptualization of coherence between chemistry and biology as depicted in Fig. 24.1. The students are encouraged to identify the subsystems, acknowledge interactions between these subsystems, and describe those interactions and their effects in terms of propositions. Specific LT activities are designed for each set of propositions; this enables the students to stepwise compose a model consisting of concepts and propositions that is in accordance with Fig. 24.1 and Table 24.1.

The designed LT sequence was divided into four distinct phases (Lijnse and Klaassen 2004). A schematic representation of this LT sequence is shown in Table 24.2.

In phase 1, a casus is introduced concerning the contamination of tap water with *Legionella pneumophila* at a school, and the students are asked to formulate the main problem and possible solutions. By addressing the problem and thinking about possible solutions, the students develop a need for specific knowledge on *Legionella pneumophila* and methods of eradication. This provides for a motive to look at the problem from another perspective, from which the leading question can be answered. In phase 2, the students look deeper into the problem presented in phase 1 from the perspective of an eradication advisor, in order to find an answer to leading question 1. The students are provided with information on methods of eradication that represents the specific knowledge of the advisor on this matter. They are asked to select a method for eradication that meets the criteria described in leading question 1. The method that the students probably select (copper-silver ionization) meets the criteria but offers a potential risk: the accumulation of heavy metals in organisms. The students acknowledge the potential risk and develop a motive for a change of perspective from advisor to researcher in order to answer leading question 2. In phase 3, the students explore the effects of the biocide on *Legionella pneumophila* from the perspective of a scientific researcher, by explaining the effects presented as empirical data in graphs and tables, in terms of interactions between the biology and chemistry subsystems. The students summarize the interactions in a schematic representation and reflect on their findings. Additionally, they reflect on an abstract representation of the interactions between levels of organization and micro and macro levels. In phase 4, the students perform an assessment specifically designed for this purpose, consisting of a posttest in which students apply the learned strategy discussed in phase 3 in a new context, based on an authentic practice concerning cadmium uptake by plants.

3.2 Data Collection and Data Analysis

The LT strategy has been tested at three schools in the Netherlands. In this paper, we report on the results of one case study. A biology teacher and his group of upper secondary, preuniversity students (grade 11, age 16–17) were involved in the case study ($n = 17$). During the enactment of the LT sequence, the students worked together in five groups (three groups of three students and two groups of four students).

Data collection took place by audio- and videotaping of the LT process, classroom observations, interviews with teachers and students, and by collecting student materials which contained specific assignments in which students were asked to

Table 24.2 Learning phases, descriptions, corresponding subquestions, and propositions in the LT sequence

Phase in the LT sequence	Questions that guided the LT sequence	Propositions (Table 24.1) addressed per phase
Phase 1. An orientation on the context by introducing a casus concerning the contamination of tap water with *Legionella pneumophila*. By addressing the problem, a need for specific knowledge is induced. This need for knowledge provides for a motive to change perspective from student to advisor	Subquestions	
	– What is *Legionella*?	
	– What are the possible consequences of *Legionella* contamination of tap water?	
	– What are the possible solutions for the problem addressed by this casus?	A_1; A_2; B_1; B_2
	Leading question 1	
	How can *Legionella* be eradicated in a safe and effective way?	
Phase 2. Exploration of the context from the perspective of the advisor provides knowledge that enables leading question 1 to be answered. The answer given to leading question 1 gives rise to leading question 2. Leading question 2 induces a new need for specific knowledge. This need for knowledge provides a motive for a second change of perspective, from the perspective of advisor to the perspective of scientific researcher	Subquestions	
	– Which methods are available for the eradication of *Legionella*?	A_1; A_2; B_1; B_2
	– What is a biocide?	
	– What are effective, safe, and sustainable methods to eradicate *Legionella* in the school tap water system?	1_1; 1_2
	– What are the potential risks of the use of heavy metals as biocides?	1_1; 1_2
	Leading question 2	
	Why do you need copper and silver in the biocide and can these substances be replaced?	
Phase 3. Exploration of the context from the perspective of the researcher provides information on explanations for phenomena emerging at different levels of the system. Students develop a model based on their findings and use the model to answer leading question 2	Subquestions	

(continued)

Table 24.2 (continued)

Phase in the LT sequence	Questions that guided the LT sequence	Propositions (Table 24.1) addressed per phase
A general reflection on the first three phases takes place by discussing an abstract model derived from Fig. 24.1	– What are the effects of the biocide's interactions with colonies, bacteria, and the cell membrane?	$1_{1\text{-}3}$; $A_{1\text{-}3}$; $B_{1\text{-}3}$; $C_{1\text{-}3}$; $I_{1\text{-}2}$; $II_{1\text{-}2}$; $III_{1\text{-}2}$
	– What are the effects of the interactions of copper and silver ions with phospholipids, proteins, DNA, and RNA?	$1_{1\text{-}3}$; $D_{1\text{-}3}$; $E_{1\text{-}3}$; $I_{1\text{-}2}$; $II_{1\text{-}2}$; $III_{1\text{-}2}$
	– How do these effects explain phenomena occurring at other levels of organization?	W; X; Y; Z
	– How can coherence between chemistry and biology be described in terms of the interactions between macro/micro levels and levels of organization?	$1_{1\text{-}3}$; $A_{1\text{-}3}$; $B_{1\text{-}3}$; $C_{1\text{-}3}$; $D_{1\text{-}3}$; $E_{1\text{-}3}$; $I_{1\text{-}2}$; $II_{1\text{-}2}$; $III_{1\text{-}2}$; W; X; Y; Z
Phase 4. In a test specifically designed for this purpose, students apply the strategy used in phase 3 in a new context, based on the authentic practice concerning cadmium uptake by plants	Subquestion – What are the effects of cadmium uptake by plants? Leading question 3 Is it safe for humans to eat lettuce plants that were grown on soil contaminated with cadmium?	

explicate coherence between the domains chemistry and biology, e.g., written answers, essays, schematic representations, concept mapping at various stages of the LT process, and finally a test specially designed for this purpose.

Analysis and interpretation of the data were performed according to the following procedure: video and audio fragments of the actual execution concerning critical stages in the scenario were transcribed verbatim by the first author and were compared with the intended outcomes of those stages; the additional data were used for triangulation when necessary; and the qualitative descriptions were verified by a second researcher (the second author of this paper) and further discussed by the research team.

The systematic analysis of the data took place according to the following steps:

A. *Identification of propositions present in data.* By analyzing the verbatim transcriptions of the discourse of the students within a group, between the students and the teacher and in class discussions, and by analyzing the worksheets, ideas, arguments, and reasoning by the students concerning the selected propositions.
B. *Interpretation of propositions present in data.* The nature of the propositions brought forward was identified for each group of students. These results were compared with the propositions described in Table 24.1.
C. *Analysis of student reasoning.* The students' arguments and reasoning about propositions were mapped. At this stage of the data analysis, the attained learning outcomes were compared with the intended learning outcomes of the LT activities as described in detail in the scenario.
D. *Recording of deviations between attained and intended learning outcomes.* The recording of the deviations of the attained LT process in relation to the intended LT process as explicated in the scenario, e.g., the students encounter unforeseen problems or in comprehensions, take a detour, or come up with an unexpected solution or argument in discussions. These deviations give arguments for redesign of the LT strategy.

4 Results

In phases 1, 2, and 3 of the LT sequence, the students were asked in specific assignments to elaborate and explicate the propositions described in Table 24.1. In Table 24.3, the outcomes of the data analysis of the learning outcomes in terms of realized propositions and deviations from the intended learning outcomes are presented at the level of the student groups.

4.1 Learning Outcomes in Terms of Realized Propositions in Phases 1–3

In phases 1–3 of the LT sequence, all five groups were able to describe the propositions concerning the interactions between chemical and biological entities and the effects of the interactions at that particular level (propositions A_1, A_2, B_1, B_2, C_1, C_2).

Table 24.3 Realized propositions and deviations from the intended learning outcomes in phases 1–3 of the LT sequence

Proposition	Number of groups that realized proposition	Exemplary student utterances	Deviations from the intended learning outcomes
Propositions concerning chemical subsystems acting on or interacting with biological subsystems			
A_1	5	"Copper silver ionization acts on Legionella"	Groups of bacteria were not denoted as colonies
A_2	5	"Legionella dies as a consequence"	Decrease of bacteria was mentioned
A_3	2	"Level of organism" "Molecular level, because of the copper and silver ions"	After this activity the teacher explained the levels of organization once again
B_1	5	"Copper and silver kill the bacterium"	
B_2	5	"It kills the bacterium and the number decreases"	
B_3	2	"Population, organism, and ions"	Three groups mentioned the organism level; two groups also mentioned the cellular level
C_1	5	"The copper opens the cell membrane and causes it to leak"	The damage done to the cell was often described as "leaking"
C_2	5		The term "permeability" was only used after introduction by the teacher
C_3	3	"Level of organelles of the cell" "Molecular level, because of the phospholipids"	Students were rather unfamiliar with the specific term "subcellular level"
D_1	0	"The negative phosphate group attracts the Cu^{2+}; this causes an interaction"	The interaction of copper ions with proteins was not mentioned
D_2	0		
D_3	5		Students gave no explanation for the interaction of copper ions with proteins
E_1	5	"Ag^+ acts on DNA, RNA, and proteins in the cell"	One group thought that DNA and RNA were proteins, until the teacher explained the difference; two groups thought that DNA and RNA contain amino acids
E_2	3	"DNA and RNA consist of amino acids; the Ag^+ binds to the amino acids, that is why DNA and RNA do not function properly anymore"	
E_3	5		

Propositions concerning the coherence between levels of organization in biology			
I_1; II_1; III_1	5	"I think we have to zoom in"	One group did not discuss the connection between sublevels
I_2; II_2; III_2	4		
Propositions concerning the effects of interactions between subsystems on other subsystems			
W	5	"Because of the Cu^{2+}, the lipids and proteins change and so the cell membrane changes. This is why other ions can get into the cell"	
X	4	"The silver ions enter the cell through channels in the membrane"	One group did not connect increased permeability with silver ions entering the cell
Y	4	"DNA and RNA become damaged because particles can enter the cell now"	One group did not connect silver ions with damage to DNA and RNA
Z	5		
Propositions concerning the coherence between the macro and micro level in chemistry			
1_1; 1_2	5	"When you perform electrolyses on solid copper, you can create copper ions"	
1_3	0	"I cannot remember us talking about this 'macro-micro' aspect during chemistry lessons"	The students were unfamiliar with the terms macro and micro in chemistry; the teacher explained the terms to the students

Although the students were able to identify chemical and biological entities, the determination of the biological level of organization proved to be more difficult for the majority of the groups (propositions A_3, B_3, C_3, D_3, E_3). After the teacher explained how to think in terms of levels of organization in a classroom discussion, propositions A_3, B_3, D_3, and E_3 were denoted by all five groups but proposition C_3 remained unclear. Propositions D_1 and D_2 were not mentioned at all by students. The propositions concerning the effects of the interactions between subsystems on other subsystems (propositions I–III and W–Z) were denoted by the majority of the groups (4–5). Propositions 1_{1-3} were denoted by students in terms of substance particle, instead of macro-micro. Taken together, it can be concluded that on average 70 % of the relations between the chemical and biological concepts emerging from the LT sequence in phases 1–3 were expressed by the students.

4.2 Deviations from the Intended Learning Outcomes

In phase 1, in which students oriented themselves on the casus, students kept focusing solely on the biological aspects of the problem, considering mainly *why* eradication is necessary, leaving out the *how* aspect of the eradication. During phase 2, the students developed a greater interest in the chemical aspects of *Legionella* eradication in their role as advisor. Students combined these aspects with the biological aspects, which gave rise to the leading question for phase 3. At the beginning of phase 3, the students tended to descend rapidly from population level to molecular level, leaving out the organism, cell, and subcellular levels because the word *ions* was mentioned in the student materials. After a classroom discussion about the meaning of biological levels of organization and the effect of the biocide at population level, the students were motivated to zoom in on the next levels of organization step by step. Although the students were unfamiliar with the terms "macro" and "micro," they recognized the terms "substance," "molecules," "ions," and "particles" from their chemistry lessons, and they were able to fill out a schematic representation of the interactions between biological and chemical subsystems and to discuss the interactions between chemical and biological entities.

In phase 4, the students performed the test on cadmium uptake by plants individually. Analysis of the written answers of the students showed that the students were capable of applying the strategy that was learned in phase 3 of the LT sequence in a new context. The students identified the subsystems, denoted levels of organization, and described interactions between subsystems in terms of propositions, in the same way as they had done in phases 1–3. On average, 80 % of the propositions were described by the students (the total number of propositions present divided by the number of realized propositions). These results indicate that the students were able to apply the procedure learned in phase 3 and showed that they were capable of identifying and describing the majority of the propositions.

5 Conclusion and Discussion

In previous sections, we described coherence between chemistry and biology as relations between chemical and biological concepts that were derived from an authentic practice in which chemical and biological knowledge and processes are applied. The specific relations between these concepts or chemical and biological entities (subsystems) that are part of a system were described as propositions. The major results indicate that the LT strategy appears to be promising. The subsequent LT sequence, in which this coherence was conceptualized and operationalized in terms of LT activities, encouraged students to experience and express the coherence between chemistry and biology. The LT strategy enabled students to (1) identify the chemical and biological subsystems, (2) acknowledge interactions between these subsystems, and (3) describe the interactions and effects of interactions between chemical and biological subsystems in detail.

In this study, the coherence between chemistry and biology from a students' perspective was measured through the identification and analysis of propositions put forward by the students during the enactment of the LT sequence. Through the use of the scenario, in which the concepts and propositions derived from the authentic practice were described in detail, it was possible to compare the attained learning outcomes with the intended learning outcomes (Lijnse 1995). The comparison indicated the propositions that were either realized, partly realized, or not realized by students, and it also indicated the specific moment in the LT process that the proposition was or was not realized (Table 24.3).

Because of the nature of the propositions, e.g., relations between chemical concepts, between biological concepts, and their interrelations, the students were encouraged to discuss the coherence within the domains as well as the coherence between the domains. By adopting a systems thinking approach, in which thinking in terms of levels of organization and macro-micro thinking were combined, students analyzed the empirical data presented in the LT sequence in a systematic way. For each level of organization, the students determined (1) which chemical entity acted on which biological entity, (2) what the effects of these interactions are at that specific level, (3) the consequences of those interactions at other levels, and (4) how the effects of the interactions can be explained. Utterances 3 and 4 induced thinking back and forth between the different levels of the system (Knippels 2005).

The central research question addressed in this study is to what extent students experience and express coherence between chemical and biological concepts by attending an authentic practice-based curriculum unit embodying the domains of chemistry and biology. The findings show that students do experience coherence between chemical and biological concepts. This conclusion is empirically based on the students' expressed relations between chemical and biological entities in terms of propositions. Although the results of the first enactment are promising, the findings also give rise to a partial redesign of the LT sequence for use in a second enactment. With regard to the applied systems thinking approach, our findings indicate that the approach structures and promotes the expression of relations between chemical and biological concepts, but it also gives rise to some learning difficulties. With respect to the thinking in terms of levels of organization in biology, it appeared that these remain implicit from a students' perspective. This implies that the thinking in terms of levels of organization will have to be emphasized and explained at some stage during the enactment of the unit. Additionally, in chemistry education the terms "macro" and "micro" are often represented by the terms "substance" and "particle." The latter terms are recognized by students from their chemistry lessons. Furthermore, it became clear that students were not used to studying bacteria in detail. This led to difficulties in the determination of the organism level, the cell level, and especially the subcellular level. Students were familiar with the cell organelle level, but a bacterium does not contain cell organelles. Therefore the subcellular level in particular will have to be explicated in the LT activities. Propositions D_1 and D_2 remained unclear to the students, because of the way in which the empirical data on interactions between copper ions and the cell membrane were presented in the LT materials. Although the students described the major groups of molecules present in the membrane of the cell (phospholipids and proteins), the LT activities focused mainly on the interactions between copper ions and phospholipids.

This study revealed that coherence between chemistry and biology in secondary education from a students' perspective can be achieved by synthesizing the prominent thinking strategies applied in both domains. The macro-micro thinking in chemistry and the thinking in terms of levels of organization in biology might both be regarded as special forms of systems thinking. Further research is needed to explore the benefits of combining both thinking strategies with respect to achieving coherence between chemistry and biology in secondary science education, by means of an authentic practice-based curriculum unit.

References

Bardeen, M. G., & Lederman, L. M. (1998). Coherence in science education. *Science, 281*(5374), 178–179. doi:10.1126/science.281.5374.178.

Boersma, K. T., van Graft, M., Harteveld, A., Hullu, E., de Knecht-van Eeckelen, A., Mazereeuw, M., et al. (2005). *Vernieuwd biologieonderwijs van 4 tot 18 jaar [New biology education from 4 to 18 years old]*. Utrecht: CVBO.

Boersma, K. T., van Graft, M., Harteveld, A., Hullu, E., de Knecht-van Eeckelen, A., Mazereeuw, M., et al. (2007). *Leerlijn biologie van 4 tot 18 jaar vanuit de concept-contextbenadering [Biology curriculum for ages 4 to 18 based on the concept-context approach]*. Utrecht: CVBO.

Boersma, K. T., & Waarlo, A. J. (2009). On the theoretical in- and output of 'design research' in biology education. In M. Hammann, K. T. Boersma, & A. J. Waarlo (Eds.), *The nature of research in biological education: Old and new perspectives on theoretical and methodological issues*. Utrecht: FIsme-Press.

Bulte, A. M. W., Westbroek, H. B., De Jong, O., & Pilot, A. (2006). A research approach to designing chemistry education using authentic practices as contexts. *International Journal of Science Education, 28*(9), 1063–1086. doi:10.1080/09500690600702520.

Cañas, A. J., Valerio, A., Lalinde-Pulido, J., Carvalho, M., & Arguedas, M. (2003). *Using WordNet for word sense disambiguation to support concept map construction*. In M. A. Nascimento, E. S. de Moura, & A. L. Oliveira (Eds.), String processing and information retrieval. 10th international symposium, SPIRE 2003, Manaus, 8–10 Oct, 2003. Proceedings. Berlin/Heidelberg: Springer

Engeström, Y. (1987). *Learning by expanding: An activity-theoretical approach to developmental research*. Helsinki: Orienta-Konsultit.

Geraedts, C., Boersma, K. T., & Eijkelhof, H. M. C. (2006). Towards coherent science and technology education. *Journal of Curriculum Studies, 38*(3), 307–325. doi:10.1080/00220270500391589.

Gilbert, J. K. (2006). On the nature of 'context' in chemical education. *International Journal of Science Education, 28*(9), 957–976. doi:10.1080/09500690600702470.

Johnson, D. K., & Ratcliff, J. L. (2004). Creating coherence: The unfinished agenda. *New Directions Higher Education, 125*, 85–95. doi:10.1002/he.141.

Klaassen, C. W. J. M. (1995). A problem-posing approach to teaching the topic of radioactivity. Dissertation, University of Utrecht (CD-β series Vol. 18). CD- β Press, Utrecht.

Knippels, M. C. P. J., Waarlo, A. J., & Boersma, K. T. (2005). Design criteria for learning and teaching genetics. *Journal of Biological Education, 39*(3), 108–112. doi:10.1080/00219266.2005.9655976.

Lave, J. (1988). *Cognition in practice*. Cambridge: University Press.

Lijnse, P. L. (1995). 'Developmental research' as a way to an empirically based didactical structure of science. *Science Education, 79*, 189–199. doi:10.1002/sce.3730790205.

Lijnse, P. L. (2005). Reflections on a problem-posing approach. In K. T. Boersma, M. Goedhart, O. De Jong, & H. M. C. Eijkelhof (Eds.), *Research and the quality of science education*. Dordrecht: Springer.

Lijnse, P. L., & Klaassen, C. W. J. M. (2004). Didactical structures as an outcome of research on teaching-learning sequences? *International Journal of Science Education, 26*, 537–554. doi:10.1080/0950069031000161 4753.

Novak, J. D., & Cañas, A. J. (2007). Theoretical origins of concept maps: How to construct them, and uses in education. *Reflecting Education, 3*(1), 29–42.

Ogborn, J. (2005). 40 years of curriculum development. In K. T. Boersma, M. Goedhart, O. De Jong, & H. C. M. Eijkelhof (Eds.), *Research and the quality of science education*. Dordrecht: Springer.

Osborne, J., & Dillon, J. (2008). *Science education in Europe: Critical reflections*. London: The Nuffield Foundation.

Osborne, J., & Collins, J. (2001). Pupils' views of the role and value of the science curriculum: A focus-group study. *International Journal of Science Education, 23*(5), 441–467. doi:10.1080/09500690010006518.

Prins, G. T., Bulte, A. M. W., Van Driel, J. H., & Pilot, A. (2008). Selection of authentic modelling practices as contexts for chemistry education. *International Journal of Science Education, 30*(14), 1867–1890. doi:10.1080/09500690701581823.

Rudduck, J., Harris, S., & Wallace, G. (1994). 'Coherence' and students' experience of learning in the secondary school. *Cambridge Journal of Education, 24*, 197–204.

Schmidt, W. H., Wang, H. C., & McKnight, C. C. (2005). Curriculum coherence: An examination of US mathematics and science content standards from an international perspective. *Journal of Curriculum Studies, 37*(5), 525–559. doi:10.1080/0022027042000294682.

Van Aalsvoort, J. (2004). Activity theory as a tool to address the problem of chemistry's lack of relevance in secondary school education. *International Journal of Science Education, 26*(13), 1635–1651. doi:10.1080/0950069042000205378.

Van den Akker, J. J. H., Gravemeijer, K. P. E., McKenny, S., & Nieveen, N. (2006). Introducing educational design research. In J. J. H. Van den Akker, K. P. E. Gravemeijer, S. McKenny, & N. Nieveen (Eds.), *Educational design research*. London/New York: Routledge.

Venville, G. J., Wallace, J., Rennie, L. J., & Malone, J. A. (2002). Curriculum integration: Eroding the high ground of science as a school subject? *St Science Education, 37*, 43–84. doi:10.1080/03057260208560177.

Verhoeff, R. P., Waarlo, A. J., & Boersma, K. T. (2008). Systems modelling and the development of coherent understanding of cell biology. *International Journal of Science Education, 30*, 543–568. doi:10.1080/09500690701237780.

Westbroek, H. B. (2005). Characteristics of meaningful chemistry education: The case of water quality. Dissertation, University of Utrecht. CD-ß Press, Utrecht.

Chapter 25
The Relationship Between Teaching and Learning of Chemical Bonding and Structures

Ray Lee and Maurice M.W. Cheng

1 Introduction

Problematic issues about the teaching and learning of chemical bonding at the senior secondary level have been widely reported in the literature. For example, students mixed up the constituent particles and forces in different structures (Coll and Taylor 2001). Students often ignored the electrostatic nature of chemical bonding (Taber 1997). They tended to see it as a process, i.e., electron transfer or sharing (Taber 1998), or a physical entity, i.e., a sea of electrons (de Posada 1997). In the explanation of macro phenomena, students often merely recited the properties of different substances – they were unable to explain them in terms of the interactions of chemical species at the submicro level (Ben-Zvi, Eylon, and Silberstein 1986). Some common alternative conceptions on metals, ionic, and covalent substances are categorized below.

1.1 Reported Alternative Conceptions

1.1.1 Metals

The electrostatic nature of metallic bonding was ignored. Some regarded the "sea of electrons" as the metallic bonding (de Posada 1997). Some learners disregarded metallic bonding as a kind of bonding but "just a force" or "improper bonding" (Taber 1998, 2001, 2003b), because they related bonding to "electron sharing" and/or "electron transfer."

R. Lee (✉) • M.M.W. Cheng
The University of Hong Kong, Hong Kong
e-mail: waishing312@gmail.com; mwcheng@hku.hk

C. Bruguière et al. (eds.), *Topics and Trends in Current Science Education: 9th ESERA Conference Selected Contributions*, Contributions from Science Education Research 1,
DOI 10.1007/978-94-007-7281-6_25, © Springer Science+Business Media Dordrecht 2014

1.1.2 Ionic Substances

Students treated ionic substances as molecular in various ways. Butts and Smith (1987) reported that some students conceptualized sodium chloride as sodium and chlorine atoms being held together by covalent bonds and often named the substance as "molecules of sodium chloride."

Taber (1994) found that many students neglected the electrostatic nature of ionic bonding and overemphasized the process of electron transfer. Learners regarded ionic bonding as the transfer of electrons, rather than the attraction between cations and anions (that could or could not result from the transfer of electrons). Students also paid undue attention to the "history of electrons" – they considered there was an ionic bond only between the pair of ions where electron transfer had occurred. Each sodium ion was only bonded to one chloride ion and attracted to five others just by "forces," not bonds; the valency of an atom determined the number of ionic bonds formed (e.g., sodium can only donate one electron, so it can only form one ionic bond with the chlorine atom) (Taber 1997).

1.1.3 Covalent Substances

Students simply regarded covalent bonding as the sharing of electrons, rather than the electrostatic attraction between the nucleus and the sharing electrons (Taber 1998). Besides, students tended to believe all molecules were formed from atoms, and atoms had a need to fill their shells. So covalent bonds were formed in order to produce filled shells. Students also paid undue attention to the history of electrons concerning covalent bonding. They suggested that each atom would get its own electrons back when a covalent bond was broken, and attraction force only existed between a nucleus and the atom's own electrons. In addition, they did not differentiate between intermolecular forces and covalent bonding.

1.2 Pedagogical Learning Impediments

Taber (2001) conducted a comprehensive review of the literature on the teaching and learning of chemical bonding and identified "pedagogical learning impediments." It was believed that these impediments related to the way chemical bonding was taught in school chemistry. Four key pedagogical learning impediments are summarized:

1. *Dichotomous classification of bonding* Students only regarded bonding as either covalent or ionic. Bonds only formed after an electron transfer or electron sharing. Metallic bonding or the van der Waals force was not ("proper") bonding.
2. *The overemphasis on the octet rule* Students tended to use the octet rule as an explanation for why reaction occurred. For example, two chlorine atoms would

share electrons to form a covalent bond because both of them "needed" or "wanted" to achieve an octet.

3. *The use of anthropomorphic language* Students reasoned chemical changes as a desire of chemical species rather than as a result of electrostatic interactions.
4. *Atomic ontology and the initial atomicity* Students often assumed that atoms were the only fundamental units of substances. When conceptualizing chemical reactions such as the formation of sodium chloride, they often reasoned the formation of the substance from individual sodium and chlorine atoms. It in some ways triggered students' use of anthropomorphic language and an overgeneralization of the octet rule.

1.3 A Curricular Model of Chemical Bonding

A curricular model of chemical bonding was proposed by Taber (2001); this model aimed to tackle these pedagogical learning impediments and associated alternative conceptions. It was proposed that the sequence of the teaching and learning of solid structures/bonding starts with metallic structures (addressing impediment (1) above). It would then go on to ionic structures, to giant covalent structures, and finally to simple molecular structures. In addition, the curricular model emphasized the following: firstly, it highlighted molecules and ions (rather than atoms) as the basic unit of matter so as to avoid the assumption of initial atomicity (addressing impediment (4)); secondly, the curricular model stressed that chemistry should be based on physical principles – the nature of bonding, structures, and properties of substances was explained in terms of electrostatic forces, but not the octet rule (addressing impediment (2)), nor the desires of atoms, while addressing bonding in terms of electrostatics would also serve as a good foundation for subsequently learning about electronegativity, bond polarity, hydrogen bonds, and solvent-solute interactions; and lastly, Taber warned that metaphors should be used carefully – when they are used, they should be reiterated with scientific descriptions or explanations, the aim being to avoid students treating anthropomorphic language as an explanation of chemical phenomena (addressing impediment (3)).

1.4 Our Teaching Plan

Based on Taber's curricular model and the literature on scientific modelling (for instance, Gilbert and Boulter 2000), we devised a teaching plan of chemical bonding for a Grade 10 chemistry class (see Table 25.1). We emphasized how different models of structures explained the variety of properties of different solid substances. We envisaged that the students would be more confident with macro phenomena (Nakhleh and Krajcik 1994). In this teaching plan, each topic started by addressing the properties of substances at a macro level. Structures at the submicro level and

Table 25.1 The suggested teaching plan

Topic	Students should learn	Students should be able to
Giant metallic structure	Properties of metals	State some properties of metals
		Point out some limitations of the particle model
	Metallic bonding	Locate metallic bonding in the electron-sea model
		Describe the nature of metallic bonding
		Explain the properties of metals in terms of their structures and bonding
Giant ionic structure	Electrostatic interactions	Describe how ions attach to form an ionic lattice in crystallization
	Ionic bonding	Locate ionic bonds and repulsion in an ionic lattice
		Describe the nature of ionic bonding
		Compare the strength of ionic bonds and repulsion in an ionic lattice
	Properties of giant ionic structures	State some properties of ionic substances
		Explain the properties in terms of their structures and bonding
Giant covalent structure	Properties of giant covalent structures	State some properties of diamonds
		Draw the electron diagrams of diamonds
	Covalent bonding	Locate the covalent bonds in diamonds
		Describe the nature of covalent bonding
		Explain the properties of diamond in terms of their structures and bonding
Simple molecular structure	Properties of simple molecular substances	State some properties of simple molecular substances
	Nature of the van der Waals force	Use electrostatic principles to explain the meaning and nature of the van der Waals force
		Explain the properties of simple molecular substances in terms of their structures and bonding
	Molecules	Locate the van der Waals force and covalent bonding in diagrams
		State the differences between the van der Waals force and covalent bonding

the electrostatic interactions of the structural constituents were then introduced to explain the properties. The overarching idea in the teaching was "electrostatic forces" rather than the events that occurred to electrons in bond formations (i.e., the transfer of electrons, the sharing of electrons, or the formation of a sea of electrons). Also, use was made of visual representations at the submicro level to facilitate the students' learning.

Due to space limitations, we will describe in detail some ways in which the students were taught metallic bonding and ionic bonding. It is to illustrate how we realized the design principles in this plan. The teaching of covalent bonding followed the same principles, but is not described here.

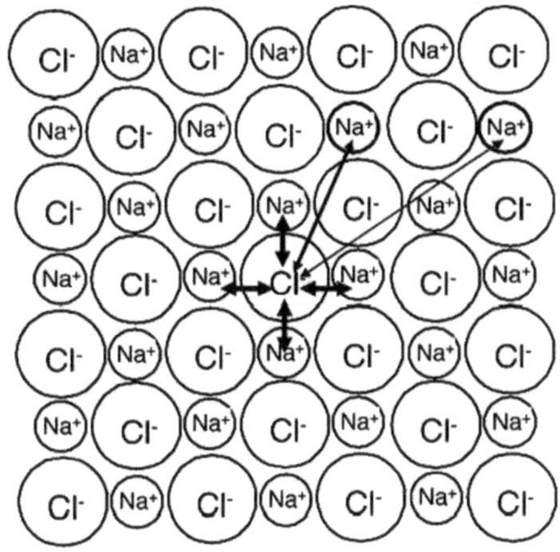

Fig. 25.1 Ionic bonding as electrostatic forces

In this teaching plan, the students started by recalling the macro properties of metals, with which they were familiar. The students were required to explain those properties with a submicro representation of metals based on the simple particle model. The electron-sea model was brought out when the students found that some of the properties (e.g., electrical conductivity, high boiling point) could not be explained sufficiently by the simple particle model. Electrostatic forces between delocalized electrons and metal cations were highlighted and were explicitly represented by lines in diagrams. The malleability, ductility, electrical conductivity in solid and liquid states, and boiling point of metals were discussed based on electrostatics.

Traditionally, ionic bonding has been represented as a transfer of an electron from a sodium atom to a chlorine atom, which resulted in an ion pair of sodium chloride. In this teaching plan, the teaching and learning of ionic bonding focused on the electrostatic forces between cations and anions (see Fig. 25.1). Such forces were represented by double arrows connecting sodium and chloride ions. In this visual model, focusing on the chloride ion in the center, there were electrostatic forces between it and its surrounding sodium ions. This was to avoid the development of an "ion pair" conception (one chloride ion formed one ionic bond with one sodium ion, possibly with the electron-donating sodium ion) and to facilitate the students' learning based on Coulombic electrostatics. That is, a negative charge particle would attract to any positive charge particle, and vice versa. In order to model the strength of the electrostatic forces with respect to distance, use was made of arrows of different thicknesses (see the arrows in Fig. 25.1): the thinner the lines, the weaker the forces. During the teaching, such a submicro representation was used to address the properties of table salt, i.e., hardness; electricity conductivity in the solid, liquid, and aqueous states; and (non-)malleability.

The teaching plan was delivered to a class of 15-year-old Grade 10 students in Hong Kong. The teaching lasted for 8 weeks with 2 h of lesson time per week. In this study, we aimed to seek answers to the following research questions:

1. What would the students' learning outcomes be based on this curricular model? More specifically, we would like to investigate if common alternative conceptions reported in the literature would still be developed.
2. What would the impediment be to the students' learning if the students were taught based on this curricular model?

2 Methodology

To answer the research questions, three students – Jenny, Kim, and Dave – were selected for interviews. Based on a prediction by the first author, who taught the students, the achievement of the three students in chemistry would be the first 20th, 50th, and 70th percentile of the student population in Hong Kong. The prediction was a rough estimation by comparing the interviewees' test and exam scores with the first author's past cohort of students. During the 8 weeks of teaching, individual interviews were conducted after each topic (in Table 25.1) had been taught. Hence, each student was interviewed four times.

The interview protocol consisted of four main parts (see Table 25.2). The first part required the students to describe each structure, which was printed on a

Table 25.2 The interview protocol in this study

	Questions	Possible follow-up questions
Structure	Can you use your own words to describe this structure?	Can you describe the particles involved? Is there any other representation for this particle?
Electrostatics	What kind of force is there in this structure?	Why do these forces exist? Do any other forces exist?
Bonding	Can you explain the meaning of metallic bonding/ionic bonding/covalent bonding/van der Waals force?	Can you indicate the bonding in this structure? Do any other kinds of bonding exist?
Malleability and ductility	This diagram (Fig. 25.2) represents a substance that has changed its shape (Fig. a/c to Fig. b/d) as a result of an external force (the arrows). Are such transitions possible for giant metallic/giant ionic/giant covalent structures?	Can you add something to the diagram to make it more detailed? How would you use these diagrams to support your claim?

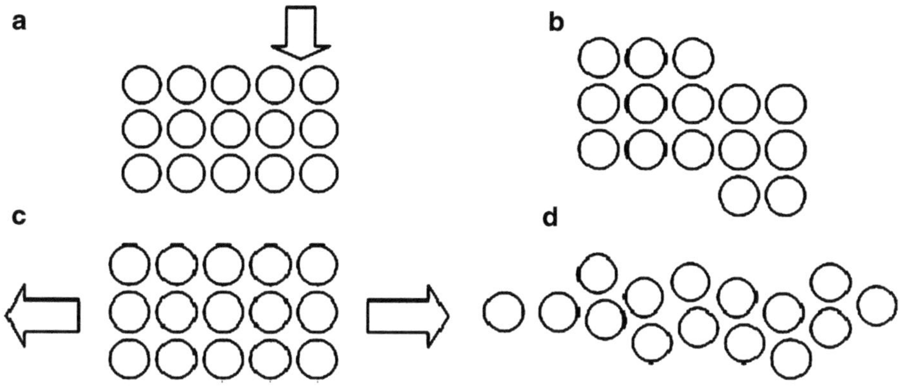

Fig. 25.2 Probing card for malleability and ductility

card, with their own words. Clarification of terms such as the name/identifier of particles was sought whenever needed. Secondly, the interviewer asked if there was any force in the diagrams shown to them. Questions concerning the nature of electrostatic forces were asked. Thirdly, we asked the students to explain the meaning of the type of bonding in different structures. Electrostatic forces and bonding were separated for questioning purposes, as we wanted to see how the students related them. Then we used diagrams (see Fig. 25.2) to probe the students' explanation of macro phenomena. The whole protocol was applied to the four rounds of interviews, with the exception that the part on malleability and ductility was not questioned in the interviews for the simple molecular structure, as it was not on the curriculum in Hong Kong.

The interviews were conducted by the first author. They were videotaped and transcribed for analysis. We compared the data with (1) the accepted scientific views, (2) alternative conceptions, and (3) pedagogical learning impediments reported in the literature. The comparison formed the basis of the evaluation of the teaching.

3 Results and Discussions

The findings are tabulated in Tables 25.3 and 25.4.

From the data, the students did not define ionic bonding as the "transfer of electrons" and did not definitely regard covalent bonding as the "sharing of electrons" (see Table 25.4). They had no apparent difficulties in understanding and applying Coulombic electrostatics in simple situations (see Table 25.3). The high-achieving student, Jenny, was able to represent her ideas through creating diagrams which were not found in her textbook and were not used in classroom teaching (Fig. 25.3). The middle-achieving student, Dave, also demonstrated similar capability (Fig. 25.4). They were able to relate chemical bonding to electrostatic interactions.

Table 25.3 A summary of the findings concerning the scientific concepts ("✓" represents a clear understanding was demonstrated by the student; "×" represents responses that involve alternative conceptions; "✓ ×" means that the student demonstrated understanding of a particular concept with some minor mistakes)

Structures	Concepts	Dave	Kim	Jenny
Giant metallic structure	Structure	✓ ×	✓	✓ ×
	Electrostatics	✓	✓	✓
	Bonding	×	×	✓
	Malleability and ductility	×	✓ ×	✓
Giant ionic structure	Structure	✓	✓	✓
	Electrostatics	✓	✓	✓
	Bonding	✓	✓	✓
	Malleability and ductility	✓	✓	✓
Giant covalent structure	Structure	✓ ×	✓ ×	✓ ×
	Electrostatics	✓	✓	✓
	Bonding	✓ ×	✓ ×	✓
	Malleability and ductility	×	×	✓
Simple molecular structure	Structure	✓	✓ ×	✓ ×
	Electrostatics	✓	✓	✓
	Bonding	✓	✓	✓

Table 25.4 A summary of the findings concerning the pedagogical learning impediments ("M" represents that the problem appears in the interview on giant metallic structure. "I" represents that the problem appears in the interview on giant ionic structure. "C" represents that the problem appears in the interview on giant covalent structure. "S" represents the problem appears in the interview on simple molecular structure. "Nil" represents no problems in that area concerning the pedagogical learning impediments)

Pedagogical learning impediments	Dave	Kim	Jenny
Dichotomous classification of bonding	Nil	Nil	Nil
Initial atomicity	I,C,S	C	C
Overemphasis on octet rule	I,C,S	C	Nil
Use of anthropomorphic languages	C,S	C	Nil

3.1 *Students' Learning of Electrostatics*

The plan emphasized electrostatics. From the tabulated result (Table 25.3), all the students did well in problems related to electrostatic principles. It is likely that the emphasis on electrostatic principles helped tackle the dichotomous classification of bonding (see Table 25.4). The students showed a sound understanding of both the direction and strength of electrostatic forces between charged particles. Jenny could apply concepts of electrostatics to explain the meaning of the "sea of electrons."

All of the students did not limit ionic bonding to the attraction between neighboring ions. They noted the existence of ionic bonding between oppositely charged ions even far away. They recognized not only the attractive force in the structure but

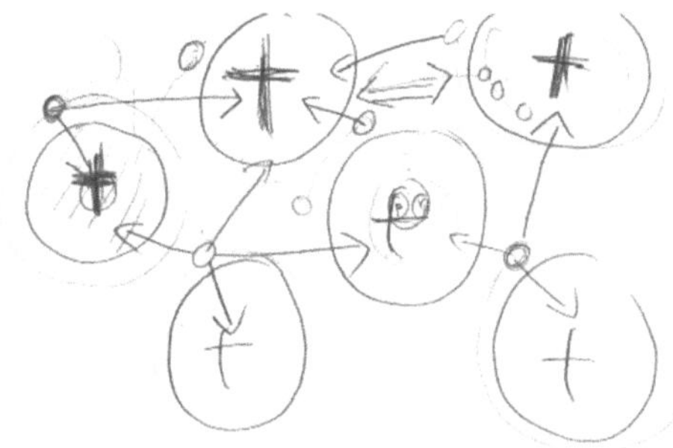

Fig. 25.3 Jenny's drawing of metallic bonding

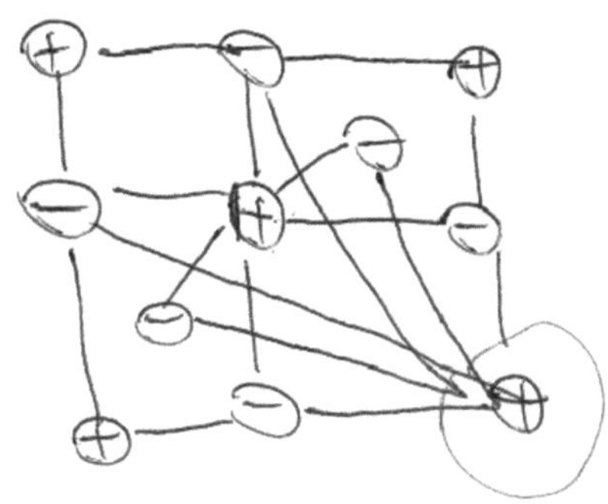

Fig. 25.4 Dave's drawing of ionic bonding

also the repulsion in a giant ionic structure. They also explained the brittleness of ionic substances. For example, in Jenny's interview concerning malleability and ductility of ionic substances:

I: interviewer S: student

I: Can you explain whether an ionic substance can maintain the shape in Fig. 25.2b after a force is applied?

S: When there is a shift in position of the ions, there is repulsion between same charges and hence the structure breaks down.

I: After a shift in position, does the structure consist of attractive forces?

S: Yes, there is attraction between oppositely charged ions.

I: So why does the structure break down even if there is attraction in the structure?

S: Before the shift in position, the ions align in such a way that attraction is stronger than repulsion. After a shift in position, the positives get close, as do the negatives. The distance between same charges is shorter than that between opposite charges. So the repulsion is larger than the attraction in this case.

Dave and Kim provided similar responses to Jenny's. They were able to analyze the distance between ions and compare the forces involved in order to draw a conclusion.

The results showed that the learning relating to giant ionic structure benefited most from our teaching plan. All the students were able to describe giant ionic structures in some sense; the meaning of ionic bonding was well understood by the students. They

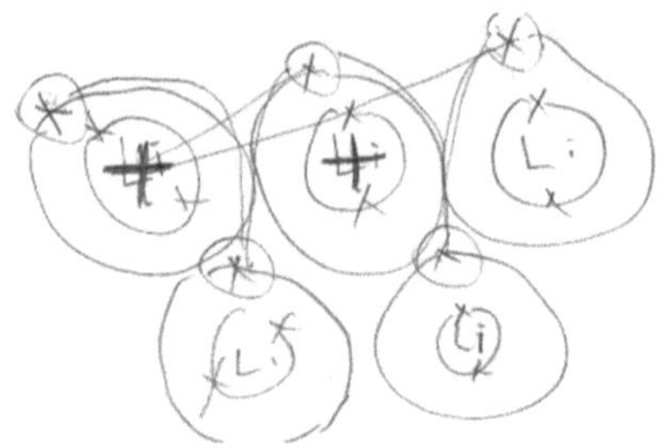

Fig. 25.5 Dave's representation of metallic bonds

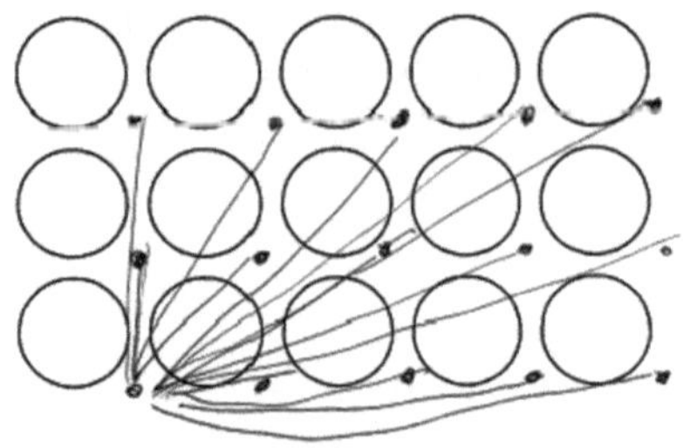

Fig. 25.6 Kim's representation of metallic bonds

were also able to explain the brittleness of ionic substances. All these responses were well linked up with concepts of electrostatics. None of the participants referred to ionic bonding as "electron transfer."

3.2 Did Electrostatics Help?

The results showed that an understanding of electrostatics helped to facilitate the understanding of the structure and bonding of giant ionic and simple molecular substances. Nevertheless, the students' learning was not without challenges. While the low-achieving student could state some electrostatic principles (such as the interaction of particles with the same and different charges), applying them in making sense of chemical bonding was still problematic.

Dave viewed metallic bonding as being equivalent to a sea of electrons (as reported in the literature). When he was asked to use lines to represent the metallic bonds, he used some curly lines to link up the outermost electrons (Fig. 25.5) but added that "there is no attraction between electrons." Kim provided similar responses when he was asked about the malleability and ductility of metals. On the one hand, he drew lines to link up electrons and stated that they were metallic bonds (Fig. 25.6). On the other hand, he was able to state that there was an electrostatic attraction between metal ions and delocalized electrons (Fig. 25.7). It could be concluded that even if he understood the basis of electrostatics, the understanding did not link up with or transfer to an understanding of the nature of metallic bonding.

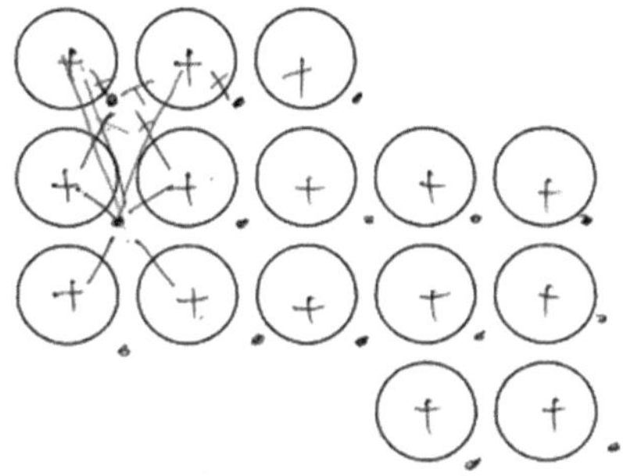

Fig. 25.7 Kim's representation of electrostatic attractions

Our findings indicated that while the use of electrostatics as the overarching idea could be useful for some students, the success in transferring the ideas of electrostatics to an understanding of chemical phenomena was not always guaranteed.

3.3 Domination of the Octet Framework

The same problem existed in the case of covalent bonding. Some students had grasped the concepts about electrostatics but were unable to utilize it in learning about covalent bonding. While all the participants were able to define covalent bonding as "the electrostatic forces between nucleus and bonding electrons," their understanding of electrostatics was not a panacea for the overuse of the octet rule.

In the formation of a covalent bond, Kim (the middle achiever) described "particles combined because of... the octet rule." Another piece of evidence for this inference was found in a later part of the interview as he indicated that electrons of carbon atoms would share because they could not carry out electron transfer:

> It doesn't have eight electrons on its own, and they cannot get electrons from other atoms, so they share.

Kim demonstrated an adequate understanding of electrostatics (Table 25.3). Yet he preferred to use the octet rule to explain the cause of bonding as shown in his first response. Both the octet and electrostatic framework coexisted in Kim's cognitive structure.

While Dave (the lowest achiever) was able to describe the structure and the meaning of covalent bonding in terms of electrostatics, he made use of the octet rule to explain various phenomena. He indicated that covalent bonds would form again between atoms in Fig. 25.2b because "they can have octets." His explanation regarding the poor ductility of giant covalent structures was also based on the octet rule:

> It is because the arrangement of atoms (in Fig. 25.2d) will not allow an octet arrangement of each atom, so there won't be any covalent bonds formed.

Dave used the octet rule as a tool for explaining the formation of bonding. Malleability and ductility were determined by whether there were bonds that could form again in the diagram. In another instance, the interviewer asked questions

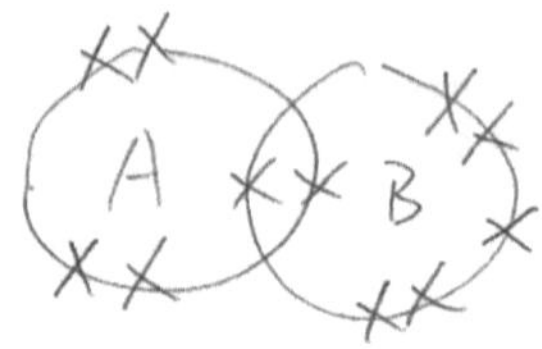

Fig. 25.8 Atom A and B for probing understanding of covalent bonding

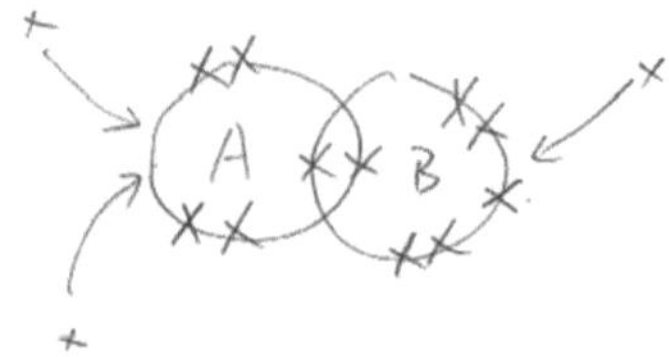

Fig. 25.9 Atom A and B with more electrons added

concerning a picture of two atoms, A and B, which were covalently bonded without fulfilling the octet structure (Fig. 25.8). Dave's response regarding atom A and B was as follows:

I: Do you find any covalent bond in this picture (Fig. 25.8)?
S: No. It does not have an octet structure. […] It is because it is not stable. If we have some method to make them achieve an octet, like adding an electron to atom B, then it will have a covalent bond (Fig. 25.9).

The octet framework was clearly indicated in Dave's response. Even if he was able to state the meaning of covalent bonding in light of electrostatics, he preferred the octet rule as an explanation for the formation/existence of covalent bonding.

We believe that the students' overuse of the octet rule is closely associated with their assumption of initial atomicity; they tended to start from a situation with atoms before they brought out the octet rule. We observed that the students could identify covalent bonds in terms of electrostatics in the structure of diamonds. But an undue emphasis was also given to an explanation of the formation of diamonds from individual atoms. Actually, we are doubtful of the need for teachers to explain the process of diamond formation from individual carbon atoms. Such an explanation of process promotes the assumption of initial atomicity and hence a misuse of the octet rule (Taber 2003a).

3.4 Why Did the Octet Rule Dominate?

The students' preference for the octet framework over the electrostatic framework is not surprising. When students are asked to draw the Lewis structure of a chlorine molecule – as is required and often asked in public examinations – they indeed do not have to focus on electrostatics. Such a focus would not offer any help in working out correct representations of chlorine (e.g., its Lewis structure, its chemical formula). At the beginning of their learning, students tend to use trial and error to work out the representations, until they can draw a diagram in which all atoms achieve the octet. It can be noted that students utilize this chain of thought whenever they need to work out the molecular formula. Their goal is very clear – to achieve the octet (and hence scores in examinations). Indeed, working out the molecular formula from two different elements is a very common type of question in public examinations. Once they have mastered such "skills," the process would be rapid, as they would know that a group VI atom would "need" or "want" two electrons to achieve the octet, so it can form one double bond or two single bonds. As time goes by, students are exposed to intensive examination training which promotes the use of the octet rule. The role of the electrostatic nature of bonding would only diminish.

3.5 Do Initial Atomicity and Anthropomorphism Matter?

The students' understanding of electrostatics did not solve the use of anthropomorphic language either. Although the teacher did not rely on anthropomorphic language or analogies, results showed that the students could still make use of such language in the interviews.

Initial atomicity and the overuse of the octet framework were found in Dave's interview, when he was asked about the differences between ions and atoms. We are not sure whether students need to work out the complicated process of how atoms change to ions based on the octet rule before they could give an answer to the question. Yet, initial atomicity and anthropomorphism did not affect Dave's understanding of bonding as electrostatic forces. It is acceptable as long as the "history" of ions does not hamper learning, and the use of anthropomorphic language does not overwrite students' electrostatic framework. But clearly, we should devise more ways to stress that ions and molecules are also fundamental entities that exist in nature. We should put our emphasis on comparing the differences between ions, molecules, and atoms, rather than how ions and molecules are formed from discrete atoms.

4 Conclusions and Implications

The case studies revealed that the curricular model proposed by Taber (2001) was helpful to direct the students' learning of chemical bonding from the "transfer of electrons" or "sharing of electrons" to the electrostatic interactions of chemical species. Also, it was found that the learning of some basic Coulombic electrostatics principles was unproblematic. However, applying those principles to the understanding of chemical bonding was not always successful. It was especially evident in the lowest-achieving student's learning of covalent bonding. It is postulated that while the student learned covalent bonding as electrostatic attractions, the undue emphasis on the octet rule in the curriculum and assessment might have steered the students' learning focus.

The students started by thinking in terms of a compound or a molecule from discrete atoms and then the combination of these atoms based on the octet rule. Such a combination was further justified by anthropomorphism. It was a chain reaction of several learning impediments, which resulted in a well-connected alternative framework of ideas in the students' cognitive structure. To solve this problem, we may prevent the overuse of the octet framework by avoiding initial atomicity. There are several reasons for this:

1. Causation exists in the problem: it is the initial atomicity that triggers the overuse of the octet rule and anthropomorphic language, but not the other way round (i.e., the use of anthropomorphic language would not lead students to think in terms of single atoms, as does the octet rule).
2. Anthropomorphic language is hard to avoid even when the teacher seldom uses such language, as is shown in our data.
3. The octet rule is an inevitable content in the curriculum; though its overuse is a learning impediment, it is still beneficial for the prediction of stable species.

As a result, we should devise more ways to stress that ions and molecules are also entities that already exist in nature. We suggest that, when dealing with the topic of chemical bonding, the idea of the formation of ions or molecules should be skipped. It is not advisable to ask "how/why do ions in sodium chloride form," "how do diamonds form from carbon," or "how/why do chlorine molecules form," since these questions would guide students towards referring to single atoms at an introductory stage, just because they have not learned different structures yet. While it is very important to ask why and how in the learning of science, it is time to ask ourselves why do we need to ask this "why"? The salt in the ocean does not come from the reaction of sodium solid (or sodium atoms) and chlorine gas (or chlorine atoms) nor is our water necessarily the result of an explosion of hydrogen gas and oxygen gas. The same idea could be applied to giant covalent structures. There is a need to explain why diamonds are stable and strong in terms of their bonding and structure, but it is redundant to explain how diamonds form from discrete atoms. If asking the question brings a distorted view of science that is hard to change in later learning, we would do better to tactfully avoid that question. We may shift our emphasis to the comparison of ions, molecules, and atoms (rather than how ions and molecules form from atoms). The formation of ions or molecules from elements could then be taught when students start learning about chemical reactions, once they have mastered the structure of different substances.

The results of the study showed that the learning of giant covalent structures least benefited from the teaching plan. After the above reflections, the following teaching order of giant covalent structures is proposed:

1. Recall some macro properties of diamonds, such as "hard" or "poor conductor of electricity."
2. Simply present an electron diagram (a submicro representation) of a diamond, without mentioning discrete atoms or asking questions that may lead to thinking of discrete atoms.

3. Point out the electrostatic forces in the diagram/structure.
4. Explain the meaning of covalent bonding in terms of the electrostatic forces mentioned.
5. Explain its properties with its bonding and structure.
6. Comment on its stability with the octet rule.

Such a flow of ideas avoids the initial atomicity and emphasizes covalent bonding as an electrostatic force. The octet rule is positioned last to avoid students associating it with covalent bonding, so that its role as a guideline does not turn into an explanation. Submicro ideas of bonding and structure would also be applied to explain macro phenomena. This seems to be a better alternative compared with the sequence in the curriculum and our teaching plan.

References

Ben-Zvi, R., Eylon, B.-S., & Silberstein, J. (1986). Is an atom of copper malleable? *Journal of Chemical Education, 63*(1), 64–66.

Butts, B., & Smith, R. (1987). HSC chemistry students' understanding of the structure and properties of molecular and ionic compounds. *Research in Science Education, 17*, 192–201.

Coll, R. K., & Taylor, N. (2001). Alternative conceptions of chemical bonding held by upper secondary and tertiary students. *Research in Science & Technological Education, 19*(2), 171–191.

de Posada, J. M. (1997). Conceptions of high school students concerning the internal structure of metals and their electric conduction: Structure and evolution. *Science Education, 81*(4), 445–467.

Gilbert, J., & Boulter, C. (Eds.). (2000). *Developing models in science education*. Dordrecht: Kluwer.

Nakhleh, M., & Krajcik, J. S. (1994). Influence on levels of information as presented by different technologies on students' understanding of acid, base, and pH concepts. *Journal of Research in Science Teaching, 31*(10), 1077–1096.

Taber, K. S. (1994). Misunderstanding the ionic bond. *Education in Chemistry, 31*(4), 100–103.

Taber, K. S. (1997). Student understanding of ionic bonding: Molecular versus electrostatic framework? *School Science Review, 78*(285), 85–95.

Taber, K. S. (1998). An alternative conceptual framework from chemistry education. *International Journal of Science Education, 20*(5), 597–608.

Taber, K. S. (2001). Building the structural concepts of chemistry: Some considerations from educational research. *Chemistry Education: Research and Practice in Europe, 2*(2), 123–158.

Taber, K. S. (2003a). The atom in the chemistry curriculum: Fundamental concept, teaching model or epistemological obstacle? *Foundations of Chemistry, 5*, 43–84.

Taber, K. S. (2003b). Mediating mental models of metals: Acknowledging the priority of the learner's prior learning. *Science Education, 87*(5), 732–758.

Chapter 26
Blending Physical and Virtual Manipulatives in Physics Laboratory Experimentation

Georgios Olympiou and Zacharias C. Zacharia

1 Introduction

Given the importance of laboratory experimentation for science education, many researchers have attempted to investigate and document the value of using Physical Manipulatives (PM; real-world physical/concrete material and apparatus) and Virtual Manipulatives (VM; virtual apparatus and material, which exist in computer-based simulations) in science laboratory experimentation (Finkelstein et al. 2005; Hofstein and Lunetta 2004; Jaakkola et al. 2010; Toth et al. 2009; Triona and Klahr 2003; Winn et al. 2006; Zacharia 2007; Zacharia and Anderson 2003; Zacharia and Constantinou 2008; Zacharia and Olympiou 2011; Zacharia et al. 2008). Comparative studies have also been undertaken in order to identify which of these two modes of experimentation (PM or VM) is the most preferable across several science subject domains (Finkelstein et al. 2005; Klahr et al. 2007; Toth et al. 2009; Triona and Klahr 2003; Zacharia 2007; Zacharia et al. 2008, 2012; Zacharia and Constantinou 2008; Zacharia and Olympiou 2011).

The findings from these studies revealed instances where the use of VM would appear to be as beneficial to student learning as PM (Klahr et al. 2007; Triona and Klahr 2003; Zacharia and Constantinou 2008; Zacharia and Olympiou 2011), more beneficial to student learning than the use of PM (Finkelstein et al. 2005; Zacharia 2007; Zacharia et al. 2008), and vice versa (Marshall and Young 2006). Olympiou and Zacharia (2012) after studying the material and methods used in these studies, attributed the differences in outcomes to the differing affordances (qualities of PM or VM that offer the possibility of an interaction relative to the ability of a learner to interact) that the PM and VM carried in each of these studies. As a result, they argued in favor of combining the two modes of experimentation rather than using them alone. Similar argument was made by several other researchers (Jaakkola et al. 2010;

G. Olympiou (✉) • Z.C. Zacharia
University of Cyprus, Nicosia, Cyprus
e-mail: olympiog@ucy.ac.cy

C. Bruguière et al. (eds.), *Topics and Trends in Current Science Education: 9th ESERA Conference Selected Contributions*, Contributions from Science Education Research 1,
DOI 10.1007/978-94-007-7281-6_26, © Springer Science+Business Media Dordrecht 2014

Toth et al. 2009; Winn et al. 2006; Zacharia et al. 2008; Zacharia and Constantinou 2008; Zacharia and Olympiou 2011). However, up to date, it was only Olympiou and Zacharia (2012) that proposed a detailed framework depicting how PM and VM could be blended. This framework takes into consideration the PM and VM affordances and specifically targets the content of each lab experiment separately. In other words, the PM and VM are combined and used in conjunction in the context of each experiment in a way that they match the needs of each experiment separately (Olympiou and Zacharia 2012).

The purpose of this chapter is to present the Olympiou and Zacharia (2012) framework and test its reliability in the physics domain on *Light and Color*. In doing so, we conducted two studies that involved the implementation of a PM and VM blended combination according to our proposed framework and compared it to conditions that involved the use of PM and VM alone. The overarching goal was to check whether students' conceptual understanding was enhanced more in the case of the blended condition than using PM or VM alone in both studies.

2 PM and VM Affordances

Science education literature on PM and VM laboratory experimentation has pointed to the fact that PM and VM have a significant overlap in terms of the affordances that they could offer in laboratory experimentation (e.g., manipulation of material, perceptual grounding for concepts that might otherwise be too abstract to be easily understood). On the other hand, the literature of this domain revealed that PM and VM carry certain unique affordances that differ between them. Thus, their presence during laboratory experimentation results in a different effect on student learning, in favor of the manipulatives that carry these additional affordances/advantages (Finkelstein et al. 2005; Olympiou and Zacharia 2012; Winn et al. 2006; Zacharia 2007; Zacharia et al. 2008).

In the case of PM, actual and active touch of concrete material is reported as one such unique affordance (Zacharia and Olympiou 2011), whose learning effect could be even more enhanced when combined with other human senses, especially visual and auditory (Han and Black 2011). A second affordance of PM is that it is something that originates from the physical world and hence the only set of manipulatives that reflects the true nature of science (e.g., measurement errors are always present). A third affordance is that only PM involve the development and acquisition of psychomotor skills through setting-up and running experiments and learning how to use human senses for making observations and measurements (Triona and Klahr 2003).

On the other hand, the use of VM, unlike the use of PM, could (a) allow students to visualize abstract/conceptual objects and processes that are normally beyond perception (Winn et al. 2006); (b) provide capabilities for altering the natural time scale and simplifying real-world models, thus making phenomena more visible to learners (deJong and Njoo 1992); (c) allow students to change variables which would be impossible or unrealistic to change in the natural world (Windschitl 2000); (d) provide students with immediate feedback about errors in the experimental setup and thus the opportunity to repeat the same experiment immediately (Huppert and Lazarowitz 2002); (e) allow students to perform a wide range of experiments faster and more easily and thus experience more examples (Huppert and Lazarowitz 2002); (f) provide a dynamically linked, information-rich, and multiple representation environment (Hsu and Thomas 2002); (g) facilitate learning by focusing students' attention more directly on the targeted phenomena (deJong and Van Joolingen 1998); and (h) enable students to experience what might be too expensive or difficult to carry out with PM and permit experiments to be performed repeatedly in a safe environment (Doerr 1997).

Given all these affordances, it becomes obvious that VM carry many unique affordances that could complement PM in surpassing many of their inherent deficiencies within the context of school science experimentation. The attempt was for VM to "match" the experimental affordances provided by PM and to exceed them by providing even more affordances than PM.

A number of studies involving the use of PM and VM (which carry a number of the aforementioned advantageous affordances) showed both that the use of VM-enhanced student learning more than the use of PM and vice versa (Finkelstein et al. 2005; Marshall and Young 2006). Thus, highlighting the added value that each mode of experimentation, PM or VM, brings into a learning experience and the essence of combining PM and VM rather than using them alone. Moreover, it is important to note that the aforementioned PM and VM unique affordances were found to be conducive to students' learning across several subject domains and age groups (e.g., Finkelstein et al. 2005; Huppert and Lazarowitz 2002; Toth et al. 2009; Zacharia 2007). Someone could reasonably argue that these affordances are content independent, which provides us with the opportunity to use this knowledge across the science subject domains. On the other hand, research has shown that these affordances are learning-objective dependent. In other words, research has associated all these affordances with specific learning objectives (Zacharia et al. 2008). This means that PM or VM should be used when their affordances, unique or not, serve the objectives of an experiment (Olympiou and Zacharia 2012).

3 Blending PM and VM

Zacharia et al. (2008) suggested that the development of a framework that portrays how PM and VM should be blended, to enhance students' conceptual understanding more than when PM and VM are used alone, is to take the learning objectives of an experiment/activity and carefully analyze them in terms of what the student should be introduced/exposed to (e.g., an authentic real experience; an experience of measurement errors; an experience that involves observation of reified objects, such as atoms). This implies that the pedagogical and didactical parameters of an experiment (e.g., content, collaboration, cognitive skills), which are reflected through its learning objectives, are better served through a research-based, targeted use of PM and VM. Of course, such a framework presupposes knowledge of what PM and VM could offer, particularly, in terms of unique affordances.

Given these ideas, Olympiou and Zacharia (2012) proceeded with the development of a framework blending PM and VM, while aiming at enhancing students' conceptual understanding. Right below we present analytically the Olympiou and Zacharia framework.

3.1 The Olympiou and Zacharia Framework for Blending PM with VM

The Olympiou and Zacharia framework involves a series of steps that need to be followed in order to reach a fine blending of PM and VM (see Fig. 26.1). First, it requires the analysis of the study's teaching material in order to identify the overarching general learning objective (e.g., the promotion of conceptual understanding), as well as the specific learning objectives of each one of the teaching material's experiments. Second, it requires the consideration of the characteristics (e.g., prior knowledge) of the student group that would be involved in the learning process. Third, it requires the identification of PM and VM unique affordances through the relevant literature review and matching them with the learning objectives at task. Fourth, it entails creating blended combinations of PM and VM per experiment, while considering at the same time a set of parameters. In particular, blending PM and VM requires knowing the PM and VM affordance that are available, whether students could switch mode of experimentation (from PM to VM and vice versa) and whether the students have the knowledge and skills required for using the VM and PM.

4 Methodology

For the purposes of validating our framework for blending Physical and Virtual Manipulatives, we conducted two studies which followed the same teaching material and experimental design but differed in the number of participants. Right below we provide all the information related to the methods applied in both studies.

4.1 Sample

The participants of the first study were 70 undergraduate students and of the second study were 114 undergraduate students. All students were enrolled in an introductory physics course that was based upon the *Physics by Inquiry* curriculum (McDermott and The Physics Education Group 1996), intended for preservice

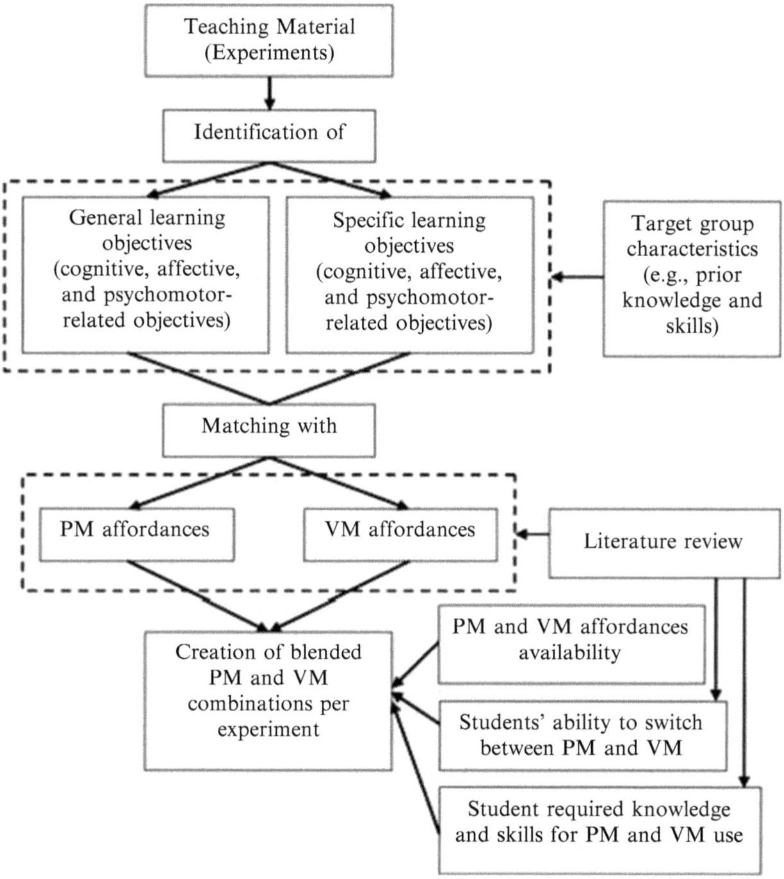

Fig. 26.1 The Olympiou and Zacharia framework for blending PM and VM (Olympiou and Zacharia 2012, p. 29)

elementary school teachers. The course took place at a university in Cyprus. The participants of both studies were randomly separated into three groups, namely, two control groups (henceforth called CG) and one experimental group (henceforth called EG). In the case of the first study, the CG1 was assigned 23 students to use PM, the CG2 was assigned 23 students to use VM, and the EG1 was assigned 24 students to use the blended combination, whereas in the case of the second study, the CG3 was assigned 42 students to use PM, the CG4 was assigned 36 students to use VM, and the EG2 was assigned 36 students to use the blended combination.

The students in all groups were randomly assigned to subgroups (three persons in each subgroup) as suggested by the curriculum of the study (McDermott et al. 1996). This particular curriculum is grounded upon a social constructivist framework that facilitates a constructive, situated, and collaborative learning process that assures that the engagement is truly collaborative and helps all students make explicit their ideas. Knowledge and understanding is co-constructed among peers through complementing and building on each others' ideas (Duit and Treagust 1998). However, all the measurements taken in both studies targeted the individuals and not their groups as a whole.

4.2 Curriculum Materials: Physics by Inquiry

The selection of the *Physics by Inquiry* curriculum was based on the fact that through numerous studies it appeared to enhance undergraduate students' conceptual understanding across physics subject domains (Zacharia et al. 2008) including the subject domain of *Light and Color*.

For the purposes of both studies, three parts of the module of *Light and Color* were used (McDermott et al. 1996). The first section (Sect. 1) focuses on an introduction to light, light sources, masks, screens, and shadows, the second section (Sect. 2) focuses on colored paint, and the third section (Sect. 3) focuses on colored light. In Sect. 1, students are encouraged to develop a mental model that enables them to account for complicated phenomena, such as the formation of images and shadows from extended sources. In Sects. 2 and 3, the students conduct experiments with colored paints and colored light, in an attempt to understand how to mix paints of different colors to obtain a particular color of paint and how to combine light of different colors to obtain a particular color of light. Moreover, on the basis of their observations, the students are encouraged to develop a mental model that enables them to predict the color an object will be when viewed under light of different colors.

4.3 Material: Physical and Virtual Manipulatives

PM involved the use of physical instruments (e.g., rulers), objects (e.g., metal rings), and materials (e.g., lamps) in a conventional physics laboratory. During PM experimentation feedback was available to the students through the behavior of the actual system (e.g., shape of a shadow on a screen) and through the instruments that are used to monitor the experimental setup (e.g., rulers).

VM involved the use of virtual instruments (e.g., rulers), objects (e.g., metal rings), and materials (e.g., lamps) to conduct the study's experiments on a computer. Most of these experiments were conducted through the Virtual Lab Optilab shown in Fig. 26.2 (Hatzikraniotis et al. 2007). For a very small number of experiments of the curriculum, the software could not provide all the material needed for the experimental setup; hence, interactive simulations were developed and used to complement the *Optilab* software.

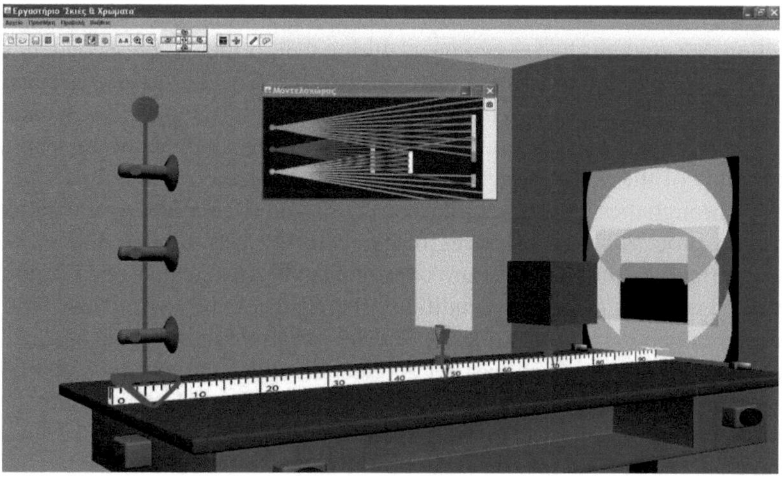

Fig. 26.2 The Optilab environment

Optilab was selected because of its fidelity and the fact that it retained the features and interactions of the domain of *Light and Color* as PM does. In its open-ended environment, students of the CG2, CG4, EG1, and EG2 were able to design and conduct the experiments mentioned in the module of *Light and Color* by employing the "same" material as the ones used by the students using PM. In the *Optilab* environment, students were provided with a virtual workbench on which experiments can be performed, virtual objects to compose the experimental setup, virtual materials whose properties are to be investigated, and virtual instruments (e.g., rulers) or displays (e.g., screen) as illustrated in Fig. 26.2. No feedback was provided by the software during the set up of the (virtual) experiment. The level of feedback was analogous to what is routinely available to students through PM experimentation.

4.4 *Experimental Design and Procedure*

A pre–post comparison study design was used for the purposes of each study that involved three groups in each case. The first study involved three conditions, namely, CG1, CG2, and EG1, whereas the second study involved the same three conditions accordingly, CG3, CG4, and EG2.

Given this experimental design, all participants were administered the L&C pretest before getting engaged in the treatment of the condition they belonged to. After a week, (pre-)test 1 was administered and a brief introduction that aimed to familiarize all students with the material they were about to use. The introduction to the routines and procedures of the *Physics by Inquiry* curriculum was very important because they differ from those involved in the more traditional, passive modes of instruction that students had experienced during their primary and secondary school years. For example, the role of the instructors in the *Physics by Inquiry* curriculum is quite different from that in a traditional instruction. It is supportive in nature and requires instructors' engagement in dialogues with the students of a group at particular points of the activity sequence, as specified by the *Physics by Inquiry* curriculum. Through these dialogues, the instructors aim to encourage reflection across the inquiry processes and practices involved in the activities of the *Physics by Inquiry* curriculum and not to lecture or provide ready-made answers/solutions. For the purposes of this study, all conditions shared the same instructors, who were previously trained in implementing the *Physics by Inquiry* curriculum and had experienced its implementation at least for 2 years. Finally, along with the instructional part of each section, conceptual tests were also administered both before and after each section. The duration of the study was 13 weeks. Students met once a week for one and a half hour. The time-on-task was the same for all conditions.

4.5 Data Collection

Both studies involved the collection of data through the use of conceptual tests before, during, and after the study. Specifically, the same conceptual test (*Light and Color* test or L&C test) was administered to assess students' understanding of *Light and Color* concepts concerning light, shadows, colored paint, and colored light, both before and after the study. Additionally, tests were administered before and after introducing each section (tests 1, 2, and 3), with each test being identical before and after each section. The tests were developed and used in previous research studies by the Physics Education Group of the University of Washington (McDermott et al. 1996). All tests (1, 2, 3, and L&C test) contained open-ended items that asked conceptual questions all of which required explanations of reasoning. The tests 1, 2, and 3 included four open-ended items each that were used for the assessment of Sects 1, 2, and 3, respectively, whereas the L&C test included five open-ended items assessing all sections of the study's curriculum. The L&C test targeted both the specific concepts introduced in each section as well as the interconnections and interdependencies of these concepts. No identical items were included in the L&C test and the rest of the tests.

Each item of each test was scored separately; however, a total correct score was derived from each test and used in the analysis. All tests were scored and coded blind to the condition in which the student was placed. A rubric table was used that specified different criteria for the responses to each item separately. Each response was scored on each criterion. The minimum score on each test was 0, and the total maximum score for all criteria of all items on each test was 100. Two independent raters scored about 20 % of the data. The reliability measures (Cohen's Kappa) for scoring of the L&C test (pre- and posttest) and tests 1, 2, and 3 (pre- and posttests) were above 0.9 across all tests.

4.6 Data Analysis

The data analysis involved both quantitative and qualitative methods in both studies. In particular, all tests were scored though the use of scoring rubrics and the resulted student performance scores were analyzed by using (a) one-way ANOVA for the comparison of the pretest scores of the three conditions on each test, (b) paired samples *t*-test for the comparison of the pretest scores to the posttest scores of each condition across all tests, and (c) one-way ANCOVA for the comparison of the posttest scores of the three conditions on each test. For the latter procedure, the students' scores in the corresponding pretests were used as the covariate.

The aim of the first procedure was to determine whether the three conditions of the study were comparable with regard to the sample's entry understanding of physics concepts from the subject domain of *Light and Color*, before the study and before each section. The aim of the second procedure was to investigate whether the use of the blended combination of PM and VM and the use of PM or VM alone, within the context of the *Physics by Inquiry* curriculum, improved students' conceptual understanding. The aim of the third procedure was to investigate whether the three conditions of the study had differences on the outcome measures (understanding of physics concepts in the domain of *Light and Color*) of each test. For the ANOVA and ANCOVA analyses, the effect size, η^2, is also reported (Cohen 1988).

The qualitative analysis focused on identifying and classifying students' scientifically acceptable (SAC) or scientifically not acceptable (SNAC) conceptions concerning *Light and Color* within their answers on the tests. This analysis followed the procedures of open coding (Strauss and Corbin 1998), in which the researchers first underlined the most important sentences in each student's pre- and posttests and marked keywords that characterized the student's conceptions with respect to *light*, *shadows*, *colored paint*, and *colored light*. By comparing the sentences underlined and the keywords derived from the tests, the content-specific similarities and differences in students' test responses about the aforementioned four categories of conceptions were explored and summarized. Then, the researchers constructed "qualitatively different" subcategories of description, essentially across rather than within the responses, that were used to classify the conceptions of *light*, *shadows*, *colored paint*, and *colored light* held by students for each condition separately. By comparing the similarities and differences between the students of each condition, subcategories of conceptions emerged for each of the aforementioned four categories of conceptions (for an example see Table 26.3). The purpose of this analysis was to reveal the subcategories of description that could characterize the qualitatively different perspectives in which *light*, *shadows*, *colored paint*, and *colored light* were conceptualized or experienced by the students of each condition. Finally, the prevalence for each of the resulting subcategories for each test was calculated across all the conditions. The aim of the latter calculation was to compare whether students' conceptions, both within and between conditions, changed over the course of the study.

5 Results

5.1 Conceptual Test Performance

5.1.1 Study 1

At the first study the one-way ANOVA procedure indicated that the three conditions did not differ in pretest scores across all of the study's tests ($F < 1$, $p > 0.01$ across all cases). The paired-sample *t*-test analysis showed that the blended combination of PM and VM and PM and VM alone improved students' conceptual understanding after each section of the first study ($p < 0.01$ across all cases).

The ANCOVA procedure revealed differences among the study's three conditions across all tests (see Table 26.1). Bonferroni-adjusted pairwise comparisons suggested that students' posttest scores in the PM alone and VM alone conditions were significantly lower than those of the students in the PM&VM condition across all tests ($p < .01$ across all PM&VM versus PM or VM comparisons; lower than the 0.016, which is the lowest *p*-value given by the Holm – Bonferroni method), whereas the same pairwise comparisons did not show any significant difference between the students' posttest scores of the PM and VM alone conditions across all tests.

5.1.2 Study 2

At the second study the one-way ANOVA procedure indicated that the three conditions did not differ in pretest scores across all of the study's tests ($F < 1$, $p > 0.01$ across all cases). The paired-sample *t*-test analysis showed that the blended combination of PM and VM and PM and VM alone improved students' conceptual understanding after each section of the first study ($p < 0.01$ across all cases). Both of these analyses revealed similar results as in study 1.

The ANCOVA procedure applied in this study also revealed differences among the study's three conditions across all tests (see Table 26.2), and the Bonferroni-adjusted pairwise comparisons also suggested that students' posttest scores in the PM alone and VM alone conditions were significantly lower than those of the students in the PM&VM condition across all tests ($p < .01$ across all PM&VM versus PM or VM comparisons; lower than the 0.016, which is the lowest *p*-value given by the Holm – Bonferroni method). Moreover, the pairwise comparisons did not show

Table 26.1 One-way ANCOVA results and p-values of the comparisons concerning the scores of the student responses across tests in study 1

Test	F	df	P	η^2
Test 1	5.104	2, 66	<0.05	0.13
Test 2	6.07	2, 66	<0.05	0.15
Test 3	5.89	2, 66	<0.05	0.15
L&C test	10.95	2, 66	<0.05	0.25

Table 26.2 One-way ANCOVA results and p-values of the comparisons concerning the scores of the student responses across tests in study 2

Test	F	df	P	η^2
Test 1	4.558	2, 110	<0.05	0.08
Test 2	4.34	2, 110	<0.05	0.07
Test 3	9.55	2, 110	<0.05	0.15
L&C test	13.53	2, 110	<0.05	0.2

any significant difference between the students' posttest scores of the PM and VM alone conditions across all tests.

As shown in Table 26.2, the findings of study 2 are in accordance with the findings of study 1 and suggest that the PM&VM condition enhanced students' understanding of the light and color concepts that were introduced through the curriculum material of this study, more than the PM or VM conditions did. Furthermore, as in study 1, study 2 revealed that the use of PM alone and the use of VM alone were equally effective in promoting students' understanding in *Light and Color*.

5.2 *Understanding of Light and Color Conceptions in Both Studies*

The qualitative analysis revealed that the PM alone and VM alone conditions shared mostly the same conceptions across the light and color concepts studied (*light*, *shadows*, *colored paint*, and *colored light*), as either SAC or SNAC, both before and after the L&C test was administered, in both studies. The same regarded tests 1, 2, and 3 both before and after the introduction of each section of the study's curriculum. The PM&VM condition appeared to share the same SAC and SNAC conceptions with the other two conditions only before the study (at the pretest of each section), in both studies.

After all students were exposed to the treatments of their conditions, they managed to surpass many of their SNACs and adopt SACs, with the PM&VM condition having the highest increase in percentage across all SACs and the higher decrease in percentage across all SNACs, in both studies. The latter implies that the blended combination had greater impact on students' transition from SNAC to SAC than the PM and VM alone conditions did, in both studies (for an example see Table 26.3).

6 Discussion

The findings of both studies revealed that the use of a blended combination of PM and VM, according to the Olympiou and Zacharia framework, was more conducive to students learning than the use of PM and VM alone. This shows that the application

Table 26.3 A sample of SAC and SNAC concerning the white light as they emerged from the qualitative analysis of test 3

	Study 1 conditions						Study 2 conditions					
	PM[a]		VM[a]		PM&VM[a]		PM[a]		VM[a]		PM&VM[a]	
Conceptions[b]	Pre %(n)	Post %(n)	Pre %(n)	Post %(n)	Pre %(n)	Post %(n)	Pre %(n)	Post %(n)	Pre %(n)	Post %(n)	Pre %(n)	Post %(n)
SAC[c] 1[e]	0 % (0)	74 % (17)	0 % (0)	87 % (20)	0 % (0)	100 % (24)	2 % (1)	69 % (29)	0 % (0)	97 % (35)	0 % (0)	94 % (34)
SNAC[d] 2[f]	30 % (7)	0 % (0)	40 % (9)	0 % (0)	21 % (5)	0 % (0)	19 % (8)	2 % (1)	17 % (6)	6 % (2)	28 % (10)	6 % (2)
SNAC 3[g]	70 % (16)	26 % (6)	60 % (14)	13 % (3)	79 % (19)	0 % (0)	67 % (28)	21 % (9)	72 % (26)	0 % (0)	94 % (34)	0 % (0)

[a]PM condition (PM), participants used PM alone; VM condition (VM), participants used VM alone; PM&VM condition (PM&VM), participants used a blended combination of PM and VM

[b]Conceptions of the subcategory titled "white light"; a subcategory of the category "combination of light of different colors"

[c]SAC, scientifically acceptable conception

[d]SNAC, scientifically not acceptable conception

[e]SAC 1, the white light is produced when all the light colors of the spectrum of light are combined before they reach our eyes

[f]SNAC 2, the white light is a pure color and not a combination of colored light

[g]SNAC 3, the white light is produced by combining the same colors as in colored paint

of the Olympiou and Zacharia framework caused the same outcomes in both studies, which depicts that the outcomes of the framework are reliable. It should be noted that both studies were carried out in the context of a normal course in physics, thus offering ecological validity to the findings.

Needless to say, further research is needed to reach to a general conclusion about the reliability of this framework. In particular, this framework needs to be tested across different ages, subject areas, and subject domains, as well as wider sample sizes and different types of PM and VM. Moreover, further research is needed on how to optimize the PM and VM blends. For example, are there any particular PM and VM affordances that should coexist or never be combined? Additionally, more frameworks (of the same or different rationale as the one followed in this study) that focus on blending PM and VM affordances in order to enhance students conceptual understanding should be developed and tested out (in the same and in different subject domain), as well as frameworks that target learning objectives besides conceptual understanding oriented ones. For instance, how such a framework should look like if the study's learning objectives were focusing on aspects of the nature of science (e.g., what is science and scientific knowledge, distinction between observations and inferences, human error).

It is important to understand, though, that any future implementation of the Olympiou and Zacharia framework (or another similar framework) requires sound understanding of the learning objectives of each experiment at task and of the affordances that each manipulative carries, particularly those unique to only PM or VM. Additionally, it requires good knowledge of the characteristics, prior knowledge, and skills of the students involved. Our experience suggests that failure to consider any of the aspects of the framework will result in misleading outcomes.

Finally, the fact that the use of a blended combination of VM and PM appears to be more conducive to learning through laboratory experimentation than the use of PM and VM alone challenges the already established norms concerning experimentation in the science classroom. Specifically, it challenges the laboratory experimentation as we experienced it through PM or VM, in a way that calls for its redefinition and restructuring, in order to include blended combinations of VM and PM in science learning environments. As the research in both worlds continues, we expect further refinement of the frameworks aiming to blend PM and VM and better learning outcomes among learners.

References

Cohen, J. (1988). *Statistical power analysis for the behavioral sciences* (2nd ed.). Hillsdale: Erlbaum.

de Jong, T., & Njoo, M. (1992). Learning and instruction with computer simulation: Learning processes involved. In E. de Corte, M. C. Linn, H. Mandl, & L. Verschaffel (Eds.), *Computer-based learning environments and problem solving* (pp. 411–427). Berlin: Springer.

de Jong, T., & van Joolingen, W. R. (1998). Scientific discovery learning with computer simulations of conceptual domains. *Review of Educational Research, 68*, 179–202.

Doerr, H. (1997). Experiment, simulation and analysis: An integrated instructional approach to the concept of force. *International Journal of Science Education, 19*, 265–282.

Duit, R., & Treagust, D. F. (1998). Learning in science – from behaviorism towards social constructivism and beyond. In B. J. Fraser & K. J. Tobin (Eds.), *International handbook of science education* (pp. 3–25). Dordrecht: Kluwer.

Finkelstein, N. D., Adams, W. K., Keller, C. J., Kohl, P. B., Perkins, K. K., Podolefsky, N. S., et al. (2005). When learning about the real world is better done virtually: A study of substituting computer simulations for laboratory equipment. *Physical Review Special Topics – Physics Education Research, 1*, 1–8.

Han, I., & Black, J. B. (2011). Incorporating haptic feedback in simulation for learning physics. *Computers in Education, 57*, 2281–2290.

Hatzikraniotis, E., Bisdikian, G., Barbas, A., & Psillos, D. (2007). Optilab: Design and development of an integrated virtual laboratory for teaching optics. In C. P. Constantinou, Z. C. Zacharia, & M. Papaevripidou (Eds.), *Proceedings of the 7th International Conference on Computer Based Learning in Science.* Crete: Technological Educational Institute of Crete.

Hofstein, A., & Lunetta, V. (2004). The laboratory in science education: Foundations for the twenty-first century. *Science Education, 88*, 28–54.

Hsu, Y.-S., & Thomas, R. A. (2002). The impacts of a web-aided instructional simulation on science learning. *International Journal of Science Education, 24*, 955–979.

Huppert, J., & Lazarowitz, R. (2002). Computer simulations in the high school: Students' cognitive stages, science process skills and academic achievement in microbiology. *International Journal of Science Education, 24*, 803–821.

Jaakkola, T., Nurmi, S., & Veermans, K. (2010). A comparison of students' conceptual understanding of electric circuits in simulation only and simulation-laboratory contexts. *Journal of Research in Science Teaching, 48*, 71–93.

Klahr, D., Triona, L. M., & Williams, C. (2007). Hands on what? The relative effectiveness of physical versus virtual materials in an engineering design project by middle school children. *Journal of Research in Science Teaching, 44*, 183–203.

Marshall, J. A., & Young, E. S. (2006). Preservice teachers' theory development in physical and simulated environments. *Journal of Research in Science Teaching, 43*, 907–937.

McDermott, L. C., & The Physics Education Group. (1996). *Physics by inquiry*. New York: Wiley.

Olympiou, G., & Zacharia, Z. C. (2012). Blending physical and virtual manipulatives: An effort to improve students' conceptual understanding through science laboratory experimentation. *Science Education, 96*, 21–47.

Strauss, A., & Corbin, J. (1998). *Basics of qualitative research. Techniques and procedures for developing grounded theory*. Thousand Oaks: SAGE.

Toth, E. E., Morrow, B. L., & Ludvico, L. R. (2009). Designing blended inquiry learning in a laboratory context: A study of incorporating hands-on and virtual laboratories. *Innovative Higher Education, 33*(5), 333–344.

Triona, L., & Klahr, D. (2003). Point and click or grab and heft: Comparing the influence of physical and virtual instructional materials on elementary school students' ability to design experiments. *Cognition and Instruction, 21*, 149–173.

Windschitl, M. (2000). Supporting the development of science inquiry skills with special classes of software. *Educational Technology Research and Development, 48*, 81–95.

Winn, W., Stahr, F., Sarason, C., Fruland, R., Oppenheimer, P., & Lee, Y.-L. (2006). Learning oceanography from a computer simulation compared with direct experience at sea. *Journal of Research in Science Teaching, 43*, 25–42.

Zacharia, Z. C. (2007). Comparing and combining real and virtual experimentation: An effort to enhance students' conceptual understanding of electric circuits. *Journal of Computer Assisted Learning, 23*, 120–132.

Zacharia, Z. C., & Anderson, O. R. (2003). The effects of an interactive computer-based simulations prior to performing a laboratory inquiry-based experiments on students' conceptual understanding of physics. *American Journal of Physics, 71*, 618–629.

Zacharia, Z. C., & Constantinou, C. P. (2008). Comparing the influence of physical and virtual manipulatives in the context of the physics by inquiry curriculum: The case of undergraduate students' conceptual understanding of heat and temperature. *American Journal of Physics, 76*, 425–430.

Zacharia, Z. C., & Olympiou, G. (2011). Physical versus virtual manipulative experimentation in physics learning. *Learning and Instruction, 21*, 317–331.

Zacharia, Z. C., Olympiou, G., & Papaevripidou, M. (2008). Effects of experimenting with physical and virtual manipulatives on students' conceptual understanding in heat and temperature. *Journal of Research in Science Teaching, 45*, 1021–1035.

Zacharia, Z., Loizou, E., & Papaevripidou, M. (2012). Is physicality an important aspect of learning through science experimentation among kindergarten students? *Early Childhood Research Quarterly, 27*(3), 447–457.

Chapter 27
Becoming a Health-Promoting School: Effects of a 3-Year Intervention on School Development and Pupils

Steffen Schaal

1 Introduction

Healthy adolescents generally cope better with the challenges of puberty than such with a deficient health. The average of German children and adolescents is reported to be in good health status, but some subgroups are partially giving cause for concern, and especially children living in disadvantaged sociocultural situation exhibit an exceeding poor health behaviour (cf. Hurrelmann et al. 2006; Ravens-Sieberer et al. 2008; Kurth and Schaffrath-Rosario 2007). Recent research in large-scale surveys showed that only each fourth German boy and each sixth girl from 11 to 17 years meet the recommended daily activity level (Lampert et al. 2007), the consumption of fruits and vegetables decreases massively from childhood to adolescence (Mensink 2007) and one fifth of German school-aged children (N=23.111) reports psychosomatic disorders at regular intervals (Hurrelman et al. 2003). The health behaviour of young people is influenced by several factors (family, peer group, media, environment, etc.), and school massively affects the adolescents' well-being (Hascher 2007, 2008). Schools increasingly are called on to counteract challenging educational demands in the life of young people (Tang et al. 2008). Schools' role is to improve different literacy, to enhance specific knowledge construction and to support personal development. In the case of science education, the development of scientific literacy, education for a sustainable development and health promotion are central issues (cf. St. Leger et al. 2010). Zeyer and Odermatt (2009) highlight the link between health and science education to promote health literacy. According to them, the health literacy acquisition should be a facet of science education that deals with biomedical concepts of health to enable a critical reflection. Thus, students are

S. Schaal (✉)
Ludwigsburg University of Education, Reuteallee 46, 71634 Ludwigsburg, Germany
e-mail: schaal@ph-ludwigsburg.de

C. Bruguière et al. (eds.), *Topics and Trends in Current Science Education: 9th ESERA Conference Selected Contributions*, Contributions from Science Education Research 1,
DOI 10.1007/978-94-007-7281-6_27,

supported to reflect science education concepts under the perspective of their applicability in everyday-life contexts, and this deems to empower them in decision-making and behaviour. But health-related behaviours such as physical activity or nutrition behaviour are difficult to change (Schwarzer 2007), health literacy might not be sufficient for sustainable health promotion and school reform has to cope with several health-related challenges and risk behaviours (CDCP 2010). Health promotion (HP) is a more comprehensive and holistic approach that addresses determinants of health in the school environment as well as individual prerequisites (Rowling and Jeffreys 2006), and it necessitates a whole-school development.

Addressing young people in school settings for HP activities:

1. Requires linkages to youth's experiences and vitality rather than normative advices for behaviour and interdictions.
2. Should consider the discrepancy between pedagogic values and role models of youth's real-life surroundings. These normative and sociocultural prerequisites influence the success of health-related interventions notably.
3. Ideally fits into age-appropriate lifestyles, stage of development and cultural influences. Younger children profit more from informative interventions, for instance, while adolescents need more participative approaches (Jensen and Simovska 2005; Lohaus and Lissmann 2005; Simovska 2007).

The concept of health-promoting school (HPS) aims to empower schools to develop adequate curricula and to concentrate on healthy circumstances for working and learning within the whole-school setting, complemented by approaches to reduce individuals' risk behaviour. The HPS involves pupils, teachers, parents and other school staff in the development and implementation process. Hereby different fields of action and principles are described in literature (Weare 2000; St Leger 2000). Within the HPS development process, science education plays a multilayered and decisive role: For instance, the curriculum is as important as the construction of comprehensive and holistic concepts of health and a healthy school environment. Biology takes a leadership in health education (cf. Turner et al. 1999; Webb 1995), and health-related contexts could be found in several science education curricula.

But a systematic and obligatory engagement in HP is missing in southwestern Germany secondary school curricula. Therefore it rests on each single school's principal and school's conference to route an individual pathway to pupils' health. Thus, HP on the one hand can be individualised and adapted to each school's requirements. On the other hand, it rests arbitrarily if HP or health education becomes a part of school's curriculum and everyday life. However, for further development it is crucial to elaborate, to implement and to evaluate evidence-based, age-appropriate health education and health promotion in German school reform and in different school science education or life science curricula. But in this field of research, again, the evidence about the effects of HP activities on the school development status and on each student's health knowledge, attitudes or beliefs is rare.

The aim of this 3-year study was to initiate and to attend lower secondary schools during their school development and the adoption of health promotion in southwestern Germany. The research question was to what extend an external support could initiate and sustain the adoption of health promotion approaches in school under a leadership of biology and science education. The research focuses on process- and outcome-based evaluation of the school development at the organisational and on the individual pupil's level.

2 Research on the Adoption of Health Promotion in School

Several obstacles often impair the implementation of health promotion in school, and it is crucial to support the systematic development of resources and frameworks in the school environment at the same time. Gugglberger (2011) describes five different strategies of supporting HPS in Austria that are validated by a body of evidence in literature (e.g. Hoyle et al. 2008; Jourdan et al. 2008; Aldinger et al. 2008):

1. *Exchange among schools*: networking, exchange of knowledge, best practices and resources
2. *Certification and quality control*: indicators for HPS, visualisation of efforts
3. *Consultation and information*: workshops, trainings, expert meetings, manuals
4. Implementation of *specific HP programmes*
5. *Coordination* of HP actors and information

Deschesnes and colleagues (2010) examined a prediction model for the adoption of the HPS approach. They highlighted the presence of health promotion leaders in school, the perceived school contextual barriers, the past investment in healthy lifestyles in school and the beliefs in collective efficacy within HP as most important predictors for sustainable health promotion in school. Bessems and colleagues (2011) conducted a study for the evaluation of the adoption of a healthy diet programme for secondary schools. Teachers' reasons to reject the programme participation were mainly due to inflexible biology curriculum, and the authors conclude that, again, the involvement of school management and the development of a school's health policy might facilitate the adoption of health promotion programmes. Jourdan and colleagues (2010) worked on the contribution of staff to health education in French schools, and they identified three main roles: Staff mediates the health education process (i) as educators in everyday-life issues, (ii) as individual supporters and (iii) as facilitators and partners in collective projects. The authors conclude health to be a whole-school approach which involves teachers as well as nonteaching staff and which needs to be supported to build a professional identity concerning health education (the term health education is used in this paper synonymously with the acquisition of health literacy). Carlson and Simovska (2012) conducted a school-based HP project helping pupils to explore critically and to improve their health-related conditions at school and in the community. They focused on the learning outcomes like for instance health-related knowledge, skills, visions, critical thinking or decision making. The cross-cultural study reveals pupils' improved health-related action competences, which could also be seen as the successful acquisition of health literacy and to be conducive to pupils' heath-related action competence.

St. Leger and colleagues (2010) describe several evidence-based recommendations for the adoption of health promotion in school, such as:

- Participation of the students and parents in the life of school
- Adequate time for class-based and out-of-class (health-related) activities
- Implementation of a whole-school approach rather than a classroom learning approach
- Exploration of health issues within the context of the students' lives and community
- Consistency of health promotion across the school and between the school, home and the wider community
- Resources to complement the fundamental role of teachers grounded on a theoretical and factual base

Stewart-Brown (2006) reported about evidence of health promotion in schools and stated that beneath the prevention of violence, programmes for healthy eating and physical activities are the most sophisticated and effective approaches to HP in school. In general, effective HPS programmes were complex and multifactorial and involved curriculum development, school environment and/or community. The authors also complain about the poor empirical quality of numerous studies which might be caused by the HPS approach itself: Schools are intended to develop their own profiles, and they conduct activities which impedes controlling all factors within large cluster trials for RCT studies.

Dealing with these hits, recent work focuses on physiological effects of HP activities, for instance, within physical activity and dietary interventions (Magnusson et al. 2012). The results add some slight evidence to the evaluation of the effectiveness of HP programmes, but again the validity and the reliability of a single measurement are not sufficient to stand on their own. Therefore, the evaluation of the complex HPS approach requires a triangulation of research methodology that is sometimes still missing in recent studies (cf. Stewart-Brown 2006). As a conclusion, HP approaches in school involve different fields of action and follow specific principles listed in Table 27.1.

3 Purpose of the Study and Rationale

The purpose of the 3-year research project was to adapt, implement and evaluate a comprehensive concept of school-based health promotion in the areas of nutrition, physical activity and stress management. The programme GUT DRAUF for young people aged 12–18 years is conducted by the Federal Centre for Health Education (BZgA) in Germany since the 1990s (Mann-Luoma et al. 2002). The GUT DRAUF[1]

[1] "GUT DRAUF" is a popular German expression for "in shape".

Table 27.1 Principles and fields of action of HPS according to Paulus (2003)

	Principles
Salutogenesis as conceptual understanding of health (Antonovsky 1979)	Focusing on resources and factors that support health and well-being rather than on factors that cause disease
Participation and empowerment of students, teachers, parents and staff	Involves all school members in decision-making and strengthens their self-confidence that results from the perception that they can influence change (cf. Brodie et al. 2009)
Internal and external networking	In-school teamwork and collaboration with external partners
Acquisition of integral health concepts and health literacy	Understanding health and well-being as process of successful resource deployment
	Fields of action
Teaching, learning and curriculum development	Learning about health in meaningful contexts
School culture and environment	Developing a healthy setting and a culture of appreciation
Health management at school	Creating a healthy school environment (school meals, ergonomy, rhythmicisation, vaccination, dental services, etc.)
Health services and cooperation	Collaboration with community health services and public health institutions

programme aims to increase pupils' individual *health literacy*, to influence *health-related attitudes* and *beliefs* as well as to provide guidance for school development process towards an HPS without strictly guided or standardised interventions in *physical activity*, *stress management* and *nutrition*. The intervention mainly focuses on the training of programme multipliers empowering them to implement HP in classroom learning, to create spaces where pupils get the opportunity to participate in the construction of a healthy environment according to their needs as well as to initiate and to sustain school development towards HPS. GUT DRAUF provides further support for the latter offering a standardised certification process.

The implementation of the GUT DRAUF programme within the study described in this chapter was threefold:

- Pre-service teacher training: In cooperation with the BZgA, university students are trained to the GUT DRAUF criteria and carry out actions autonomous as assistant teachers in intervention schools.
- Development, proving and implementation of GUT DRAUF modules in grades 5–9 (11–16 years) which are oriented on the specific curricula of science education (nutrition, apparatus concerning physical activity, physiology, consequences and avoidance of stress).
- Process consulting towards HPS with the intervention schools' supervision teams (teachers, parents, pupils, staff) under an academic leadership of the researchers.

The general research questions are related to longitudinal effects over the 3-year intervention period and to cross-sectional effects comparing intervention and control group. In detail, the following research areas are of particular interest:

- Achievement of concepts about nutrition and physical activity
- Health-related well-being
- Psychophysiological stress symptoms
- Health-related self-efficacy
- Attribution of nutrition and physical activity
- Food frequencies, physical activity behaviour
- Indicators for health-promoting school

Special interests in this study were the relationships between the development of the HP status of the involved intervention schools compared to the control schools, the functional health literacy (cf. von Wagner et al. 2007) and the pupils' attitudes and/or other health-related properties and behaviours. Especially the relationship between the development of an adequate curricular funding of HP and pupils' attitudes, properties and behaviours would point towards the link between health promotion and science education as an interdisciplinary challenge.

4 Methods and Material

4.1 Study Design and Sample

The study was conducted as a field experiment (cf. Randler 2012) in longitudinal, cross-sectional design with four annual data collections (t_1–t_4) among five lower secondary schools (grades 5–9, age range 10–17 years) in southwestern Germany. The whole-school community mandated the participation in the research project, and at the beginning of the intervention, a school HP supervision team consisting of teachers, parents and pupils was installed at each school. The participating schools were randomly assigned to the treatment. Three schools (intervention schools N=1120 pupils, age M=13.2±1.9) received the threefold school development support according to the research schedule ((1) HP activities carried out by pre-service teachers, (2) GUT DRAUF modules within science education curricula and (3) process consulting towards HPS). The GUT DRAUF project team and pre-service teachers specifically supported the intervention schools in the phase of curricular implementation of the health promotion modules within regular teaching. The results of the baseline survey of the research project are published in Schaal (2011).

The intervention consisted of activities and support for the HP teams, pupils, parents and teachers which are indicated in Fig. 27.1. Two schools served as control group (N = … […]); the mean age of control school pupils was significantly lower (ANOVA (F)=31.1, $p<0.001$), while gender distribution did not differ.

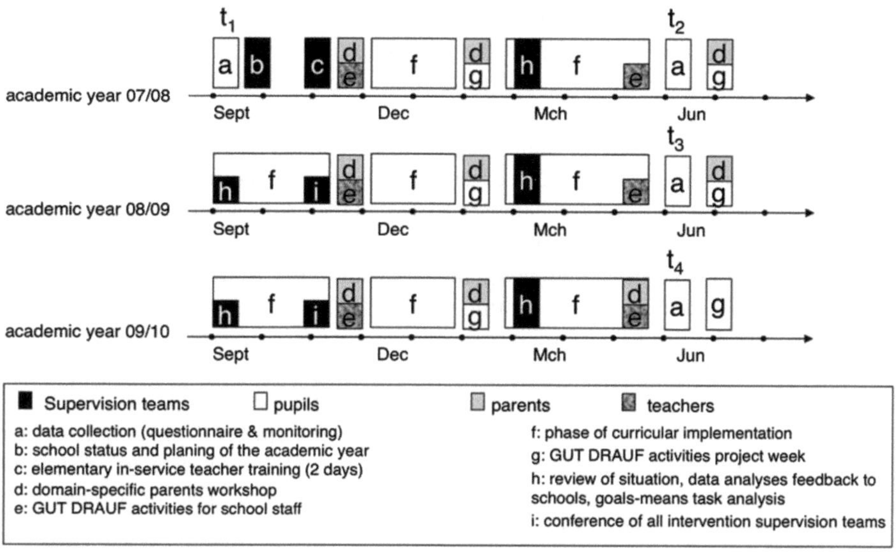

Fig. 27.1 Project and research schedule for the intervention schools. At the control schools, only steps a and h were conducted; further activities were in the schools' own responsibility

University students, especially trained for the survey, conducted the empirical data collection within regular school courses. The data collection started in September 2007 with a baseline survey, followed by an annual survey just 1 week after the schools' spring holidays in May or June (t_1–t_4). Pupils answered, for instance, the questionnaire four times during the 3-year period, and thus the data can be analysed longitudinally. The data of all survey were analysed and published within specific report (∅ 60 pages) at the beginning of each academic year analysing the status of the HP development of the school and the pupils' health-related outcomes. The aim of the reporting was to highlight the school-specific fields of action within the HP development process and a recommendation of HP activities for the academic year specified for each form.

The control schools just participated in the annual data collection, and their school steering comities received a specific report. The further school development process was a task for the control schools' supervision teams without any further external support or consulting of the project team.

4.2 *Measures and Instruments*

The research methodology of the study was twofold: First, the HPS development process was analysed using structured interviews according to Parsons and Stears (2002) with the responsible groups of the school community separately (principal, parents' representative, HP supervision group). Second, pupils completed a questionnaire consisting of heath-relevant scales described in Table 27.2. In total, more than 2,000 pupils at the five schools in 89 classes participated in the data collection, data analysis was conducted using SPSS.

Using interviews and large-scale methodology considers the requested triangulation to get insight into the school development process and into the intervention outcomes at the pupils' level (cf. Stewart-Brown 2006). All scales are either adapted from existing instruments or they are developed explicitly for the survey in this research project providing adequate quality indicated by Cronbach's α.

The indicators for a health-promoting school (HPS) were assessed using strongly structured, standardised interviews with fixed responses according to the German grading system (1 ≙ highly established aspect, 6 ≙ poorly or not at all established, e.g. "*Students have possibilities to participate in school development initiatives*" [*one of six questions within the indicator for pupils'/teachers' participation*] or "*School offers a forum to discuss and to implement recent aspects of health promotion as well as their integration in the school curriculum*" [*one of six questions within the indicator for HP curriculum*]). Respondents were at least two representatives for each focus group at all schools (the supervising teams, the principals and the parents' representatives) within a face-to-face interview. The same person conducted all interviews to avoid an interviewer bias (Bash 1987); subjects' further queries were answered following a standardised catalogue of responses. The average duration of an interview was in total 30 ± 16 min, and in total N = 42 subjects were interviewed. Every HPS indicator consisted of at least five questions, and the mean of these answers is used for data analyses.

The statements of the different responding focus groups were compared and accumulated by a three-step procedure:

1. Calculating HPS indicators: The mean rating of each focus group (HPS supervising team, principal, parents) for each indicator for HPS was calculated separately (e.g. six questions concerning the establishment of a health promotion curriculum).
2. Focus group comparison: The mean values of the different focus group answers for each HPS indicator are compared (e.g. principals' view versus parents' representative's perspective). As the systemic insight of teachers and principals in the school development process is higher than the parents' one, their vote counts higher in the case of disagreement. For example, the principal's answers lead to a mean of 3.5 within the HPS indicator *curriculum*, while the mean of the parents' representative's answers is 4.1; the principal's perspective is weighted stronger.
3. Accumulation: The mean values of each HP indicator (e.g. participation, HP work conditions, HP curriculum, bindingness of HP) are assigned to three levels of establishment (∅ 1.0–2.0 ≙ well established, ∅ 2.1–3.9 ≙ established, ∅ ≥ 4.0 poorly or not established).

Table 27.2 Description of the questionnaire and the guided interview (C's α $\triangleq$ Cronbach's α)

Domain of interest	Scale description	Respondents
Concepts about nutrition and physical activity	Nutrition knowledge (10 items)	*Questionnaire*: pupils (N = 1.958)
	Physical activity knowledge (3 items)	
Health-related well-being	KIDSCREEN inventory (Ravens-Sieberer et al. 2006)	
	In total 27 items, 4-point rating Likert scale (C's α 0.90)	
	Subscale *school-related well-being*, 4 items (C's α 0.78)	
Psychological stress symptoms	6 items, 3-point rating Likert scale (Lohaus et al. 2006) (C's α 0.84)	
Nutrition and physical activity self-efficacy	Nutrition: 5 items, 4-point rating Likert scale (adapted from Lach 2003) (C's α 0.65)	
	Physical activity: 12 items, 5-point rating Likert scale (adapted from Giebel 1999) (C's α 0.84)	
Food frequency protocol	In total 20 items to food/beverage intake, 6-point rating Likert scale	
	4 subscales (positive/negative food consumption, positive/negative beverage, C's α 0.61–0.76)	
Physical activity protocol	Hours per week, 4 activity domains, 4-point rating Likert scale each	
Indicators for a health-promoting school (*according to the fields of action in Table* 27.1)	Structured interviews to the developmental fields of action (acc. to Parsons and Stears 2002) (C's α overall 0.77):	*Guided interview* (*N=42*):
	Pupils'/teachers' participation (6 questions)	Supervising teams
	School environment (7 questions)	Principal
	Health-promoting work conditions (10 questions)	Parents' representative
	Health promotion curriculum (6 questions)	
	Teacher's support in health promotion (7 questions)	
	Bindingness of health promotion (4 questions)	
	Participation of parents (6 questions)	
	Connectedness to social networks (5 questions)	

4.3 Statistical Procedures

Descriptive analysis is used for the interview data, while inferential statistical procedures are used for the large-scale pupils' questionnaire data. The variables were measured four times (t_1–t_4) during the research project, and the general linear model (GLM) with repeated measures was used for the analysis of covariance (Rutherford 2004). This procedure evaluates whether the sample means of dependent variables are equal while other influencing variables and their influence are controlled.

5 Results

5.1 Indicators for HPS as a Result of Structured Interviews

The indicators for HPS were analysed longitudinally for each school, and the progress during the 3-year period of the study was compared. The data analysis revealed a veritable improvement of the establishment in different fields of HP at the intervention schools from the baseline (t_1) to the final survey (t_4). Two of the intervention schools improved

- In eight cases one level (indicator for HPS: not established → established or established → well established)
- In three cases two levels (indicator for HPS: not established → well established)

A decrease was not reported in any intervention school. One intervention school started at the baseline survey t_1 with an advanced HPS status, and thus there was nearly no space for further improvement.

In contrast, at the control schools an improvement was detected only in one case for one level. A graphical overview is provided in Figs. 27.2 and 27.3; the columns represent one indicator for HPS.

As constraint of the expressiveness of the interview data, it has to be mentioned that the results are based upon personal perceptions of the different groups' representatives involved in the schools, and thus the interpretation should be conducted carefully. But the status or the improvement of HPS indicators can be seen as covariates that influence the pupils' self-reported health measures. Therefore, the schools' HPS indicators are used with their discrete values as influencing variables within the ANCOVA.

5.2 Pupils' Self-Reported Health Measures: Quantitative Survey

The response rate of the quantitative questionnaire survey was above 80 %, and in total N = 1.958 pupils participated at all four measuring times (t_1–t_4). Pupils' socioeconomic status was controlled using four items of the international HBSC

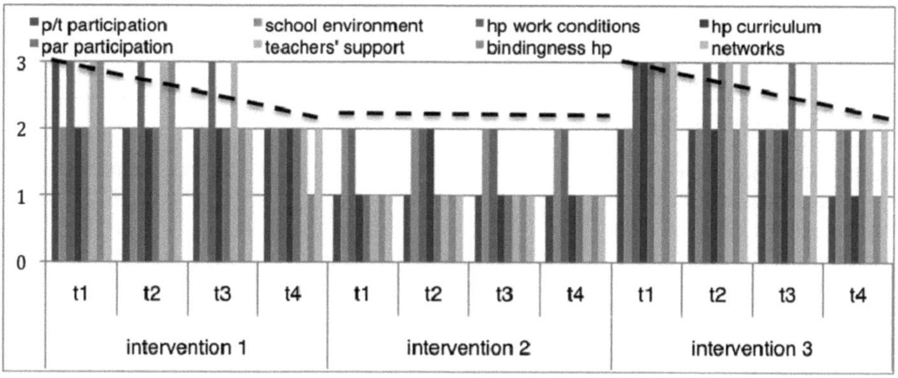

Fig. 27.2 Graphical overview of the development of the different indicators for HPS at the intervention schools. The *dashed lines* indicate the tendencies of improvement during the intervention (scale: 1 ≙ HPS field well established, 2 ≙ HPS field established, 3 ≙ HPS field poorly/not established)

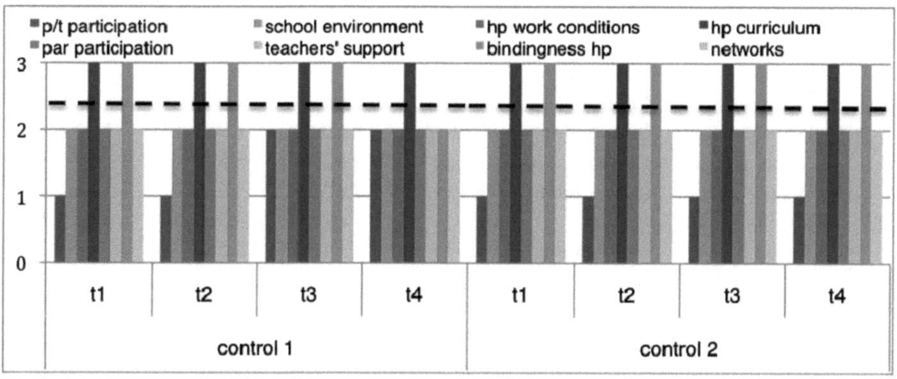

Fig. 27.3 Graphical overview of the development of the different indicators for HPS at the control schools. The *dashed lines* indicate the tendencies of improvement during the intervention (scale: 1 ≙ HPS field well established, 2 ≙ HPS field established, 3 ≙ HPS field poorly/not established)

large-scale study (*Health Behaviour of School-aged Children*, Hurrelman et al. 2003). As the sample size in this study was insufficient for multilevel modelling and hierarchical analyses (Maas and Hox 2005), repeated measures general linear modelling (GLM) were applied for multivariate analyses and controlled by residual analysis. An overview of the dependent variables, the factors and the covariates within a repeated measures ANCOVA are listed in Table 27.3.

All factors and covariates were implemented in a complete repeated measures ANCOVA, and during a stepwise procedure according to Milliken and Johnsons (2002), the final model was selected eliminating nonsignificant covariates. The socioeconomic status as well as gender distribution did not influence any model and thus it was not used for further data analyses. As indicators for HPS, only the

Table 27.3 Variables introduced into the GLM analysis for repeated testing

Dependent variables (questionnaire for pupils)	Factors	Covariates
Concepts about nutrition and physical activity		
Health-related well-being (general and subscale school)	Age	Socioeconomic status
Psychological stress symptoms		
Nutrition and physical activity self-efficacy	Treatment (intervention vs. control)	Eight indicators for HPS (see Table 27.2)
Food frequency protocol		
Physical activity protocol		

Table 27.4 Selected variables unaffected from the intervention and eliminated from the final ANCOVA model

Dependent variables		t_1 (mean ± SD)	t_2 (mean ± SD)	t_3 (mean ± SD)	t_4 (mean ± SD)
Socioeconomic status	Control	7.7 ± 2.2	–	–	–
	Intervention	7.9 ± 2.3	–	–	–
Concepts about physical activity (max. 3)	Control	1.06 ± 0.7	0.8 ± 0.6	1.03 ± 0.7	1.04 ± 0.7
	Intervention	1.07 ± 0.7	0.8 ± 0.5	1.03 ± 0.7	1.02 ± 0.7
Concepts about physical activity (max. 10)	Control	5.8 ± 1.5	7.9 ± 2.2	5.5 ± 1.6	5.5 ± 1.5
	Intervention	5.4 ± 1.9	7.7 ± 2.3	5.6 ± 1.7	5.8 ± 1.6
Physical activity-related self-efficacy (1 ≙ low – 5 ≙ high)	Control	3.0 ± 0.9	3.7 ± 1.0	3.5 ± 1.1	3.6 ± 1.1
	Intervention	3.0 ± 0.8	3.7 ± 1.1	3.5 ± 1.1	3.5 ± 1.1
Psychological stress symptoms (max. 18)	Control	10.5 ± 3.3	10.8 ± 3.5	10.8 ± 3.8	10.5 ± 3.8
	Intervention	10.9 ± 3.4	10.8 ± 3.4	10.8 ± 3.4	10.5 ± 3.9

status of HP work conditions (hpwc) at school and the existence of an HP curriculum (curr) showed a significant influence in all models, and on that account they were used for the final model. The concepts about physical activity and nutrition, the activity-related self-efficacy, the self-reported physical activity as well as the psychological stress symptoms did not change significantly in any school during the intervention at all (Table 27.4).

These results were quite surprising, and due to the limited space in this chapter, further reporting considering these domains will be limited to some aspects. Concerning the knowledge acquisition in the domains of nutrition and physical activity, the pupils did not significantly improve their knowledge during the 3-year period (grades 5–7, grades 6–8, grades 7–9). The self-efficacy related to physical activity was not influenced, even though physical activity during school breaks was an important aspect for intervention schools and the improvement of physical

Table 27.5 Overview of the generalised linear model (GLM) for the dependent variables (* indicates the interrelationship between two variables)

Dependent variable (R^2)	Factors and covariates	F	p	η^2
Health-related well-being in school	Treatment	$F_{2,763}=15.1$	*<0.001*	0.02
(R^2=0.19)	Age	$F_{2,763}=30.4$	*<0.001*	0.05
	Treatment*curr	$F_{2,763}=20.5$	*<0.001*	0.03
	Treatment*hpcw	$F_{2,763}=13.9$	*<0.001*	0.02
Unfavourable food consumption	Treatment	$F_{2,902}=1.4$	*n.s.*	–
(R^2=0.13)	Age	$F_{2,902}=3.1$	*< 0.05*	0.007
	Treatment*curr	$F_{2,902}=1.1$	*n.s.*	–
	Treatment*hpcw	$F_{2,902}=1.4$	*n.s.*	–
Unfavourable beverage consumption	Treatment	$F_{2,459}=26.4$	*<0.001*	0.04
(R^2=0.15)	Age	$F_{2,459}=33.8$	*<0.001*	0.07
	Treatment*curr	$F_{2,459}=2.3$	*n.s.*	–
	Treatment*hpcw	$F_{2,459}=37.7$	*<0.001*	0.06
Nutritional self-efficacy	Treatment	$F_{1,107}=4.2$	*<0.01*	0.01
(R^2=0.19)	Age	$F_{1,107}=3.6$	*<0.05*	0.01
	Treatment*curr	$F_{1,107}=1.1$	*n.s.*	–
	Treatment*hpcw	$F_{1,107}=1.4$	*n.s.*	–
Physical stress symptoms	Treatment	$F_{2,457}=3.1$	*<0.05*	0.01
(R^2=0.18)	Age	$F_{2,457}=11.9$	*<0.001*	0.03
	Treatment*curr	$F_{2,457}=7.1$	*<0.001*	0.02
	Treatment*hpcw	$F_{2,457}=2.3$	*n.s.*	–

activity behaviour was a goal of every school development process. For the other dependent variables, the results are indicated in Table 27.5. The model-fit R^2 (between 0 and 1.0, predicts the future outcome of the applied model for the dependent variable) and the proportion of explained variance η^2 (estimates to what extend the variance of the dependent variable is explained by a specific variable) were adequate for a field study (*c.f.* Cohen 1988). As health-related behaviour is quite stable and the parents' influence as well as the influence of the peer group and other factors could not be controlled under ecological conditions, even a low η^2 allows drawing conclusions when comparing the intervention versus the control schools.

All dependent health-related variables were mainly influenced by age, and except the food consumption, the treatment (intervention vs. control) had a measurable influence. For instance, the age explains up to seven percent of the individuals' differences within the consumption of unfavourable beverage, which means that older pupils tend to prefer lemonade, coke or beer. Six percent of the variance in beverage behaviour can be explained by treatment in combination with the HP work conditions at school (hpcw): If a pupil visited an intervention school in which, for instance, the assortment of drinks is explicitly part of the HPS development process, his/her behaviour is more favourable.

As a conclusion, intervention school pupils report a more favourable health-related well-being, a less unfavourable nutritional behaviour, a higher nutritional self-efficacy and less physical stress symptoms. The interaction effect between the intervention and the two fields of HP (HP curriculum and HP work conditions) indicates the relationship between the general school development and the pupils' perception.

6 Discussion and Conclusion

The way to become a health-promoting school is a bumpy road, and school development is crucial. As indicated by Gugglberger and Dürr (2011) capacity building in and for schools is necessary at all levels, for teachers, staff and school administration. The authors pointed towards the institutionalisation of health promotion and the empowerment of the school staff to construct adequate resources. The results in this study are in line with this recommendation: An external and intensive support of the HPS development and implementation process leads to an improvement of school development status of the intervention schools and resulted in measurable effects on the pupils' level. In further research the impact of professional development compared to the general school development process seems to be an interesting aspect.

However, as the study was conducted under low-invasive, field-research conditions, the effect sizes of the specific factors are quiet low and the results have to be interpreted carefully. It has to be considered that the intervention at every specific school was not standardised and the outcomes of the 3-year programmes could be assigned to the specific health promotion approach supporting the systemic development of all school members rather than to implement standardised programmes. This may be one reason why the knowledge construction in the domains of nutrition and physical activity did not succeed. This is disillusioning especially against the background of regular science teaching which failed in these two domains. This seems to be an argument to implement aspects of health explicitly into the science teaching and to relate it to relevant contexts.

The interaction effects of the treatment and the bindingness of HP as well as its insertion into the school curriculum point towards the necessity to bindingly implement health promotion into school development processes. The data reveals that not only the planning of a school development process conducts to the implementation of health promotion. It rather seems to be important to set development goals, to implement specific activities and to evaluate them together with (external) mediators and supporters as recommended in systematic development approaches (e.g. Fidler 2002; Rolff 2007; Senior 2012). One step towards this direction could also be an external certification process and quality control described by St. Leger and colleagues (2010) to increase the commitment to the HP.

The influence of age on health-related measures is well known in literature (cf. Jacobs et al 2002; Frost and McKelvie 2004; Hampel and Petermann 2005), and thus the results of this study are not surprising. As age plays an important role within the support of young people towards adequate health behaviour, it should be undoubted to find age-appropriate contexts to relate health promotion activities to adolescents' everyday life. As example, the nutritional behaviour of youth would be scarcely influenced by just teaching biomedical concepts (cf. Schaal 2009). In fact, the cognitive aspects of health have to be supplemented by the perception of enjoyment and well-being in a healthy setting to be sustainable and relevant for health behaviour. Thus, school should be a place to practise and to perceive health, which was one aspect of the GUT DRAUF health promotion programme. Combined with the Zeyer and Odermatt's health literacy approach (2009), health promotion could be a part of the science curriculum, and the path from concept construction to health behaviour could be facilitated through participation in changing the personal school environment (Brodie et al 2009, Bessems et al. 2011).

As a conclusion, it can be stated that the school development programme in this study partially succeeded and that further research has to be conducted on the interrelation of health literacy construction in science education (Harrison 2005), the health promotion within a developing school and the health behaviour of young people. Coherent models for cognitive, emotional and organisational aspects of health promotion in school could facilitate interventions to be planed to the point.

Acknowledgements We are grateful to all school members who participated in the study. A special thank goes to Gabriele Topaltzis who fed the data into the computer and supported all the administrative tasks. The research project was funded by the Ludwigsburg University of Education, grant # 1430 54771 *Gesundheitsförderung und Schulentwicklung*.

Literature

Aldinger, C., Zhang, X.-W., Liu, L.-Q., Guo, J.-X., Hai, Y., & Jones, J. (2008). Strategies for implementing health-promoting schools in a province in China. *Global Health Promotion, 15*, 24–29.

Antonovsky, A. (1979). *Health, stress and coping*. San Francisco: Jossey-Bass.

Basch, C. (1987). Focus group interview: An underutilized research technique for improving theory and practice in health education. *Health Education Quarterly, 41*(14), 1–48.

Bessems, K., van Assema, P., Paulussen, T., & de Vries, N. (2011). Evaluation of an adoption strategy for a healthy diet programme for lower vocational schools. *Health Education Research, 26*(1), 89–105.

Brodie, E., Cowling, E., Nissen, N., Paine, A., & Warburton, D. (2009). Understanding participation: A literature review. Pathways through participation. Retrieved from http://pathwaysthrough-participation.org.uk/wp-content/uploads/2009/09/Pathways-literature-review-final-version.pdf. Accessed 25 May 2012.

Carlsson, M., & Simovska, V. (2012). Exploring learning outcomes of school-based health promotion – a multiple case study. *Health Education Research, 27*(3), 437–447.

Centers for Disease Control and Prevention. (2010). Youth risk behaviour surveillance – US 2009. Surveillance Summaries. MMWR, 59.

Cohen, J. (1988). *Statistical power analysis for the behavioral sciences*. Hillsdale: Erlbaum.

Deschesnes, M., Trudeau, F., & Kébé, M. (2010). Factors influencing the adoption of a Health Promoting School approach in the province of Quebec, Canada. *Health Education Research, 25*(3), 438–450.

Fidler, B. (2002). *Strategic management for school development*. Thousand Oaks: Sage.

Frost, J., & McKelvie, S. (2004). Self-esteem and body satisfaction in male and female elementary school, high school, and university students. *Sex Roles, 51*(1/2), 45–54.

Giebel, M. (1999). Diagnostik der Motivation zur Verhaltensänderung kardiovaskulären Risikoverhaltens bei Bundeswehrsoldaten unter Berücksichtigung des transtheoretischen Stufenmodells der Verhaltensänderung nach Prochaska. Dissertation, University of Marburg.

Gugglberger, L. (2011). Support for health promoting schools: A typology of supporting strategies in Austrian provinces. *Health Promotion International, 26*(4), 447–456.

Gugglberger, L., & Dür, W. (2011). Capacity building in and for health promoting schools: Results from a qualitative study. *Health Policy, 2011*(101), 37–43.

Hammig, B., Ogletree, R., & Wycoff-Horn, M. (2011). The relationship between professional preparation and class structure on health instruction in the secondary classroom. *The Journal of School Health, 81*, 513–519.

Hampel, P., & Petermann, F. (2005). Age and gender effects on coping in children and adolescents. *Journal of Youth and Adolescence, 34*(2), 73–83.

Harrison, J. (2005). Science education and health education: Locating the connections. *Studies of Science Education, 41*(1), 51–90.

Hascher, T. (2007). Exploring students' well-being by taking a variety of looks into the classroom. *Hellenic Journal Psychology, 4*, 331–349.

Hascher, T. (2008). Quantitative and qualitative research approaches to assess student well-being. *International Journal of Educational Research, 47*(2), 84–96.

Hoyle, T., Samek, B., & Valois, S. (2008). Building capacity for the continuous improvement of health-promoting schools. *The Journal of School Health, 78*, 1–8.

Hurrelmann, K., Klocke, A., Melzer, W., & Ravens-Sieberer, U. (2003). *Jugendgesundheitssurvey. Internationale Vergleichsstudie im Auftrag der Weltgesundheitsorganisation WHO (HBSC)*. Weinheim: Juventa.

Hurrelmann, K., Albert, M., & Infratest, T. N. S. (2006). *Jugend 2006. Eine pragmatische Generation unter Druck – 15. Shell Jugendstudie [Youth 2006 – A pragmatic generation under pressure]*. Frankfurt: Fischer.

Jacobs, J., Lanza, S., Osgood, D., Eccles, J., & Wigfield, A. (2002). Changes in children's self-competence and values: Gender and domain differences across grades one through twelve. *Child Development, 73*(2), 509–527.

Jensen, B., & Simovska, V. (2005). Involving students in learning and health promotion processes–clarifying "why?" "what?" and "how?". *Promotion and Education, 12*, 150–157.

Jourdan, D., Samdal, O., Diagne, F., & Carvalho, S. (2008). The future of health promotion in schools goes through the strengthening of teacher training at a global level. *Promotion and Education, 15*, 36–38.

Jourdan, D., Mannix Mc Namara, P., Simar, C., Geary, T., & Pommier, J. (2010). Factors influencing staff contribution to health education in schools. *Health Education Research, 25*(4), 519–530.

Kurth, B.-M., & Schaffrath-Rosario, A. (2007). The prevalence of overweight and obese children living in Germany. Results of the German Health Interview and Examination Survey for Children and Adolescents. *Gesunheitsbl-Gforsch-Geschutz, 50*, 736–743.

Lach, J. (2003). Gesundheitsförderndes Ernährungsverhalten von Jugendlichen und die Konsequenzen für die schulische Praxis. *Empirische Pädagogik, 17*(1), 39–56.

Lampert, T., Mensik, G., Romahn, N., & Woll, A. (2007). Körperlich-sportliche Aktivität von Kindern und Jugendlichen in Deutschland. Ergebnisse des Kinder- und Jugendgesundheitssurveys (KiGGS). *Gesunheitsbl-Gforsch-Geschutz, 50*, 634–642.

Lohaus, A., & Lissmann, I. (2005). Entwicklungsveränderungen und ihre Bedeutung für die Gesundheitsförderung. In A. Lohaus, M. Jerusalem, & J. Klein-Heßling (Eds.), *Gesundheitsförderung im Kindes- und Jugendalter*. Göttingen: Hogrefe.

Lohaus, A., Eschenbeck, H., Kohlmann, C.-W., & Klein-Hessling, J. (2006). *SSKJ 3–8. Fragebogen zur Erhebung von Stress und Stressbewältigung im Kindes- und Jugendalter*. Göttingen: Hogrefe.

Maas, C., & Hox, J. (2005). Sufficient sample sizes for multilevel modeling. *Methodology, 1*(3), 86–92.

Magnusson, K., Hrafnkelsson, H., Sigurgeirsson, I., Johannsson, E., & Sveinsson, T. (2012). Limited effects of a 2-year school-based physical activity intervention on body composition and cardiorespiratory fitness in 7-year-old children. *Health Education Research, 27*(3), 484–494.

Mann-Luoma, R., Goldapp, C., Khaschei, M., Lamersm, L., & Milinski, B. (2002). Integrierte Ansätze zu Ernährung, Bewegung und Stressbewältigung. Gesundheitsförderung von Kindern und Jugendlichen. *Gesunheitsbl-Gforsch-Geschutz, 45*, 952–959.

Mensik, G. (2007). Lebensmittelverzehr bei Kindern und Jugendlichen in Deutschland Ergebnisse des Kinder- und Jugendgesundheitssurveys (KiGGS). *Gesunheitsbl-Gforsch-Geschutz, 50*, 609–623.

Milliken, G., & Johnson, D. (2002). *Analysis of messy data. Volume III: Analysis of covariance*. New York: Chapman and Hall.

Parsons, C., & Stears, D. (2002). Evaluating health-promoting schools: Steps to success. *Health Education, 1*(102), 7–15.

Paulus, P. (2003). Schulische Gesundheitsförderung – vom Kopf auf die Füße gestellt. In K. Aregger & U. Lattmann (Eds.), *Gesundheitsfördernde Schule – eine Utopie?* Luzern: Sauerländer.

Randler, C. (2012). Field experiments in learning research. In: N. Seel (Ed.), *Encyclopedia of the Sciences of Learning* (pp. 1293–1297). New York/Dodrecht/Heidelberg/London: Springer.

Ravens-Sieberer, U., & BELLA-group. (2008). Prevalence of mental health problems among children and adolescents in Germany: Results of the BELLA study. *Europe Child & Adolescent Psychology, 17*(1), 22–33.

Ravens-Sieberer, U., et al. (2006). *The KIDSCREEN questionnaires – Quality of life questionnaires or children and adolescents*. Lengerich: Pabst Science.

Rolff, H.-G. (2007). *Studien zu einer Theorie der Schulentwicklung*. Weinheim: Beltz.

Rowling, L., & Jeffreys, V. (2006). Capturing complexity: integrating health and education research to inform health-promoting schools policy and practice. *Health Education Research, 21*(5), 705–718.

Rutherford, A. (2004). *Introducing ANOVA and ANCOVA: A GLM approach*. London: Sage.

Schaal, S. (2008). Biological education and health promoting school – The implementation and evaluation of a school development process. In D. Raichvarg et al. (Eds.), *Sustainable development, ethics and education for the 2020s: What challenges for biology?* Burgundy: BioEd proceedings IUBS-CBE, UNESCO, LDES University of Geneva, CIMEOS and Mission Culture Scientifique, University of Burgundy.

Schaal, S. (2009). Weil's schmeckt – Fast Food & Co. *Unterricht Biologie, 341*, 8–15.

Schaal, S. (2011). Gesundheitsförderung in Schulentwicklungsprozessen. In R. Mann, B. Schulz, & S. Streif (Eds.), *GUT DRAUF – Zwischen Wissenschaft und Praxis*. Köln: BZgA.

Schwarzer, R. (2007). Modeling health behaviour change: How to predict and modify the adoption and maintenance of health behaviours. *Applied Psychology, 57*(1), 1–29.

Senior, E. (2012). Becoming a health promoting school: Key components of planning. *Global Health Promotion, 19*, 23–31.

Simovska, V. (2007). The changing meanings of participation in school based health education and health promotion. *Health Education Research, 22*, 864–878.

St Leger, L. (2000). Developing indicators to enhance school health. *Health Education Research, 15*, 719–728.

St Leger, L., Young, I., Blanchard, C., & Perry, M. (2010). *Promoting health in schools – From evidence to action*. St. Denis Cedex: IUHPE.

Stears, D. (2000). Profiling the health promoting school: Valuing assets and evaluating the management of change. *Health Education, 100*, 74–80.

Stewart-Brown, S. (2006). *What is the evidence on school health promotion in improving health or preventing disease and, specifically what is the effectiveness of the health promoting schools approach*. Copenhagen: WHO Regional Office for Europe's Health Evidence Network (HEN).

Tang, K., Nutbeam, D., Aldinger, C., St Leger, L., Bundy, D., Hoffmann, A., et al. (2008). Schools for health, education and development: A call for action. *Health Promotion International, 24*(1), 68–77.

Turner, S., Öberg, K., & Unnerstad, G. (1999). Biology and health education. *European Journal of Teacher Education, 22*(1), 89–100.

Von Wagner, C., Knight, K., Steptoe, A., & Wardle, J. (2007). Functional health literacy and health-promoting behaviour in a national sample of British adults. *Journal of Epidemiology and Community Health, 61*(12), 1086–1090.
Weare, K. (2000). *Promoting mental, emotional and social health. A whole school approach.* London: Routledge.
Webb, G. (1995). Sleeping position and cot death: Does health promotion always promote health? *Journal of Biological Education, 29*(4), 279–285.
Zeyer, A., & Odermatt, F. (2009). Health literacy – A link between health education and science education. *ZfDN, 15*, 265–285.

Chapter 28
Disagreement in "Ordinary" Teaching Interactions: A Study of Argumentation in a Science Classroom

Ana Paula Souto-Silva and Danusa Munford

1 Introduction

Seeking to better understand how knowledge is socially constructed in classrooms, educational researchers have turned to the study of discourse in school context. Some of the implications of this perspective are that language is perceived as a mediational means for thinking and that the role of the teacher is of a "language user" who supports conceptual understanding (Edwards and Mercer 1987). Researchers and policy makers consider argumentative discourse as essential in science learning (e.g., Driver et al. 2000; Jiménez-Aleixandre and Erduran 2007; NRC 2000; Sadler 2006; Sandoval and Milwood 2007; Schwarz 2009; Scott et al. 2007; Zohar 2007). Thus, since the late 1990s and early 2000s, there have been great efforts from the community to conceive and develop curriculum materials to implement argumentation in science classrooms. In parallel, argumentation gradually became a prevalent theme in science education research, with an increasing number of articles published in journals and presented at conferences. Consequently, in the recent literature there is a great diversity of focuses, approaches, definitions, and understanding of argumentation (Osborne and Patterson 2011). Nevertheless, it is possible to identify some trends in this research (see, for instance, Jimenez-Aleixandre and Erduran 2007; Ozdem and Erduran 2011). It is beyond the scope of the present work to present a detailed review of the knowledge produced in the field. We will focus on articles and book chapters of interest to discuss aspects related to the study of prospective teachers' and teachers' argumentation practices.[1]

[1] We will rely on previous reviews or edited books and focus mainly on articles related to teachers' discourse or teacher education. The first author conducted a search for articles published between 2002 and 2008 in three main journals in the field (Science Education and JRST). A research group from our university identified articles for the period between 2008 and 2012 in the same journals. (This group included both authors; Professor Dr. Marina Tavares; graduate students Cláudia Starling, Simone Estevez, Vanessa Capelle, and Margareth Lovisi; and undergraduate student

A.P. Souto-Silva • D. Munford (✉)
Faculdade de Educação, Universidade Federal de Minas Gerais, Belo Horizonte, Brazil
e-mail: danusa@ufmg.br

C. Bruguière et al. (eds.), *Topics and Trends in Current Science Education: 9th ESERA Conference Selected Contributions*, Contributions from Science Education Research 1,
DOI 10.1007/978-94-007-7281-6_28,

Research into argumentation in education has had two major foci: learning to argue and learning science through argumentation (Schwarz 2009). Initially, research in science education emphasized issues related to the development of argumentation abilities and only later emerged an interest for understanding relationships between learning science and argumentation (e.g., Andriessen 2006; Baker 2009; Chin and Osborne 2010; Schwarz 2009). Studies that focused on students' learning contributed to better understanding the importance of teachers' discourses and strategies, as well as to exploring relationships between teaching and learning science through argumentation/learning to argue. More recently, there has been great criticism around the oversimplification and overvalorization of argumentation in science education (McDonald and Kelly 2012). Some authors tried to consider classroom community/context in their investigations, examining classroom discourse – including attention to teachers' practice (e.g., Berland and Reiser 2011; McNeil and Pimentel 2010; Zembal-Saul 2009). Thus, currently, there is an effort to portray complexities of argumentation in science classrooms. Results indicate that scientific argumentation involves multiple goals (e.g., sense making and persuading) (Berland and Reiser 2011), requires that teachers assume multiple roles (e.g., constructor, critiquer, authority) (Berland and Reiser 2011; McNeil and Pimentel 2010), and that they use multiple strategies (McNeil and Kraijick 2008). Moreover, there is a great deal of variation in argumentation from classroom to classroom. This variation is related to each community's practices as well as to the interpretations of argumentation which their members hold (Berland and Reiser 2011; McNeil and Pimentel 2010).

In spite of these important contributions to knowledge about argumentation, it is important to be aware that most studies in our field have involved a design approach, and the analytical tools adopted are still centered in the Toulmin model, with some adaptations. In this sense, concerns discussed by McDonald and Kelly (2012) are not overcome and deserve the attention of the field. Moreover, although scholars acknowledge the teacher's role when trying to understand learning through argumentation/to argue, there are still relatively few studies on aspects of teacher education and teachers' practice in argumentative contexts (e.g., McNeil and Krajcik 2008; Osborne et al. 2004; Sandoval and Reiser 2004; Zembal-Saul 2009; Zohar 2007). A significant challenge when exploring relationships between teaching and learning through argumentation/learning to argue is that there are no longitudinal studies on teachers' practices. Moreover, most research with experienced teachers takes place in a school classroom context, whereas most research with prospective teachers does not examine science teaching practices at school, and it occurs in the context of university level courses in preservice teacher education.[2]

Many studies have shown limitations in prospective teachers' and experienced teachers' abilities to construct arguments (Zohar 2007), but there is evidence that

Rafael Alves.) The works discussed in this chapter were selected from these two lists, considering its relationships with the issues addressed here.

[2] Exceptions would be studies like Zembal-Saul (2009) and Aavramidou and Zembal-Saul (2005).

prospective teachers are able to construct arguments (e.g., Zembal-Saul et al. 2002; Sadler 2006). The quality of prospective teachers' practices has been related to their experiences during teacher education, in courses that address argumentation in the context of science teaching (Sadler 2006, Avraamidou and Zembal-Saul 2005; Zembal-Saul 2009). Elementary prospective teachers participating in explicit instruction about "teaching science as argument" can have different understandings of science teaching and learning practices (i) "based on hands-on activities"; (ii) "based on investigation," involving the use of testable questions and the design of fair tests; (iii) "based on evidence," involving supporting claims with evidence; and (iv) "based on argument," considering how claim and evidence are related, as well as examining alternative explanations/points of view (Avraavmidou and Zembal-Saul 2005; Zembal-Saul 2009). Prospective teachers with an argument-based perception not only incorporated in their teaching at school various strategies that were modeled during the course but also were able to reflect critically on their own practice in an elementary classroom (Zembal-Saul 2009).

Studies with experienced teachers indicate that even when working with the same curricular materials, their instructional strategies and discourse vary significantly (McNeil and Krajcik 2008; McNeil and Pimentel 2010; Simon et al. 2006). For instance, depending on the teacher, argumentation can occur more often during whole-class discussions, or when students were working in small groups, or even in role-playing activities (Simon et al. 2006). Empirical evidence indicates also that, often, teachers define explanation, but they address some components (i.e., claims and evidence) more than others. In some cases, reasoning is not even mentioned. These results show that the challenges for experienced teachers are similar to those for prospective teachers who engage in teaching science as argument. Furthermore, strategies like "making the rationale of scientific explanation explicit" are more effective for promoting teachers' learning to argue than others (e.g., modeling scientific explanation), even though they are rarely adopted, and are not included in curricular materials. Moreover, some instructional strategies that are extensively used, like defining explanation and its components, are effective only when combined with others (in this case, making the rationale explicit) (McNeil and Krajcik 2008). There is also evidence that open questions have a key role in supporting students' argumentation and interactions among students (McNeil and Pimentel 2010), as also observed among prospective teachers (Zembal-Saul 2009). However, not all teachers use open questions as often as it would be desirable. Finally, establishing connections with everyday life is a strategy that is less often used, and there are contradictory data about its contribution to learning (McNeil and Krajcik 2008; McNeil and Pimentel 2010).

In this study, to learn about a prospective novice teacher's practices in argumentative contexts, we examined discursive interactions in a science classroom. In this case, argumentation was not related to the use of curricular materials, but emerged in the classroom as part of "ordinary" events in science lessons. Moreover, we tried to develop a methodological approach that creates more opportunities to characterize aspects of argumentation from the classroom community perspective. We intend to contribute to research on the *processes* involved in supporting science learning through argumentation. We investigated the following research questions: *(1) In what aspects does argumentation differ in various instructional contexts of a science classroom? (2) How does a science teacher use language in different argumentative contexts in the classroom?*

2 Methods

2.1 Research Setting

The study took place in a middle school science classroom for adult learners attending evening classes. This school was part of an outreach university project in a major city in Brazil. Prospective teachers from different areas were responsible for teaching the school discipline in which they were majoring. They worked with two classes, and also participated in meetings coordinated by professors from the College of Education, discussing planning/evaluation and reflecting on their teaching practices.

The teacher, Mr. Sunday, had graduated very recently, but had worked as a teacher for about two years before the data were collected. He was very committed to promoting students' learning, and he was very attentive to learners' demands. Although he often taught through lectures, he interacted a lot with students, mainly due to his ability to encourage them to participate, and in asking questions to check understanding.

The study was conducted in a class that initially comprised 25 students, with ages between about 20 and 75, most of them female. They were very participative and kept systematic notebook records of what was addressed in science classes.

These adult learners are at the beginning of the process of schooling, and, like younger students, they face challenges of participating in science classrooms, in particular, sharp differences between everyday knowledge and discourse, and school science knowledge and discourse. These common challenges create conditions for the occurrence of differences of opinion. Thus, the study of adults can contribute to a better understanding of science learning in general.

2.2 Methodological Orientations

A naturalistic design utilizing qualitative research methods (Lincoln and Gubba 1985) was employed in the present study. It can be characterized as a case study (Stake 2000; Merriam 1998), informed by ethnography in education (Castanheira et al. 2001; Dixon and Green 2005). The adopted theoretical framework reflects an emphasis on participants' perspectives on the phenomena under investigation, as well as an understanding of science learning as socially constructed and mediated by language.

If we want to deepen our knowledge about teachers' role and practices, it is also important to examine methodological approaches adopted in research on argumentation in science education (Erduran 2007). In a 2007 review, Erduran identified analytical frameworks that were more significant: (i) evidence and justification, (ii) epistemic practices and criteria, (iii) arguers and the nature of arguments, and (iv) participation in discussions. The same author also pointed out that Stephen Toulmin's model is the most prevalent analytical approach. Bricker and Bell (2008) noted that argumentation research in science education has been limited to only a few approaches. The emphasis on Toulmin's model resulted in a robust knowledge about the quality of arguments, in particular about the characterization of the structure of the argument and its components. However, we believe that this model and some of adaptations that were proposed tend to make little contribution to providing descriptions of ways of arguing that were constructed within a science classroom culture through everyday interactions.

In our study, we chose to use the Pragma-Dialectical theory of argumentation (Van Eemeren et al. 1996, 2002) to learn about *argumentation processes* in learning contexts. We focused on the teacher's discourse throughout argumentation, and the procedures involved in resolving differences of opinion, always considering the broader context in which argumentative interactions took place. We chose to adopt tools for argumentation analysis that had greater potential to enable us to learn about argumentation phenomena from the participants' perspective. Thus, we chose to adopt an alternative perspective to evidence-based models that could broaden our understanding about learning in science classrooms from an argumentation perspective (Bricker and Bell 2008).

Initially, the transcripts were analyzed, the main reference being the "analytic overview" framework and evaluation procedures from the Pragma-Dialectical theory of argumentation (Van Eemeren et al. 1996, 2002). This theory combines aspects of discourse analysis, rhetoric, and dialectics to analyze argumentation. We focused on the following: (i) identifying points of view, and establishing what the main and subordinate differences of opinion were and whether they were explicit or implicit; (ii) identifying protagonist or antagonist positioning; and (iii) characterizing the structure of arguments.

3 Data Collection

Over 8 months, the first author conducted participant observation (Spradley 1980) with narrative recording in field notes, combined with audio and video recording (Green et al. 2001). We also conducted three 2 h-long interviews with the teacher to have insights into participants' perspectives. Moreover, we worked collaboratively with the teacher; we thus had records of informal conversations and of meetings. Finally, we had access to the teacher's written records about the lessons.

4 Analysis

The first phase of analysis involved transforming the audio and video recordings into some type of written or graphic representation of data (Cameron 2001). We used tools from interactional ethnography such as event maps and storylines with different grain sizes (Castanheira et al. 2001; Dixon and Green 2005). These representations made it possible to situate the lessons and the events in the history of the classroom. Moreover, through this analysis we were able to identify lessons in which argumentation occurred. We selected lessons involving argumentation during whole-class discussions, in which the resolution of the difference of opinion took a significant amount of time (approximately between 5 and 20 min). Following a contrastive perspective (Castanheira et al. 2001), we focused in this study on argumentation in three lessons with different characteristics (see Table 28.1). For each lesson, argumentative events were transcribed word-by-word.

In the second phase, we used the "analytic overview" tools (Van Eemeren et al. 1996, 2002) to analyze the word-by-word transcriptions. Moreover, based on Van Eemeren and colleagues' work, we developed three new different types of representation to better visualize how arguments were constructed through interaction and how argumentation differs in the various contexts: first, representing relationships between argumentation components and discourse (Fig. 28.1); then, representing the whole structure of argumentation in detail at the end of the events, considering the multiple points of view at stake (Fig. 28.2); and, finally, representing the whole structure of argumentation, but in a way that makes visible broader characteristics of argumentation in the events of a lesson (e.g., the relationships between main and subordinate differences of opinion) (Fig. 28.3).

Table 28.1 Major characteristics of the lessons selected for analysis

	Lesson 1: blood types	Lesson 2: solid waste investigation	Lesson 3: ecological relationships
Theme of the lesson	Blood types (ABO): antigens/antibodies	Choosing units to measure solid waste	Ecological relationships between living beings
Instructional context	Lesson is part of a series of lessons that address the circulatory system	Lesson is part of an inquiry unit on solid waste	Lesson is part of a series of lessons on ecological interactions.
	Lesson is designed to prepare students for a lab activity	Lesson is designed to prepare students for collecting data on their own waste	Lesson is designed to prepare students for reading a book on ecology and doing a group presentation
Learning focused on	Scientific concepts	Scientific practices	Scientific concepts
Differences of opinion involving…	Students' knowledge versus school science knowledge	Teacher's and students' doubts	Different points of view from school science knowledge

Graphical Representation	Turn	Word by word transcription with analysis
PV1' Use of an unconventional unit to measure waste; 1.1 Possible unit; 1.2 Possible unit	1	**Teacher:** The quantity of fruit peel is hard to define; we have to figure out an unit to measure it... *[**Protagonist** to PV1; it is implicit that the conventional unity to measure solid waste, that is kilo, is not appropriate. They have to choose a different unit.]*
	2	**Lucas:** Oil bottle *[Possible unit]*
	3	**Teacher**: yes! We will have to create a unity. To fill in a plastic bag [possible pattern] with peel, and you define this as number 10, for instance. We have to discuss and see what is the best way. *[**Protagonist** to PV1]*
PV2' Conventional unit; 2.1' Kg would be a good unit	4	**Giovana**: we could measure its weight *[**Antagonist** to PV1; it is implicit that the conventional unity would be the best way to measure the waste]*

Fig. 28.1 An example (from lesson 2 on solid waste) of the first type of representation created to use Pragma-Dialectics in classroom-based studies of argumentation. It is generated *during transcript analysis* with the goal being to establish direct relationships between participants' talk and elements of argumentation (points of view and supporting elements), how they are constructed through time, what the implicit/explicit elements are, and what antagonist/protagonist utterances are

5 Results

The results show differences in the nature of argumentation in the three lessons. In the "blood types" lesson, we identified greater complexity in the *relationships between differences of opinion.* In this lesson, there was one main difference of opinion and two *levels* of subordinate differences of opinion. In each level, we identified two differences of opinion. In short, the argumentation in lesson 1 involved complex *hierarchical relationships among the various differences of opinion* (Fig. 28.4).

Argumentation started when the teacher drew a table on the blackboard to represent antibodies and antigens present in each blood type (considering the AB system) (Fig. 28.5). He explicitly related the symbols in the table to structures that were present in the blood and to observable phenomena like agglutination and thrombosis. However, even before Mr. Sunday filled the rows for the AB and O groups, a student asked whether people could receive blood from a different type in case of

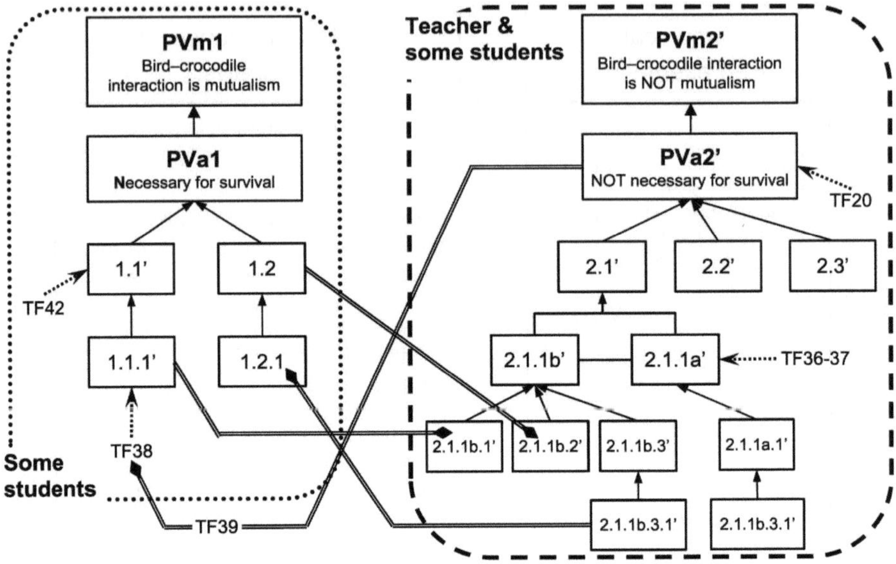

Fig. 28.2 An example (from lesson 3 on mutualism) of the second type of representation created to use Pragma-Dialectics in classroom-based studies of argumentation (simplified). It is a summary of the transcript analysis results. It makes it possible to characterize the structure of the argumentation, identify points of view and supporting elements of argumentation (described and associated with turns/utterances), as well as providing evidence on how (and by whom) arguments are generated in interaction (mapping turns/utterances) (*Dotted lines* indicate support to element. *Double lines* indicate confronting element, prime indicates implicit element, and PVm means "point of view in the main difference of opinion." PVa or PVb indicates points of view in the subordinate differences of opinion, numbers indicate arguments supporting point of view, and TF indicates speak turns)

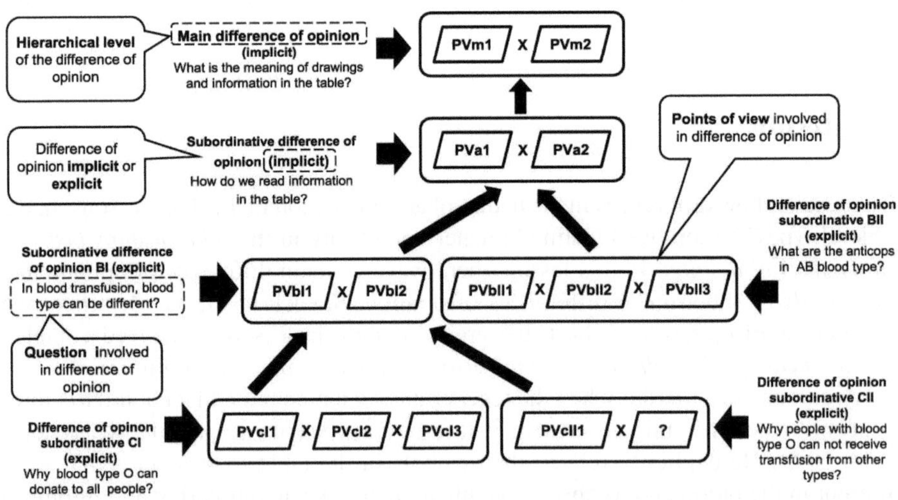

Fig. 28.3 An example (from lesson 1 on blood type) of the third type of representation created to use Pragma-Dialectics in classroom-based studies of argumentation. This representation makes it possible to characterize the broad structure of Argumentation, that is, the hierarchical relationships between main differences of opinion and secondary differences of opinion

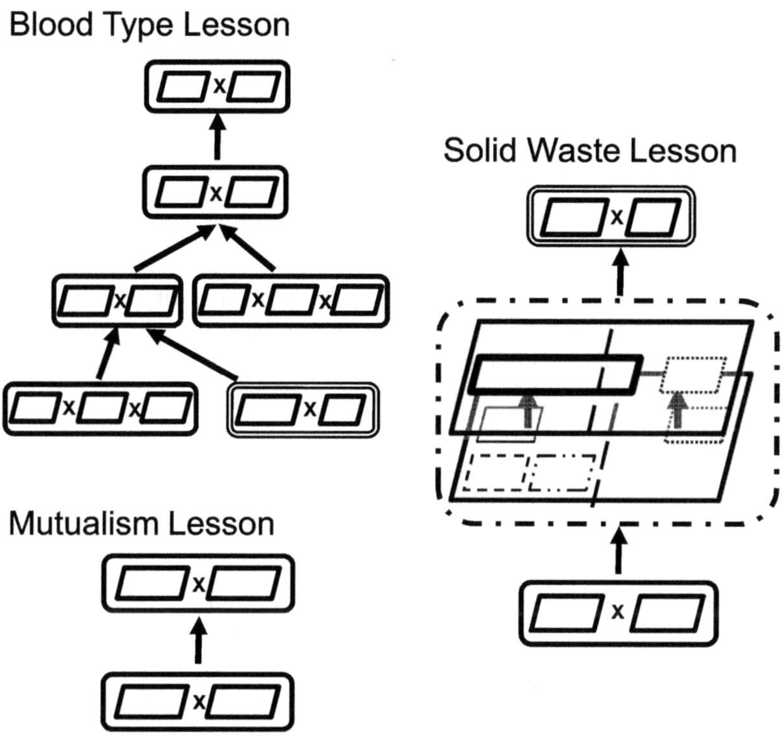

Fig. 28.4 Representations of hierarchical relationships between various differences of opinion and the nature of the difference of opinion in the three lessons. The contrast shows that there is a greater complexity in hierarchical relationships in the blood type lesson. The contrast also shows that the most diverse and complex types of differences of opinion were present in the solid waste lesson. In this lesson, there were multiple mixed (*dashed-dotted lines*), simple mixed (*continuous lines*), and simple non-mixed (*double lines*) differences of opinion. In the other lessons, there were only simple differences of opinion, predominantly mixed (SM) (*small quadrilateral shapes* represent "points of view"; "differences of opinion" are represented as *rectangles around* points of view)

Fig. 28.5 Representation of the table that the teacher drew on the blackboard to help students to understand antibodies and antigens present in each blood type (considering the AB system)

Blood type	Antigen	Antibody
A	A	Anti-B
B	B	Anti-A
AB	AB	?
O		

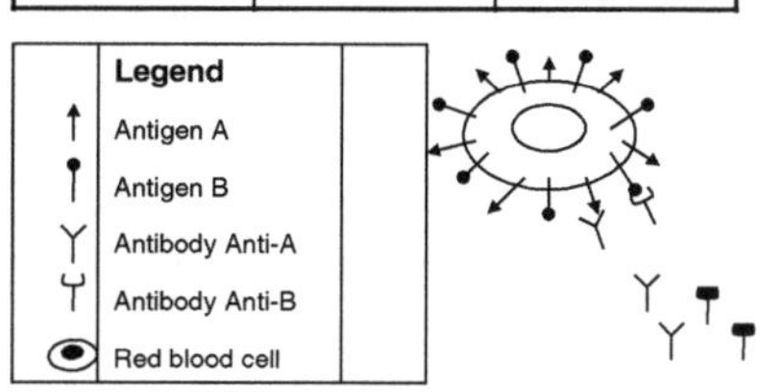

transfusion. As the discussion progressed, teacher and students debated around which antigens and antibodies are present in the AB blood type:

Teacher:	(…) What kind of antibodies are present in the plasma of people who have the AB blood type?
Bianca:	A and B.
Teacher:	Ah!
Margareth:	Isn't it both?
Erica:	It's gonna have both.
Teacher:	Antibodies A and B? But then, look, if it has antibodies A and B. (…) These two types of antibodies are going to bond to their red blood cells and are going to recognize their own red cells as a threat.
Gabriela:	It always has to be the opposite, isn't?

The main difference of opinion (MDO) was related to a conflict regarding different ways of "reading" the table. For the teacher, based on scientific ways of knowing, the table represented structures from the blood in a living organism. Antibodies and antigens interacted with each other in accordance with principles that could be established through observation. For students, the table represented some type of abstract principle or "school rule" that determines that anti-B antibodies are present whenever A antigens are present, and anti-A antibodies are present whenever B antigens are present. All the subordinate differences of opinion resulted from these different ways of understanding the information in the table. The teacher did not address this MDO, but instead, focused on more specific problems/questions that potentially could be solved more easily, and would result in resolution of the MDO. However, he did not succeed in resolving the subordinate differences of opinion, possibly because both teacher and students were not conscious of the nature of the disagreement (i.e., MDO).

In the solid waste investigation lesson, we identified greater complexity in the *nature of the difference of opinion*. This lesson was part of an inquiry-based unit on solid waste (11 one-hour classes). Students addressed the question "How can we solve the problem of solid waste?" collecting and analyzing data about the solid waste they produced at home and contrasting it with data from their city and from different parts of the state (rural and urban regions).

Based on an analysis of interactions, we concluded that what was at stake in this lesson was "What does it mean to adopt a unit?" (the main difference of opinion). On the one hand, the teacher defended the use of *principles* that orient scientists' practices – that is, all should adopt the same unit, in this case, units that everyone could use at home to measure solid waste. For the teacher, what was important was to create a common "language" to quantify solid waste. On the other hand, students tended to argue in favor of the use of standardized units that were present in their everyday lives (e.g., kilograms, liters). They were not concerned with the *rationale* behind choosing/using certain units, and they defended choices based on "useful" *examples* that were already established. Because the kilogram was used to measure everything in their everyday lives, they wanted to use that, although many of them did not have a scale to make the measurements. Finally, during the discussion a debate emerged around using units to measure specific materials/objects (e.g., counting egg shells, a spoon to measure coffee) versus using units for categories of materials (bags or boxes to measure organic matter, metal cans).

The lesson involved three differences of opinion of different types (Fig. 28.4). First, the main difference of opinion was implicit and involved the most elementary form of difference of opinion, called single non-mixed. It involves only one proposition and only one standpoint is defended and then questioned by the other party. Second, a subordinate difference of opinion related to "How can we measure our solid waste?" was explicit in interactions. In this subordinate difference of opinion, more than one proposition was under dispute *and* various points of view were involved (Fig. 28.5). Finally, there was another difference that was subordinate to the first subordinate difference of opinion: "What are we measuring: types of material or specific objects?"

In short, the whole lesson involved quite complex argumentation in relation to differences of opinion. Moreover, there was a process of reformulation of these points of view and a search for conciliation between them. This complexity was associated with a context of engaging students in scientists' practices. It was the first time Mr. Sunday had conducted this type of investigation in his classroom. Because he was a novice teacher, the question of deciding which unit to use to measure the solid waste was a genuine question in this class. Thus, there is evidence that the fact that the teacher was not sure about the answer to the problem posed produced divergence and complexity in relation to the nature of the difference of opinion.

Finally, in the ecology lesson, we identified greater complexity in *argumentation structure* (see Fig. 28.2). This lesson took place after a sequence of lessons in which the teacher lectured on ecological relationships, providing an overview of the biology categories to students. The episode that we analyzed started with the teacher recalling the definition of "mutualism" and asking students to give an example of these relationships. A student suggested that the relationship between crocodiles and spur-winged plover (*Vanellus spinosus*) represented an instance of mutualism. The teacher problematized the example, leading to an explicit difference of opinion on whether it was mutualism or not. As a consequence of the discussion, following the definition of mutualism, the class started to debate whether the living beings involved depended on the relationship to survive or not. In this case, both Mr. Sunday's and the students' argumentation structure were more complex, combining various features described in Pragma-Dialectics Theory (Van Eemeren et al. 2002) (Fig. 28.2): (i) the participants presented independent and multiple defenses to the same point of view (multiple argumentation) and (ii) they combined interdependent arguments to support their point of view (coordinative argumentation) and they used layers of arguments to support a single point of view (multiple subordinative argumentation). Contrasting the structure of argumentation in the mutualism lesson 3 (Fig. 28.2) with the structure in the blood type lesson (Fig. 28.6), it is possible to observe that the second is much more linear (more vertical distribution of its components) than in the mutualism lesson (Fig. 28.2) because in this case, teachers and students only used layers of arguments to support a single point of view (multiple subordinative argumentation).

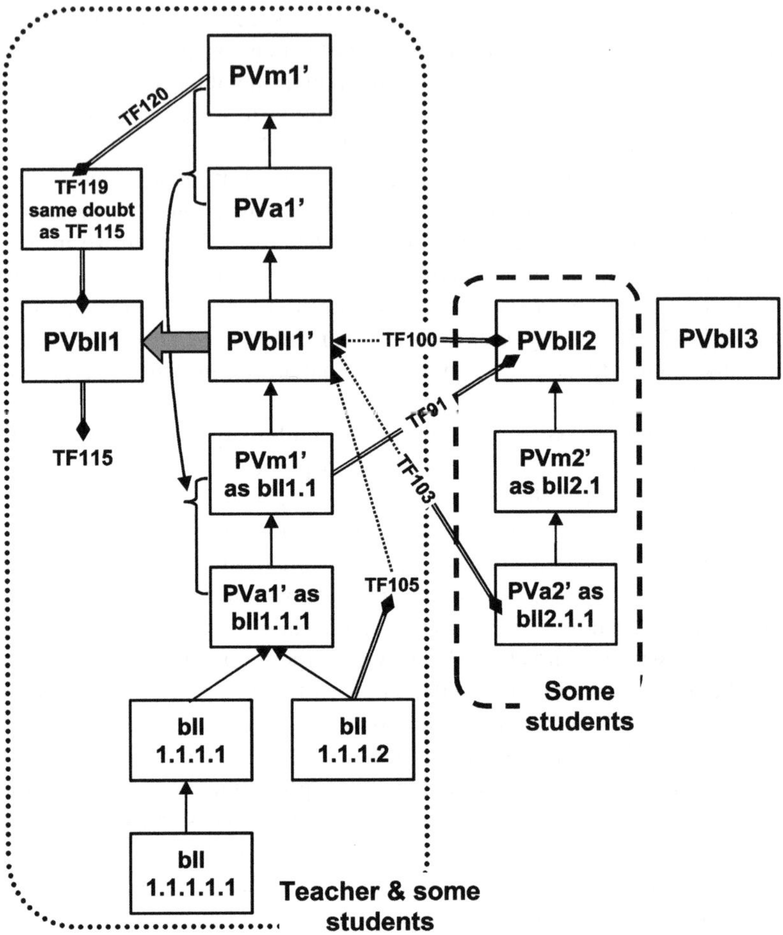

Fig. 28.6 Representations of argumentation structure for the blood type lesson. The structure of argumentation is much more simple (in this case, linear), and the contribution of students to argumentation is less significant compared to the structure in the mutualism lesson 3, represented in Fig. 28.2 (*Dotted lines* indicate support to element; double lines indicate confronting element; prime indicates implicit element; PVm means "point of view in the main difference of opinion"; PVa or PVb indicates points of view in the subordinate differences of opinion; numbers indicate arguments supporting point of view; TF indicates speak turns)

Our analysis evidenced the complexity of argumentative discourse, supporting the notion that this complexity is constructed through social interaction, and depends on a broad variation in the teacher's argumentative discourse. In this study, the teacher's repertoire of argumentative discourse involved not only helping students to provide more supportive elements for a point of view, making the structure of argumentation more complex (like in the mutualism lesson, when students argue about scientific concepts) – an aspect that is usually emphasized in the literature. His repertoire also included engaging students in multiple subordinative "argumentations" to link scientific knowledge and everyday knowledge (like in the blood type lesson). Finally, he better defined which points of view were under debate, and he explored various dimensions of these points of view, when teacher and students were experiencing new types of approaches and activities in science class (like in the solid waste lesson).

In all three lessons, Mr. Sunday tended to act both as protagonist of his own point of view and as antagonist of others' points of view. An exception is in lesson 2 on solid waste, when he had doubts about what was the best unit to be used to measure organic solid waste. In this case, he defended multiple points of view, acting as protagonist and antagonist. The students, on the other hand, tended to act only as protagonists of their points of view. The premises involving the (main and subordinate) differences of opinion in lessons 1 and 2 are, in general, implicit. Only in lesson 3 is the main difference of opinion explicit, making it possible, for the participants to be conscious of which issue is leading to disagreement.

We identified significant variation in Mr. Sunday's language use in two aspects: how he raised questions and how he made explicit/implicit his points of view or his arguments. Our analysis indicated that the teacher's discursive strategies were influenced by his goals, as well as by the relationships between the students' knowledge and scientific school knowledge. Another interesting aspect of Mr. Sunday's practice was that, through language and with a single utterance, he simultaneously kept discussion open and oriented discussion to a close, sustaining dialogic discourse as described in Mortimer and Scott (2003).

6 Implications and Conclusions

Our study supports the notion that argumentation occurs in science classrooms in different forms depending on the context and certain aspects seem to be particularly influential. The complexity of argumentation tends to be present in different contexts but involves different aspects: elements of the structure of argumentation and the nature of the difference of opinion (at a more "micro" level of analysis) and the hierarchical relationships between differences of opinion (at a more "macro" level). We believe that these different types of complexity were made visible because we adopted an analytic framework that is not restricted to one aspect or level of argumentation. Thus, this research makes evident that the field of theory of argumentation can contribute to the development of new methods to investigate argumentation and science learning.

References

Aavramidou, L., & Zembal-Saul, C. (2005). Giving priority to evidence in science teaching: A first-year elementary teacher's specialized knowledge and practice. *Journal of Research in Science Teaching, 42*, 965–986.

Andriessen, J. (2006). Arguing to learn. In R. K. Sawyer (Ed.), *The Cambridge handbook of the learning sciences* (pp. 443–460). New York: Cambridge University Press.

Baker, M. (2009). Argumentative interactions and the social construction of knowledge. In N. M. Mirza & A. N. P. Clermont (Eds.), *Argumentation and education: Theoretical foundations and practices* (pp. 127–144). London: Springer.

Berland, L. K., & Reiser, B. (2011). Classroom communities' adaptations of the practice of scientific argumentation. *Science Education, 95*, 191–216.

Bricker, L. A., & Bell, P. (2008). Conceptualizations of argumentation from science studies and the learning sciences and their implications for the practices of science education. *Science Education, 92*, 473–498.

Cameron, D. (2001). *Working with spoken discourse*. Los Angeles: SAGE.

Castanheira, M. L., Crawford, T., Dixon, C., & Green, J. (2001). Interactional ethnography: An approach to studying the social construction of literate practices. *Linguistics and Education, 11*(4), 353–400.

Chin, C., & Osborne, J. (2010). Students' questions and discursive interaction: Their impact on argumentation during collaborative group discussions in science. *Journal of Research in Science Teaching, 47*(7), 883–908.

Dixon, C., & Green, J. (2005). Studying the discursive constructions of texts in classrooms through interactional ethnography. In R. Beach, J. Green, M. Kamil, & T. Shanahan (Eds.), *Multidisciplinary perspectives on literacy research* (2nd ed.). Santa Barbara: Hampton Press Cresskill.

Driver, R., Newton, P., & Osborne, J. (2000). Establishing the norms of scientific argumentation in classrooms. *Science Education, 20*, 1059–1073.

Edwards, D., & Mercer, N. (1987). *Common knowledge. The development of understanding in the classroom*. London: Methuen/Routledge.

Erduran, S. (2007). Methodological foundations in the study of argumentation in science classrooms. In Erduran, S. & Jiménez-Aleixandre, M. P. (Eds.), *Argumentation in science education: Perspectives from classroom-based research* pp. (47–69). New York: Springer.

Green, J., Dixon, C., & Zaharlick, A. (2001). Ethnography as logic of inquiry. In J. Flood, D. Lapp, J. R. Squire, & J. Jensen (Eds.), *Research on teaching the English language arts* (pp. 201–224). Mahwah: Lawrence Erlbaum.

Jimenez-Aleixandre, M. P., & Erduran, S. (2007). Argumentation in science education: An overview. In M. P. Jimenez-Aleixandre & S. Erduran (Eds.), *Argumentation in science education: Perspectives from classroom-based research* (pp. 3–25). New York: Springer.

Lincoln, Y. S., & Gubba, E. G. (1985). *Naturalistic inquiry*. Beverly Hills: SAGE.

McDonald, S. P., & Kelly, G. (2012). Beyond argumentation: Sense-making discourse in the science classroom. In M. S. Khine (Ed.), *Perspectives on scientific argumentation: Theory, practice and research* (pp. 265–281). New York: Springer.

McNeil, K. L., & Pimentel, D. S. (2010). Scientific discourse in three urban classrooms: The role of the teacher in engaging high school students in argumentation. *Science Education, 94*, 203–229.

McNeill, K. L., & Krajcik, J. (2008). Scientific explanations: Characterizing and evaluating the effects of teachers' instructional practices on student learning. *Journal of Research in Science Teaching, 45*, 53–78.

Merriam, S. B. (1998). *Qualitative research and case study applications in education*. San Francisco: Jossey-Bass.

Mortimer, E. F., & Scott, P. (2003). *Meaning-making in secondary science classrooms*. Berkshire: Open University Press.

NRC. (2000). *Inquiry and the National Science Standards: A guide for teaching and learning*. New York: National Academy Press.

Osborne, J., & Patterson, A. (2011). Scientific argument and explanation: A necessary distinction? *Science Education, 95*, 627–638.

Osborne, J., Erduran, S., & Simon, S. (2004). Enhancing the quality of argument in school science. *Journal of Research in Science Teaching, 41*(10), 994–1020.

Ozdem, Y., & Erduran, S. (2011). The development of an argumentation theory in science education. Paper presented at the 2011 ESERA Conference, Lyon.
Sadler, T. D. (2006). Promoting discourse and argumentation in science teacher education. *Journal of Science Teacher Education, 17*(4), 323–346.
Sandoval, W. A., & Reiser, B. J. (2004). Explanation-driven inquiry: Integrating conceptual and epistemic scaffolds for scientific inquiry. *Science Education, 88*, 345–372.
Sandoval, W., & Millwood, K. (2007). What can argumentation tell us about epistemology? In S. Erduran & M. P. Jiménez-Aleixandre (Eds.), *Argumentation in science education: Perspectives from classroom-based research* (pp. 71–88). New York: Springer.
Schwarz, B. B. (2009). Argumentation and learning. In N. M. Mirza & A. N. P. Clermont (Eds.), *Argumentation and education: Theoretical foundations and practices* (pp. 91–126). London: Springer.
Scott, P., Asoko, H., & Leach, J. (2007). Student conceptions and conceptual learning in science. In S. K. Abell & N. G. Lederman (Eds.), *Handbook of research in science education* (pp. 31–55). Mahwah: Lawrence Erlbaum.
Simon, S., Erduran, S., & Osborne, J. (2006). Learning to teach argumentation: Research and development in the science classroom. *International Journal of Science Education, 28*, 235–260.
Spradley, J. (1980). *Participant observation*. New York: Holt, Rinehart; Winston.
Stake, R. E. (2000). Case studies. In N. K. Denzin & Y. S. Lincoln (Eds.), *Handbook of qualitative research* (pp. 435–454). Thousand Oaks: SAGE.
Van Eemeren, F. H., Grootendorst, R., Henkemans, F. S., Blair, J. A., Johnson, R. H., Krabbe, E. C. W., et al. (1996). *Fundamentals of argumentation theory: A handbook of historical backgrounds and contemporary developments*. Mahwah: Lawrence Erlbaum.
Van Eemeren, F. H., Grootendorst, R., & Henkemans, A. F. S. (2002). *Argumentation: Analysis, evaluation, presentation*. New Jersey: Lawrence Erlbaum.
Zembal-Saul, C. (2009). Learning to teach elementary school science as argument. *Science Education, 93*, 687–719.
Zembal-Saul, C., Munford, D., Crawford, B., Friedrichsen, P., & Land, S. (2002). Scaffolding preservice science teachers' evidence-based arguments during an investigation of natural selection. *Research in Science Education, 32*, 437–463.
Zohar, A. (2007). Science teacher educational and professional development in argumentation. In M. P. Jimenez & S. Erduran (Eds.), *Argumentation in science education: Perspectives from classroom-based research* (pp. 245–268). New York: Springer.

Chapter 29
Analysis of Teaching and Learning Practices in Physics and Chemistry Education: Theoretical and Methodological Issues

Patrice Venturini, Andrée Tiberghien, Claudia von Aufschnaiter, Gregory Kelly, and Eduardo Mortimer

1 Introduction

Over the past decade, classroom activities have been studied with the aim of describing and modeling teaching and learning practices. Mainly based on the analysis of videotaped teachers' and students' discourses and actions, the corresponding studies relate to particular domains of research in science education. We have chosen to focus on three of them because of their importance in the field: conceptual change, meaning-making, and inquiry-based science education (IBSE). The conceptual change approach considers individual students facing concepts and their difficulties, first from a constructivist perspective and far later paying attention to social interactions and to cultural context in which meaning emerges. Instead, the meaning-making approach emerged as a consequence of the application of a sociocultural perspective to the teaching and learning processes and has a strong root in Vygotsky theory. Lastly, the chosen perspectives of inquiry-based learning environments connect teaching, learning considered from a socioconstructivist perspective and knowledge, referring to the nature of scientific knowledge and scientific practice. In spite of their differences, all of these approaches deal with science learning and

P. Venturini (✉)
University of Toulouse, Toulouse, France
e-mail: patrice.venturini@univ-tlse2.fr

A. Tiberghien
UMR ICAR, CNRS, University of Lyon 2, ENS-Lyon, Lyon Cedex, France
e-mail: andree.tiberghien@univ-lyon2.fr

C. von Aufschnaiter
University of Giessen, Giessen, Germany

G. Kelly
Pennsylvania State University, University Park, PA 16802, USA

E. Mortimer
Federal University of Minas Gerais, Minas Gerais, Belo Horizonte, Brazil

C. Bruguière et al. (eds.), *Topics and Trends in Current Science Education: 9th ESERA Conference Selected Contributions*, Contributions from Science Education Research 1,
DOI 10.1007/978-94-007-7281-6_29,

teaching processes, focusing more or less on one of them or on both. Our purpose is to give an example of studies related to each of these three approaches in order to present some theoretical frameworks and methodologies used in video-based analyses of classroom activities and discuss their interest and their limits. In each case, focusing on physics and chemistry education, we first outline the main research trends in the domain over the last decade to situate the studies on which we report after this overview. A final discussion on this kind of analyses closes the chapter.

2 Analyses of Teaching and Learning Practices Related to Conceptual Change

2.1 Overview of Research Trends

Student thinking and conceptual change have been studied since the 1970s, but studies have changed in emphasis over recent decades (Vosniadou 2008). At first, researchers mainly focused on the characterization of students' thinking (misconceptions, mental models, etc.). From the beginning of the 1990s, some science education researchers considered student thinking as a basis for a learning process leading to scientifically accepted ideas. It is only since the 2000s that research has been influenced by Vygotskian ideas and has focused more on classroom discourse. Thus, the theoretical frameworks supporting these investigations have a strong sociocultural dimension. These analyses are mainly carried out from audio or video recordings of whole-class talks and address three different (but often interrelated) issues. A first set of studies details how students develop an understanding of particular concepts and how their conceptions of these concepts change during learning, in relation with the social discussions in the classroom (e.g., Havu-Nuutinen 2005, about floating and sinking; Rincke 2011, about the concept of force). A second set of studies provides more general information on the conceptual change process. For example, Eschach (2010) reveals "conceptual flow patterns" or structures of whole-class discussions occurring within physics lessons; von Aufschnaiter and Rogge (2010) present conceptual development as "a process that develops from explorations to intuitive rule-based and then to explicit rule-based understanding." A third group of studies deals with the links between teaching practices and conceptual change. Thus, particular learning environments including teaching strategies designed to persuade children of the usefulness and validity of the target scientific concept (Loxley 2009) or chosen sets of learning goals (Beeth and Hewson 1999) influence and shape conceptual changes and student thinking. Fruitful learning environments can also encompass computer simulations (Suzuki 2005, about the concept of force) or other material artifacts (like ambiguous objects used in a sorting activity linked to the concept of matter, Varelas et al. 2008). The former (environment including computers) leads to a dialog among students who have different perspective and to a process of mutual changes in thinking, whereas the latter (environment including ambiguous objects) induces discursive spaces where there is no specific answer or way of thinking.

2.2 *Analysis of Learning Practices from a Conceptual Change Perspective*

Conceptual change research carried out *during* instruction, focusing on the processes by which "change" comes about, is a relatively young discipline which seems to have its roots in interviews or teaching experiments (e.g., Riemeier and Gropengießer 2008). Researchers who address conceptual change in science classrooms typically have begun to focus their attention toward classroom discourse (e.g., work done by Eshach 2010, or Rincke 2011) in which all students are talking to or with a teacher. "Conceptual change" can then have two meanings: (1) changes in how the entire class conceptualizes different science concepts, thus learning among the *community* of students, or (2) how an *individual student* develops his/her own conceptual understanding while participating in the community. For (1) a focus on entire class discussions is a useful means to investigate how different and/or new meanings of a concept evolve. Video-based research in this focus typically uses one camera to follow the teacher, whereas another camera is used to document the entire class (e.g., Seidel et al. 2005). However, the discourse of a community does not give information about individual progress. Thus, students may participate without understanding the meaning of the discussion. In order to distinguish between the community and the individual, Eshach (2010, pp. 470f.) introduces the idea of an "individual space" and a "collective space." Basing on this idea, we therefore use video in a slightly different way than it is normally employed. Within a classroom setting, we focus on student groups of three to four students. We use two to four cameras, each of which documents one group. This approach makes it possible to follow individual contribution to classroom discourse, which can be assessed completely through group cameras, but also to investigate how students make use of outcomes of this discourse for their own progress in conceptual understanding. Following Eshach's views of interactional spaces, this approach makes it possible to connect the collective space, the social and material learning opportunities, with the individual space. Importantly, it is not only an individual student's participation in classroom discourse that matters but also how he/she works on, for instance, experiments that are carried out during group work or worksheets which have to be completed and so on. These activities which occur outside whole-class discussions are a good indicator of conceptual understanding. Here, a student is typically much more active than during class discussions, which in turn makes it easier to reconstruct conceptions from verbal and nonverbal activities.

In all our research projects, we focus on student groups which are composed of two to four students, either in classroom settings or in laboratory settings in which only one group is present at a time. We do not have a camera which is directed toward the teacher but can hear the teacher on at least one group camera clearly; the same is true for all student participation in classroom discourse even if a particular student is not assessed by a group camera. Videos are coded by trained raters especially to identify the social settings (e.g., teacher talk, classroom talk, single student work, work in pairs or group work) and individual student activity (e.g., verbal content-specific activity, nonverbal content-specific activity, reading/writing, organizational activity, raise hand, observation/listen, emotional utterance, or off-task activity). In order to investigate conceptual change processes, all sequences are transcribed in which we applied codes associated with content-specific individual activity, either verbal or nonverbal. These transcripts are then investigated utterance by utterance or activity by activity to assess conceptual understanding and the progress of it. When analyzing transcripts, we focus, for instance, on whether ideas are "correct," which level of conceptualization these ideas have (e.g., v. Aufschnaiter and Rogge 2010), how many content elements are interrelated, and how long coherent activities take. We also assess how students experience their current activities and their learning.

As an example, Table 29.1 shows the discussion of three students (grade 11, about 16 years old) about how warm different objects feel. In the right column, our analysis of the students' utterances is presented with reference to the different categories we typically employ (underlined). In the excerpt, the students seem to realize for the first time that even though different materials (here: the metal blades of the scissors and the plastic handle) feel differently warm, they can have the same temperature. Overall, it takes them about 10 min with repeated own activity to realize the difference between sensory experience and measured temperature. This experience helps them to establish a conceptual understanding of the zeroth law of thermodynamics in the further course of the instruction.

Results of our work can inform science education in different ways:

(a) From differences and similarities in different students' conceptual change processes, we can generalize patterns which describe how concepts are formed and stabilized through individual activities. Especially, our results indicate that own experiences play a crucial role in concept formation and that conceptual change processes seem to be spiral rather than linear; establishing new conceptual understanding thus requires "back-and-forth movement" (e.g. v. Aufschnaiter and Rogge 2010).
(b) From relating social and material learning opportunities on the "collective space" to individual conceptual change processes, we can conclude which kind of learning opportunities promote conceptual change processes or may hinder them. Previous research has revealed that it seems to be very important for "information" – for instance, a teacher explanation or an explanation given by another student – to match what the individual student already knows (e.g., v. Aufschnaiter 2003). Even though this result sounds trivial taking into account Vygotsky's (1978) idea of the zone of proximal development, it is harmed repeatedly in everyday teaching by teachers asking questions which students cannot answer, giving explanations which are far away from students' current experiences and understanding, and so on.
(c) The methodological approach taken broadens current video-based classroom research as it moves away from videoing whole-class settings toward documenting individual student activity while taking part in instruction. The benefit of this

Table 29.1 Analysis of students' discussions about how warm objects feel

Transcript	Analysis (references to categories used are underlined)
Scene 1 [7:00–7:35]	Students explore phenomena
(*The scissors plastic handle is measured*)	
S2:24. Oh [surprised]	Students are astonished about the result of their exploration
S1:We do not know the room's temperature	Students start to cross-connect content, but it is not clear what this cross-connection is about to mean
(*The scissors metal blades are measured*)	Exploration
S1:I reckon 16. What does the thermometer say?	Students have an intuition about the phenomenon. This intuition is not correct
S3: (Looks at thermometer) 23	Exploration
S2: Point 8	Exploration
(Students giggle)	Indicates students' experiences (astonishment/ irritation)
S2: Point 7.	Exploration
That cannot be true. I'll have a look on the other side (turns scissors around). 23.7. It is freezing cold [ironically]	Intuition about the outcomes which is not correct. Further exploration. [Turning around the scissors doesn't make sense here and indicates how strong the intuition is and how demanding the physics concept is for the students]
(Students giggle) S1: Strange	Indicates students' experiences (astonishment/ irritation)
Scene 2 [8:35-8:40]	Exploration
S2: (Touches blades and handle). That is so differently warm, that cannot be true	Intuition about the outcomes which is not correct
S1: (Touches blades)	Exploration
Scene 3 [18:18-18:32]	Exploration
S3: (Measures temperature of the air) 22.7	
S1: Oh [surprised]	Experiences
S2: Well, all objects... S3:are warmer... S2: All are warmer than the air. Including those we thought of as cold	Elaboration indicates an intuitive understanding of a certain concept (but is not yet expressed conceptually)
S3: Yes. Awesome, really	Experiences

approach is information on what constitutes "optimal" teaching. The indicator for teaching which is "in situ" optimal is how it promotes content-specific individual activity in the class, for as many students as possible. A camera which focuses on the entire class cannot give enough information on individuals as these appear as small "spots" on the video for which we can often not say in detail what the students are doing or discussing.

3 Analyses of Teaching and Learning Practices Related to Students' Meaning-Making in a Sociocultural Context

3.1 Overview of Research Trends

Two research trends based on a sociocultural perspective directly address the question of students' meaning-making in socially shared practices. The first one, rooted in Swedish research, is also inspired by pragmatism and the late works of Wittgenstein. Its focus is "to use a formal theory of meaning-making in illuminating the connection between how people produce meaning and what meaning is produced in a specific practice" (Wickman 2004). It considers learning as a dynamic process where relationships are constructed in encounters between individuals and between individuals and the world (Wickman and Östman 2002). Learning proceeds when people notice gaps and fill them with relations they build to what stands fast in these encounters (i.e., is not questioned in talks or acts). This theoretical mechanism for learning provides a methodological approach for analyzing video-recorded students' talk and action, for noting the details of how discourses change and how students become participants in new practices, and thus for describing people's ways of making meaning in action in a particular sociocultural context (e.g., Lidar et al. 2010a, b, about gravity and the shape of the Earth; Wickman 2004, about practical work in inorganic chemistry). The students' "practical epistemology analysis" is completed by an "epistemological moves analysis" to categorize the actions that a teacher takes with the aim of helping students to learn. Using both analyses allows the researcher to investigate the relation between the epistemological moves that teachers make in teaching and the practical epistemology that the students use in their learning process (Lidar et al. 2010a, b), shedding light on the interplay between these intertwined activities (Lidar et al. 2006).

The second trend (Mortimer and Scott 2003) is mainly focused on the teacher as its concern is to characterize "the various ways in which the teacher acts to orchestrate the talk of science lessons in order to support student learning" (p. 24). To reach this aim, Mortimer and Scott have developed a discursive analytic framework based on five linked aspects (e.g., the communicative approach, interactive vs. noninteractive, authoritative vs. dialogic). Their methodology involves the video recording of a set of lessons. Videos are transcribed, and data are mapped into episodes characterized by a specific function in the flow of discourse and then analyzed. The analytic framework empowers the authors to examine, for example, the movements between authoritative and dialogic discourse in the set of lessons and argue that shifts between communicative approaches are necessary to support meaningful learning in science (Scott et al. 2006). Other scholars have borrowed this framework or some of its elements for a discursive analysis regarding science meaning-making in particular cases (e.g., Chin 2006, to analyze teacher questioning and feedback, or Yeo and Chee Tan 2010, to analyze the use of authoritative sources).

3.2 Analysis of Particular Events Related to Students' Meaning-Making: The Turning Points

An inevitable consequence of classroom discourse being characterized as an alternation of dialogic and authoritative discourse is that transitions between the two types of discourse will be critical for planning teaching sequences. If transition is the rule, one of the most important events in the classroom should be the *turning points*, which first and foremost are identified in terms of a change in communicative approach – from dialogic to authoritative or vice versa – during the staging of a teaching intervention.

In terms of methodology of research, we are going to discuss how we could characterize what we are calling the turning point entries and exits. We assume that turning points can be planned by the teacher, even if he/she has never heard about the expression "turning points," or can be spontaneous, in the sense that shifts between authoritative and dialogic discourse (or vice versa) happen independently of the teacher's will. For example, a student's question may restart a dialogic sequence when the teacher decides to answer it. In this article we are going to discuss only the planned turning points, trying to find out which are the cues that we can look for in the ongoing classroom talk to determine a turning point entry or exit.

The data analysis involved examining the videotapes using Videograph® and Transana® to identify all instances of turning points. A typography of turning points was then developed based on the structure and functions of those points. A key part of this development involves identifying unambiguous classes of types of turning points and then being able to allocate examples of turning points to the appropriate classes.

The teacher whom we are going to analyze to exemplify what is a turning point develops a teaching sequence (about 5 h) focusing on the topic of forces, with grade 7 students in a secondary school in a rural area of the North of England.

The turning point is part of this sequence and focuses on teaching and learning about the normal force: that a table exerts an upward force on a cup which is placed on it. The starting point here is that children of this age typically find it difficult to believe that inanimate objects such as a table can exert a force on a cup. Jonathan the teacher in this case started by organizing his class into pairs and giving each pair of students a concept cartoon showing four possibilities for the forces acting on a bottle on a shelf: (a) the bottle is not moving, there are no forces on it; (b) the only force on the bottle is the force of gravity pulling it downwards; (c) there are two forces on the bottle – the force of gravity and the push of the shelf upwards, which balances it; and (d) a shelf cannot push. It is just in the way of the bottle and stops it falling. The students talk in pairs about each of these statements indicating whether they "agree" or "disagree" or are "not sure" about each one. Each pair then works with another pair to compare views and to reach a consensus within the group of four. Finally, Jonathan calls the class around the table at the front of the room.

Jonathan has opened up a *dialogic space* (Wegerif 2007), where students are able to express their ideas. The concept cartoon prompts the students to talk through the various possible models, and it is clear that differences in point of view have been created. In fact, as Jonathan comments, the students *really didn't agree at all*. Josie expresses clearly the view that "a table can't push up," which is in disagreement with the likes of Ryan who thinks that there are two forces on the bottle. Jonathan summarizes this situation using a noninteractive/dialogic approach. There is clear evidence here that the concept cartoon exercise has been effective, in engaging at least some students in thinking about whether or not tables can push. Having opened up these differences in student thinking, the question is: What might be the next step?

We return to the class for the next lesson, which was taught 2 days later. Jonathan starts the lesson and refers back to the debate from the previous lesson: "I'd like to get you to think about one of the ideas that you really *argued* about on Monday." At this point Jonathan has gone as far as is possible in terms of opening up the different ways of modeling the bottle on the shelf. He and the class have reached the *turning point* for this particular intervention. Jonathan now states quite clearly what is to come next:

Teacher: What I want to do… I want to leave you this morning… with a picture of something that might help you to believe that *that* [*knocking on the table*] can push up. Now this is a very logical little argument, so you're going have to follow it through.

With the help of one of the students, Sam, Jonathan presents an argument in favor of the table pushing, using a balloon. He gets Sam to put his hands on either side of this balloon and gently squeeze it together, changing the shape of the balloon. After this, he gets Sam to push the balloon over the table and asked Sam what happens with the shape of the balloon. Sam answered that with his action, the balloon flattened. Jonathan then asked the whole class to answer the question, "Where on Earth is the other force that's changing the *shape*?" Different students answer that the table was pushing.

In this way Jonathan enlists the help of one of the students, Sam, and presents an argument to suggest that the table can push up, focusing attention on the forces acting on a balloon. He achieves this by taking an interactive/authoritative communicative approach, played out through I-R-E patterns of three (initiation-response-evaluation, Mehan 1979). This pattern of interaction continues up to the point when Holly provides the correct response that the other force is "from the table." Jonathan then conducts rapid confirmatory exchanges (Edwards and Mercer 1987) with Levi and Penny prior to concluding the episode with an authoritative statement. In this way Jonathan exits the turning point by "presenting" a logical argument centered on the analogous case of the balloon.

One key characteristic of this turning point is that here the impetus for learning comes from the differences in the students' views about the *models* presented in the concept cartoon, whereas in some case the impetus for learning came from observing a *phenomenon*. Here there is no phenomenon which is open to dispute. There is no arguing about whether or not bottles stand on tables. The point at issue lies with how that situation is modeled in terms of forces. In this case, therefore, the impetus for learning is generated by the students engaging in the modeling task with the concept cartoon. Here, creating differences involves setting up differences in students' views about possible models.

A well-elaborated transition from dialogic to authoritative discourse (and vice versa) becomes an extremely sensitive point in the planning of a teacher. We sought to demonstrate that the transition between dialogic and authoritative discourses can occur anywhere in a teaching sequence, not only at the beginning, when students are still imbued with their everyday ideas. In advanced stages, when students have mastered the scientific tools that allow them to risk a hypothesis to solve a problem, different points of view may again appear in this process, but most of these points of view are well grounded in scientific discourse.

That this dialectic between authoritative and dialogic discourses can occur in classrooms seems to be fundamental for students to progress in their understanding of science. The stronger in articulating authoritative discourse the students become, the greater the chance that they take the risk of offering different points of view in solving problems, reinstalling the dialogic discourse.

4 Analyses of Teaching and Learning Practices Related to IBSE

4.1 *Overview of Research Trends*

Inquiry was a major focus for the reform efforts of the 1960s in the United States (Yager 1997) and of the 1970s in Europe, and it has been studied for a long time in science education. Among the studies in the last 10 years involving classroom observations with video data, several perspectives have emerged. Some studies mainly focus on the nature of inquiry, such as its authenticity (Chinn and Malhotra 2002; Schwartz et al. 2004); tackle the nature of science dimension (Sandoval 2005); concern socioscientific issues (Walker and Zeidler 2007); or give room to links between inquiry and argumentation (Sampson et al. 2011). Other studies deal with collaborative aspects during inquiry activities (Watson et al. 2004), including with the support of software (computer-supported collaborative learning, Makitalo-Siegl et al. 2011). The way in which teachers implement inquiry in their classroom practices after that inquiry has been added to the official curriculum constitutes another research theme (Blanchard et al. 2010; Smithenry 2010), as do the teacher's role and the characteristics of interactions between the teacher and the students when doing inquiry activities.

In this brief review, we now give an overview of the studies that have the same focus as our own to the extent that they carry out an analysis of teachers' actions when classroom activities concern inquiry. Several analyses related to this trend concern particular types of actions like the teacher's questioning and answering and the students' expectations, whereas others deal with both the teacher's and the students' related actions, either as being specifically linked to inquiry, like developing hypotheses (Shimoda et al. 2002), or as being more general, like assessment (Ruiz-Primo and Furtak 2007). Lastly, investigations approach scientific inquiries dealing with the "social structures that constitute a classroom community" (Enyedy and Goldberg 2004) or with the need for the teacher to withhold the answers and be reticent (van Zee 2000; Furtak 2006). More generally, they show the importance of studying together the teacher's actions and the way in which the classroom situation is designed and enacted to interpret students' actions and vice versa.

4.2 Analysis of IBSE Teaching Practices with the Joint Action Theory in Didactics

This study deals with the analysis of teaching practices in grade 8 related to the voltage law in a series circuit. Our aim was to provide information on how a young male physics teacher coped with inquiry-based physics teaching. Our analysis was supported by the Joint Action Theory in Didactics where the didactic action is considered as a joint action by teacher and students (JATD – Sensevy 2009, 2011). Its use is in line with the previous conclusion according to which it is important to analyze at the same time both the teacher's and the students' action.

JATD regards the situations that make up teaching and learning practices as a set of collaborative didactic games. This allows us, for each game, to consider what is at stake (related to knowledge), what are the definitory rules (to play), what are the strategic rules (to win), the players' investment, etc. To analyze the classroom situations in terms of games, two concepts are used: (1) the milieu which is made up of all the conceptual and material elements in the environment that act on the students and on the teacher and on which the students and the teacher act and (2) the didactic contract which consists of a set of perennial and local knowledge-related rules governing the teacher's and the students' actions, defining their rights and duties, and sharing and limiting their respective responsibilities. Three dimensions thus account for the dynamics of each game: (1) the evolution of knowledge as it unfolds throughout the game; (2) the evolution of the milieu, which is its continuous reorganization due to the students' and the teacher's interventions; and (3) the evolution of the teacher's and the students' respective responsibilities for the progress of knowledge in the class. The teacher intervenes during the didactic game in different ways. He/she can first define the game. He/she can also devolve the game, acting in such a way that the students agree to play the game properly and on their own. Usually the teacher regulates the game, intervening in order that the students modify their actions to become more relevant regarding the stakes of the game. Lastly, the teacher can institutionalize the knowledge at stake, pointing out to the students that their activity has reached a part or the totality of that knowledge.

The grain size of the games can be related to the temporal span of the analysis. Thus, we consider that the modeling of classroom practices necessitates choosing levels of analysis related to different timescales (Lemke 2001): macroscopic (teaching sequence: several weeks, months), mesoscopic (less than a session:

about 10 min), and microscopic (couple of seconds to a couple of minutes). The emergence of one phenomenon at a particular level depends on what unfolds in both the lower and higher levels.

Our analysis was based on a particular methodology. The video of the lesson constituted our main data. The teacher was also interviewed about what he/she intended to do and what he/she actually did during this lesson and why. These talks, the worksheets distributed during the lesson, and the curriculum formed our second set of data. These data were analyzed in three consecutive steps but also by going back and forth between them.

We first made an a priori analysis of the given tasks in relation with the curriculum from an epistemological point of view in order to identify possible variants of the lesson development. We also synthesized the teacher's comments and characterized the students' worksheet according to the curriculum requirements. Lastly, using Transana software (Woods and Fassnacht 2010), we transcribed the classroom interactions.

Second, analyzing the lesson at a mesoscopic level, we divided it according to the social classroom organization, the thematic content of classroom discourse, and the evolution of the milieu, the responsibility for the progress of knowledge, and the meaning of the knowledge involved in the discourse, that is to say according to different didactic games. Each of these parts was indexed in Transana with the corresponding keywords. From an analysis at microscopic level, we characterized sub-games with keywords related to JATD concepts and knowledge at stake using categories of facets. Facets are little components of knowledge whatever it may be (Minstrel 1992) and are defined on the basis of "knowledge to be taught" (curriculum, textbook), students' misconceptions, and classroom productions. This qualitative analysis at micro- and mesolevels provided us, on the one hand, with a static, global view of the lesson with the amount of time linked to the different keywords allowing quantitative calculations and, on the other hand, with a dynamic view of the lesson development given by the succession of keywords over time.

Finally, getting back to the detailed interactions and their meaning regarding the knowledge at stake and supported by Transana outputs, we analyzed the successive didactic games to precisely map out the dynamics of the lesson, including in particular the characterization of knowledge development (continuity or discontinuity in the chronogenesis). Moreover, we inferred some of the determinants of these dynamics. The uncertainty of these inferences was reduced by cross-checking our interpretations with what happened in other games and with information coming from the second set of data.

Both theoretical and methodological frameworks led us to findings, of which we give only some very limited excerpts now to illustrate their nature.

The stake of the first game of the observed lesson was to recall knowledge from the previous lesson orally about nominal voltage and voltage measurement. The stake of the second game was to understand the problem presented by the teacher. This presentation was the first step of the inquiry process, as specified in the French curriculum. The first part of the game [min 3:22] sounded like a story: "A child afraid of the dark wants to light his hut and for this, he uses five bulbs with bases indicating 0.1 A and 3.5 V, 5 bulbs with bases indicating 0.2 A and 6 V, and a 12 V battery in a series circuit. The bulbs light up badly or not at all"; the teacher then asked the students to help the child. In the second part of the game, the teacher asked the students to summarize the problem [min. 5:32]. In short game 3, the students were asked "to propose a solution" [min 6:21], then "an explanation" [min 6:53]. The devolution process (by which the teacher acts in such a way that the students accept the responsibility of constructing knowledge) mainly took place on the practical aspects and the students suggested adding a battery or removing some lamps. One of them hypothesized that the sum of the whole nominal voltage of the bulbs was higher than the battery voltage [min 7:12], but the teacher's epistemological point of view about the knowledge construction prevented him from using this student's hypothesis. As he said in the interview, "knowledge must come from practical experiments [and not from preconceived opinions]."

Then, he asked the students (game 4) to work in pairs and write a question summarizing the problem [min 7:34]. To obtain the scientific question he was looking for, the teacher made many regulations, taking more and more space in the interactions and giving more and more prompts on the content. For example, less than one minute after the students started working in pairs, the teacher spoke to the whole class and strongly guided the students toward proposing a question concerning the bulb voltage in the circuit that would not be appropriate because it was different from the nominal voltage. The facets in the following part of the game (belonging to the category "relations between the functioning and the characteristics of an element") confirmed that the students did that. We consider the teacher's intervention as a breakdown, a discontinuity in the didactic contract: Before the intervention, the students had the responsibility for knowledge; after it, they had only to follow the teacher's request. After the students stopped working in pairs and orally gave their proposals, in agreement with the teacher's proposal, he introduced a new proposal:

1. T: (*Teacher*) Why is the voltage across the lamps not the right one, and what other question might we ask? If it is not the right one, there must be another one. [*min 12:53*]
2. S: (*Student*) Yes.
3. T: What could you try to do with your own knowledge? (*student's murmurs*) Yann, what could you try to do with your knowledge? If the voltage is not the right one, then in reality there is another one!
4. S: Check with a voltmeter.
5. T: Here it is, we could try to make the measurement. Do all of you agree with me?
6. S: Yes.
7. P: So, we want to know how the… (*T writes on the blackboard and waits for an answer*)
8. S: The voltage
9. T: Here it is, the voltage is shared out in the circuit. [*min 13:15*]

In this excerpt, the teacher modifies the knowledge on which the question is focused. Whereas turn 1 deals with the comparison between nominal and actual voltage across bulbs and thus with the local characteristics of one component of the circuit, turn 9 deals with the way the voltage is shared in the circuit and thus necessitates thinking about this circuit as a whole, as a system. In fact, many studies have shown that it is very difficult for students to take this viewpoint. They rather interpret a circuit in terms of a causal relationship between the generator and the receivers involving the dipoles in a sequential reasoning (Closset 1983). Thus, we consider that there is a significant discontinuity of knowledge between the beginning and the end of this exchange. Moreover, as students could not build meaningful relations with the milieu regarding the knowledge at stake, in turn 3, we see that their sole resource was a usual school practice according to which the knowledge of the previous lesson (voltage measurement) must be useful in the following one, that is to say, a perennial rule of the didactic contract.

These short excerpts of our findings indicate that the JATD framework enables us to account for didactic phenomena occurring in a physics lesson. Moreover, as shown here with the use of facets, patterns in the interactions, and timescales, this theory can work with other frameworks produced in science education or in neighboring fields. To go further and obtain a more integrated view on teaching and learning practices, we think it would be interesting to combine the use of JATD with the analyses of students' practical epistemologies (Wickman and Östman 2002). We also consider that software like Transana is very valuable for analyzing such lessons. It gives the researcher some quantitative information linked to qualitative analysis, helping him/her to objectivize his/her conclusions. But it requires keywords to be associated with video episodes. At a mesoscopic level, we used several ways of structuring the session, according to the social organization, the theme at stake, and the game played. Whereas the first two are mainly associated with one aspect and are rather easy to determine, the last one implies taking the milieu, the contract, and the knowledge at stake into account and considering the rules of the game. It therefore needs much more interpretation from the researcher. Lastly, the study shows that the continuity of knowledge can be analyzed using JATD at a mesoscopic level (e.g., how responsibility for knowledge is shared or relations between themes) as well as microscopic level (e.g., construction of the meaning of interactions, in particular from a student's perspective).

5 Final Discussion

The three cases presented offer the field of science education ways of examining theories and methods for researching pedagogical approaches in science education. Whether looking at conceptual change, meaning-making, or inquiry in science, a set of common methodological themes emerges from the theoretical commitments of the research teams presenting each of the three cases. We discuss three of these common themes.

First, the cases show how everyday life in schools is interactionally accomplished. By opening up the processes of conceptual change, the first case focused on student conceptual change evinces the relationship of the commonly constructed knowledge of the group and the variation among the individual knowledge among the students and teacher constituting the group. Conceptual change among students occurs through the opening of a collective space where students can engage and make sense of the concepts in question. Changes in understanding are accomplished in and through discourse, as meanings are constructed, modified, corrected, and taken up among students (Kelly et al. 2012). In the second case, focused on meaning-making, a turning point was constructed through the shared knowledge and joint recognition between the teacher and students that this significant discourse event occurred and that a shift in the conversation is needed to accomplish the next emerging goal. Thus, a turning point becomes recognized among participants through interaction. The third case examined an inquiry event in which the discursive moves by the teacher led to discontinuity of knowledge among students. The case shows how a brief interactional episode can change the dynamics of the classroom conversation and the meanings made available to students.

Second, communicative approaches and interactive spaces are constructed intertextually through dialog, each offering different opportunities for learning. Across the three cases, the different interactive spaces for communication – individual and collective space in case 1, dialogic space in case 2, and reorganization of the milieu in case 3 – are constructed through discourse and make use of previous knowledge evoked to stimulate student discussion. By making intertextual references to previous knowledge, that is, ways of talking in the particular milieu evoking previous discourse, the teachers in each case use various communicative approaches to engage students in substantive talk about science ideas. By opening up the conversation for meaning-making, the teachers make choices about how to situate students in dialog about science. These choices have consequences for student learning. Thus, analysis of classroom discourse needs to examine both the use of previously established meanings of the conceptual knowledge of science and the reevoking of common ways of being in the collective action. This shows the connection across conceptual understanding, common discourse, and social practices.

Third, classroom norms and practices are established over time. In each of the cases, examples of micro-moment interactions occur within broader social contexts where local norms for such interaction frame events as they are interactionally enacted among participants. For example, in the third case focused on inquiry, the learning game of the classroom is framed within a set of perennial and local knowledge-related rules governing the teacher's and the students' actions. Such local rules are continuously constructed, evoked, enacted, and renegotiated through the everyday discourse processes of this, or any, classroom. Understanding the nature of inquiry, student conceptual change, or key turning points in instructional conversations requires methodological approaches that move across timescales and interactional spaces to examine how what counts as science in a milieu is constructed over time (Kelly 2008). To understand the consequences of different pedagogical approaches, research methods need to examine both the moment-to-moment interactions where meanings are constructed and the overtime practices that stabilize meaning through conversation within a collective.

References

Beeth, M.-E., & Hewson, P. W. (1999). Learning goals in an exemplary science teacher's practice: Cognitive and social factors in teaching for conceptual change. *Science Edition, 83*, 738–760. doi:10.1002/(SICI)1098-237X(199911)83:6<797::AID-SCE9>3.0.CO;2-Y.

Blanchard, M. R., Southerland, S. A., Osborne, J. W., Sampson, V. D., Annetta, L. A., & Granger, E. M. (2010). Is inquiry possible in light of accountability? A quantitative comparison of the relative effectiveness of guided inquiry and verification laboratory instruction. *Science Edition, 94*, 577–616. doi:10.1002/sce.20390.

Chin, C. (2006). Classroom interaction in science: Teacher questioning and feedback to students' responses. *International Journal of Science Edition, 28*, 1315–1346. doi:10.1080/09500690600621100.

Chinn, C. A., & Malhotra, B. A. (2002). Epistemologically authentic inquiry in schools: A theoretical framework for evaluating inquiry tasks. *Science Edition, 86*, 175–218. doi:10.1002/sce.10001.

Closset, J. L. (1983). Sequential reasoning in electricity. In Research on Physics Education. Proceedings of the First International Workshop. La londe des maures, pp. 313–319.

Edwards, D., & Mercer, N. (1987). *Common knowledge: The development of understanding the classroom*. London: Routledge.

Enyedy, N., & Goldberg, J. (2004). Inquiry in interaction: How local adaptations of curricula shape classroom communities. *Journal of Research in Science Teaching, 41*, 905–935. doi:10.1002/tea.20031.

Eshach, H. (2010). An analysis of conceptual flow patterns and structures in the physics classroom. *International Journal of Science Edition, 32*, 451–477. doi:10.1080/09500690802635247.

Furtak, E. M. (2006). The problem with answers: An exploration of guided scientific inquiry teaching. *Science Edition, 90*, 453–467. doi:10.1002/sce.20130.

Havu-Nuutinen, S. (2005). Examining young children's conceptual change process in floating and sinking from a social constructivist perspective. *International Journal of Science Edition, 27*, 259–279. doi:10.1080/0950069042000243736.

Kelly, G. J. (2008). Inquiry, activity and epistemic practice. In R. Duschl & R. Grandy (Eds.), *Teaching scientific inquiry: Recommendations for research and implementation* (pp. 99–117). Rotterdam: Sense Publishers. 288–291.

Kelly, G. J., McDonald, S., & Wickman, P. O. (2012). Science learning and epistemology. In K. Tobin, B. Fraser, & C. McRobbie (Eds.), *Second international handbook of science education* (pp. 281–291). Dordrecht: Springer.

Lemke, J. L. (2001). The long and the short of it: Comments on multiple timescales studies of human activities. *Journal Science of Learning, 10*, 29–43.

Lidar, M., Lundqvist, E., & Ostman, L. (2006). Teaching and learning in the science classroom: The interplay between teachers' epistemological moves and students' practical epistemology. *Science Edition, 90*, 148–163. doi:10.1002/sce.20092.

Lidar, M., Almqvist, J., & Ostman, L. (2010a). A pragmatist approach to meaning-making in children's discussions about gravity and the shape of the earth. *Science Edition, 94*, 689–709. doi:10.1002/sce.20384.

Lidar, M., Lundqvist, E., & Ostman, L. (2010). Comparative studies of manners of teaching. Communication presented at ECER 2010 *Education and Cultural Change*. University of Helsinki, 25–27 August 2010.

Loxley, P. M. (2009). Evaluation of three primary teachers' approaches to teaching scientific concepts in persuasive ways. *International Journal of Science Edition, 31*, 1607–1629. doi:10.1080/09500690802150114.

Makitalo-Siegl, K., Kohnle, C., & Fischer, F. (2011). Computer-supported collaborative inquiry learning and classroom scripts: Effects on help-seeking processes and learning outcomes. *Learning & Instruction, 21*, 257–266. doi:10.1016/j.learninstruc.2010.07.001.

Mehan, H. (1979). *Learning lessons: Social organization in the classroom*. Cambridge, MA: Harvard University Press.

Minstrell, J. (1992). Facets of students' knowledge and relevant instruction. In R. Duit, F. Goldberg, & H. Niedderer (Eds.), *Research in physics learning: Theoretical issues and empirical studies* (pp. 110–128). Kiel: IPN.
Mortimer, E., & Scott, P. (2003). *Meaning-making in secondary science classrooms*. Maidenhead: Open University Press.
Riemeier, T., & Gropengießer, H. (2008). On the roots of difficulties in learning about cell division: process-based analysis of students' conceptual development in teaching experiments. *International Journal of Science Edition, 30*, 923–939. doi:10.1080/09500690701294716.
Rincke, K. (2011). It's rather like learning a language: Development of talk and conceptual understanding in mechanics lessons. *International Journal of Science Edition, 33*, 229–258. doi:10.1080/09500691003615343.
Ruiz-Primo, M. A., & Furtak, E. M. (2007). Exploring teachers' informal formative assessment practices and students' understanding in the context of scientific inquiry. *Journal of Research in Science Teaching, 44*, 57–84. doi:10.1002/tea.20163.
Sampson, V., Grooms, J., & Walker, J. P. (2011). Argument-driven inquiry as a way to help students learn how to participate in scientific argumentation and craft written arguments: An exploratory study. *Science Edition, 95*, 217–257. doi:10.1002/sce.20421.
Sandoval, W. A. (2005). Understanding students' practical epistemologies and their influence on learning through inquiry. *Science Edition, 89*, 634–656. doi:10.1002/sce.20065.
Schwartz, R. S., Lederman, N. G., & Crawford, B. A. (2004). Developing views of nature of science in an authentic context: An explicit approach to bridging the gap between nature of science and scientific inquiry. *Science Edition, 88*, 610–645. doi:10.1002/sce.10128.
Scott, P., Mortimer, E., & Aguiar, O. (2006). The tension between authoritative/dialogic discourses: A fundamental characteristic of meaning-making interactions in high-school science lessons. *Science Edition, 90*, 605–631. doi:10.1002/sce.20131.
Seidel, T., Prenzel, M., & Kobarg, M. (2005). *How to run a video study*. Muenster: Waxmann. Technical Report of the IPN Video Study.
Sensevy, G. (2009). Outline of a joint action theory in didactics. In Proceedings of the Sixth Conference of European Research in Mathematics Education, Lyon. http://ife.ens-lyon.fr/publications/edition-electronique/cerme6/wg9-12-sensevy.pdf. Retrieved on 5 Jan 2012.
Sensevy, G. (2011). *Le sens du savoir. Éléments pour une théorie de l'action conjointe en didactique (The meaning of knowledge. elements for a joint action theory in didactics)*. Brussels: De Boeck.
Shimoda, T. A., White, B. Y., & Frederiksen, J. R. (2002). Student goal orientation in learning inquiry skills with modifiable software advisors. *Science Edition, 86*, 244–263. doi:10.1002/sce.10003.
Smithenry, D. W. (2010). Integrating guided inquiry into a traditional chemistry curricular framework. *International Journal of Science Edition, 32*, 1689–1714. doi:10.1080/09500690903150617.
Suzuki, M. (2005). Social metaphorical mapping of the concept of force "CHI-KA-RA" in Japanese. *International Journal of Science Edition, 27*, 1773–1804. doi:10.1080/09500690500206507.
van Zee, E. (2000). Analysis of a student-generated inquiry discussion. *International Journal of Science Edition, 22*, 115–142. doi:10.1080/095006900289912.
Varelas, M., Pappas, C. C., Kane, J. M., Arsenault, A., Hankes, J., & Marmotes Cowan, B. (2008). Urban primary-grade children think and talk science: Curricular and instructional practices that nurture participation and argumentation. *Science Edition, 92*, 65–95. doi:10.1002/sce.
von Aufschnaiter, C. (2003). Interactive processes between university students: Structures of interactions and related cognitive development. *Research in Science Edition, 33*, 341–374. doi:10.1023/A:1025452430958.
von Aufschnaiter, C., & Rogge, C. (2010). Misconceptions or missing conceptions? *Eurasia Journal of Mathematics, Science & Technology Edition, 6*, 3–18.
Vosniadou, S. (2008). *Handbook of research on conceptual change*. Hillsdale: Erlbaum.
Vygotsky, L. S. (1978). *Mind in society. The development of higher psychological processes*. Cambridge, MA: Harvard University Press.
Walker, K. A., & Zeidler, D. L. (2007). Promoting discourse about socio-scientific issues through scaffolded inquiry. *International Journal of Science Edition, 29*, 1387–1410. doi:10.1080/09500690601068095.

Watson, R. J., Swain, J. R. L., & McRobbie, C. (2004). Students' discussions in practical scientific inquiries. *International Journal of Science Edition, 26*, 25–45. doi:10.1080/0950069032000072764.

Wegerif, R. (2007). *Dialogic education and technology: Expanding the space of learning*. New York: Springer.

Wickman, P. O. (2004). The practical epistemologies of the classroom: A study of laboratory work. *Science Edition, 88*, 325–344. doi:10.1002/sce.10129.

Wickman, P. O., & Östman, L. (2002). Learning as discourse change: A sociocultural mechanism. *Science Edition, 86*, 601–623. doi:10.1002/sce.10036.

Woods, D., & Fassnacht, C. (2010). *Transana v2.42*. Madison: The board of regents of the university of Wisconsin system. http://www.transana.org.

Yager, R. (1997). Science education a science? *Electronic Journal of Science Education, 2*, 1–4. Retrieved 2 April 2012, from http://wolfweb.unr.edu/homepage/jcannon/ejse/yager.html.

Yeo, J., & Chee Tan, S. (2010). Constructive use of authoritative sources in science meaning-making. *International Journal of Science Edition, 32*, 1739–1754. doi:10.1080/09500690903199564.

Part VI
Teaching Resources, Curriculum

Chapter 30
Designing a Learning Progression for Teaching and Learning About Matter in Early School Years

Andrés Acher and María Arcà

1 Introduction

The atomic-molecular model of matter is one of the most important scientific models that permeate the science curriculum beginning in the early school years. Due to its prominent role in learning science, anchoring in this expert model, we want to support young students, from pre-k to 4th grade, to gradually develop increasingly complex models of the internal structure of materials that make sense for them while they engage in interpreting properties and changes in these materials. Learning progressions (LP), intended as "descriptions of the successively more sophisticated ways of thinking about topics that can follow one another as children learn about and investigate a topic over a broad span of time" (National Research Council, USA 2007), offers a working platform to think about how to accomplish this instructional goal. **The aim of this short article is to present** our proposed learning progression organized around three elements of design: (1) children's articulations of intuitive ideas, (2) materials that contextualize these articulations, and (3) teacher acts promoting and sustaining children's gradual articulations of ideas. We present illustrations of classroom work representing the interplay between these design elements drawn from several empirical investigations involving pre-k and elementary teachers and children. Some of these investigations have been already reported in our previous work (Arcà and Acher 2005; Acher, Arcà and Sanmartí 2007; Acher and Arcà 2009a, b). We use excerpts from these data, which include classroom discourse, children's drawings and explanations of these drawings, pictures of classroom events that include mimic games, or children's artifacts to identify aspects of the interplay of the designed elements that are evidenced in the classroom work. The

A. Acher (✉)
Martin-Luther-Universität Halle-Wittenberg, Halle-Saale, Germany
e-mail: andres.acher@paedagogik.uni-halle.de

M. Arcà
CNR, Rome, Italy

C. Bruguière et al. (eds.), *Topics and Trends in Current Science Education: 9th ESERA Conference Selected Contributions*, Contributions from Science Education Research 1,
DOI 10.1007/978-94-007-7281-6_30,

data help elucidate the meaning of these elements of the progression and help distinguish different refinements along the progression. The data are not intended as a test of a particular activity or teaching approach, or as an overall validation of the progression. Rather, the data help illustrate what kinds of classroom work are possible with appropriate support and how this work can be characterized using the three selected elements of design. Furthermore, although our proposed LP has been elaborated from data coming from 3- to 9-year-old children's classrooms, being constrained by length limitations, we have chosen to concentrate on illustrations that come from 3- to 4-year-old children's classrooms.

This chapter starts first by explaining our rationale for the three selected elements of our LP design representing also our units of analysis; second, we briefly describe the context of our work; third, we report examples to illustrate some *refinements* in our progression with some lessons learned in the design process. We plan to conclude this chapter by delineating our idea of *progression* and possible ways to continue with our work to finally identify some coincidences with other LP frameworks in similar areas of research.

2 A Rationale for Our Design Work/Design Commitments

Our overall aim is to engage young students in gradual constructions of increasingly complex models that work for them, to explain aspects of the behavior of and changes in materials. We want to create instructional support for this gradual engagement based on understanding what *refinements* along these constructions may look like when emerging from classrooms where multiple experiences with materials are complemented with guided reflections. It is in this context of engagement in in-depth knowledge constructions that we aim to present our work on designing a LP for teaching and learning about matter in early school years. We develop our work as formative research (Collins, Joseph and Bielaczyc 2004), testing and refining our design as a continuous process reported in prior research (op. cit. pg.1). Through this research, we are engaged in refining three elements of design that guide our ongoing work.

Our first element of design entails defining *ideas* to guide students' model constructions. We explore how children communicate articulations of two complementary *intuitive ideas* or *ways of understanding* (Arcà 1984) the properties and changes in materials when they interact with these materials and reflect with others in real classroom situations. We select two complementary *ideas* that are useful for them to make sense of everyday experiences with materials:

(a) The understanding of the macroscopic behavior of materials through invented microscopic elements that are parts of these materials. This reflects our instructional goal of encouraging young children to practice interpretations of materials through what is not evident at perceptual level. We think that the idea of *parts and a big quantity of parts* offers children the basis to develop an abstraction upon their own capabilities to break an entire piece of something and being challenged to think how to recompose interpretations of materials' behaviors through these parts. All this falls on the "standard problem" of explaining macroscopic properties of matter in terms of submicroscopic forms in what are meaningful and understandable ways for children. Much research has shown this to be a didactically difficult challenge (e.g., Wiser and Smith 2008). While we do not claim to solve this problem, we do argue for including this idea as a gradual instructional way out to start dealing with the problem.

(b) The understanding of these "invented parts" as elements of structures, in which not just the characteristics of the parts but the way they are connected and organized play a role in understanding the properties of and changes in materials. In this case, we understand that children can also be challenged to develop interpretations about material behavior upon different forms of relationships among the invented parts.

The above *ideas* or *ways of understanding* are not only chosen based on capitalizing on what children can do with them but also with what they can *potentially do* with them – i.e., the *generative character of these ideas* (Acher et al. 2007). This means that even when children communicate simple articulations of ideas to explain how materials behave, these simple articulations keep the potential to evolve and generate more complex articulations to explain in depth the same or other related behaviors. For example, if teachers work with children to start imagining the "inside of water" as many "tiny parts of water," they can continue working to encourage them imagining "how these parts are bonded to each other" when they "break" water into drops by different means. The ***generative character of these ideas*** is also exploited in modern scientific interpretations of the behavior of materials. Scientists generate new knowledge upon the idea of "particles organized in a structure." For example, within material science, the idea of "mesoscopic organization" has been implemented to define the minimum unit inside each material that can explain the properties of a perceptible chunk of material (e.g., Lhen 1995). So, we also selected the two ideas upon a *resonance* or a *productive match* with scientists' ways of knowing and doing (Arcà and Acher 2005; Acher et al. 2007). Through these two ideas, we like to think that epistemology and children's cognitive abilities share a fruitful space to examine how different articulations emerge in classrooms, from simple to complex interpretative articulations, characteristic of a scientific-domain-specific area. We want to inform our LP design by examining these articulations.

The second element of our design incorporates the *material contexts* in which the selected ideas are articulated and developed. These articulations emerge – and are also stimulated – from students' continuous interactions with the material world. For instance, when children start interpreting material behaviors through "invisible parts," they may be supported to think about these parts as *organized into structures*. This includes involving them in "doing things with these materials" like breaking them while discussing possible ways to explain how materials react to these actions through *imagining how the invisible parts are connected.* Through this second design element, we want to build on the school tradition of children manipulating materials, re-signifying the purpose of these manipulations through a "material" aspect of a scientific practice that deals with its interpretations (Tiberghien 1994). We want to reinforce the importance for teaching and learning science of the "doing and thinking" as two sides of the same coin. We want to inform our LP by examining the extent to which different materials and their manipulations provide a fruitful phenomenological anchorage to support – or to limit – children in generating, using, and revising the different forms of articulated ideas.

The third element of our design includes the broad range of *teacher acts* to promote children's articulations of ideas through reflections about connections between doing and thinking with these ideas (i.e., to create a fruitful "cognitive environment," Selmi 1982). This includes paying attention to teachers generating opportunities to strengthen and communicate children's views on materials, such as promoting reflections when comparing ideas or building on each other's ideas during group discussions. It also involves paying attention to manipulations of materials that "perceptively support" the emergence of ideas or to how to facilitate the articulation of ideas through drawings or gestures that are particularly important at early ages. The choice to include *teaching acts* in our LP responds to our view of the instructional responsibility teachers have to engage students in processing and reflecting about what children are doing and thinking. In our classrooms, we work with teachers to discuss how to *act* upon a twofold challenge: promoting and identifying articulations of ideas during children's manipulations of materials *and* understanding that reflecting on these articulations are aspects of a long-term intellectual process that children need to go through. We do not know what these reflections may look like, but as part of our design process, we learn from teachers *how they act* and create opportunities for this to happen. So we want to inform our LP design by examining *acts* that contextualize the articulation of ideas and children's manipulations of these materials from the point of view of what teachers do.

The above three design elements constitute the interwoven aspects we want to see "progressing." This means that we want to examine how young children in continuous interaction with the material world learn to gradually construct more complex interpretative abstract forms within their reach, with embodied meaning that transcends their tendency to think concretely about matter and its transformations. And we also want to examine how teachers contribute to making these gradual constructions reflective moments in their classrooms. From the many occasions on which we have worked with teachers to implement our proposal, we know that this requires many *refinements* that, in particular with young students, require a multiplication of experiences with materials to repetitively practice abstract interpretations from perceptive sources. In order to obtain "better pictures" of these *refinements* of our LP, we decided to "flip-flop" between the three units of analysis shown in Fig. 30.1 below. At the center of the figure lie the selected *ideas* or *ways of understanding*. Forms generated by articulating these *ideas* – including the very simple ones – are *models*, as children use those to interpret the materials that they manipulate. However, in our work with teachers, we decided not to hide these forms in the word "models" (already defined forms) as we prioritize the aim of obtaining insights into the process of understanding how these models are *generated* – something that

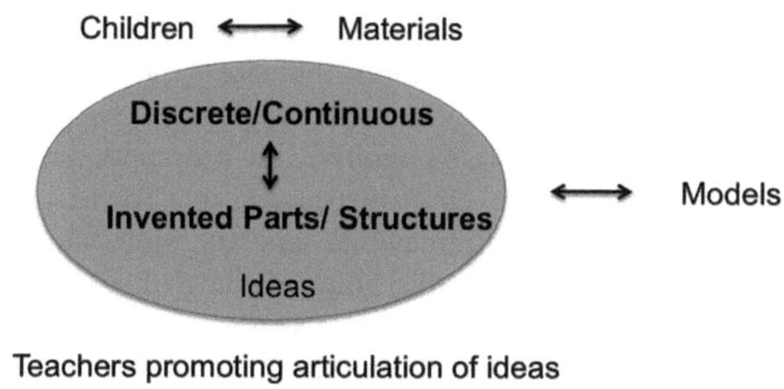

Fig. 30.1 Framework for designing our LP for teaching and learning about matter in early school years. At the center, the two *ideas* designed to guide children's interpretations of their interactions with materials from their physical environment (at the *top*). At the bottom, the identification of teacher acts promoting these interpretations and, on the right-hand side, models that emerge from this classroom work

is intrinsic to the modeling practices but that does not make modeling a particular learning goal as we contribute to clarifying in other LP work about scientific modeling with older students (Schwartz et al. 2009, 2012).

Our decision to "flip-flop" between the above units of analysis responds to the search for a balance between sometimes competing perspectives and the integration of diverse systems of knowledge. This integration is not easy and implies making compromises, balancing losses and gains. We accept these challenges in our LP design, guiding our decisions upon the commitment to contribute to building long-term, in-depth, and coherent thinking in children's science classrooms that respects children's capabilities and generates learning environments which stimulate and challenge these capabilities. So, our main goal in framing our research as an iterative design process involves examining "progression" in instructional real environments to identify classroom examples that "freeze" *refinements* in the progression, capturing classroom events that show some form of ideas and the models that emerge there from, and descriptions of *material contexts* and *teacher acts* contextualizing those events, with the hope of informing teachers, curriculum designers, and other researchers in our field.

3 Context for the Learning Progression Design Work

We based our iterative process of design on children's science classrooms on recreating a "cognitive environment." These *environments* emphasize the way both children and teachers – although with very different responsibilities – view their own participation as a challenge for understanding how to better articulate *ideas* to reach increasingly complex interpretations of the material world. In this case, the material world concentrates on material behaviors and its transformations. These challenges work as driving forces for teachers, who continuously work with children to generate curiosity and to look for explanations under a broad spectrum of long-term exploratory-perceptible ventures that are typical of young children. Furthermore, rather than thinking only in terms of activities, we work with teachers to understand how to implement these cognitive environments with opposite structures of partitioned teaching modules of instruction that focus on what we know (e.g., facts and skills). Within this framework, teachers motivate children to interpret their everyday experiences from the point of view of selected ideas/ways of understanding, for example, going on walks on a snowy and on a muddy day, seeking to make children interact with "different kind of waters" that they experience and then encouraging them to communicate their ideas about how these waters behave; also implementing different ways of representing ideas and the articulation of ideas such as drawings, body expressions, and gestures; encouraging children to "imagine" and to represent what may happen at an invisible level inside materials and when different materials interact between each other; assessing with children the organization of their ideas; and getting involved in explanations and in the reconstruction of new aspects of their manipulations with materials. Through this process, teachers promote flexible modes of participation involving children in whole-group and small-group discussions and considered collaborative aspects of knowledge construction. With greater teacher experience and stronger student capabilities, sometimes related with older ages, teachers work on more specific instructional challenges such as on extending the articulation of ideas emerging from one material context to explain a wide range of materials or comparing and assessing some forms of articulated ideas upon explaining different behaviors of the same materials. Teachers and children participate in experimental classrooms (n=20–28, in different classrooms, ages from 3 to 9 years old). Classroom sessions vary in length, from about 35 to 90 min, and in number, from 8 to 20, depending on age differences and the types of experiences in which children are engaged. As we stated in the Introduction section (p.1), we obtain children's data from this kind of classroom. We also collect excerpts from teachers' interventions and teachers' plans that help us to elucidate some *teacher acts* contextualizing children's articulations of ideas. These data help us to illustrate what kinds of *refinements* in articulating the selected ideas young children engage in when immersed in classrooms where the teacher plays a supportive role and how these *refinements* can be characterized to contribute to tracing the evolution of children's understanding of a material and its transformations. Through these illustrations, we also seek to improve the meaning of the selected design elements of our proposed LP as others – and ourselves – try to adjust them to new educational experiences.

4 Illustrations of the Proposed Learning Progression

In order to illustrate *refinements* in our progression within the space limitations of this chapter, we chose first to summarize the LP through the *articulation of ideas* and *transitions between these articulations* contextualized by one characteristic *teaching action and one associated challenge* (see Fig. 30.2 below). Our intention is to offer the

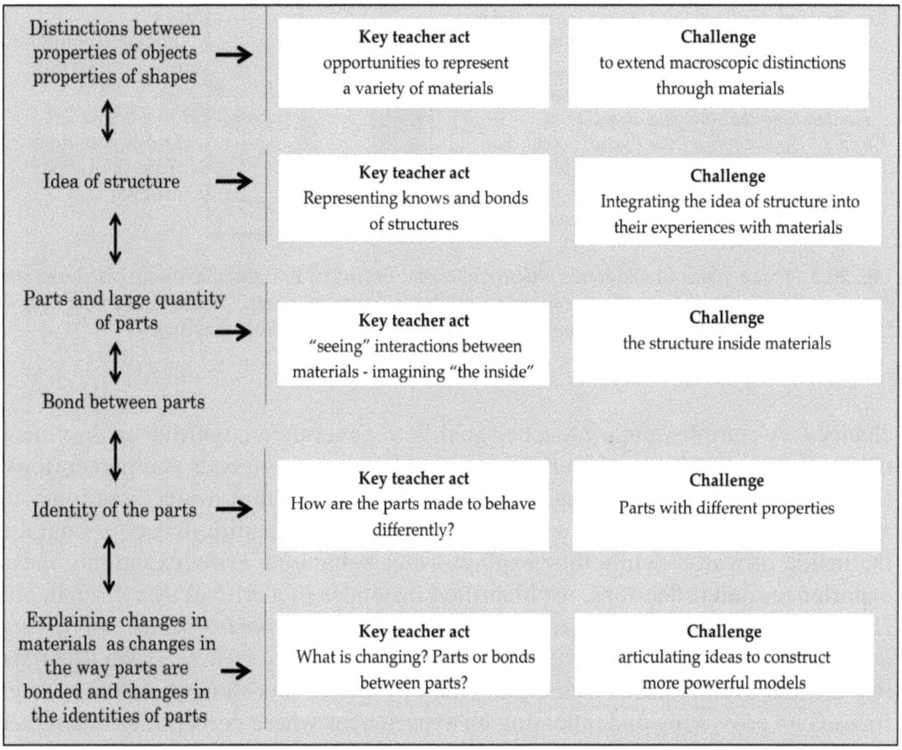

Fig. 30.2 Summary scheme of the whole proposed learning progression based on Acher and Arcà (2009a). Left-hand column: *transitions* in the *refinements* of the designed *ideas*. Arrows in both directions represent iterations of *refinements* that do not necessarily respond to a rigid longitudinal sequence. Right-hand column: key teacher acts and challenges in supporting children's *refinements*

whole LP picture, to then focus on some *refinements* based on classrooms of 3- and 4- year-old children as a way to show the kind of challenges that emerged from our examinations. We report these *transitions*, first by briefly describing the environment in which the **material context** to develop the first ideas is organized; we then comment on **teacher acts** generating opportunities to provoke and use **students' ideas** and end by showing examples of **students' models** as a way to assess the forms of organization of ideas that students use to interpret a material's behaviors.

5 Illustrating a Transition: From the Idea of Parts to Ideas of an Internal Structure of Materials

In our classrooms of 3-year-old children, teachers engage the children in a wide range of experiences with water, with the first goal of widening their perceptions regarding this material. The children experience water movements and

Teacher *[while observing children's expressions about different 'sizes of water' coming out from the syringes]: What is the difference between a drop and a trickle of water? How do I understand that this is a drop?*
Child 1*: The drop is a little one, all round. You can count them: 1, 2... A trickle is long... with many drops.*
Teacher*: Can we draw these differences?*

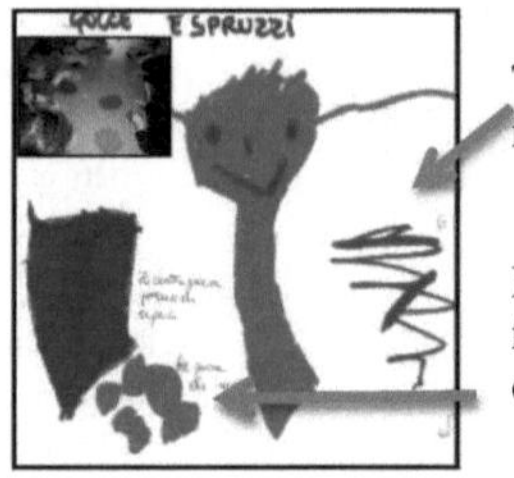

Fig. 30.3 Three-year-old children's classroom event. From left to right: classroom discourse and drawings and explanations when comparing different "parts of water." "Little and round drops" falling from the rubber drop compared with a "continuous trickle" from a syringe

changes. A complementary teacher goal is to generate a cognitive environment that supports the children's reflections about these experiences and perceptions, for example, by encouraging them to compare different water reactions, to anticipate what may happen with water under certain conditions, or "to imagine the inside of water" while they explain water behaviors. From examining these experiences and reflections, we identified instances of a critical *transition* in our LP: *promoting the idea of parts – and a big quantity of parts – while interacting with a whole chunk of material to understand its internal structure*. We illustrate two critical events of this transition. One of these events shows how the teacher focuses on provoking and reflecting on experiences where perceptions of smaller parts of water are potentially easy to obtain. For example, when the children engage in transporting water with hands, glasses, spoons, plates, and other artifacts, the teacher directs the attention to moments where the water "seems to break." She takes these moments as opportunities to start describing the "broken parts of water" as drops of an entire chunk. Soon thereafter, she motivates the classroom to find ways to "hunt for these drops" and takes advantage of a child's proposal to use syringes to do so. She sets experiences to widening the children's perceptions of drops of water through collecting, ejecting, and transporting water with syringes and rubber droppers. She captures the children's gestures and expressions and takes those as opportunities to continue experiencing the many ways to break water and to characterize "parts of water." For instance, parts of water are "drops" that fall and come apart from each other easily but become dribbles when water drips from a sponge or "continuous trickles" with syringes. Figure 30.3 below exemplifies this kind of moment, with the teacher steps of comparing two ways to obtain and characterize "parts of water": drops and trickles.

Refinements towards practicing the habit of "looking at" materials like water through the *idea of parts* emerge from the multiple opportunities that teachers generate to identify and describe these parts while interpreting a wide range of material behaviors. We illustrate a teacher's awareness of this need with classroom events in which she encourages children to continue comparing different water behaviors that include "drops of water into water." We observed in many classrooms

Teacher: *We have already played at making big and little drops. Let's try now to put colored drops in cold and warm water. Let's do that many times and let's also try to draw what we see.*

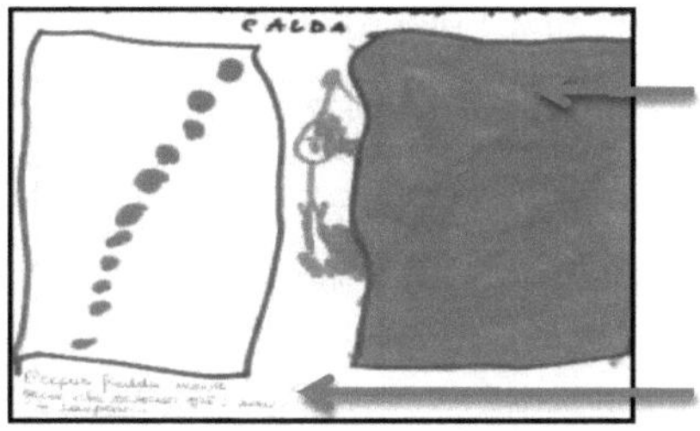

"This is warm water, when the colored drop gets there, it quickly becomes colored".

"Cold water with colored drops that fall in and do not break".

Fig. 30.4 3-year-old children's classroom event when discrete-continuous structures of water are experienced and characterized. *Long arrow* shows little drops of water falling in cold water. *Short arrow* shows colored drops coloring water

that teachers engage children in making drops with colored water and encourage them to observe and compare what happens when these drops fall in cold and warm water. This particular experience offers children the opportunity to continue finding ways to communicate *the idea of parts* either when colored water drops keep their parts structure in cold water or when those come back to a continued structure into warmer water. Figure 30.4 below illustrates this kind of classroom moment.

The other critical event which we chose in order to illustrate *refinements* towards the *idea of internal structures from an idea of parts and a big quantity of parts* consists in showing a teacher generating opportunities to ***use the ideas of parts*** to interpret material behaviors by imagining an *internal structure* of these materials. In the context of the same kind of experiences with water from the above examples, while children break water with their hands, the teacher encourages them to explain further, from the idea of parts emerging from these breaking experiences, what water is made of. For example, we found the teacher guiding children to imagine the internal structure of water by posing a problem: "...but if a whole chunk of water can be broken into drops, can a drop still be broken in parts?" She lets a drop fall onto the plastic surface of a table and encourages the children to observe how it breaks into many "little drops" and then poses another problem: if there are many "little drops," what makes them stay together in one entire drop? The children develop their explanations based on imagining a "special glue" that "holds the little drops together" as if they were inside the drop. Figure 30.5 below shows a classroom event when internal structures of drops of water emerge.

Figure 30.5 above shows how children are guided to *use the ideas of parts* to construct their first models that explain the behavior of a drop of water – a chunk of material – through its internal structure. A drop made of "many little drops" held together by an invented "drop glue" reveals the ideas which they articulate to explain what happens when a drop of water falls onto a table. These are 3-year-old children's models that offer us partial proof of children's use of the designed ideas of our LP.

Teacher: *How do we understand that water is made all of little parts like some of you have said ?*
Child 1: *Blow, blow, and then you will see that little parts come out from the water and those get apart.*
Child 2: *I made these little parts with my fingers, the water moves.*
Teacher: *Is it ok to say that those are 'little parts of water'?*
Child 1: *Yes, because the little parts of water are called "drops of little parts".*
Child 3: ***'Little drops - little parts'.***
Teacher: ***And how would 'little drops - little parts' be made** inside?*
Child 4: *There are still little 'little parts' . Look, you can see them.*

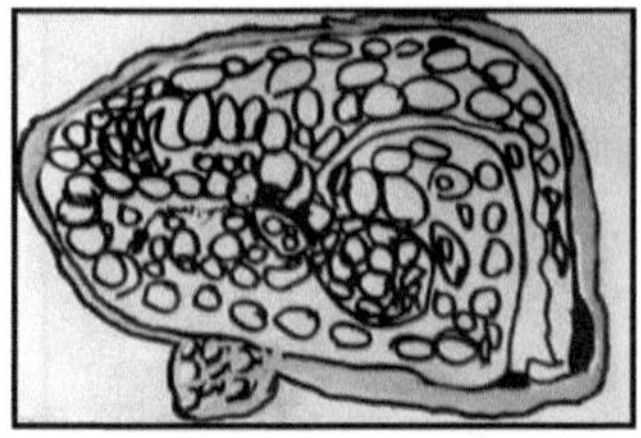

Elisa [explaining her drawing]: I made a big drop made of many little drops. Then, I also made the 'drop glue' that makes drops stay together .

Fig. 30.5 Three-year-old children's classroom event in which models of internal structures of a drop of water emerge

6 Extending the Idea of Parts to Different Materials and Refining the Idea of Structure

Another *transition* of our designed LP entails extending the *idea of parts* to different materials and refining the *idea of internal structures* with first *ideas of bonds between parts* to explain material behaviors. We illustrate this *transition* through events in a classroom of 4-year-old children, where the teacher, upon developing the idea of parts as components of different materials, first guides children to look inside these materials and then to focus on refining the idea of internal structure. The event which we present comes from similar classroom contexts to those shown in the previous examples. This time, one of the teacher's goals is to engage children in a wide range of experiences with different materials, encouraging them to perceive and to interpret "whole chunks" of those materials as composed by many parts. Through materials intentionally selected according to their easily perceptible discretization, like chunks of butter, bread, water, or chalk, children describe with words and drawings the visible structure of these materials, having broken them by different means while looking for "different powders" of those materials. Once the children have gained experience with macroscopic discrete views, the teacher engages them in experiencing interactions between water and the materials. For example, chalk-water interactions offer the children the possibility to experience decomposition-composition of chunks of chalk in many parts. It is during this kind of experience that the teacher guides the children's view to the inside of materials, supporting the construction of first models to explain internal structures of materials. Figure 30.6 below represents this kind of classroom event. On the left-hand side, there is a classroom dialog with an explicit teacher request to look inside a chunk of chalk during chalk-water interactions. On the right-hand side, there is a child's drawing with his own explanation on this material interaction.

Interactions between materials after working with discrete views of those materials constitute a critical *material context* to guide children's interpretations towards

Teacher: *Let's look inside [the chunk of chalk]. How does water get into there?*
Child 1: *The water in the chalk got very far inside....*
Teacher: *How do you know that?*
Child 1: *Because when you cut the piece of chalk, we could see the water going very far inside. It was all wet.*
Child 2: *No, the water makes everything wet. If you are soft, weak... it gets very far inside, but if the inside is very hard, the water can't get in and stays outside like little drops.*

"This is water getting into th e chalk because it finds holes and gets through... ".

Fig. 30.6 Four-year-old children's classroom event in which the teacher proposes that the children explain the "inside of chalk" after observing their experiences of putting chalk into water

Fig. 30.7 Four-year-old children's classroom event in which children pretend to be parts of materials. They get tight or loose together depending on the characteristic of the material they want to represent

more refined models of internal structures. We found, however, that *refinements* towards these models need more work from teachers. We examined many ways in which teachers intend to refine models of internal structures through *ideas of bonds between parts* in our classrooms of 4-year-old children. We found teachers engaging children in mimic games in which they play at being "children materials" while explaining how materials are made inside, allowing the other material to get inside them or not. For example, we observed some "chalk children" hugging and holding themselves tight as if they were "little parts of chalk together till the water comes," as they say. We also observed "water children" who fight to get inside the "chalk children," getting them all wet, as they also say. Figure 30.7 below represents this kind of classroom event.

Through mimic games, we illustrate how teachers compose their acts using the power of metaphors as fruitful resources, especially at these ages, to re-signify the children's experiences and support them in tackling the *refinements* of more complex models of internal structures of materials that go beyond an idea of parts to integrate an idea of bond between these parts. Paying attention to the many ways these parts are held together constitutes partial evidence of children refining their ideas of internal structures of materials. We found further evidence when teachers intend to make the children use the incipient *idea of internal structures*. We observed teachers encouraging children to compare different materials interacting with water, guiding their interpretations of these experiences to "look inside materials." They

Teacher: *Let's put the duplo into the water. Does the same thing happen? [like with chalk]*
Child 1*: No, because the chalk wants the water and the duplo doesn't want it at all. And it keeps it outside. And you can see it's all wet when you touch it. Inside, the duplo is made of very hard parts that hold together very tight like we have already done.*

This is the chalk with little parts inside. These are a bit together but don't hug tight. And there is the water. It gets inside and fills it.

The duplo has little parts inside that are very close together, like children hugging tight. Over here, I've drawn the water drop that can't get in.

Fig. 30.8 From left to right, 4-year-old children's classroom discourse, drawings, and explanations of water-chalk and duplo-water comparisons after experiencing breaking different materials

take advantage of the different levels of permeability of materials to refine the idea of internal structures. Figure 30.8 illustrates this kind of classroom experience. Children compare chalk-water and duplo (plastic)-water interactions first by having group discussions, then drawing what they think happens between the materials, and finally, explaining their drawings.

Children's models with *refined* internal structures of materials that explain what materials are made of as well as material behaviors during interactions with water emerge upon *teacher acts* that include the ones shown above. Teachers compose their acts integrating critical material contexts that, for example, support a potentially easily perceptible idea of parts, to then facilitate visible-abstract passages towards the inside of materials while interacting with colored water. But teachers also display a battery of making-sense tools like metaphorical mimic games or analogical comparisons to guide children in using the idea of parts in novel contexts and imagining how these parts are linked together, to explain the structure of novel materials or some material behaviors.

What we want to highlight through the above illustrations is the way we like to see *refinements* during young children's *constructions of models* that work for them to explain different aspects of the changes in and of the behavior and internal structures of materials. For us, while focused on children's articulations of *ideas*, the characterizations of *refinements* need to also include *a material-context component* and a *teacher-act component* as critical for learning environments in children's science classrooms. Our decision to integrate these three components is in line with the instructional goal that we pursue with our design. We want to capture instances of the way in which we work with our teachers. We explain to them the potential of our designed ideas, and at the same time, we challenge them to understand how they can enrich children's perceptual experiences with materials to recreate their ideas of internal structures made of parts in different materials. We learn from them how they work on making these ideas emerge through a sensitive guidance of classroom talk, free drawings, and other kinds of representations that trigger children's imaginations. Complementarily, we also work with them on the potential of using emerging children's models as their own guidance for checking "how things are going."

Water is a critical material to accomplish these goals. We observed that teachers focus on exploring the structure of water, encouraging the children to develop analogies between drops of water that "perceptively" get together with solid materials from which it is easy to make parts but which are not easy to put back together.

7 A Key Point in Revisiting the Meaning of Our Learning Progressions

Our idea of *learning progression* builds upon the kind of *refinements* that we like to see occur not only in an incremental *vertical way* represented by complex articulations of ideas but also in a *transversal way* represented by the ability to use these articulations to interpret different material contexts and aspects of these contexts. This *vertical and transversal* growth in the progression emerges from *iterative refinements* in which children develop the habit of constructing and using different articulations of *ideas* to understand the behavior of and changes in materials. This entails them revisiting the idea of internal parts as many times as needed to then think through incrementally complex articulations as their own abilities – and the teacher's ability to make this happen –allow. At an early age, starting with classrooms of 3- and 4-year-old children, we aim then to practice "ways of looking at" materials that retain the potential for further elaboration in other school years, when learning about molecular structures and chemical or physical transformations.

We continue the process of understanding what *refinements* look like and "how far" it is possible to go in the progression. For us, researchers and science educators, this idea of progression allows us to rethink what would be needed to support the construction of children's models that include other aspects with generative anchorages on the atomic-molecular model of matter, which are necessary in different school years. This may include ideas of ratio between weight and density (Wiser and Smith 2008), built upon the idea internal parts structures, or may include creative ways to stress reflective aspects of scientific modeling such as more explicit ways to reflect on the selected ideas as generative ideas (i.e., aspects of metamodeling knowledge).

8 Some Coincidences with Other LP Frameworks

In our ongoing iterative design cycle of our LP for teaching and learning about matter in early childhood, we commit ourselves to coupling learning aspects involved in deepening children's *domain-specific ways of knowing* upon children's own potential. We also commit to broadening this knowledge with aspects of its construction and use, including *material contexts* and *teaching acts* as key aspects of learning environments. Furthermore, we constantly obtain our evidence from real classrooms, where we look for insights into habits of mind that, either for students or for teachers, take years to become established classroom norms. These commitments are aligned with what has recently been called a "requisite of a learning progression" (Duschl et al. 2011).

Our idea of *refinements* in the progression, although driven by epistemological decisions, does not concentrate on constantly testing alignments to a target canonical step of the atomic-molecular theory of matter. We rather commit to the idea of the *generative character* of our designed ideas and the many ways of making this happen in the classrooms – *the instructional element*. In that sense, we are proposing and examining how to fine-tune *resonances* between children's and scientists' ways of knowing and doing can be part of children's science classroom culture. Few or no studies have been done in early childhood from the perspective that we are bringing here. However, our narrative of *progression*, through its *vertical* and *transversal* dimensions of growth, is aligned with other European perspectives that, although based on older students, pushes for attending to long-term didactical processes that seriously consider alternatives which integrate general skills with content knowledge, the points of view of students and the ability of teachers to build upon them (e.g., Lijnse 2007). Engaging in designing, investigating, and revising long-term teaching-learning processes in children's science classrooms entails different challenges. While many of these challenges do not differ from those that part of our community identifies as main challenges in the learning progressions in science (Alonzo and Gotwals 2012), we like to choose one core challenge: the ability to view key *cognitive children's abilities*, the *material context* in which these abilities develop, and *teacher acts* not as competing forces but as integrated goals to work with in conceptualizing a learning progression for teaching and learning about matter from a very early age. Our learning progression is an argument, supported by evidence, about the need to pay attention to the integration of those three forces in designing possible pathways and uncovering their associated challenges.

Acknowledgments We wish to offer our heartfelt thanks to teachers Anna Aiolfi and Marika Quaglietta, Scuola dell' Infanzia Andersen 1 Circolo di Spinea, Italy. We also wish to thank Lisa Kenyon and Maria Evagorou for first revisions of this manuscript and anonymous reviewers for their suggestions.

References

Acher, A., & Arcà, M. (2009). Children's representations in modeling scientific knowledge construction. In A. Teuval (Ed.), *Representational systems and practices as learning tools in different fields of knowledge* (pp. 109–133). New Jersey: Sense Press. Chapter VIII.

Acher, A., & Arcà, M. (2009). Modeling materials in early school years. Paper presented at the annual meeting of AERA, San Diego, USA. Symposium: Developing and refining a learning progression for matter from pre-K to Grade 12: Commonalities and contrasts among four current projects. Chair: Carol Smith. Discussants: Joe Krajcik & Clark Chinn.

Acher, A., Arcà, M., & Sanmartí, N. (2007). Modeling as a teaching learning process for understanding materials. A case study in primary education. *Science Education, 91*, 398–418.

Alonzo, C., & Gotwals, A. W. (2012). *Learning progressions in science: Current challenges and future directions*. Rotterdam: Sense Publishers.

Arcà, M. (1984). Strategies for categorizing change in scientific research and in children's thought. *Human Development, 27*, 335–341.

Arcà, M., & Acher, A. (2005). Children's Models in Scientific Knowledge Construction. In: *Children's drawing: Its relation to learning and instruction in kindergarten and primary Education*. Symposium. N. Scheuer (Chair). 11th EARLI Conference Proceedings. Nicosia, Cyprus.

Collins, A., Joseph, D., & Bielaczyc, K. (2004). Design research: Theoretical and methodological issues. *Journal of the Learning Science, 13*(1), 15–42.

Duschl, R., Maenga, S., & Sezenb, S. (2011). Learning progressions and teaching sequences: A review and analysis. *Studies in Science Education, 47*(2), 123–182.

Lhen, J. M. (1995). *Supramolecular chemistry*. Weinheim: VCH.

Lijnse, P. (2007). Didactical structures as an outcome of research on teaching-learning sequences? *International Journal of Science Education, 26*(5), 537–554.

National Research Council – USA. (2007). *Taking science to school: Learning and teaching science in grades K-8*. Washington, DC: The National Academies Press.

Schwarz, C. V., Reiser, B. J., Davis, E. A., Kenyon, L., Acher, A., Fortus, D., et al. (2009). Developing a learning progression for scientific modeling: Making scientific modeling accessible and meaningful for learners. *Journal of Research in Science Teaching, 46*, 632–654.

Schwarz, C. V., Reiser, B. J., Acher, A., Kenyon, L. O., & Fortus, D. (2012). Issues and challenges in defining a learning progression for scientific modeling. In A. Gotwals & A. Alonso (Eds.), *Learning progressions for science*. New Jersey: Sense Press.

Selmi, L. (1982). *La Sezione dei cinque anni. [The classroom of 5-year-old children]*. Milano: Fabbri Eds.

Tiberghien, A. (1994). Modeling as a basis for analyzing teaching-learning situations. *Learning and Instruction, 4*, 71–87. 1994.

Wiser, M., & Smith, C. (2008). Teaching about matter in grades K-8: When should the atomic molecular theory be introduced? In S. Vosniadou (Ed.), *International handbook of research on conceptual change* (p. 205). Mahwah: Lawrence Erlbaum.

Chapter 31
'Realistic-Fiction Storybooks' as a Resource for Problematic Questioning of Living Being with Pupils in Primary School

Catherine Bruguière and Eric Triquet

1 Introduction

Since the 2000s, scientific education has been part of the development of an integrated approach to science and literacy which is based on a common disciplinary epistemology (Millar and Ogborne 1998). In this context, the aim in scientific literacy is to ensure that the pupils acquire more than knowledge about the basic concepts in science but also a vision of *how* such knowledge relates to other events, *why* it is important and *how* this particular view of the world came to be (p. 174, Osborne 2007). The hypothesis supported by scientific literacy is that the link between the reading and writing of learning texts is a necessary condition for scientific conceptualisation (Butzow and Butzow 1998). By suggesting a broader approach to literacy, Street (2001) extends the practices of reading-writing to diverse semiotic systems which offer new tools for interpreting the world; such is the implicit conjecture contained in the term literacy used in many English-language research studies.

Different studies on science curriculum plans structured around the triumvirate of 'speaking, reading and writing' (i.e. French, Australian, UK curricula) show improvements in pupils in primary school. Chall (1996) and Saul (2004) show that pupils who use reading and writing to acquire knowledge about scientific subjects develop, at the same time, more reading and writing skills. Nevertheless, this transition from 'learning

C. Bruguière
EA 4148 S2HEP, University of Lyon 1, ENS-Lyon, University of Lyon, Villeurbanne Cedex, France

E. Triquet (✉)
EA 4148 S2HEP, Université Lyon 1-ENS de Lyon , Université de Lyon, Villeurbanne 69622, France

Université Grenoble 1, Saint-Martin-d'Hères 38041, France
e-mail: eric.triquet@ujf.fr

C. Bruguière et al. (eds.), *Topics and Trends in Current Science Education: 9th ESERA Conference Selected Contributions*, Contributions from Science Education Research 1,
DOI 10.1007/978-94-007-7281-6_31, © Springer Science+Business Media Dordrecht 2014

to read' to 'reading to learn' is not without difficulty for primary-school pupils because of the different nature of the texts to which they are exposed (Fang 2006).

Whereas they learn to read using narrative texts, they learn sciences from expository texts that are problematic for them. Consequently, the choice of curricular materials becomes all-important in the development of scientific literacy in primary school: Which texts should be chosen among the growing range of books for children? Should informational texts or fictional texts be used? How does one combine the use in class of different types of texts that do not belong to the same field of application?

Numerous studies (Rice 2002; Schussler 2008) focus on spotting inaccuracies or the gaps inherent in children's literature. More interesting are those studies that analyse the role of a type of literary resource such as the 'Dialogue Concerning the Two Chief World Systems, Ptolemaic and Copernican' (Piliouras et al. 2011) and 'The Panda's Thumb' by Gould (1980) (Straits and Nichols 2007) in teaching scenarios that prioritise knowledge about the nature of science. When reading scientific texts is combined with writing realistic stories, some authors (Ritchie et al. 2008; El-Hindi 2003) reveal that primary-school pupils show a considerable improvement in the use of canonically accurate knowledge of concepts and the use of concepts becomes more familiar. These mixed-genre writing activities would appear to be a particularly relevant resource in primary school because they involve pupils in a constant work of critical back-and-forth between reality and fiction.

In this chapter, we propose to focus on mixed-resource, realistic-fiction storybooks (Bruguière et al. 2012; Bruguière and Triquet 2012). It is a matter of identifying to what extent such a resource can enable pupils not only to question but also to test their representations of 'reality'. By 'reality', we mean a (scientific) reality constructed by scientists. It represents the current scientific knowledge of what surrounds us.

2 Theoretical Framework

2.1 Narrative in Science Education

Studies that look at the narrative dimension in science education and more particularly those that consider narrative as a valued component of scientific literacy (Strube 1994) are connected not only with the narrative turn of the 1990s in the United States where storytelling originated but also with the vision of scientific education espoused by UK education specialists of that period, who advocated 'greater use of the most powerful and persuasive way of conveying ideas: the narrative form' (Reiss et al. 1999).

The foundation of these approaches can be linked to the cognitive perspective of narrative envisaged by Greimas (1966), who consider narrative to be the telling of sequential events, whether real or imaginary, in a way that portrays a meaningful, coherent whole. From Strube's work, we draw out three arguments for the usefulness of narrative in science education that can also partly be found in Bruner:

- The consideration of curriculum integration
- Narrative as a pedagogical tool
- Narrative as the reflection of the way in which the human mind orders experience

Narrative structure, Strube tells us, not only explicitly links disparate understandings into a coherent whole but also gives value to those links 'because they form and allow narrative to occur'. Moreover, narratives contribute to the construction of a common culture that makes it possible to establish curricular coherence with science; students need to share in the stories that scientists tell each other.

Narrative is an effective teaching tool because it places concepts in acceptable, easily assimilable and memorable form. On this subject, Fisher (1987) talks about *Homo narrans*, whose particular feature is to recall the objects in an artificial story in order to be able to remember them. Beyond that, it is through narrative that pupils' thoughts are ordered and structured, whatever the discipline in question. Therefore, a command of narrative seems essential in a science curriculum plan. Nevertheless, the explanatory capacity of narrative is put into perspective because more often than not it does not enable one to transcend a linear mode of reasoning that is causal (Viennot 1996) or sequential (Orange-Ravachol 2005). But we will focus on the way in which narrative deals with the world (which exists) but is directed not at how things are but rather as how they might be or might have been (Bruner 1997, 2002).

2.2 Narrative Realistic Fiction as Hybrid Literature

It is because we consider that fictional narrative has a close connection with reality that we can describe certain realistic-fiction storybooks as being likely to present a knowledge value for readers. We thus reject the position which removes fiction from all external references, either because it considers fiction to be strictly self-referential, anchored only in the imagination of its author, or because it considers fiction as a substitute for evading reality, as is Salmon's criticism (2007) of storytelling. In the latter case, we understand the risk for pupils of transforming reality into fiction and of attributing a real value to fiction. On the other hand, we adopt the position developed by U. Eco (1979, trad; 1985) where he shows that the worlds constructed through fiction are not by any means fanciful worlds because nothing there is entirely contingent and unrelated to the world that we know, or rather the world that we construct.

Due to the fact that it extends its reference beyond the internal world that it describes to the external world of natural phenomena, fictional narrative can offer us scientific knowledge about the real world.

In children's storybooks, the fictional world is recounted through a 'narrative' structured around a plot that is often developed according to a quinary system (Larivaille 1974): initial state/complication/dynamic/resolution/final state. Through these different transformations made necessary for the purposes of the story, a meaningful coherence is constructed. It is this cognitive function of the story which interests us, because it partly ties up with the activity of scientific problem-posing.

2.3 Narrative Realistic Fiction as Problematic Questioning in Science

The importance of problems in the functioning of science is now a matter of consensus. Therefore, the production of scientific knowledge cannot be detached from the problems on which it is based (Bachelard 1927, reprinted 1976). For Michel Fabre (1999), 'identifying a problem presupposes a background knowledge considered, at least provisionally, to be reliable'. When this background knowledge is proven wrong in a particular situation, this will give rise to the problem. The underlying idea is therefore that the mastery of a problem is not limited to its resolution but also includes its construction, that is to say, a phase of making connections between the elements identified as significant and pinpointing what is essential compared with what is of secondary importance.

The meaning of the fictional narrative is achieved in the conflict between the elements of fiction on the one hand and the elements of reality on the other, between therefore what is within the realms of possibility and what reflects scientific validity.

A story begins therefore when a complication appears in the order of things as we expect them, which disrupts our representations of the world. There we find a similarity with the scientific problem as we have envisaged it. These shared characteristics lead us to form the hypothesis that in realistic-fiction stories, the plot is a potential lever for a problematic questioning of the reality to which the story refers. In fact, reading such a story does not lead readers directly to wonder about the world, but rather to wonder about the understanding of the logic of the plot. Therefore, we seek to highlight how readers are led to question themselves about elements of reality in order to construct meaning, based on their own knowledge. Such questioning arises from the understanding of the story's message, most often crystallised in a statement described as the 'golden sentence' by Tompkins (2003), to which we accord a problematic dimension because its meaning is not obvious to pupils from the outset.

The plot resolution, meanwhile, corresponds to the sequence of actions that aim to resolve the initial complication. According to Bruner (1997), it is by 'domesticating the unexpected and the extraordinary' that stories give shape and meaning to the world around us and lead readers 'into the realms of possibility, of what could be, of what could have been, of what will be perhaps one day' (Bruner 2002, p. 16). Consequently, one question is knowing what knowledge of the real-world readers will need in order to construct a possible world that is as close as possible to the world of reference in the story (Bruguière et al. 2007). But more interesting still is the question of how this type of story can lead readers to a reconstruction of reality based on the possible world of the fiction. Here, there is a reversal which we think is particularly interesting to explore.

3 Research Question

Following Bruner (1997, 2002), we therefore attribute an epistemic aim to realistic-fiction stories. We argue that by calling upon readers to engage in 'thought experiments', under certain didactical conditions these stories will lead them into a double process of questioning of reality and of 'projections' of their own representations. Our discussion will explore two questions.

The first question is knowing how the reading of a fictional story such as *Fish is Fish* (Lionni 1974, translated into French in 1981 under the title *Un poisson est un poisson*) can enable pupils to ask themselves questions about the real-world phenomena that are at issue. Our hypothesis is that interpretative work on the problematic statement *'un poisson est un poisson'* ('a fish is a fish') should help pupils aged 7–8 years to simultaneously question the notion of species and that of classification criteria.

The second question is how to define what knowledge of the real world these pupils need in order to understand the logic of the plot. Therefore, we seek to highlight how they are led to construct meaning by testing their own representations.

4 Methodology

The analysis focuses on a case study relating to the question-based reading of a realistic-fictional storybook *Fish is Fish* (Lionni 1974, translated into French in 1981), in the course of two didactical situations that arise from the storybook's actual narrative. They were put into practice by a teacher with French pupils aged 7–8 years (21 pupils) as part of a science sequence within the syllabus about the unity and diversity of living things.

Below is a summary of the storybook *Fish is Fish* (Lionni 1974, translated 1981):

> Two inseparable friends – a minnow and a tadpole – swim in a pond. But one morning, the tadpole notices that he has grown two little legs and proudly announces that he is a frog, which his friend the minnow disputes. Once he has become an adult, he goes off to explore the terrestrial world. Back in the pond, he tells his friend about all the extraordinary things he has seen. The fish imagines birds, cows and men just like him. Finding himself alone again, he decides to go out in search of the other world. There, on the banks of the pond, he almost suffocates. Fortunately, his friend comes to his aid by pushing him back into the water. He then realises that a fish is a fish.

Our pupils' corpus is made up of recorded exchanges as well as drawings done during one of these situations. The chosen storybook, which can be considered emblematic of realistic-fiction stories because it enables an integrated approach between science and literature (Butzow and Butzow 1998) or the arts (Hansen 2006), is not written by a scientific author. Presumably, its aim is therefore not to make science more accessible, as is the case, for example, with the Uncle Albert trilogy by the physicist Stannard (2001) or the illustrated books written by Hawking et al. (2007). This storybook belongs to realistic fiction in the sense that the story is subject to certain reality constraints, the understanding or at least the questioning of which is akin to understanding the fictional narrative. Another aspect of this contemporary storybook is the relationship of interdependence between the text and the image, where the image is neither illustrative nor denotative but connotative, since it functions as understanding in relation to the text (Defourny 2009). What is at stake in the narrative is situated in neither the text nor the image but in an intermediary space, an intelligible whole termed 'chronotrope' by Bakhtine (1984), who links the two inseparably. The storybook very much appears as a scripto-visual object (Jacobi 2005) or as offering multimodal texts (Hasset and Curwood 2009), the narrative unit of which is the double page because it represents for readers, whatever its format (portrait or landscape), the space of inscription of the text and the image. In fact, the movement from one double page to the other sets the pace of the narrative by creating the effect of suspense. We will therefore take the double page as our analytical unit, which leads us to split the storybook under study into 15 double pages.

The analysis of the corpus is guided by two lines of analysis, which are driven by the didactical analysis of the storybook:

- A line of analysis that tackles the problematic dimension of the storybook, crystallised in the paradoxical expression *'Fish is Fish'* and the way in which the pupils take that on board. *'Fish is Fish'* represents the problematic statement insofar as the first *Fish* refers to a particular individual, whereas the second *Fish* refers to fish as a group. Indeed, it raises the following question: in what way is a (particular) fish a (generic) fish? The proposition whereby the same word, 'fish', designates different references constitutes a semantic paradox which is a driving force, as Clément et al. (2004) demonstrated in reference to language-exchange students putting in place an argumentation on an epistemological level. The whole story tries to develop the relationship between the two characters depending on the understanding that they share this statement or not. Therefore, the tadpole, by asserting that 'frogs are frogs and a fish is a fish; that's just the way it is', expresses a universal knowledge according to which the specific identity of each individual is immutable and that this identity changes over the course of events inscribed in its development, an idea disputed by the minnow.

  The activity in which the pupils are invited to engage consists of oral exchanges on the meanings taken on by the statement *'Fish is Fish'* in the course of the story.
- A line of analysis driven by the characterisation of a species and the way in which the pupils select criteria and assess their pertinence. This supposes that the criteria are not considered as having been given or taken at random, but as problematised criteria, that is to say, criteria for which thought has been given as to why it is possible for them to be chosen and as to why it is impossible that any

other criteria be selected. In the storybook, it is the fish that, in striving to visualise the species of the aerial world that the frog describes to him, is employed in a work of characterisation. He is going to combine known characters with unknown characters to imagine new animals (cows, birds, people). The proposed activity is modelled on the thought experiment in which the fish engages. It consists in each pupil drawing a representation of how the fish imagines the cow to be, from the description given him by the frog.

p. 9. 'Cows', said the frog. 'Cows! they have four legs, horns, eat grass, and carry pink bags of milk'.

Instructions: draw *how the fish imagines the cow according to what the frog tells him*

5 Data Analysis

The two teaching situations which we developed are situated differently with regard to the fictional story but are part of the same work of constructing the problem relating to what an animal species is. The first concerns the statement 'Fish is Fish' and takes place after the full reading of the storybook, whereas the second, which consists in doing imaginary drawings, takes place after reading the first seven double pages.

5.1 Plot and Problem-Posing

In the first didactical situation, the pupils have the whole of the storybook text on the board in front of them and are invited by the teacher to pick out when the problematic statement 'Fish is Fish' appears and the different meanings that it assumes depending on the character who utters it and the moment it occurs in the story.

It is a matter of identifying to what extent these pupils grasp the problematic dimension of this statement to engage in biological questioning on what a fish species is compared to a frog species. Is the fact that there are convergences between these two species, which live in the same environment at the larval stage, enough or not enough to consider them as belonging to the same species? In what way is this scientific questioning coupled with epistemological questioning?

5.1.1 Analysis of Two Key Moments of Discussion During the Didactical Situation

The first moment of discussion consists of a comparison of the meanings assumed by the statement *'Fish is Fish'* when it is uttered by the tadpole (p. 3) and by the fish (p. 15).

p. 3: They argued and argued until finally the tadpole said, 'Frogs are frogs and fish is fish and that's that!'

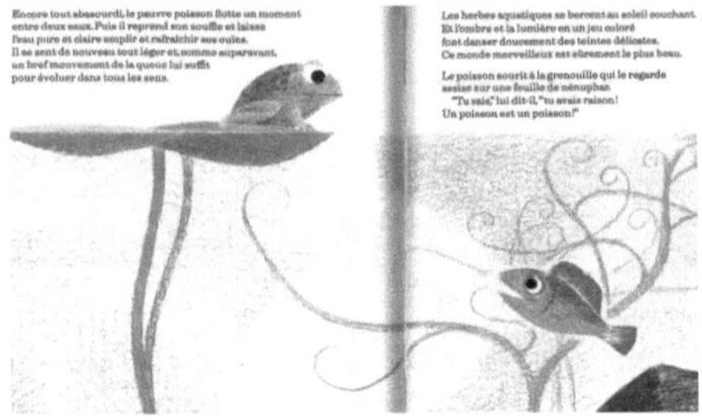

p. 15: He smiled at his friend the frog, who sat watching him from a lily pad. 'You were right', he said. 'Fish is fish'.

It is the status of the statement *'Fish is Fish'* that is under discussion. For pupil L, the difference in meaning stems from the difference in speaker, in the first sentence it is the tadpole speaking and in the second sentence it is the fish, whereas for pupil A the difference lies in 'how' these two characters say it: in one case, the tadpole explains or rather asserts that a fish is a fish, whereas in the other case, the fish admits because he has the proof (the reason) for which he is a fish and belongs to the fish species.

L Well, yes, a bit but…. Well, in fact, there it's the tadpole who says a fish is a fish, and that's just the way it is.

M It's the tadpole who says that, okay.

L And then there.

M Who is speaking there?

L There, it's the fish.

M Yes, okay. So there, we have, it's the tadpole who says that and there it's the fish. Does that change anything? Go on A….

A Well, it doesn't really change the difference, it doesn't change because it doesn't matter what he says, but what changes is that there he explains it to him.

M Where do you mean, there?

A Well, when the tadpole speaks there, he explains it to him.

M The tadpole explains.

A Whereas there, compared to there he doesn't explain there, he doesn't explain, he doesn't explain, er you know, er, you were right, a fish is a fish, there er, he means that he admits, he admits that he is right.

Whilst pupil A here outlines questioning about the status of a statement by shifting pupil L's argument from the speaker to the validity of the statement, she does not go so far as to provide the reason why the fish admits *a fish is a fish*.

During the second moment of discussion, the teacher goes back over the meaning of the part of the sentence: 'You know' he said to him, 'you were right!' uttered by the fish before he declares 'A fish is a fish!' p. 15.

For the pupils, it is difficult to associate the use of the term *raison* (in the French translation, meaning 'right' or 'reason') by the fish with knowledge, with the fact of knowing. Only pupil C understands the fish's experience of suffocation on the bank as a reason enabling him to have learned that he could not live out of water. This pupil manages to link the fish's experience to the fact that he can recognise that the frog is right without however, she says, the fish being able to deduce from this that he is a fish. The confrontation with reality through experience is taken as a foundation for knowledge, as an element of proof for moving from belief to scientific knowledge.

M Now, I'm going to ask the question again: what does it mean 'you know you were right'?

Ma It means that *you* were right.

M What does it mean that she was right?

L It means that a fish is a fish and a frog is a frog.

M That doesn't answer my question.

C He was wrong to underestimate him because....

M Very good.

C Because he had an experience.

M So the fish had an experience and went on land, is that it?

C He says that he can't live there, whereas his friend can, so at the end of the story he has learned that a fish can't live out of water.

M Has he learned what a fish is?

C No.

M He has learned, I agree with you, he has learned, er, that a fish can't, that he can't...

C Live out of water.

M Has he learned that he is a fish?

C No, not yet.

Here begins an attempt at rationalising what makes it possible to say whether or not such a living being belongs to such or such a species but, more broadly, about the necessity of firmly anchoring one's knowledge in actual grounds.

5.2 *Plot and Putting Representations to the Test*

The second situation focuses on the double page of the storybook that corresponds to the moment when the tadpole-turned-frog describes to his friend the cow he came across on land. The illustration is concealed so as to enable the pupils to draw this animal as the fish can imagine it, from the frog's description.

Afterwards, the drawings are displayed on the board, and the pupils compare and discuss them in order to work out what is necessary or not in depicting a cow and to give the reasons for this. This is about enabling the pupils to set out the characteristic attributes of cows and, from there, to question the notion of classification criterion.

In the storybook, the fish tries to visualise the terrestrial world such as he is able to reconstruct it based on the environment that is his own, the aquatic world of the pond. As a result, he will combine known characteristics with unknown characteristics in order to imagine the cow described by the frog. This piecing-together process amounts, to some extent, to a classification process that centres on the determination of relevant characteristics.

5.2.1 Analysis of the Pupils' Drawings

The drawings show very different degrees of variation. Thus, the majority of the 'cow' drawings include characteristics known to pupils (eyes, tail, hoof, spot, ear, etc.) but not to the fish because they are not mentioned by the frog, as well as the general morphology of a real cow. The pupil's point of view predominates over the fish's point of view, offering drawings that are hardly out of kilter with reality. That is why the teacher asks the pupils if they have depicted all the cow's characteristics and to justify those they have depicted. The pupils acknowledge that they have only depicted *important things*, and when listing the depicted characteristics, *horn, udder, black spots*..., they realise that the spot characteristic is present on almost all the drawings even though it is not mentioned by the frog. Furthermore, when the teacher asks the pupils '*do all cows have spots?*' some work then begins on the relevance of the 'spot' characteristic.

Presence of two eyes, two horns, four legs with hooves, a tail, an udder... With a general 'cow' type morphology

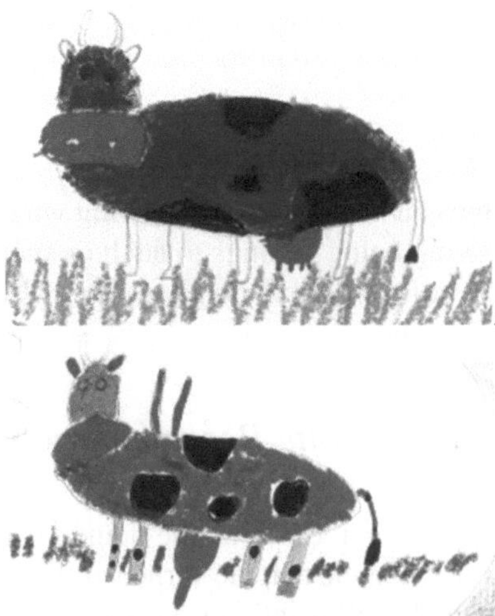

Drawing A

Drawing B

Certain drawings are more representative of the fish's point of view (drawings C, D, E) because they move away from what the pupils know about the cow and strive to interpret the frog's description, whilst allowing themselves variations on the colour, number and positioning of the characteristics.

A horn on its head and another on its flank
Two tails (one black, the other dark grey) at the end of its back

Drawing C

Two horns on its flank and not on its head

Drawing D

Three udders, including two on either side of its neck

Drawing E

The pupils very easily justify the origin of variations in the characteristics by the imprecisions in the frog's description. Therefore, it is the fact that the frog does not say where the cow's horns are located which makes it possible to imagine that the horns could be on its flank. The pupils realise that the indication of criteria is not enough to characterise a cow, they still need to be situated in a certain organisational structure and their number needs to be exact.

The space for interpretation opened up by the implicit aspects of the description seems to be moderately invested by the pupils, but the oral justification for what they depicted in their drawing leads them to ask themselves questions about the contingent nature of certain attributes such as the spots or the common or necessary nature of other attributes. This piecing-together process amounts, to some extent, to a classification process that centres on the identification of relevant characteristics, that is to say, elements that are significant for characterising an animal species.

6 Conclusion

The question-based reading of the realistic-fiction storybook 'Fish is Fish' combined with didactical activities written down in the storybook's actual narrative is a potential lever in pupils aged 7–8 years for problematic questioning of the notion of species in relation to that of criterion. The variations in meaning taken on by the problematic statement 'Fish is Fish' over the course of the story trigger the search for reasons that enable the fish character to move from belief to knowledge about an individual's belonging to a species. Furthermore, the creation of an imaginary drawing which forces the pupils to adopt a character's point of view makes it possible to imagine new possible characteristics for a species but, above all, to consider a species' criteria not as given by the characteristics themselves but to be constructed on the basis of some characteristics. This exercise seems particularly effective in terms of enabling pupils to challenge what they think they know about what identifies a species. The characterisation of living species is thus considered as a necessary part of working on the problematisation of the notion of criterion, which has been situated here mainly at the problem construction phase. Consequently, realistic-fiction storybooks can be a fertile resource for coping with scientific questions in the context of reading activities based on understanding the plot. But for the teacher, this requires an in-depth analysis of such storybooks, in particular in terms of the fiction/science coupling around the elements of scientific knowledge evoked in the story.

References

Bachelard, G. (1927, reprinted 1976). *Essai sur la connaissance approchée Thèse principale*. Éditions Vrin.

Bachelard, G. (1976/1927). *Etude sur l'évolution d'un problème de physique*. Paris: Vrin.

Bakhtine, M. (1984). *Esthétique de la création verbale. Bibliothèque des idées*. Paris: Gallimard.

Bruguière, C., Héraud, J.-L., Errera, J.-P., & Rembotte, X. (2007). Mondes possibles et compréhension du réel. La lecture d'un album en cycle 2 comme source de questionnement scientifique. *Aster, 44*, 69–106.

Bruguière, C., & Triquet, E. (2012). Des albums de fiction réaliste pour problématiser le monde vivant. *Repères, 45*, 1–22.

Bruguière, C., Triquet, E., Héraud, J.-L., & Soudani, M. (2012). Fictional storybook as a scientific and epistemological question – building tool for primary school. In C. Bruguière, A. Tiberghien, & P. Clément (Eds.), *E-Book proceedings of the ESERA 2011 conference: Science learning and citizenship. Part 14 (co-ed. Costas Constantinou and Jane Johnston)* (pp. 12–19). Lyon: European Science Education Research Association. ISBN 978-9963-700-44-8.

Bruner, J. (1997). *L'éducation, entrée dans la culture* (p. 255). Paris: Retz.

Bruner, J. (2002). *Pourquoi nous racontons-nous des histoires?* Paris: Retz. 112 p. [Original US edition: *Making Stories: Law, Literature, Life*. Farrar, Strauss & Giroux, New York, 2002].

Butzow, C. M., & Butzow, J. W. (1998). *More science through children's literature. An integrated approach* (p. 245). Englewood Cliffs: Teacher Ideas Press.

Chall, J. S. (1996). *Learning to read*. New York: McGraw-Hill.

Clément, P., Héraud, J.-L., & Errera, J.-P. (2004). Paradoxe sémantique et argumentation: analyse d'une séquence d'enseignement sur les grenouilles au cycle 2. *Aster, 38*, 123–150.

Defourny, M. (2009). *Le livre et l'enfant, Recueil de textes de Michel Defourny*. Bruxelles: De Boeck.
Eco, U. (1985 translation). *Lector in fabula. Le rôle du lecteur ou la coopération interprétative dans les textes narratifs*. Grasset, Livre de poche.
Eco, U. (1979, trad. 1985). *Lector in fabula*. Paris: Grasset.
El-Hindi, A. E. (2003). Integrating literacy and science in the classroom: From ecomysteries to readers theatre. *Reading Teacher, 56*, 536–539.
Fabre, M. (1999). *Situations-problèmes et savoirs scolaires*. Paris: PUF.
Fang, Z. (2006). The language demands of science reading in middle school. *International Journal of Science Education, 28*(5), 491–520.
Fisher, W. R. (1987). *Human communication as narration: Toward a philosophy of reason, value, and action*. Columbia: University of South Carolina Press.
Gould, S. J. (1980). *The Panda's thumb*. New York: W. W. Norton.
Greimas, J. (1966). *Sémantique structurale*. Paris: Larousse.
Hansen, C. (2006). The art of author study: Leo Lionni in the primary classroom. *The Reading Teacher, 60*(3), 276–289.
Hassett, D. D., & Curwood, J. S. (2009). Theories and practices of multimodal education: The instructional dynamics of picture books and primary classrooms. *The Reading Teacher, 63*(4), 270–282.
Hawking, S., Hawking, L., & Galfard, C. (2007). *Georges et les secrets de l'univers*. Pocket Jeunesse.
Jacobi, D. (2005). *Les sciences communiquées aux enfants*. Saint-Martin-d'Hères (Isère): P.U.G.
Larivaille, P. (1974). L'analyse morphologique du récit. *Poétique, 19*, 368–388.
Lionni, L. (1974). *Fish is fish*. New York: Dragonfly Books – Alfred A. Knopf.
Lionni, L. (1981 translation: *Un poisson est un poisson*). *Fish is fish*.
Millar, R., & Ogborne, J. (1998). *Beyond 2000: Science education for the future: A report with ten recommendations*. King's College London, School of Education.
Orange-Ravachol, D. (2005). Problématisation fonctionnaliste et problématisation historique en sciences de la Terre chez les chercheurs et les lycées. *Aster, 40*, 177–204.
Osborne, J. (2007). Science education for the twenty-first century. *Eurasian Journal of Mathematics, Science & Technology Education, 3*(3), 173–184.
Piliouras, P., Siakas, S., & Seroglu, F. (2011). Pupils produce own narrative inspired by the history of science: Animation movies concerning the geocentric-heliocentric debate. *Science Education, 20*, 761–795.
Reiss, M. J., Millar, R., & Osborne, J. (1999). Beyond 2000: Science/Biology education for the future. *Journal of Biological Education, 33*(2), 68–70.
Rice, D. C. (2002). Using trade books in teaching elementary science: Facts and fallacies. *The Reading Teacher, 55*(6), 18–22.
Ritchie, S., Rigano, D., & Duane, A. (2008). Writing an ecological mystery in class: Merging genres and learning science. *International Journal of Science Education, 30*(2), 143–166.
Salmon, C. (2007). *Storytelling. La machine à fabriquer les images et à formater les esprits*. Paris: La Découverte.
Saul, E. W. (Ed.). (2004). *Crossing borders in literacy and science instruction: Perspectives on theory and practice*. Newark: International Reading Association.
Schussler, E. (2008). From flowers to fruit: How children's books represent plant reproduction. *International Journal of Science Education, 30*(12), 1677–1696.
Stannard, R. (2001). Communicating physics through story. *Physics Education, 36*(1), 30–34.
Straits, W. J., & Nichols, S. E. (2007). Using historical non-fiction and literature circles to develop elementary teachers' nature of science understandings. *Journal of Science Teacher Education, 18*, 901–912.
Street, B. (2001). Ethnographic perspectives on literacy. In B. V. Street (Ed.), *Literacy and development. Ethnographic perspectives* (pp. 1–17). London: Routledge.
Strube, P. (1994). Narrative in the science curriculum. *Research in Science Education, 24*(1).
Tompkins, G. (2003). *Literacy for the 21st century*. Upper Saddle River: Prentice Hall.
Viennot, L. (1996). *Raisonner en physique: la part du sens commun*. Bruxelles: De Boeck.

Chapter 32
Nature of Science as Portrayed in the Physics Official Curricula and Textbooks in Hong Kong and on the Mainland of the People's Republic of China

Ka Lok Cheng and Siu Ling Wong

1 Background and Purpose

It is widely agreed that the nature of science (NOS) should gain access to science classrooms (e.g. Arriassecq and Greca 2007; Driver et al. 1996). The increasing attention to the philosophical and social aspects of science is evident from the inclusion of these aspects in the science education curriculum specifications in many Western countries (McComas and Olson 1998).

The attention to NOS in the science curricula of Hong Kong and mainland China started at the beginning of the twenty-first century. In Hong Kong, the shift from the predominantly content-focused goals to the promotion of scientific literacy is evident in the new set of Curriculum and Assessment Guides devised for senior secondary level science subjects (CDC and HKEAA 2007). Promotion of scientific literacy is stated as the overarching aim in the Physics, Chemistry and Biology Curriculum and Assessment Guides. For example, in the Physics Curriculum, the following statement of aim could be found:

> The overarching aim of the Physics Curriculum is to provide physics-related learning experiences for students to develop scientific literacy, so that they can participate actively in our rapidly changing knowledge-based society, prepare for further studies or careers in fields related to physics, and become lifelong learners in science and technology. (CDC and HKEAA 2007, p. 4)

An understanding of NOS is a key component for achieving scientific literacy. As such, enabling students to appreciate and understand NOS is also explicitly spelt out as an aim in the Physics Curriculum and Assessment Guide. Similar expansion in the goals of science education has also occurred in mainland China. The newest official Physics Curriculum, which has been piloted since 2004, is aimed at enhancing students' scientific literacy.

K.L. Cheng (✉) • S.L. Wong
Faculty of Education, The University of Hong Kong, Hong Kong
e-mail: chengkla@hku.hk

C. Bruguière et al. (eds.), *Topics and Trends in Current Science Education: 9th ESERA Conference Selected Contributions*, Contributions from Science Education Research 1,
DOI 10.1007/978-94-007-7281-6_32, © Springer Science+Business Media Dordrecht 2014

Table 32.1 NOS ideas to be examined in the current study

1.	Laws as generalizations and theories as explanations of the generalizations
2.	Creative elements of the scientific processes
3.	Tentative and developmental nature of science
4.	Distinction and relationship between science and technology
5.	Theory-laden nature of scientific processes
6.	Empirical nature of scientific knowledge
7.	Different ways of performing scientific investigations
8.	Interactions between science, technology and society
9.	Moral and ethical dimensions of science
10.	Scientists as a community

Despite such a similarity in the curriculum intents, the specific contents of the official curricula are dissimilar. These differences are expected to be translated into the corresponding textbooks. The differential extent and manner of the official curriculum-textbook alignment further complicate the picture and affect how NOS ideas are presented in the textbooks.

Textbooks, in turn, are vital intermediaries for the translation of official curricula into implemented curricula. The study of textbooks is particularly important in Chinese societies in which the reliance on textbooks is high. With such consideration in mind, the NOS ideas in the textbooks take the foreground of the current study. In this study, we compare the NOS elements in the official Physics Curricula with the corresponding textbooks. The extent of the alignment between the official curricula and textbooks is also studied.

In the current study, the term 'NOS' is defined in light of existing literature, especially Lederman (2007), in which NOS is confined to the nature of scientific knowledge only, and McComas and Olson (1998) and Osborne et al. (2003), in which the term is defined in a broader and more inclusive sense. The NOS ideas identified for the study include not only those that provide descriptions on the nature of scientific knowledge but also those relevant to the methods of scientific inquiry and the social dimensions of science. Through synthesizing the lists of NOS ideas in the above studies and performing content analysis with the official curricula and textbooks used at the two study sites, NOS ideas in the following list (Table 32.1) were identified.

The current study serves to provide authentic materials to illustrate how NOS ideas are presented in the official curricula and textbooks used in the Chinese contexts. These materials could also function as the background for studying whether a common set of NOS ideas exist, as asserted by McComas and Olson (1998), when Asian perspectives are taken into account. This study also paves the way for further studies on whether the textbooks function to create the cultural barriers that adversely affected non-Western students' learning of science, as asserted by Sutherland and Dennick (2002).

2 Methods

The most recent official Physics Curricula used in Guangzhou, the capital of the Guangdong Province of the People's Republic of China (MoE 2003, hereafter known as 'CHN Standards'), and in Hong Kong (CDC and HKEAA 2007, 'HK Guide') were examined in the current study. Study was made of two sets of corresponding Physics textbooks from each city, including the one published by People's Education Press (Centre for Research and Development on Physics Curriculum Resources of PEP 2004, 'PEP') and the one published by Guangdong Education Press (Editorial Group for Physics Textbooks, Guangdong 2004, 'GEP') from Guangzhou and the sets published by Oxford University Press (Wong and Pang 2009, 'OUP') and by Longman (Tong et al. 2009, 'Longman') in Hong Kong. Table 32.2 summarizes the type and site of deployment of these official curricula and textbooks.

The qualitative method was used in the current study. The full text of the official curricula and the above sets of textbooks, including texts for both the core and elective modules, were inspected for the presence of NOS content. Unlike the Turkish biology textbooks studied by Irez (2009), the NOS content in the official curricula and textbooks examined in the current study were found in various places of the official curricula and textbooks instead of being well grouped under a single teaching unit. As such, the development of an inventory of NOS ideas found in the official curricula and textbooks was the key task in the early stage of the current study.

We made reference to the existing literature on NOS in the identification process while keeping in mind that NOS ideas in the lists developed in the West may not include all NOS ideas found in the examined official curricula and textbooks. The text was repeatedly read and the NOS content was coded using the items in the existing lists as far as feasible, while new codes were added if the existing codes were unable to accommodate the identified NOS content. Codes were refined by the first author after being repeatedly sensitized by the existing NOS literature. Codes were consolidated through the comparison of the content of similar meaning. The resulting list of codes, together with the references to the original texts, was validated by the second author, who was knowledgeable about NOS. The coding was reiterated until a consensus had been reached between the two authors. The finalized list of NOS ideas, which result from the consolidated codes, is given in Table 32.1.

Table 32.2 Official curricula and textbooks examined in the current study

Type of official curricula and textbooks	Guangzhou (mainland China)	Hong Kong
Official curricula	*CHN standards*	*HK guide*
Textbooks	*PEP*	*OUP*
	GEP	*Longman*

The NOS ideas identified in the official curricula and textbooks were then compared with the contemporary NOS understandings. Comparisons among the textbooks from the same site and the determination of the alignments between the official curricula and the corresponding textbooks were then performed.

3 Results

Ten NOS ideas were identified from the official curricula studied (Table 32.1). Several of them are similarly represented in the official curricula in both Hong Kong and mainland China, including 'laws as generalizations and theories as explanations of the generalizations', 'tentative and developmental nature of science', 'empirical nature of scientific knowledge' and 'scientists as a community'.

However, some of these NOS ideas, including 'theory-laden nature of scientific processes', 'different ways of performing scientific investigations', 'interactions between science, technology and society' and 'moral and ethical dimensions of science', are presented with significant differences. For any given idea, the differences in terms of their presence in the official curricula are expected to breed differences in the presentation in the corresponding textbooks. To allow the differences in terms of the NOS ideas presented in the textbooks and the extent of curriculum alignment to be illuminated within the bounds of this article, four NOS ideas are selected for discussion below.

3.1 Theory-Laden Nature of Scientific Processes

3.1.1 Official Curricula and Textbooks in Guangzhou

The theory-laden nature of scientific processes refers to the effect of theoretical commitment (including beliefs) on how scientists perform investigations. While the term 'theory-laden' is absent in *CHN Standards*, it states that students should recognize that 'experimental manipulations are to be performed under the guidance of relevant principles' (p. 52).

Correspondingly, in *PEP*, narratives on how belief affects scientists' capability to observe the occurrences in experimental settings are included in the chapter on electromagnetism to illustrate how scientific works are 'under the guidance of relevant principles':

> It is almost certain that no one among the audience could note such phenomena (deflection of magnetic needles). However, since Ørsted kept thinking about the relationship between electricity and magnetism, such phenomena delighted him greatly. (*PEP* Elective 3–2, p. 2)

> Under the limitation of such a concept (alignment of the directions of force with the objects involved), Ørsted placed the magnetic needle in the same direction as the conducting wire and failure thereby resulted. (*PEP* Elective 3–2, p. 80)

Using Ørsted as an example, the authors illustrate the influence of theoretical commitment on the making of knowledge-generating observations. The experimental design is dependent on the theoretical assumption: While a correct assumption could guide scientists' efforts to obtain the required observations, a mistaken assumption will lead to a failure to make positive observations.

GEP also uses an example of a scientist (Colladon) in the chapter of electromagnetism and states that he 'was only concerned with the situation when the electricity became steady... and therefore missed the chance to discover induced current' (Elective 1–1, p. 48), to show that a mistaken assumption could lead to a failure to make meaningful observations. It further illustrates that mathematical law was altered to accommodate the expected experimental results (presence of a constant value of elementary charge) during the determination of elementary charge by Millikan:

> The value of *e* calculated from initial experimental data increased as the size of oil droplets diminished ... He thereby amended Stokes's law to obtain reasonable experimental results. (*GEP* Elective 3–1, p. 19)

3.1.2 Official Curricula and Textbooks in Hong Kong

In contrast, in *HK Guide*, the words 'observation' and 'observe' (together with their derivatives) are mostly associated with the fostering of students' careful observations. While the use of 'physical laws in mathematical form to predict natural phenomena and [the need to subsequently] verify these predictions with careful observation and experimentation' (p. 53) in scientific investigations is stated, there is no specification regarding how the influence of theoretical commitment to scientific process should be communicated to the students. If the textbook reflects such omission faithfully, the impact of theoretical commitment on the experiments should be unidentifiable throughout the text.

In *OUP*, Röntgen's 'accidental' discovery of x-rays (Radioactivity, p. 35) and the 'surprise' in Rutherford's scattering experiment (p. 39) show scientific discoveries as randomly appearing events. Similarly, in the section on Kinetic Theory in *Longman*, Brown is described as having no idea of the meaning and significance of his experiments on pollen grains:

> The English botanist Robert Brown (1773–1858) discovered the random motion of pollen grains suspended in water under a microscope during an experiment ... Although he could not explain the phenomenon, Brownian motion was named after him. (*Longman*, Vol. 1, p. 163)

In these examples, the inadequate emphasis on the importance of scientists being observant and attentive to fine details might portray a distorted image of scientists as a passive subject waiting for the meaningful observations to appear by themselves.

3.2 Different Ways of Performing Scientific Investigations

3.2.1 Official Curricula and Textbooks in Guangzhou

There is a dedicated chapter on scientific investigations in *CHN Standards* in which the 'components of scientific investigations' are listed in a table as shown in Table 32.3. The seven components are remarkably similar to the stepwise scientific method, the presence of which is stated as one of the myths about NOS (McComas 1998).

Table 32.3 Components of scientific investigations as presented in *CHN Standards* (pp. 10–11)

Components of scientific investigations	Basic skill requirements for scientific inquiries and physics experiments
Question formulation	Able to discover physics-related problems
	Capable of expressing these problems clearly as physics problems
	Aware of the significance of problem discovery and question formulation
Speculation and hypothesis formation	Able to propose problem-solving methods and solutions to problems
	Capable of predicting the results of physics experiments
	Aware of the importance of speculation and hypothesis
Experiment planning and design	Able to develop a plan according to the experimental objectives and conditions
	Capable of selecting the appropriate experimental methods, set-ups and instruments
	Able to consider the experimental variables and their control
	Aware of the role of planning
Experimentation and data collection	Capable of collecting data using a variety of methods
	Able to perform experiments according to guidelines and to use common instruments
	Capable of faithfully recording experimental data and aware of the meaning of the duplicated collection of experimental data
	Safety-conscious
	Aware of the importance of the objective collection of experimental data
Analysis and reasoning	Capable of analysing experimental data
	Able to develop conclusions according to the observations and data
	Able to explain and describe the experimental data
	Aware of the importance of analysis and reasoning to experiments
Evaluation	Capable of analysing the differences between hypotheses and experimental results
	Able to attend to the unresolved issues in the inquiries and discover new problems
	Capable of improving the inquiry plan with reference to the experience gained
	Aware of the significance of evaluation
Exchange and cooperation	Able to write reports for the experimental inquiries
	Able to uphold principles while respecting others during cooperation
	Cooperative
	Aware of the importance of exchange and cooperation

In response, *PEP* nominates a particular sequence of experimental processes (as portrayed through a diagram duplicated in Fig. 32.1), which approximates the 'components of scientific investigations' in *CHN Standards*. It is also stated that such a method was developed by Galileo:

> Galileo's study of motion… created a highly useful scientific method for the modern scientific development, or can be considered as highlighting many basic elements of the process of scientific study… (*PEP* Physics 1, p. 48)

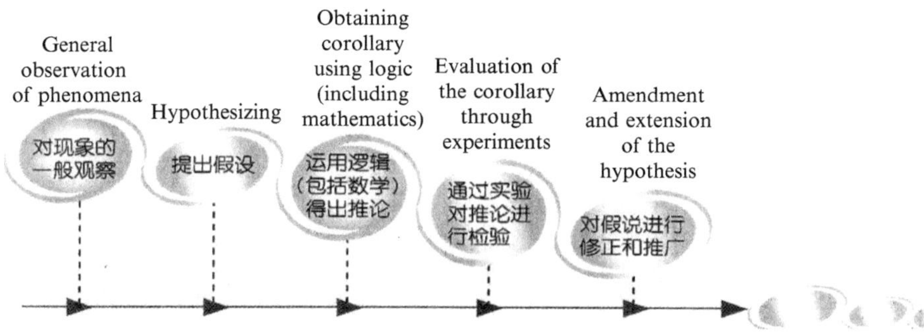

Fig. 32.1 Sequence of steps of scientific investigation in *PEP* (Physics 1, p. 48)

While an explicit message that describes the sequence of 'components of scientific investigation' as 'the scientific method' is absent in *GEP*, procedures of the experiments like the one on the factors that affect resistance (Elective 3–1, pp. 36–38) are organized under headings like 'Experiment Design', 'Experiments and Data Acquisition', 'Analysis and Reasoning', 'Evaluation and Exchanges', etc. Students might infer that such a sequence of steps is '*the* scientific method'.

3.2.2 Official Curricula and Textbooks in Hong Kong

Physics teachers in Hong Kong are not given a table of 'components of scientific investigations' as are their counterparts in Guangzhou. However, in the 'Scientific Investigation' part of the 'Learning Targets', they are given a similar checklist of official expectations:

Students are expected to:

- Ask relevant questions
- Propose hypotheses for scientific phenomena and devise methods to test them
- Identify dependent and independent variables in investigations
- Devise plans and procedures to carry out investigations;
- Select appropriate methods and apparatus to carry out investigations
- Observe and record experimental observations accurately and honestly
- Organize and analyse data, and infer from observations and experimental results
- Use graphical techniques appropriately to display experimental results and to convey concepts
- Produce reports on investigations, draw conclusions and make further predictions
- Evaluate experimental results and identify factors affecting their quality and reliability
- Propose plans for further investigations, if appropriate

(*HK Guide*, pp. 9–10)

While different methods used in science are briefly covered in both official curricula, the existence of a universal scientific method, which was considered

Table 32.4 Interactions between science, technology and society as depicted in the official curricula

CHN standards	HK guide
The roles of the widespread use of heat engines in bringing about the changes towards science, social development and the mode of living (p. 22)	How major breakthroughs in scientific and technological development that eventually affect society are associated with a new understanding of fundamental physics (p. 49)
… aware of the social problems that are brought about by the technological applications of physics (p. 2)	… controversial issues about the effects of microwave radiation on the health of the general public through the use of mobile phones (p. 38)

as a myth by McComas (1998), is implied in *HK Guide* just as in *CHN Standards*: The table in *CHN Standards* and the bullet list in *HK Guide* showing chronological sequences of scientific processes are implying the existence of such a universal method.

The textbooks in Hong Kong seem to be immune from the influence of the official curriculum in this respect. Both *OUP* and *Longman* include mostly 'cookbook' experiments. The only content in each experimental section is a very short sequence of imperative statements that usually spans no more than a third of a page, which is accompanied by a photograph that shows the experimental set-up (detailed procedures are usually found in the 'experimental workbooks'). Such succinctness (if not incompleteness) precludes any possible presentation of a fixed sequence of actions for scientific inquiries as found in *HK Guide*.

3.3 Interactions Between Science, Technology and Society

3.3.1 Official Curricula in Both Sites

The comparison of the official curricula in both sites (Table 32.4) reveals that, while the discussion on both the pros and cons of the technological applications of science are specified in both texts, *CHN Standards* attend to a larger extent to the societal development brought about by technological advances, while *HK Guide* pays more attention to the controversies that could result from the use of technologies.

3.3.2 Textbooks in Guangzhou

The focus on the role of scientific and technological advances in bringing about societal development in *CHN Standards* is translated into the text of *PEP*. In Table 32.5, we can see in how much detail the effects of the Industrial Revolution are depicted in *PEP*. Particular attention can be paid to the statements regarding the

Table 32.5 Societal effects of the Industrial Revolution as depicted in *PEP* (Elective 1–2, p. 77)

Impacts	Relevant text segments
Rise of manufacturing and service industries	Human lives changed from agriculture-based to industry-based...
	... the dense population and development of the cities gave birth to service industries... In 1840, among the working population in Britain, 39 % worked in manufacturing industries, 38 % in service industries, while those who worked in agricultural sectors decreased to 23 %
Centralization of production	The venue of production changed from dispersed family production [houses] for agriculture and handcrafts to factories for centralized production
	The centres of production shifted from villages to cities...
Rise of the market economy and capitalism	A self-sufficient natural economy had evolved into a market economy in which production and consumption are separated
	The modes of production underwent comprehensive and deep-rooted changes and the historical advancement resulting from the replacement of feudalism by capitalism was realized
Progress in other aspects of life	The development of cities facilitated the advancements of science, technology, education and cultural industries. Human society was entering the era of industrial civilization

'rise of manufacturing and service industries' and the 'rise of the market economy and capitalism' – such emphasis may be due to the influence of Marxist ideas on historical periodization. A similar influence of Marxism in relation to the science teacher educators in mainland China can also be found in Wan et al. (2013).

While *CHN Standards* also attend to the societal problems brought about by scientific developments, the negative consequences of the applications of science are downplayed in *PEP*. For instance, the safety issues that could result from the use of nuclear energy are addressed by simply stating that such issues 'are being attended to on a continuous basis' (Elective 1–1, p. 58), while two major nuclear meltdowns, the Three Mile Island accident and the Chernobyl disaster, are just briefly touched on (pp. 59–60) without any consideration of the resulting controversy. The authors express their wishful thinking instead:

> In the twenty-first century, we may look forward to the development of safer and more economical nuclear reactors and to the maturation of nuclear fusion technology. Nuclear energy is providing more electric energy for human use. (*PEP* Elective 1–2, p. 59)

The nuclear controversy is silenced. Students are not even told that the use of nuclear power is a controversial issue, and thus they are not likely to engage in the debate on the future of nuclear power. Such coverage is already the most comprehensive reference to the safety aspect regarding the use of nuclear energy in the *PEP* set of textbooks: the only remaining remark regarding the safety of nuclear plants in the other parts of the textbooks is on the measures to prevent the leakage of nuclear waste (Elective 3–5, p. 86).

GEP also covers the societal effect of the Industrial Revolution although the coverage is briefer. However, it is less silenced on the controversies regarding the use of nuclear energy, despite its attempt to contain students' worries:

> .. the nuclear leakage in Chernobyl in the former USSR occurred and the radioactive pollution affected the north-western part of Europe. The incident caused people to worry about the use of nuclear power. (*GEP* Elective 2–3, p. 89)

> Usually, nuclear reactors use low-enriched uranium… As such, reactors will not explode like nuclear bombs. However, there are some hazards of nuclear leakage. (*GEP* Elective 1–2, p. 50)

The authors of *GEP* point out the difference between the current state of scientific research in China and that in the developed countries and admit that 'compared to the developed countries, our country [China] lags behind to some extent' (*GEP* Elective 3–3, p. 34). To explain the difference of the extent of technological advancements between China and other countries, *GEP* points out the influence of contextual and political factors on technological advancement, as it states:

> In the present century and due to harassment by foreign powers and other factors, the standard of the advanced industries of China is much lower than that of the developed countries. Since the 1980s, as China has implemented the 'Reform and Opening Up' policy, foreign advanced experiences are being enthusiastically absorbed and innovation has continued to allow the standard of the advanced industries to improve significantly, and thus the gap in relation to the developed countries is gradually diminishing. (*GEP* Elective 2–2, p. 37)

The textbook is thus used as a medium to induce students' appreciation of how the policy of the government helps societal and economic development. The discussion on the recent development of microprocessors (Elective 3–4, p. 65) functions as another example to exemplify the implications of the 'Reform and Opening Up' policy. Similarly, the text on the successful launching and return of *Shenzhou V* (Physics 2, p. 52) is also meant to exhibit the role of such policy in terms of technological development.

3.3.3 Textbooks in Hong Kong

For the societal implications of science and technology, the controversy on the use of nuclear energy is included in *OUP* and this corresponds to the specifications in *HK Guide* well. In the coverage on the controversy over the use of nuclear power, *OUP* uses a cartoon (Fig. 32.2) that depicts the demonstrations organized by pro- and anti-nuclear camps to represent the diametric views on the issue (Radioactivity, p. 90). Similarly, in the discussion of the possible harmful effects brought about by overhead power cables, students are encouraged 'to find out [about] … action taken by pressure groups and the response of electricity companies' (Electricity, p. 280).

Fig. 32.2 A picture showing the views of pronuclear power and anti-nuclear power groups in *OUP* (Radioactivity, p. 90)

These examples exhibit the authors' intention to highlight the socio-scientific conflicts among different stakeholders and groups.

In *Longman*, the impacts of science and technology on society are superficially covered. Moreover, the socio-scientific controversies are mostly replaced by scientific findings on the possible health effects, as in the case of the overhead power cables. Socio-scientific controversies are reduced to scientific colloquia:

> Some researches indicate that exposure to changing electric and magnetic fields from transmission lines may incur a number of health problems. These include leukaemia, cancer, miscarriage, clinical depression, etc. However, these conclusions are solely based on statistical studies … The detailed mechanism is still unclear. (*Longman* Vol. 4, p. 440)

3.4 Moral and Ethical Dimensions of Science

3.4.1 Official Curricula and Textbooks in Guangzhou

In *CHN Standards*, there is no specification regarding the moral and ethical dimensions of science. The only remotely connected discussion is on the character of scientists. The textbooks are required to 'use vivid information to exhibit their scientific spirit and their determined devotion to science' (p. 58). Scientists are to be honoured but not criticized.

This results in the narratives in *PEP* that describe scientists as extraordinarily moral, such as having the *courage* in making the difficult decision to suggest the use of atomic bombs (which is unlikely to be presented as such in the official curricula and textbooks of the West) and their contempt for wealth and status:

> Some scientists, especially those who had escaped from Fascist persecution, had a premonition about the threat of atomic bombs especially after they had heard that Germany had accelerated the research into nuclear chain reactions… On July 1939, nuclear physicist Szilard and others found Einstein and wished he could use his status to urge the United States to produce atomic bombs before Germany had done so… (*PEP* Elective 3–5, pp. 84–85)

> Faraday rejected the positions and titles like the President of the Royal Society, President of the Royal Institute of Great Britain, Professor of the University of London, and was not willing to receive any title of nobility… (*PEP* Elective 3–2, p. 5)

Similar narratives of scientists' advocacy of the use of atomic bombs to counter possible threats can also be found in the chapter on atomic physics in *GEP*. However, a more reflective stance is taken. The text initiates a discussion on whether scientists should be free from the moral consequences of their work:

> Should people develop nuclear technology if both the pros and cons are taken into account? Some say scientists should care about science *per se*, while the impacts of science and technology on society should be a matter for sociologists. Do you agree? (*GEP* Elective 3–5, p. 88)

3.4.2 Official Curricula and Textbooks in Hong Kong

Compared to *CHN Standards*, a more reflective perspective is adopted by *HK Guide*, in which students are requested to examine 'the roles and responsibility of scientists and the related ethics in releasing the power of nature' and 'the moral issues of using various mass destruction weapons in war' (p. 52) during their study of nuclear energy. Scientists are no longer simply heroes to be appreciated, they are to be examined.

Such a stance influences the authors of *OUP* as they also invite students to reflect on the moral responsibility of scientists. Rhetoric that attempts to identify scientists' responsibility regarding the use of nuclear weapons is used:

> Atomic bombs ended World War II. The weapon brought peace at a great price: 115, 000 died in the explosions and many others from cancer caused by radiation years later. Should atomic bombs be used to end a war? Did scientists have some responsibility for the destruction caused? (*OUP* Radioactivity, p. 96)

However, moral aspects are not covered in many cases. For example, in the practice question that takes stealth fighters as a context (Wave, p. 138), the technological design of stealth fighters is highlighted, but the moral implications of their use are absent.

In *Longman*, the moral and ethical aspects are pointed out but not discussed in depth. For example, while the 'international race for nano-weapons' and the 'technology of changing and even constructing DNA' are said to cause much impact on ethical values (Atomic World, p. 149) and nuclear energy is said to be able to 'destroy our civilization if it is misused' (Radioactivity, p. 108), no further discussion follows these brief references.

4 Discussion

4.1 NOS Ideas Presented in the Textbooks Used in the Two Study Sites

Due to the differential extent of emphasis in the official curricula or otherwise, the foci of textbooks in both study sites are different, as is delineated in Table 32.6. The theory-laden nature of science is included in the textbooks in mainland China, while similar discussions cannot be found in the textbooks in Hong Kong.

Table 32.6 NOS ideas in the textbooks of the two study sites

Guangzhou (mainland China)	Hong Kong
Role of 'beliefs' in terms of the practice of science	No mention of the effects of theoretical commitment on the practice of science
A fixed sequence of processes for any scientific investigation	Notion of the existence of such fixed sequence not found
Science is essential to the development of society	Pros and cons of science and technology; controversies regarding socio-scientific issues
Appreciation of the moral character of the scientists	Critical thinking on the moral responsibility of scientists

Table 32.7 Observations on curriculum alignments

Cases	Observations
Case I: Theory-laden nature of science	Presence/absence of an idea in official curricula leads to its corresponding presence/absence in textbooks
Case II: Elements of scientific investigations	Forms of presentation *may* lead to the transmission of implicit understandings
Case III: Controversies of socio-scientific issues	Relative importance of different NOS ideas in official curricula is translated to the corresponding textbooks
Case IV: Moral responsibility of scientists	Even the overt messages in official curricula can be downplayed in the textbooks

Secondly, the scientific investigations represented in the textbooks in Guangzhou are made up of a sequence of steps which can be understood as '*the* scientific method', while such an implied message on methodological universality is absent from the textbooks used in Hong Kong.

The emphases of the textbooks may be different even when the same category of NOS ideas is being presented. The science-society relationship as depicted in the textbooks of mainland China focuses more on the societal development brought about by technological advances, while those in Hong Kong focus more on socio-scientific controversies. Moreover, the texts in mainland China tend to give far greater appreciation to the morality of scientists, while a more reflective stance is taken by the textbooks in Hong Kong.

4.2 Some Observations on Curriculum Alignments

Table 32.7 presents some observations regarding the official curriculum-textbook alignments. Official curricula and textbooks are often interrelated and well aligned. The presence and absence of the coverage of the theory-laden character of scientific knowledge in the official curricula in Guangzhou and Hong Kong, respectively, is well translated into the corresponding textbooks used in both sites.

The idea of the presence of a universal scientific method is reinforced in the textbooks of Guangzhou through the arrangement of laboratory instructions using the organizational framework provided by the *CHN Standards*, in spite of the lack of explicit statement regarding methodological universality in the corresponding official curriculum. However, implicit messages, such as those transmitted through the use of the bullet list to present the steps of scientific investigations in *HK Guide*, are not always well translated: the very brief experimental instructions in the textbooks of Hong Kong is unlikely to allow any idea – contemporarily accepted or the otherwise – regarding the scientific process to be exemplified.

The extent of emphasis and foci in the official curricula may explain the inclusion of materials in the textbooks: the specific statements on the Industrial Revolution in *CHN Standards* result in the extensive coverage on how the Industrial Revolution led to the change of the mode of production and the rise of capitalism in the textbooks used in mainland China, in contrast with the few references made in the textbooks used in Hong Kong. Similarly, the requirements stipulated by the *HK Guide* as regards students' recognition of the controversial nature of socio-scientific issues result in the graphical representation of the conflicting ideas of the pro- and anti-nuclear camps, while only optimistic remarks are made regarding the safety issues associated with nuclear energy in the textbooks of mainland China: The related controversies are silenced.

However, the ideas that are found in the official curricula may still be virtually absent from the corresponding textbooks. For instance, the moral and ethical dimensions of science specified in *HK Guide* are reduced to a few brief statements in a set of textbooks used in Hong Kong.

5 Conclusions and Implications

From the brief survey of the official curricula in both sites, we can see there are differences between Guangzhou and Hong Kong regarding their inclusion of NOS ideas in the official curricula. There are also significant differences in terms of the inclusion and the focus of the NOS ideas in the textbooks. From the four illustrative cases, we can see how the (overt and covert) messages in the official curricula could (and sometimes fail to) influence the corresponding textbooks.

Given the status of textbooks in science classrooms, further interpretative attempts on how various factors, including the official curricula, play a role in shaping the NOS ideas found in the textbooks will be necessary for the practical knowledge on the strategic introduction of NOS ideas into the textbooks and classrooms to be generated.

References

Arriassecq, I., & Greca, I. M. (2007). Approaches to the teaching of special relativity theory in high school and university textbooks of Argentina. *Science Education, 16*(1), 65–86.

Centre for Research and Development on Physics Curriculum Resources, Curriculum Resources Research Institute, People's Education Press. (2004). *Physics*. Beijing: People's Education Press.

Curriculum Development Council and Hong Kong Examinations and Assessment Authority . (2007) *Physics curriculum and assessment guide (Secondary 4–6)*. Hong Kong: Curriculum Development Council and Hong Kong Examinations and Assessment Authority. Retrieved from 20 May 2012, http://www.edb.gov.hk/FileManager/EN/Content_5999/phy_final_e.pdf

Driver, R., Leach, J., Miller, A., & Scott, P. (1996). *Young people's images of science*. Bristol: Open University Press.

Editorial Group for Physics Textbooks, Guangdong Basic Education Curriculum Resources Research and Development Centre. (2004). *Physics*. Guangzhou: Guangdong Education Press.

Irez, S. (2009). Nature of science as depicted in Turkish biology textbooks. *Science Education, 93*(3), 422–447.

McComas, W. F., & Olson, J. K. (1998). The nature of science in international science education standards documents. In W. F. McComas (Ed.), *The nature of science in science education: Rationales and strategies* (pp. 41–52). Hingham: Kluwer Academic.
McComas, W. F. (1998). The principal elements of the nature of science: Dispelling the myths. In W. F. McComas (Ed.), *The nature of science in science education: Rationales and strategies* (pp. 53–70). Hingham: Kluwer Academic.
Ministry of Education, People's Republic of China (Ed.). (2003). *Physics curriculum standards for ordinary senior secondary schools (on trial).* Beijing: People's Education Press.
Lederman, N. G. (2007). Nature of science: Past, present, and future. In S. K. Abell & N. G. Lederman (Eds.), *Handbook of research on science education* (pp. 831–879). Mahwah: Lawrence Erlbaum.
Osborne, J., Colins, S., Ratcliffe, M., Millar, R., & Duschl, R. (2003). What "ideas-about-science" should be taught in school science? A Delphi study of the expert community. *Journal of Research in Science Teaching, 40*(7), 692–720.
Sutherland, D., & Dennick, R. (2002). Exploring culture, language and the perception of the nature of science. *International Journal of Science Education, 24*(1), 1–25.
Tong, S. S. W., Won, H. K., Kwong, P. K., Wong, Y. L., & Lee, L. C. (2009). *New senior secondary physics in life*. Hong Kong: Longman Hong Kong.
Wan, Z.H., Wong, S.L., & Zhan, Y. (2013). *When nature of science meets marxism: Aspects of nature of science taught by Chinese science teacher educators to prospective science teachers.* Science and Education, *22*(5), 1115–1140.
Wong, S. L., & Pang, W. C. (2009). *New senior secondary physics at work.* Hong Kong: Oxford University Press (China).

Chapter 33
On the Transfer of Teaching-Learning Materials from One Educational Setting to Another

R. Pintó, M. Hernández, and C.P. Constantinou

1 Introduction

One of the growing concerns within the community of researchers in science education refers to making research results available and useful for our society. As Sabelli and Dede (2001) pointed out: 'Decades of funded studies that have resulted in many exciting programs and advances have not resulted in pervasive, accepted, sustainable, large-scale improvements in actual classroom practice, in a critical mass of effective models for educational improvement'.

In attempts to share experience and knowledge, collaborating teachers and researchers often communicate their teaching-learning materials as one type of artefact that bears information with a potential influence on teaching practice (Méheut and Psillos 2004). We also identify teaching-learning materials as an important artefact for illustrating theoretical ideas, design rationales and classroom teaching materials. Hence, selecting materials that others have developed and entering into a process of making adaptations followed by classroom enactment can be a mechanism for introducing ideas into teaching and thereby promoting classroom experimentation and educational change.

R. Pintó (✉) • M. Hernández
CRECIM, Universitat Autònoma de Barcelona, Edifici GL 304, 08193, Bellaterra Barcelona, Spain
e-mail: roser.pinto@uab.es; mariaisabel.hernandez@uab.es

C.P. Constantinou
Learning in Science Group, Department of Educational Studies, University of Cyprus, P. O. Box 20537, Nicosia 1678, Cyprus
e-mail: c.p.constantinou@ucy.ac.cy

C. Bruguière et al. (eds.), *Topics and Trends in Current Science Education: 9th ESERA Conference Selected Contributions*, Contributions from Science Education Research 1,
DOI 10.1007/978-94-007-7281-6_33,

Educational policymaking often places emphasis on the transfer of successful or exemplary teaching practice as a mechanism for quality improvement in education (European Commission 2012). What constitutes exemplary teaching, the circumstances under which it can occur and how it can be diffused, if at all, are issues of great interest to educational policy development.

However, this issue of using educational materials as a way to gain information from others with a view to promote reformed teaching is neither trivial nor unproblematic. In particular, Fishman and Krajcik (2003) highlight two key considerations: (1.) How can we create science curriculum innovations that will be used in classrooms after the developers or researchers have left the scene? and (2.) How can we create science curriculum innovations that can be used in places that have never had any contact or only minimal contact with the developers or researchers? We consider that these issues are not only relevant for studies on the creation of sustainable and scalable science curriculum innovations but also for studies on the creation of transferrable innovations that can be reused across the boundaries of distinct educational systems.

In this study, we sought to examine a process of transferring teaching-learning materials as a tool for framing teaching practice in detail. Specifically, we organised an effort to exchange teaching-learning materials across different educational settings, and we set out to explore the issue of what is being transferred, what is being adapted and what conditions make the transfer process feasible on the understanding that the essential features of the design rationale are to be safeguarded.

We have framed this exchange effort as an international collaborative activity aiming to transfer different kinds of knowledge to classroom teaching. International collaboration has become possible, to a large extent, due to efforts of educational and research funding agencies to promote projects, reform initiatives and peer learning activities (Beernaert 2008; EU Council 2011). The collaboration that took place within the *Materials Science* project involved science education researchers and experienced teachers. This kind of partnership has been proposed as a useful mechanism for responding to the complexity that is inherent in efforts to plan for teaching, develop innovative materials and enact them in classrooms with a view to evaluate their quality in practice and to refine them when necessary (Couso 2011).

In particular, in this study we investigate the following research questions:

- Which kinds of adaptations are introduced before a classroom enactment when transferring teaching-learning materials from one educational setting to another?
- What are the reasons that are invoked in determining these changes?

By throwing light on such activities, we believe we can identify those parameters that can make or break an initiative to use established materials from elsewhere. Our study also has the potential to inform the ways in which science education communities organise evaluation, adaptation, reflection and sharing within a transfer process. In this sense, our investigation adds to the existing literature on design-based research by exploring an additional design cycle, which consists of the adaptation, enactment, evaluation and redesign of teaching-learning materials.

2 Rationale and Theoretical Underpinnings

2.1 *What Do We Mean by Transfer of Teaching-Learning Materials?*

In this chapter we use knowledge transfer (KT) to mean the selection of teaching-learning materials developed at another educational setting and their actual use in a classroom setting after a cycle of informed adaptation. This is distinct from the idea of transfer of knowledge sometimes used to describe the ability of a learner to apply a learning outcome in an unfamiliar context. It is also different from the idea of dissemination as the process where communication simply involves a sender and a receiver.

KT is understood as a process of sharing individual and collective experience (Rogers 1983). The perspective of KT has changed through the years, and, since today knowledge is no longer understood as value neutral, detached or 'existing on its own', it cannot be viewed as a commodity that can be transferred from a knowledge producer to a user. In this sense, KT is regarded as a process of reconstruction rather than a mere act of transmission and reception. Hutchinson and Habermann (1993) have shown the necessity to acknowledge this shift from one-way flow models to a constructivist perspective in which 'the user acts upon information by relating it to existing knowledge, imposing meaning and organization' (p. 2). Therefore, if we intend to make knowledge usable for potential users, then they should have an active role in reconstructing knowledge.

Within the *Materials Science* project, we have sought to establish this dialogue starting from the results of prior design research studies carried out in a particular country and bearing in mind that the future users are teachers from different countries.

2.2 *Which Kind of Knowledge Is Transferred?*

The process of KT can be different depending on the type of knowledge to be transferred since it may require different strategies and methods. Taking into account that, in our case, the object of transfer consists of the teaching-learning materials designed in one country that are to be applied in another country, we distinguish three types of knowledge to be transferred:

- Some scientific knowledge that underlies the topic of any teaching-learning sequence (TLS). A didactical transposition (Chevallard 1991) of the scientific knowledge has been carried out during the phase of design of the TLS.
- Pedagogical content knowledge (PCK) that includes designers' conceptions and strategies on different aspects of teaching and learning science such as contextualisation, inquiry, modelling, metacognition, sequencing and evaluation.

- The know-how about how to stage each lesson. This knowledge, which usually is not explicit, is considered essential for the consequences of implementing the sequence. Some examples of this type of knowledge are how to use materials and resources, how to engage and motivate students, how the teacher takes part in the teaching and learning process and how to manage student groups and support emotional involvement of every individual student. The know-how during the implementation of TLS is not analysed here.

TLS are highly specified artefacts that include teacher guidelines and the crucial materials to be used in classroom. As such, they are intended to communicate the formal but also the tacit knowledge for guiding the classroom implementation according to the initial design rationale.

2.3 How Is the Process of Knowledge Transfer? What Stages Does It Comprise?

In order to analyse the process of transferring an innovative TLS that conveys certain knowledge, we have adopted the perspective of Rogers (1983). According to him (p. 5), the diffusion of innovations, envisaged as 'the process by which an innovation is communicated through certain channels over time among the members of a social system', undergoes different phases such as adoption, adaptation and reinvention.

2.3.1 Adoption of an Innovation

Transferring a curriculum innovation to a different context requires that the innovation is perceived by the adopters as usable in the new context. The first important step involves the active selection of the TLS for adoption by the community where it will be implemented. Discontinuance (rejection after adoption through replacement or simply rejection as a result of dissatisfaction) can result if an innovation is adopted without conviction.

Ely (1990, 1999) presents a comprehensive set of eight environmental or human-related conditions that facilitate implementation of innovative processes:

- *Dissatisfaction with status quo* refers to an emotional discomfort resulting from the use of current processes perceived as inefficient or ineffective.
- *Adequate resources* refer to the availability and accessibility of resources needed to implement the innovation.
- *Rewards and incentives* refer to either intrinsic or extrinsic rewards that result from using the innovation. Rewards can include moral support or community appreciation.
- *Knowledge and skills* refers to users possessing or acquiring the needed skills and knowledge to employ the innovation. This condition also reflects users'

self-efficacy beliefs about using the innovation with training being a necessary part of the implementation plan.

- *Adequate time:* users must be provided time to develop a sense of familiarity and comfortableness with the innovation
- *Participation* refers to the involvement of stakeholders in the decision-making process to adopt and implement an innovation. This condition helps to develop ownership of the innovation and increases the stakeholders' interests in a successful implementation.
- *Institutional commitment* refers to the visible support by the upper level leaders in an organisation. The key is how the users perceive the leaders' commitment to the implementation of the innovation. Simple verbal endorsement of the innovation by leaders does not constitute adequate commitment. Visible forms of commitment include such things as personal communication, development of strategic implementation plans, changes to organisational policies, dedication of resources and active involvement in the implementation of the innovation.
- *Leadership* refers to the level of ownership and support given by the leaders who will manage the daily activities of those implementing the innovation. The enthusiasm of these leaders directly affects the motivation of the users of the innovation. Immediate supervisors must provide support and encouragement, answer questions, address concerns and serve as role models for using the innovation.

The literature on reform points to organisational culture, capability and management and policy as three areas that influence whether an educational innovation will be adopted (Blumenfeld et al. 2000). Organisational culture refers to local norms, routines and practices, such as professionalism, openness to sharing, risk-taking, reflection, communication and cooperation among teachers. Capability refers to educators' beliefs, needs, perspectives, understandings of the innovation and expertise to implement it. Management and policy refer to schedules, resources, infrastructure, assessments and allocation of responsibilities among different levels of the system. Taking into account these three dimensions, Blumenfeld et al. (2000) created a diagnostic tool to understand challenges to innovations, that is to say, a framework that can be used prospectively in its consideration of the likelihood of success and usability for an innovation. Each of the dimensions can be used to analyse gaps between existing conditions in a certain educational context and the requirements of an innovation. The greater the gap in any dimension, the more pressing it will be for the innovation to be modified or for the context to modify its practices, culture or capabilities.

2.3.2 Adaptation and Reinvention of an Innovation

Very often, making the decision to adopt an innovation once it has been communicated to the potential adopters seems sufficient to implement it as intended by the designers. However, this perspective does not take into account that the communication process should already be considered transformative since communications are not simply 'received' but are remade, reconstituted and transformed by the receiver (STTIS 1998).

Moreover, adoption of a TLS will be more successful when the users have developed ownership of the innovation (Ogborn 2002). The actual process of developing ownership requires adaptation to the new context: local culture, beliefs, priorities and resources can constrain this process. As stated by Burkman (1987), in many cases the failure of an innovation is not due to the quality of the product nor to the failure of adopters to appreciate this quality, but results from those responsible for its successful implementation failing to consider variables other than the product itself. Therefore, potential adopters of an innovation need not only to carefully select quality innovations but also to consider the environmental and human factors associated with implementation in order to adapt the innovation accordingly.

According to a constructivist perspective, we agree that enacting an exact copy or imitation of the innovation as proposed by designers is not feasible. In the past, the common process of reinvention (Rogers 1978) was generally not regarded favourable by research and development agencies, which may consider reinvention to be a distortion of their research findings. At present, we consider this requisite necessary if we take into account the differences between the reality of the context in which the innovation has been generated and the reality in which it will be used.

2.3.3 Adaptation-Fidelity: The Long Debate

Innovations that are implemented with high levels of fidelity (Mowbray et al. 2003) but fail to produce desired effects may need to be reinvented. In the opposite side, innovations that are implemented with very low level of fidelity can reach the point at which adaptation compromises the integrity of the original innovation, that is to say, these adaptations would trespass the 'zone of drastic mutation' (Hall and Loucks 1977). In that case, the innovation being implemented may no longer be what was originally intended. Some studies about implementation of innovations showed how critical these mutations can be when teachers have a passive role in designing an innovation and cannot recognise the reasons why old strategies should be left aside (Pintó 2005; Viennot et al. 2005). These studies support the idea that teachers and researchers working together and bringing different types of expertise to the design of an innovation are more likely to develop shared ownership of their outcomes in a way that can lead to enactments with higher fidelity.

The Rand report on the implementation of educational innovation (Berman and McLaughlin 1976) observed three patterns of implementation in innovative educational programmes of systemic reforms: (1) *co-optation* or adapting the programme without any changes in organisational behaviour, (2) *mutual adaptation* in which the programme is adapted at the same time there are changes in the organisation and (3) *non-implementation and non-adoption* in which neither happened. The most successful implementations occur when there is a mutual adaptation of both school practice and the demands of the innovation (McLaughlin 1990). Simply co-opting an innovation in a school can result in no change, as has often been seen when schools simply adopt the language of reform ideas adding the old to the new

(Hirn and Viennot 1999) without substantive changes. If an innovation deviates too widely from everyday practices, participants might react defensively or simply ignore it.

3 Context of Research

Within the *Materials Science* project, several TLS were designed, developed and refined through classroom implementation at a local level. In particular, a TLS on *Acoustic Properties of Materials*, addressed to 15–16-year-old students, was designed by a local working group (LWG) in Barcelona (Spain), formed by three researchers on science education and six experienced science teachers. Another TLS on *Electromagnetic Properties of Materials*, addressed to high school students (16–17-year-old students), was designed by a LWG in Nicosia (Cyprus), formed by two researchers on science education and five secondary science teachers. Involving teachers as members of the design group brought together diverse expertise about the corresponding topic, the local educational context, and allowed sharing and discussing different approaches to teaching, learning and designing educational materials.

After the design of each TLS, each home LWG organised a classroom enactment of the TLS in the home setting in order to evaluate the designed TLS. During this enactment, researchers from the host LWG (i.e. the group to whom the TLS would be transferred) participated in a study visit in the home setting that offered observational information and feedback on the design and implementation. In study visits, the two LWGs started to interact in order to lead to meaningful collaboration with the explicit intent to implement the TLS in the host educational setting. The two working groups were mindful from the beginning of the fact that they were designing with a dual purpose: (a) to develop materials as a tool for transferring knowledge to classroom practice and (b) to make those materials accessible for use by the other group.

After the evaluation and refinement of each TLS in their home context, the revised version was communicated to each host LWG along with the corresponding background information. During the adoption phase, the host LWG made an effort to abstract the *core elements* (Kelly et al. 2000) and to decode, through mutual interaction with the home LWG, the underlying pedagogical orientations embedded in the background document. Close communication was kept between the home and host groups for clarifications, advice and maintenance of the core features of each TLS. The core aspects of each TLS, which were abstracted by each host LWG and agreed with each home LWG, are described in Table 33.1. The intent at this stage was to develop a new version of the TLS adapted to the conditions of the host educational setting and classroom context.

The adapted versions of the TLS were implemented in the host settings with a view to collect evaluation data for local validation purposes. Thus, the TLS on Acoustic Properties of Materials was adapted to be implemented in the context of a vocational

Table 33.1 Core aspects of the TLS on acoustic properties of materials and the TLS on electromagnetic properties of materials

Type of knowledge	Core aspects	TLS on acoustic properties of materials	TLS on electromagnetic properties of materials
Scientific knowledge	Conceptual content	Sound attenuation as a process of energy dissipation by reflection and absorption	Magnetic interactions between magnets and other objects
		Acoustic properties of materials: porosity, rigidity, density	Magnetic field as a model that accounts for interaction at a distance
		Internal structure of materials that affects their acoustic properties	Electromagnetic interactions between a current-carrying wire and other objects
	Procedural content	Experimental design	Experimental design
		Measurements of sound intensity level using sound level meters	Using experimental data to interpret evidence and formulating conclusions
		Using experimental data to interpret evidence and formulate conclusions	Control of variables
			Technological design
Pedagogical content knowledge	Inquiry	Design of thought and real experiments to move forward from students' preliminary ideas	Multiple investigations of observable phenomena
		One central investigation: What properties of materials affect their acoustic behaviour?	Sequence of activities for active student collaboration in groups
		Students' active discussion	
		Teaching by questioning rather than by telling	
	Modelling	Emphasis on development and application of conceptual models to learn science and about science	Emphasis on design and development of a (physical) train model, applying previous knowledge
	Contextualisation	Problem of noise pollution caused by a disco	Magnetic train as a form of modern transport
	Other learning aspects	Proposing continuous students' oral and written expression	Distinction and interconnections between science and technology
		Guidance to increasingly use appropriate scientific terminology	Reflective writing and discussion
		Metacognition emphasis to guarantee students' awareness of the purposes of each task and self-regulation	Formative assessment by means of specially designed tests
		Emphasis on synthesis	

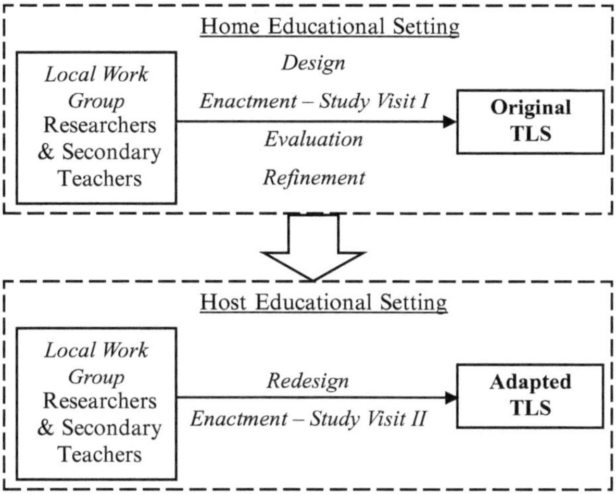

Fig. 33.1 Stages of the transfer process of each TLS from one educational context to another

school with 15-year-old students, whereas the TLS on Electromagnetic Properties of Materials was implemented with secondary school students (15–17 years old). During this implementation, members of the home LWG participated in a second study visit and offered on-site feedback on the adapted design to the host LWG in order to help abstract ideas, adapt, redesign and implement the TLS. The host LWG utilised this feedback and the data from the host enactment, in order to produce a revised version of the TLS. Figure 33.1 summarises all the processes carried out by each LWG during the transfer process of both TLS.

4 Methods

When TLSs are transferred from one context to another, there are adaptations that take place both by intent and inadvertently. In order to capture these adaptations, we analysed the following data:

- The original and the adapted version of each TLS (i.e. home and host versions of the enacted teaching sequences), which describe students' learning and assessment activities in both versions of the sequence and include teachers' notes for the implementation in both settings. Each TLS involves a brief exposition of the relevant background knowledge, a review of the main findings of the students' conceptions about the topic, descriptions of the underlying pedagogical framework and the educational context, a set of stated learning objectives, any prerequisite knowledge, a sequence of activities supported with any required teaching and learning materials or other teaching artefacts and a set of formative assessment tasks that can be used to monitor student learning.
- A background document elaborated by the host LWG, analyzing the structure and core aspects of each original TLS and specifying and justifying the changes introduced in the module when adapting it to the new educational context.

- A report elaborated by the attendees to the second study visit, describing the extent to which the core aspects of the sequence had been modified during the adaptation process and discussing the perceived factors that had facilitated or hindered the enactment of each TLS in the host context.

Analysing the aforementioned sources of data, we obtained information on the modifications performed by each host group along with their justifications for these changes. Moreover, we characterised the types of changes that were introduced and the type of knowledge that was adapted. Finally, we inferred the requirements that made the transfer process of each TLS feasible. We will consider that the process of transfer of a TLS is feasible when it is considered usable in a context different from the one in which it was designed. In this respect, we are aligned with Blumenfeld et al.'s perspective (2000) that an innovation is usable if (1) the innovation is adaptable to the organisation's context, (2) the organisation is able to enact the innovation successfully and (3) the organisation is able to sustain the innovation.

5 Results and Discussion

5.1 On the Adaptations Performed in Each TLS During the Transfer Process

During the transfer process of the TLS on Acoustic Properties of Materials from the Spanish to the Cypriot context and the TLS on Electromagnetic Properties of Materials from the Cypriot to the Spanish context, and prior to any enactment, some changes were considered necessary by each host group, apart from translating the materials into Greek and Catalan, respectively. Tables 33.2 and 33.3 present the main changes that were reported along with the justification for each change given by each host LWG.

As shown in Tables 33.2 and 33.3, both groups made changes, especially referring to the pedagogical content knowledge associated with each TLS (80 % of the modifications were related to it) albeit without radically altering their overall structure, rationale and sequencing of the modules.

5.2 Reasons Behind the Adaptations Performed in Each TLS During the Transfer Process

From the background documents prepared by each host group and the peer-review study visit interactions, observations and reports, we were able to study the reasons why each host group introduced changes to each TLS. We have identified *key driving forces for adaptations*, such as:

Table 33.2 Changes introduced in the TLS on acoustic properties of materials by the host LWG in Cyprus when adapting it to their educational setting

Modifications		Description of particular changes and justification for each change
Modifications to the scientific knowledge	Regarding the conceptual content	1. The chapter on diffraction of sound was removed since it was considered too demanding for the students of the host context. Moreover, the host LWG considered that the chapter was not an important contribution on acoustic properties of materials since the phenomenon of diffraction is affected by the geometry of objects but not by the properties of materials
		2. Extra activities were added to meet students' prerequisite knowledge (i.e. to familiarise students with sound as a wave)
Modifications to the pedagogical content knowledge	Regarding inquiry	3. More inquiry-oriented questions were posed in order to help students to elaborate on concepts and conceptual models
		4. Existing activities were modified to allow students to investigate some phenomena from different perspectives
		5. Some simulations were introduced to facilitate better student visualisation of some phenomena
	Regarding contextualisation	6. The last chapters of the material were grouped to establish a unique and perhaps more interesting context for the final project. More emphasis was placed on this project, in which students are expected to identify materials with appropriate acoustic properties for a disco
		7. A video, not foreseen in the original version, was projected to present the acoustic problems of a local disco
	Regarding other learning aspects	8. In a number of places, the text wording was simplified and shortened in order to be more accessible to students or to reduce the probability that it would be skipped
		9. Different assessment tests were designed and used, such as a framework for evaluating the posters as artefacts for presentation of conclusions and ideas

Table 33.3 Changes introduced in the TLS on electromagnetic properties of materials by the host LWG in Barcelona when adapting it to their educational setting

Modifications		Justification for each change
Modifications to the scientific knowledge	Regarding the conceptual content	1. There were some changes to the selection of contents. For instance, the concept of 'magnetic domains' was considered to be a useful idea to support students' construction of a model of magnetic materials. For this reason, this concept and a visual representation of it were explicitly introduced
		2. Some terms and concepts were also adapted or clarified to avoid conveying misleading ideas (e.g. the expression 'strength of a magnet' was avoided since it could convey the idea that the strength is a physical property of an object; the concept of magnetic field was explicitly distinguished from the idea of magnetic field lines)
	Regarding the procedural content	3. The technological design of a train model was omitted due to time constraints and teachers' perception that the train project should be tackled in the technology class rather than in the science class
Modifications to the pedagogical content knowledge	Regarding inquiry	4. Some activities, which were suggested as thought experiments in the original version of the TLS, became hands-on experiments (where feasible) to provide students with more experiences regarding a certain phenomenon (e.g. breaking a magnet into little pieces)
	Regarding modelling	5. Addition of some questions to make students' preconceptions explicit and to guide students to interpret phenomena. For example, some questions were added as a support to observe and interpret a simulation or to interpret what magnetic field lines patterns represent
		6. The progression of some concepts or ideas was changed to further support the process of building a coherent conceptual framework. For example, we decided to introduce the model of magnetic field earlier in order to apply it to the interpretation of phenomena of magnetic interactions

(continued)

Table 33.3 (continued)

Modifications		Justification for each change
		7. The activities aimed at the same goal were grouped together to better support students' coherence in the construction of certain concepts. The goals were made more explicit for the students
		8. Some visual resources were included (images, computer animations, photos, etc.) as potential facilitators of conceptual model construction
		9. Some activities were added to synthesise, apply and reflect on what had been studied in order to help students to organise new ideas
	Regarding contextualisation	10. Some statements were rephrased or further contextualised, expressing the sense of each task and the connection with the previous ones, in order to avoid students' disorientation and to help them better understand what they are doing
		11. A new introduction of the TLS was elaborated to help students recognise that magnetic and electromagnetic materials can be found commonly in their surroundings and so, to engage them in learning about 'Electromagnetic Properties of Materials'
	Regarding other learning aspects	12. Some activities were added to use and read different types of representation languages (verbal or graphic)
		13. Some activities were removed or shortened since they were considered too long or redundant
		14. The interim or follow-up tests were omitted and their questions were taken into account to elaborate the posttest

- Differences in the two *educational systems and contexts* (i.e. different student habits, school expectations and teaching/learning culture). Some changes emerged from different system perspectives on the rationale for promoting learning and also on the level of autonomy a teacher might have for innovation that might go beyond conventional practice. Other changes introduced in the TLS by the host group were justified in terms of alignment of the goals of certain activities to the level of the specific group of students who participated in the implementation of the adapted sequence in the host setting (e.g. modifications 1 and 2 in Table 33.2).

- Differences in *teachers and researchers' personal views, values and priorities*. For instance, the host members' views on how to promote students' conceptual understanding or how to assess students affected some decisions taken about the degree of guidance provided to students or about the resources selected to help students to understand some concepts or phenomena (e.g. modifications 3 and 4 in Table 33.2; modifications 5–9 in Table 33.3). We were able to identify important variation between the two groups about issues like when to introduce a conceptual model, how much emphasis to place on valid experimental designs or how explicit to make the learning objectives for the students.
- *Curricular, time and resource constraints*. For instance, time and curriculum constraints were expressed to justify the need for omitting the technological design project of the TLS on Electromagnetic Properties of Materials in the host context (i.e. modification 3 in Table 33.3). Similar arguments emerged for omitting the unit on diffraction in the module on Acoustic Properties of Materials.

From these findings we notice that, apart from common features in the two sets of modifications, there are also some significant differences. For example, whereas one group made several adaptations regarding the inquiry approach of the original TLS, the other group made several adaptations concerning the modelling approach of the original TLS. In other words, there was a different intent in the adaptations carried out by the two LWGs. This difference in intent reveals a variation in design priorities. We can interpret these differences in the adaptations carried out by the two LWG in terms of the dimensions presented in Table 33.4.

In sum, the key driving forces for adapting the TLS that are identified in Table 33.4 lead to different modifications or adaptations performed by each group. The adaptations and the driving agents mainly reflect the groups' values and principles regarding the design (and refinement) of educational materials and also reflect their individual epistemologies regarding how to teach science and how students learn science. Table 33.4 can also be interpreted as reflecting two different perspectives on the priorities and processes for inquiry-oriented approach or for modelling-based inquiry teaching and learning approach.

6 Conclusions and Implications

The analysis of the transfer process of two TLS from one educational context to another has provided insights into the types of adaptations that are carried out and the reasons that guide the rationale for such adaptations. We believe that the process we followed to adapt teaching-learning materials for use in different contexts is instructive for others who may consider similar endeavours.

Table 33.4 Driving agents for adaptations carried out by the two LWGs: dimensions of variation in the adaptations of each TLS

LWG in Cyprus	LWG in Spain
More emphasis on experimentation and inquiry learning	More emphasis on construction and application of conceptual models as interpretive frameworks
More student autonomy and extensive group work	More student support when working in groups
Fewer time constraints	More time constraints (associated with syllabus commitments)
Less emphasis on explicit metacognition	More emphasis on explicit metacognition
Less emphasis on synthesis and organisation of ideas	More emphasis on synthesis and organisation of ideas
Minimal expository text	Some expository text at the end of each inquiry and modelling process in order scaffolding the development of a conceptual model: *What does science tell us?*
High sensitivity on the students' cognitive demand	More emphasis on the precision of the scientific content
More emphasis on technological design and development in connection with science	Less emphasis on technological design and development within the science class
Individual formative assessment through presentation of specific questions for testing student thinking and understanding	Group assessment through broad questions that facilitate student verbal expression and elaboration of own understandings

Firstly, we identified that both groups felt the need to carry out adaptations in some elements of the TLS they received before any enactment could take place. Most of the changes introduced to each TLS, during the corresponding adaptation process, came from the different perspectives that each group holds around the pedagogical content knowledge that should underlie a TLS. Fewer adaptations were introduced regarding the scientific knowledge. The core scientific ideas of each TLS have not been deeply changed. The transfer of know-how (see Sect. 2.2) was not considered in this analysis. According to Blumenfeld et al.'s (2000) framework, each of the TLS was usable in the host context since it was adaptable to it and their members were able to enact the innovation successfully.

The different adaptations performed by each group reflect their values and principles regarding the design of educational materials and also reflect their individual epistemologies regarding how to teach science and how students learn science. We interpret that the exchange of both TLS was productive because both groups share basic design principles despite some differences regarding pedagogical approaches. Moreover, the scientific background was also similar in both groups, and so only small modifications have been identified. At the same time, the process of analysing the adaptations improved understandings of each others' perspective and so it was enriching for both groups.

Most of the adaptations carried out by the two local groups were made to the teaching-learning materials rather than to teachers' or school norms, routines, beliefs or practices, which remained mainly unchanged. We would like to remark that the transfer of an isolated TLS does not imply any systemic reform of an educational system but specific adaptations of the TLS to the host context.

On the other hand, transferring educational materials designed by another group in a different context implies making an effort to make one's own values, priorities and knowledge explicit and to compare them with those from the home group. In this sense, our findings, as well as other studies (e.g. Ruthven et al. 2009), support that the principles and underlying rationale of design of educational materials should be made explicit and convincing with the intent that any reader can understand it and can negotiate their meaning in order to potentially contribute to educational reform and gradual changes in teaching practice. In that way, a mutual adaptation (McLaughlin 1990) is positive regarding both school practice and demands of the TLS and, thus, can promote successful implementations.

Modifications referring to the PCK can emerge from differences in theoretical principles as well as from differences in the underlying priorities. In our case, the differences in the PCK between the LWGs, when redesigning each TLS, come from varied interpretations of the theoretical ideas. These dissimilarities across groups cannot be avoided since the sensitivities to the same issue are, and probably should be, different for each individual.

From this study, it has also been possible to infer basic requirements that need to be in place if a set of teaching-learning materials is to be transferred from one setting to another and implemented in a manner that can be deemed useful in the local setting. One of these requirements is related to Cohen and Ball's (1999) statement that, if an innovation is both highly specified and highly developed, it might be instructionally effective, but it is less likely to be adopted because it is less flexibly adaptive to the goals and needs of local teachers' contexts. As discussed by Fishman and Krajcik (2003), creating sustainable curriculum innovations requires careful balancing between specification and development in order to preserve adaptive flexibility without diminishing usability. Taking into account that the transfer process of each TLS was feasible and they were usable in each host context, we could say that the degree of specifications and explicitness provided by the materials jointly with the effort for explicit abstractions of their core aspects and their underlying rationale were key points for such positive exchanges.

Acknowledgements We would like to acknowledge the contribution of Professors Matilde Vicentini and Hans Niedderer who participated as external experts in the peer-review study visits and were able to provide valuable feedback for improving the modules. We also acknowledge Lilian C. McDermott and the Physics Education Group at the University of Washington for continued support and collaboration in the curriculum design efforts of the Learning in Science Group. Work presented in this chapter has been supported by the European Union through the European Communities Research Directorate General in the project Materials Science: University-school partnerships for the design and implementation of research-based ICT-enhanced modules on Material Properties, Science and Society Programme, FP6, SAS6-CT-2006-042942).

References

Beernaert, Y. (2008). Compendium of good practices in Mathematics, Science and Technology Education: Peer Leaning Activities (PLA) in France, the Netherlands and Sweden, 2006, 2007 and 2008.

Berman, P., & McLaughlin, M. W. (1976). Implementation of educational innovation. *The Educational Forum, 40*, 345–370.

Blumenfeld, P., Fishman, B., Krajcik, J., Marx, R., & Soloway, E. (2000). Creating usable innovations in systemic reform: Scaling up technology-embedded project-based science in urban schools. *Educational Psychologist, 35*(3), 149–164.

Burkman, E. (1987). Factors affecting utilization. In R. M. Gagne (Ed.), *Instructional technology: Foundations*. Hillsdale: Lawrence Erlbaum.

Chevallard, Y. (1991). *La transposition didactique didactical transposition* (2nd ed.). Grenoble: La PenséeSauvage.

Cohen, D. K., & Ball, D. L. (1999). Instruction, capacity, and improvement. CPRE Research Report Series RR-043, University of Pennsylvania Consortium for Policy Research in Education, Philadelphia.

Couso, D. (2011). Collaborative settings and design focus: an agenda for authentic and pragmatic science education research. *ESERA Pre-conference 2011*, Lyon.

Ely, D. P. (1990). Conditions that facilitate the implementation of educational technology innovations. *Journal of Research on Computing in Education, 23*(2), 298–305.

Ely, D. P. (1999). Conditions that facilitate the implementation of educational technology innovations. *Educational Technology, 39*, 23–27.

European Commission (2012). DG Education and culture: Management plan – 2012. http://ec.europa.eu/atwork/synthesis/amp/doc/eac_mp.pdf. Accessed 31 July 2012.

European Council (2011). Council conclusions on the role of education and training in the implementation of the 'Europe 2020' strategy. *Official Journal of the European Union, 4*(3), C70/01 – C70/03. http://eur-lex.europa.eu/LexUriServ/LexUriServ.do?uri=OJ:C:2011:070:0001:0003:EN:PDF. Accessed 31 July 2012.

Fishman, B., & Krajcik, J. (2003). What does it mean to create sustainable science curriculum innovations? A commentary. *Science Education, 87*, 564–573.

Hall, G. E., & Loucks, S. F. (1977). A developmental model for determining whether the treatment is actually implemented. *American Educational Research Journal, 14*(3), 263–276.

Hirn, C., & Viennot, L. (1999). Transformation of didactic intentions by teachers: The case of geometrical optics in grade 8 in France. In M. Komorek, H. Behrendt, H. Dahncke, R. Duit, W. Gräber, & A. Kross (Eds.), *Proceedings of the second international conference of the European science education research association*, Kiel, Vol. 2, pp. 447–450.

Hutchinson, J., & Huberman, M. (1993). Knowledge dissemination and use in science and mathematics education: A literature review, prepared for the directorate of education and human resources, division of research, evaluation and dissemination, national science foundation, Washington, DC. *Journal of Science Education and Technology, 3*, 1.

Science in Society (2010). Mobilizing and mutual learning action plans. http://ec.europa.eu/research/science-society/index.cfm?fuseaction=public.topic&id=1226&lang=1. Accessed 23 April 2012.

Kelly, J., Sogolow, E., & Neumann, M. S. (2000). Future directions and emerging issues in technology transfer between HIV prevention researchers and community-based service providers. *AIDS Education and Prevention, 12*, 126–141. Suppl. A.

McLaughlin, M. W. (1990). The rand change agent study revisited: Macro perspectives and micro realities. *Educational Researcher, 19*(9), 11–16.

Méheut, M., & Psillos, D. (2004). Teaching-learning sequences: Aims and tools for science education research. *International Journal of Science Education, 26*(5), 515–535.

Mowbray, C., Holter, M., Teague, G., & Bybee, D. (2003). Fidelity criteria: Development, measurement, and validation. *American Journal of Evaluation, 24*(3), 315–340.

National Center for the Dissemination of Disability Research (NCDDR). (1996). *A review of the literature on dissemination and knowledge utilization*. Austin: Southwest Educational Development Laboratory. http://198.214.141.98/kt/products/reviews/du/LitReview_DU.doc.

Ogborn, J. (2002). Ownership and transformation: Teachers using curriculum innovations. *Physics Education, 37*(2), 142–146.

Pintó, R. (2005). Introducing curriculum innovations in science: Identifying teachers' transformations and the design of related teacher education. *Science Education, 89*(1), 1–12.

Rogers, E. M. (1978). Re-invention during the innovation process. Presented at the *Workshop on Assessment of Current Developments on the Diffusion of Innovations*, Northwestern University, 15–16 November 1978.

Rogers, E. M. (1983). *Diffusion of innovations*. New York: The Free Press.

Ruthven, K., Laborde, C., Leach, J., & Tiberghien, A. (2009). Design tools in didactical research: Instrumenting the epistemological and cognitive aspects of the design of teaching sequences. *Educational Researcher, 38*(5), 329–342.

Sabelli, N., & Dede, C. (2001). Integrating educational research and practice: Reconceptualizing the goals and process of research to improve educational practice. http://www.virtual.gmu.edu/SS_research/cdpapers/integrating.htm. Accessed 21 May 2012.

STTIS (1998). STTIS Report RW0 outline and justification of research methodology: Work packages WP1, WP2 and WP3. http://www.crecim.cat/projectes/websttis/metoth/index.html. Accessed 14 December 2012.

Viennot, L., Chauvet, F., Colin, P., & Rebmann, G. (2005). Designing strategies and tools for teacher training: The role of critical details, examples in optics. *Science Education, 89*, 13–27.

Chapter 34
CoReflect: Web-Based Inquiry Learning Environments on Socio-scientific Issues

Andreas Redfors, Lena Hansson, Eleni A. Kyza, Iolie Nicolaidou, Itay Asher, Iris Tabak, Nicos Papadouris, and Christakis Avraam

1 Background, Framework, and Purpose

There has been an increasing acknowledgment that an informed citizenship builds on people having relevant scientific knowledge that they can use for societal decision-making. Because of this, research in the field of science education suggests that work with "socio-scientific issues" (SSI) should be included in the teaching of science (Ratcliffe and Grace 2003; Sadler 2004). Including SSI in the teaching of science prepares young people to deal with questions they meet as citizens. However, the development of this skill is not straightforward (Lee 2007). Engagement in group discussions with teacher moderation has not proved to be adequate for promoting advanced decision-making skills (Ratcliffe 1997). One of the goals of the 3-year EC project *Digital Support for Inquiry, Collaboration, and Reflection on Socio-scientific Debates* (CoReflect, www.coreflect.org) was to develop and empirically test interactive web-based learning environments (LEs) that have the potential to motivate students' inquiry-based learning in science. In this chapter we give an overview of the different learning environments and give examples of and compare the experiences from the implementations.

A. Redfors (✉) • L. Hansson
Kristianstad University, Kristianstad, Sweden
e-mail: Andreas.Redfors@hkr.se

E.A. Kyza • I. Nicolaidou
Cyprus University of Technology, Limassol, Cyprus

I. Asher • I. Tabak
Ben Gurion University of the Negev, Beersheba, Israel

N. Papadouris • C. Avraam
University of Cyprus, Nicosia, Cyprus

C. Bruguière et al. (eds.), *Topics and Trends in Current Science Education: 9th ESERA Conference Selected Contributions*, Contributions from Science Education Research 1,
DOI 10.1007/978-94-007-7281-6_34,

As part of CoReflect, seven *Local Working Groups* (LWGs) from five European countries and Israel have each designed an innovative, interactive, inquiry-based LE on the STOCHASMOS platform (Kyza and Constantinou 2007). Participatory design and design-based principles (Barab and Squire 2004; Design-Based Research Collective 2003; Penuel et al. 2007) were used to guide the design of the innovative LEs. The design-based approach seeks to bridge the often disconnected worlds of academia and theory with the realities, complexities, and constraints of educational practice. Practicing teachers were engaged in the design and implementation of these environments in order to increase ownership and motivation and test a mechanism for developing sustainable, motivating, and innovative science teaching sequences.

The design of each LE was based on a project-wide discussion of key theoretical issues, which we call *conceptual building blocks*, concerning science education, inquiry, scaffolding, motivation, understanding, reflection, collaboration, and socio-scientific issues. The scenario for each of the LEs uses a problem-based approach, presenting students with a socio-scientific dilemma, providing them with rich data, and scaffolding them in engaging in an inquiry investigation to help them answer the dilemma posed.

Students need help in order to manage the complexity of data-driven inquiries. Without appropriate scaffolding, it is reported that it might be difficult for many students to engage in high-level reasoning when dealing with data-rich environments (Kyza et al. 2011). Reflective practices, such as planning, monitoring, and evaluating one's processes and products, are especially important in inquiry-based science, where students are asked to take an inquisitive role towards learning and assume responsibility of regulating their problem-solving activities. One strategy to make science more appealing and promising for meaningful learning is to integrate new technologies in science teaching. The web-based LEs developed on STOCHASMOS were designed to scaffold students' collaboration. Collaborative learning through peer review and constructive feedback is essential for promoting the elaboration of students' ideas, while at the same time it is a highly valued activity among scientists. Collaboration, and in particular asynchronous collaboration, requires tools that can scaffold it – tools provided by the STOCHASMOS platform.

The STOCHASMOS platform supports students' reflective inquiry with rich data. Teachers can organize data and provide necessary contextual information using the STOCHASMOS teachers' authoring environment. Students have access to a learning environment, consisting of the Inquiry Environment and the Reflective WorkSpace, providing tools for capturing, organizing, interpreting, and reflecting on data. STOCHASMOS also affords synchronous communication between paired groups in the form of a chat tool and allows asynchronous communication via shared peer comments in the Reflective WorkSpace and a forum. The project management features of STOCHASMOS allow teachers synchronous and asynchronous access to their students' work. This means that a teacher can review students' work and add comments to their workspace pages, providing feedback the students can view and use at the beginning of their next session. Furthermore, the history log gives teachers information of student activities.

2 Methods

LWGs engaged in a process of iterative design consisting of three phases; the first two phases took place in their own country and the third phase concerned the enactment of the design in a partner country. During the first phase, the LEs were designed and piloted; data from these pilots were used to revise the LEs, which were then reenacted in the same country during the second phase. Peer reviewing was used for all phases, and collaborating LWGs were formed for the adaptation processes preceding the third phase, i.e., the enactment of the LE in another country. The adaptation process is described in Kyza et al. (2014).

A mixed-methods approach was followed, which included qualitative and quantitative data collection. The quality of the implementation, evaluation, and refinement process of the LEs was supported by different measures in each Local Working Group (LWG). The exchange of experiences between the LWGs was supported by templates in Google Docs. The templates provided information about the enactments, classroom conditions, encountered obstacles, suggested refinements, etc. The templates documented changes made in the revision and refinement process following the pilot enactments. They made it possible for a LWG to keep abreast of the development in other LWGs and learn from other's experiences.

For the assessment of students' knowledge development, the LWGs used domain-specific knowledge testing, assessment of reflective inquiry strategies, and micro-analysis of contrastive case groups. Data collection included pre- and posttests to assess students' worldviews, conceptual understanding of the science content, inquiry skills, and motivation. Logging of computer activities and audio/video recording of classroom activities was used.

3 Results

Seven LEs were designed and implemented in CoReflect. Information on each of the LEs can be retrieved from the CoReflect website (www.coreflect.org). Selected results from four of these, namely, *Life in the Universe*, *Genetically Modified Plants*, *Nix the Nicotine*, and *Global Warming*, are presented here. The description of the LEs is supplemented by different kinds of analyses of empirical data from the second enactment phase. These analyses serve as examples of what can be learned from the implementations of web-based inquiry learning environments on socio-scientific issues. The rationale, methods, and results from the enactment of each LE are presented here, and the results are discussed in *Conclusions and Implications*.

3.1 Life in the Universe: Students' Argumentation

The Swedish LE *Life in the Universe* (Hansson et al. 2011) caters to 9th grade students and focuses on Astrobiology. The two socio-scientific driving questions of the LE are as follows: "Should we look for, and try to contact, extraterrestrial life?" "Should we transform Mars into a planet where humans can live in the future?" This LE is available in English, Greek, and Swedish.

Students are expected to (a) demonstrate a basic understanding of essential concepts of Astrobiology, (b) discuss the nature of science, (c) link hands-on lab work to Astrobiology research, and (d) provide evidence-based answers to the driving questions, using scientific, social, economical, and ethical perspectives.

3.1.1 Rationale

The aim of this report is to shed light on how students' arguments concerning the SSI are formed during the work with the digital LE – i.e., making notes in the STOCHASMOS Reflective WorkSpace, discussing with the whole class in the forum, chatting with another group, and finally formulating final standpoints. Two student groups are discussed in more detail, while the kinds of arguments present among all student groups are presented as a background for the two cases. We report on arguments used and the presence of science content (frontier and core science) in the argumentation.

3.1.2 Methods

The students working with the LE were in their last year of compulsory school (16 years old) and worked together in groups (2–3 students) in front of one computer. Students' work included reading texts, working with activities, hands-on lab work, collecting and formulating arguments, chat and forum discussions, and the formulation of the final standpoints. Groups were paired for specific lessons and had thereby the opportunity to share documents and chat. Seven of the 14 groups were audio taped throughout their work with the LE during 5 weeks with four lessons per week (40–70 min). In addition to this, all computer activities were logged by STOCHASMOS for all student groups, i.e., arguments, final decisions, chat, and forum discussions. One or two researchers were present in the classroom during 14 of the lessons. In addition to this the students completed written individual pre- and posttests. The analysis presented here focused on students' decision-making and arguments used as presented in their written documentation – from the workspace, chat, and forum discussions; the final standpoints of the groups; as well as in the pre- and posttests. The analysis was inspired by Toulmin (1958) and had a special focus on the data part of the argument. It was based on categories established in the analysis of the pilot enactment during the first phase (Hansson et al. 2011). The study included three classes consisting of 28, 27, and 30 students, respectively – results from one class are reported here.

Table 34.1 Number of students from pre- and posttest

		Should we look for, and try to contact, extraterrestrial life?				Should we transform Mars into a planet where humans can live in the future?		
		After						
		Yes	No	N/A		Yes	No	N/A
Before	Yes	10	5	4	Yes	7	1	3
		Svea Eskil[a]	Jan Elsa[a]			Svea Eskil[a]		
	No	2	4	1	No	5	8	2
							Jan Elsa[a]	
	N/A	1	1	0	N/A	1	0	1

[a]The positions (Yes/No) of the two groups described in detail below are indicated

3.1.3 Results

Most of the participating students did not change their view on the SSI during the teaching sequence (for pre- and posttest results, see Table 34.1). The students who did change their view did this in both directions. This indicates that the teaching sequence has succeeded in presenting plausible arguments for both positions.

In the discussions within the group, the students developed their arguments and their knowledge concerning relevant aspects of the content. The students used different kinds of arguments in relation to the SSI. The analysis show that students used data related to *risks*, *chance of success and practical issues*, *technological and developmental issues*, *costs and resource prioritizing*, *curiosity*, *ethical issues*, and *peoples' views, experiences, and decision-making in the society*.

From the analysis of the data used by the groups in their arguments, we see that most groups used science-related data – both core science (e.g., distances in the universe, conditions on Mars, the importance of water for life to exist) and frontier science (e.g., consequences of contamination, possible ways to terraform Mars, the existence of water on Mars, and habitable zones).

The detailed qualitative analysis of the two groups A (Svea & Eskil) and B (Jan & Elsa) shows how these students' arguments developed in the workspace, forum, and chat and finally how their decisions were formulated and argued for in their final standpoints. The students in both groups stated we should search for extraterrestrial life in the pretest (Table 34.1). However, while the students in group A were positive also to the Mars issue, the students in group B expressed a view that we should not try to transform Mars; see Table 34.1. In their writing in the workspace, the two groups expressed very different views on risks, costs, and whether humans have the right to transform and move to another planet (ethical issues). The focus in the chat was on argumentation using data concerning risks, costs, and ethical issues – areas on which they in the workspace expressed different views. It is concerning these aspects that the students develop and add new arguments during the chat.

In their final standpoints about Mars, both groups argued for the position they had from the beginning. However, concerning whether we should search for extraterrestrial life, group B change their view during their work, and their decision was that we should not search for extraterrestrial life. We could see signs of a pending change of viewpoint towards the end of their notes in the workspace, but it became more obvious in the whole-class forum discussion, where they said "We should not disturb or destroy nature … If life is meant to be only on Earth, we should leave it at that …." In the continued discussion with the other groups, their new viewpoint became evident.

3.2 Genetically Modified Plants: Credibility Assessment

The Genetically Modified Plants LE was developed by one of the two Cypriot LWGs. It focuses on biotechnology, genetic engineering, and genetically modified organisms (GMOs) and caters to 10–12th grade high school students (15–18 years old). The socio-scientific issue addressed is GM plants' cultivation from the perspectives of health, environment, and economy. Students are asked to provide an evidence-based answer to the following problem: "Would you allow the growing of GM plants in your country?" This LE is available in Arabic, English, Greek, and Hebrew.

Students were expected to (a) develop understanding of concepts such as biotechnology, genetic engineering, and GMOs; (b) assess the credibility of evidence by applying specific criteria; and (c) make evidence-based decisions.

3.2.1 Rationale

A framework, which we call the Credibility Assessment Framework (Nicolaidou et al. 2011), guided the design of the LE with the goal of scaffolding high school students' collaborative construction of evidence-based decision-making and their assessment of the credibility of evidence. The Credibility Assessment Framework builds on the theory of situated learning, according to which LEs should provide authentic contexts and activities, multiple perspectives, coaching, and scaffolding by the teacher at critical times. Additional characteristics of such LEs refer to an authentic assessment of learning within the tasks, support for collaborative construction of knowledge, and promoting reflection and articulation (Herrington and Oliver 2000).

The research questions of this study were the following: How do 11th grade students' (a) conceptual understanding of biotechnology, (b) evidence credibility assessment skills, and (c) motivation to engage in collaborative inquiry to solve a complex socio-scientific problem change over time, as a result of students' experience with the biotechnology web-based LE?

3.2.2 Methods

The learning environment included a hands-on session during which students extracted their own DNA to help them understand the genetic modification process. Following this activity and a discussion relating to criteria for the assessment of the credibility of evidence, collaborative web-based inquiry sessions on the STOCHASMOS biotechnology learning environment were scheduled. In these sessions, students reviewed scientific data on the impact of genetically modified (GM) plants coming from sources of low, average, and high credibility and used these data as evidence to collaboratively solve the problem of whether they would allow the cultivation of GM plants in Cyprus. They organized evidence in their workspace templates and evaluated its credibility. They, then, synthesized evidence on the topic from three different perspectives (economy, environment, and health) to make their final decision as a group.

Data collection methods included pre- and posttests to assess students' conceptual understanding of biotechnology, skills in assessing evidence credibility, and the administration of two motivation instruments: FAM (Vollmeyer and Rheinberg 2003) and FKS (Rheinberg et al. 2003). Lastly, interviews with three students at the end of the enactment to assess their motivation to participate in extended scientific inquiry investigations were conducted.

The pilot enactment of the LE took place in an 11th grade class (n = 12) consisting primarily of nonscience major students at a public high school in Cyprus. Changes were made to the LE according to the research data collected and feedback from students. During the second enactment phase, the revised LE was enacted in a second 11th grade class (n = 21). Both enactments were taught over a period of 2 months during 7 weekly held 90-min lessons by a biology teacher who was part of the LWG. Students worked collaboratively in groups of three. All lessons were videotaped for further analysis.

3.2.3 Results

Statistically significant results were found for students' understanding of biotechnology concepts and students' skills in assessing the credibility of evidence. To assess students' learning gains in year 1, their pre- and posttests were compared using the Wilcoxon signed ranks nonparametric test. Students' performance increased from a mean of 16.58 (SD = 5.02) to a mean of 24.17 (maximum score = 38, SD = 6.13). The analysis indicated a statistically significant difference ($z(12) = -2.91$, $p < 0.01$) and an effect size of 0.59. To assess students' learning gains in year 2, their pre- and posttests were compared using paired sample *t*-test. Students' performance increased from a mean of 15.72 (SD = 5.69) to a mean of 20.89 (SD = 5.51). The analysis indicated a statistically significant difference [$t(17) = -5.36$, $p < 0.01$], which showed that students' conceptual understanding of biotechnology and their skills in assessing evidence credibility have improved over time.

Statistically significant results were also found for students' motivation. The analysis of students' motivation using the FAM during the pilot enactment of the LE showed statistically significant results pre and post for the "probability of success" factor. The analysis indicated a statistically significant difference $z(12) = -2.09$, $p < 0.05$, meaning that students generally thought that they were up to the difficulty of solving the inquiry-based problem and thought that they, as well as their peers, could do well in it. Another factor that yielded statistically significant results was the "absorption" factor ($z(12) = -2.03$, $p < 0.05$), meaning that students felt just the right amount of challenge, did not notice time passing, and were totally absorbed in what they were doing. Evidence from students' interviews confirmed these findings as students indicated that they were motivated to participate in scientific inquiry and that they enjoyed it.

3.3 Nix the Nicotine: Graph Synthesis

This Israeli LE addresses the socio-scientific issue of patient healthcare decisions, specifically nicotine addiction, cessation, and how to choose among alternatives based on effectiveness, risks, and individual characteristics. It catered to grades 10–12 (15–18 years old), covering topics from neurobiology and health sciences. Using clinical data, students constructed an evidence-based recommendation for one of four pharmacological smoking cessation aids. This LE is available in Arabic, English, Greek, and Hebrew.

Students are expected to (a) demonstrate a basic understanding of the neurophysiology of addiction and of health science research methods (e.g., placebo use), (b) interpret and synthesize graphs (bar, line) in order to solve a problem, and (c) employ a rational decision-making model, to be able to back claims and counter arguments with evidence.

3.3.1 Rationale

Synthesis skills may be considered a subset of procedural scientific literacy skills (Bybee 1997), because they involve credibility judgments and allocation of different weights to multiple data pieces. We focus on synthesis in the context of graph comprehension (Bertin 1983) in a process we refer to as *graph synthesis*. Graph synthesis is important, first, due to the general importance of graph analysis and, second, due to the need for and challenges specific to synthesis. Graphs are central communication devices in both scientific and lay reports, and therefore graph interpretation and production skills are emphasized in national science standards (NRC 1996). Yet, several studies have shown that full competence in reading and producing graphs is not even achieved by college and university graduates (Bowen et al. 1999; Leinhardt et al. 1990). If making sense of a single graph is a difficult task, then having to glean information from a comparative synthesis of multiple graphs is even more challenging. This activity, which we refer to as graph synthesis, is a common feature in the work of scientists (and other professionals) when they analyze their own data or when they review data published by others. A similar challenge is faced by citizens who encounter graphs when they make decisions about science-related issues based on information that they garner from news reports, web searches, or other similar sources.

3.3.2 Methods

The environment required students to compare four pharmacological smoking cessation treatments using data (graphs) modified from recently published scientific journal articles. The specific task reported here required students to extract data from two line graphs, in order to rank the treatments by their efficiency, measured as abstinence from smoking. Each graph contained data from a different clinical study, in which *only two or three* treatments were compared, in addition to a placebo control group. Thus, the overall ranking of *four* available treatments required a process of meta-analysis, involving measures of effect size.

Studies during the first phase revealed that students find these types of graph synthesis problems difficult and that most, if not all, cannot spontaneously solve these problems without explicit instruction on normalization techniques. In this study, we were interested in identifying whether explicit instruction on these techniques in the context of a rich SSI problem would enable students to apply them and reach an evidence-based decision. Our explicit instruction included an introductory unit on placebo effects and control groups in clinical trials using a whole-class discussion of case examples and prompting questions and hints in the text about the graphs in the LE. It also included "synthesis templates" in the students' computerized workspace that included questions designed to guide the analysis and synthesis of these graphs. Students were given pre-/posttests (two counterbalanced versions) that included items on graph reading skills (some from PISA), items on conceptual knowledge about nicotine addiction and the use of control groups in clinical experiments, as well as items focusing specifically on graph synthesis.

3.3.3 Results

We report on data collected from two 10th grade enactments in Israel. One class was from an average Arabic-speaking school (N=22) in a community with an average socioeconomic status (SES) and the other from a high SES Hebrew-speaking public school (N=16). The enactment in the Arabic-speaking class included ten 90–120-min sessions during a period of 2 months (a frequency of 1–2 sessions per week). The enactment in the Hebrew-speaking class included six 90-min sessions during a period of 1 month (a frequency of 1–2 sessions per week). In contrast to the pilot study during the first phase, most of the students ranked the medications correctly.

We found an overall improvement in students' ability to interpret graphs as indicated by significant pre to post gains on the PISA graph interpretation items that were included in our assessment (41.96 (SD=30.47) to 58.93 (24.73). $t=2.37$, $p<0.05$). Moreover, when compared to national statistics, these results show that students performed on par with national averages at pretest but higher on posttest. However, our results also indicate that graph synthesis is a very challenging task, because, even after the addition of specific instructional scaffolds, only ~60 % of the students in only one of the classes were able to answer the synthesis question correctly on the posttest (there were not enough pre-/post-matched pairs to obtain statistical power). Students' free-form responses justifying their answer are more promising, because they suggest advances in students' reasoning and approach to graph synthesis (analyses of the justification results are reported elsewhere). We hope that as a first step, our work will serve to include graph synthesis as an important part of the agenda for cultivating scientific literacy.

3.4 Global Warming: Understanding the Greenhouse Effect

The *Global Warming* LE catered to 11th grade students in Cyprus. Students address two driving questions: *Is climate change a man-made or a natural phenomenon? Which technology should be used in combination with the existing oil-fired power stations to cover our future needs for electricity: natural gas, wind turbines, or solar cells?* This LE is available in English, German, and Greek.

Students are expected to (a) demonstrate an understanding of the greenhouse effect; (b) develop argumentation skills, that is, the ability to provide evidence-based arguments and counterarguments; and (c) develop a specific optimization reasoning strategy for dealing with decision-making situations.

3.4.1 Rationale

The LE aims to promote understanding about the mechanism underlying the operation of the greenhouse effect, argumentation skills, and awareness of certain aspects of NOS, including the idea that the same phenomenon could be accounted for by different (and often contradictory) theories. In this report, we focus on the first of the learning objectives, the development of students' conceptual understanding of the greenhouse effect. The research question we address is: *To what extent does students' interaction with the learning materials help them improve their understanding of the mechanism underlying the operation of the greenhouse effect?*

3.4.2 Methods

The activity sequence was implemented in the context of a summer school with a group of 28 high school students (aged 15–17). The teaching intervention lasted eight 110-min sessions. Prior to and after the teaching intervention, we administered a number of open-ended tasks to assess learning gains and derive an indication of the effectiveness of the teaching materials. Students responded individually and about half of them also participated in follow-up interviews intended to provide additional insights into their reasoning. The data that emerged for each task were processed so as to describe the qualitatively different categories of students' responses. Here we discuss the results from two of the tasks. The first asked students to describe the mechanism underlying the greenhouse effect, while the second asked

Table 34.2 Categories of responses to Task I

Category of response	Pre	Post[a]
No description of the mechanism: "The greenhouse effect is when solar radiation passes through the ozone layer and causes an increase of the temperature"	15	4
Vague reference to the "trapping" of solar radiation on earth: "Some of the solar radiation is reflected by the earth's surface, but it is trapped by the greenhouse gases, and it is reflected back to the earth causing its temperature to increase"	12	4
Valid description of the mechanism: "The sun emits largely visible radiation, some of which reaches the earth's atmosphere. The atmosphere is largely transparent to the solar radiation, which is allowed to pass through and reach the surface of the earth. The earth also emits radiation (mostly invisible) to which the atmosphere is opaque. This radiation is absorbed by the atmosphere and is reemitted causing the increase in the temperature of the earth's surface"	1	16

[a] Four students were absent during the final evaluation

Table 34.3 Categories of responses to Task II

Category of response	Pre	Post[a]
Formulation of a prediction without relevant justification: "The temperature would rise because the temperature is affected by gas emissions. Thus, if there was more CO_2 and water vapor in the atmosphere, the temperature would be higher"	10	1
Formulation of a prediction and provision of a vague justification: "The temperature should rise. The CO_2 and water would not allow heat to escape"	18	7
Formulation of a correct prediction and provision of valid justification: "If it [the atmosphere] was thicker, the sun's rays would pass through but more of the invisible radiation emitted by the earth would be trapped. Therefore more heat would be trapped in the atmosphere, the greenhouse effect would become stronger, and the planet's temperature would rise"	–	16

[a] Four students were absent at the final evaluation

them to apply this mechanism to a hypothetical scenario so as to derive a (qualitative) prediction about the average temperature of the earth. This scenario involved an increase of carbon dioxide and water molecules in the atmosphere.

3.4.3 Results

Tables 34.2 and 34.3 summarize the results of the categorization of students' responses to Tasks I and II, respectively. The categories are ranked in terms of their appropriateness, and each is illustrated through a student response.

The distribution of students' responses across the three categories prior to and after the teaching intervention was compared using the Wilcoxon test, which revealed a statistically significant difference ($z(23)=-3.74$, $p<0.001$, $r=0.55$). Given that this difference stems from the increase in the frequency of the third category and the corresponding decrease in the first two categories, it could be interpreted as an indication of improvement in students' understanding of the mechanism underlying the greenhouse effect. This improvement was also evident in the interview data, which revealed that most students were able to provide detailed and articulated descriptions.

The available data for Task II (Table 34.3) suggest that the students in the first two categories did not differentiate between the solar radiation and the radiation emitted by the earth. The last category includes students that provided both a correct prediction and a valid justification that clearly distinguished between the two types of radiation and explicitly referred to the difference in terms of how each of them interacts with the atmosphere of the earth and how this impacts on the temperature of its surface.

As shown in Table 34.3, the third category was the most prevalent in the posttest while it was totally absent in the pretest data. This shift, whose statistical significance is confirmed by the quantitative analysis using the Wilcoxon test ($z=-3.84$, $p<0.001$, $r=0.54$), could be conceived as an indication of the improvement in students' understanding of the mechanism underlying the greenhouse effect.

4 Conclusions and Implications

In general students were motivated to participate in collaborative data-driven inquiry for solving problems related to society and everyday life. In this chapter we have shown four examples of analyses of students' work with web-based inquiry learning environments on socio-scientific issues. The SSI covered here range from being issues on an individual level (*Nix the Nicotine*), on societal level (*Genetically Modified Plants*), and on global level (*Global Warming*, *Life in the Universe*), even though decisions on one level also have consequences for other levels. The example analyses described in this chapter have different emphasis and show the extent to which students through their work with the LEs could:

- Develop conceptual understanding (Genetically Modified Plants and Global Warming)
- Use science as data in their argumentation (Life in the Universe)
- Develop their argumentation during their work (Life in the Universe)
- Develop skills in assessing the credibility of evidence (Genetically Modified Plants)
- Develop skills in using graph synthesis (Nix the Nicotine)

Learning gains with regard to students' conceptual understanding of the science content were reported, especially pointed out by the two Cypriot examples discussed in this chapter. Students' ability to apply the mechanism of the greenhouse effect in an alternative scenario is especially noted. The computer-based scaffolds of the inquiry process improved student learning, but there are still reasons to strive for broader and deeper learning gains. A continued exploration of how to foster the conceptual understanding and how to integrate computer- and teacher-based scaffolds should continue (Tabak 2004).

The special focus on argumentation in the Swedish study revealed that students used different kinds of data to support their arguments. Most students did not change their view on the SSI during the teaching sequence, but some did, and in both directions. Hence, the LE seemed to support both positions. It was also seen that most groups in their final standpoints, and during the chats, used science-related data. This differs from much previous research on SSI (Aikenhead 2006; Kolstø 2006; Sadler 2004). The difference between this and previous studies could depend on that Astrobiology are of interest to the students, that science content is readily available to students through the LE, and that core science (not only frontier science) is considered relevant by the students for the SSI this teaching sequence on Astrobiology focused on. That students in this study use science in their argumentation, also core science content, is promising and strengthens earlier results saying that also the learning of science content could be enhanced during work with socio-scientific issues (Sadler 2004).

Students' skills in assessing evidence credibility were discussed for the Cypriot LWG that developed a *Credibility Assessment Framework* for the acquisition of evidence credibility assessment skills. The analysis indicated a statistically significant improvement of skills in assessing evidence credibility over time.

The initial results from the Israeli LE showed that they succeeded in problematizing the issue of graph synthesis for the students. The students recognized that one cannot simply aggregate disparate graphs without some form of adjustment. However, there is still room for increasing students' understanding of the problems associated with simple aggregation and with the specific procedures required for synthesizing graphs.

The different foci of the analyses reported here mirror different interests of the researchers and teachers involved, both in the design part and in the analysis. But they probably also mirror (even though the LEs from the beginning are built from an agreed framework) different views on the reasons for using SSI in the teaching of science. Sadler (2004) stated that SSI could be used as a context for learning scientific content or the nature of science but that they could also be used to improve students' argumentation and decision-making skills. We can see that the example analyses here give support for such different outcomes from work on SSI in science classrooms.

Acknowledgments This work was funded by the "Science in Society" Initiative of the Seventh Framework Research Programme (FP7) of the European Community, under the CoReflect grant (217792) coordinated by the Cyprus University of Technology. Opinions, findings, and conclusions are those of the authors and do not necessarily reflect the views of the funding agency. We would like to thank the remaining lead CoReflect partners (Costas P. Constantinou, Marianne Cutler, Vassilis Kollias, and Hans van der Meij) and their collaborators, who all worked with the CoReflect LEs. We also appreciate the contribution of all members of the Local Working Groups in each country. The names of these LWG members are too many to enumerate here. For more information about the CoReflect project and each LWG, please visit http://www.coreflect.org.

References

Aikenhead, G. S. (2006). *Science education for everyday life. Evidence-based practice*. New York: Teachers College Press.

Barab, S., & Squire, K. (2004). Design-based research: Putting a stake in the ground. *The Journal of the Learning Sciences, 13*(1), 1–14.

Bertin, J. (1983). *Semiology of graphics: Diagrams, networks, maps*. Madison: The University of Wisconsin Press. In Berg, W. (trans.).

Bowen, G. M., Roth, W. M., & McGinn, M. K. (1999). Interpretations of graphs by university biology students and practicing scientists: Toward a social practice view of scientific representation practices. *Journal of Research in Science Teaching, 36*(9), 1020–1043.

Bybee, R. W. (1997). Towards an understanding of scientific literacy. In W. Grabe & C. Bolte (Eds.), *Scientific literacy – an international symposium*. Kiel: IPN.

Design-Based Research Collective. (2003). Design-based research: An emerging paradigm for educational inquiry. *Educational Researcher, 32*(1), 5–8.

Hansson, L., Redfors, A., & Rosberg, M. (2011). Students' socio-scientific reasoning in an astrobiology context during work with a digital learning environment. *Journal of Science Education and Technology, 20*(4), 388–402.

Herrington, J., & Oliver, R. (2000). An instructional design framework for authentic learning environments. *Educational Technology Research and Development, 48*(3), 23–48.

Kolstø, S. D. (2006). Patterns in students' argumentation confronted with a risk-focused socio-scientific issue. *International Journal Science Education, 28*(14), 1689–1716.

Kyza, E. A., & Constantinou, C. P. (2007). *STOCHASMOS: A web-based platform for reflective, inquiry-based teaching and learning*. Cyprus: Learning in Science Group.

Kyza, E. A., Constantinou, C. P., & Spanoudis, G. (2011). Sixth graders' co-construction of explanations of a disturbance in an ecosystem: Exploring relationships between grouping, reflective scaffolding, and evidence-based explanations. *International Journal of Science Education, 33*(18), 2489–2525.

Kyza, E. A., Herodotou, C., Nicolaidou, I., Redfors, A., Hansson, L., Schanze, S., Saballus, U., Papadouris, N., & Michael, G. (2014). Adapting web-based inquiry learning environments from one country to another: The CoReflect experience. In C. Bruguière, et al. (Eds.), *Topics and trends in current science education: 9th ESERA conference selected contributions* (Contributions from science education research, Vol. 1, pp. 567–582).

Lee, Y. (2007). Developing decision-making skills for socio-scientific issues. *Teaching for science literacy, 41*(4), 170–177.

Leinhardt, G., Zaslavsky, O., & Stein, M. K. (1990). Functions, graphs, and graphing: Tasks, learning, and teaching. *Review of Educational Research, 60*(1), 1–64.

Nicolaidou, I., Kyza, E. A., Terzian, F., Hadjichambis, A., & Kafouris, D. (2011). A framework for scaffolding students' assessment of the credibility of evidence. *Journal of Research in Science Teaching, 48*(7), 711–744.

NRC. (1996). *National Science Education Standards*. Washington, DC: National Academy Press.

Penuel, W. R., Roschelle, J. M., & Shechtman, N. (2007). Designing formative assessment software with teachers: An analysis of the co-design process. *Research and Practice in Technology Enhanced Learning, 2*(1), 51–74.

Ratcliffe, M. (1997). Pupil decision-making about socio-scientific issues within the science curriculum. *International Journal of Science Education, 19*(2), 167–182.

Ratcliffe, M., & Grace, M. (2003). *Science education for citizenship. Teaching socio-scientific issues.* Maidenhead: Open University Press.

Rheinberg, F., Vollmeyer, R., & Engeser, S. (2003). Die erfassung des flow-erlebens [the assessment of flow]. In J. Stiensmeier-Pelster & F. Rheinberg (Eds.), *Diagnostik von motivation and selbstkonzept [diagnosis of motivation and self-concept]* (pp. 261–279). Gottingen: Hogrefe.

Sadler, T. D. (2004). Informal reasoning regarding socioscientific issues: A critical review of research. *Journal of Research in Science Teaching, 41*(5), 513–536.

Tabak, I. (2004). Synergy: A complement to emerging patterns of distributed scaffolding. *The Journal of the Learning Sciences, 13*(3), 305–335.

Toulmin, S. (1958). *The uses of argument*. Cambridge: Cambridge University Press.

Vollmeyer, R., & Rheinberg, F. (2003). Aktuelle motivation und motivation imVerlauf [current motivation and on-line motivation]. In J. Stiensmeier-Pester & F. Rheinberg (Eds.), *Diagnostik von motivation und selbstkonzept [diagnosis of motivation and self-concept]* (pp. 281–295). Gottingen: Hogrefe.

Chapter 35
Adapting Web-Based Inquiry Learning Environments from One Country to Another: The CoReflect Experience

Eleni A. Kyza, Christothea Herodotou, Iolie Nicolaidou, Andreas Redfors, Lena Hansson, Sascha Schanze, Ulf Saballus, Nicos Papadouris, and Georgia Michael

1 Introduction

There is widespread interest in the development of innovative inquiry learning materials to respond to societal demands and improve the learning of science. One of the primary goals of science education in Europe and abroad is to support students in developing a deeper understanding of science as it relates to their own lives and help them grow into responsible citizens (National Research Council 2012; Rocard et al. 2007). Socio-scientific issues (SSI) have been discussed in the literature as having the capacity to make learning relevant to current societal needs, motivate student interest, and facilitate an integrated and meaningful approach to learning science (Hodson 2003; Sadler 2009). Nonetheless, an emphasis on the SSI approach strongly suggests that one should attend to the situated nature of learning, with researchers (e.g., Sadler 2009) indicating the importance of contextual factors to the success of learning.

Indeed, the goodness of fit of the materials to a particular learning context and the ways in which these materials are interpreted and used by teachers and students can be highly variable and can account for the quality of learning. Issues like the scalability of learning materials have been proven difficult when examining such

E.A. Kyza (✉) • C. Herodotou (✉) • I. Nicolaidou
Cyprus University of Technology, Limassol, Cyprus
e-mail: Eleni.Kyza@cut.ac.cy; c.herodotou@cut.ac.cy

A. Redfors • L. Hansson
Kristianstad University, Kristianstad, Sweden

S. Schanze • U. Saballus
Leibniz Universitat Hannover, Hannover, Germany

N. Papadouris • G. Michael
Learning in Science Group, University of Cyprus, Nicosia, Cyprus

C. Bruguière et al. (eds.), *Topics and Trends in Current Science Education: 9th ESERA Conference Selected Contributions*, Contributions from Science Education Research 1,
DOI 10.1007/978-94-007-7281-6_35,

efforts in countries like the USA, while in Europe, cultural, linguistic, and systemic differences perplex an already complicated adoption process. Thus, understanding the factors that play a role in the adaptation and enactment of reform-oriented web-based learning environments and their impact on student outcomes is important.

This chapter discusses the findings of a coordinated effort on designing and adapting inquiry web-based learning environments, each focusing on a current socio-scientific issue. This effort was conducted as part of the European, FP7-funded project "Digital Support for Inquiry, Collaboration, and Reflection on Socio-scientific Debates" (CoReflect, http://www.coreflect.org), in which university-teacher collaborating teams in six countries collaborated in designing, enacting, and exchanging web-based learning environments. Each partner team designed a web-based inquiry learning environment, which was validated locally and then adapted and enacted a second time by a collaborating team in a different national context. Through this approach, we aimed to investigate the adaptation of technology-enhanced learning materials from one country to another while gaining deeper insights into how these practices can be improved.

2 Adaptation of Inquiry Learning Materials

Not only does adaptation naturally occurs at all instances of reusing educational materials, we, along many other researchers (e.g., Luehmann 2001; Reiser et al. 2000; Squire et al. 2003), argue that adaptation is highly desirable if we are to attend to local needs and achieve an appropriate recontextualization of the original design intent. Hence, the efforts described in this chapter aim to *shed light on and describe the process of adapting* innovative, technology-supported curricula from one country to another.

Currently, few studies exist about the process of teacher appropriation and the adaptation of curriculum materials in science education in Europe. In one of these few examples, Pintó and her colleagues (Pintó 2005) examined teachers' adaptation processes in four European countries in the context of the project "Science Teacher Training in an Information Society" (STTIS), focusing on differences between the designed and enacted curriculum. Interested in the teachers' adaptation processes, the STTIS partners examined three types of curricular innovations: novel symbolic representations, computer tools, and novel teaching sequences to teach science (Pintó 2005, Pinto and Constantinou, this book). One main outcome of STTIS was the development of training materials, in the form of workshops, to guide teachers' thinking about adaptation. The CoReflect approach differs from and extends the work conducted by STTIS, since it focuses on web-based inquiry learning using socio-scientific, problem-driven scenarios; CoReflect studies the adaptation of these learning environments both as compared to the designers' intent and as they were enacted across countries. In particular, this explicit focus on exchanging curricular innovations and contrasting adaptation practices at European level is a unique characteristic of the CoReflect project.

The dissemination of learning environments should not be seen as a unidirectional process, as several factors at different levels need to be considered if the epistemic fidelity of a learning environment which was created elsewhere is to be preserved. The literature is replete with examples of such efforts leading to lethal mutations (Brown and Campione 1996), a concern that has contributed to the creation of many prescription-style curricular materials. We believe that more participatory approaches need to be adopted in order to empower teachers, maximize the potential of successful adaptations, and preserve the validity of the original learning environments. The goal of this chapter is to bring this important issue to the forefront and contribute to productive deliberations on this topic.

3 Methods

To investigate adaptation we adopted an approach guided by principles of participatory design (Carlgren 1999; Konings et al. 2007) and the design of technology-enhanced inquiry-based projects (Singer et al. 2000). These efforts were methodologically guided by design-based research (Design-Based Research Collective 2003), according to which we engaged in cycles of iterative design, enactment, and evaluation.

As part of the CoReflect project, seven Local Working Groups (LWGs) from five European countries and Israel each designed an innovative, inquiry-based learning environment (LE) using the learning and teaching platform STOCHASMOS (Kyza and Constantinou 2007; Kyza et al. 2011). STOCHASMOS (http://www.stochasmos.org) is a scaffolded, web-based platform, which includes a learning (inquiry) environment for the students as well as an authoring environment for developing new, modifying, or reusing learning environments to adapt them to new contexts.

The scenario for each of the LEs uses a problem-based approach, presenting students with a socio-scientific dilemma, such as whether one should allow the cultivation of genetically modified plants or who is responsible for global climate change. Each environment was designed to provide primary or secondary school students with rich scientific data and scaffold them in engaging in an inquiry investigation to help them answer the dilemma posed. Each LWG developed the first version of their LE individually and pilot-tested it in a classroom with students of the target age. Research data were collected during the enactments in order to assess the effectiveness of the learning environment and of the activity sequence. More details about this effort are provided in the chapter by Redfors et al. (2014).

Each LWG was partnered up with another LWG, which is called collaborating Local Working Group (cLWG). LWGs went through a process of iterative design, during which they designed, piloted, revised, and reenacted their LE. After the first revision and enactment, the cLWG peer-reviewed the LWG's LE and negotiated modifications to the environment. At the end, each LE was enacted at least twice beyond the pilot enactment in two different countries: once by the LWG which designed it and once by the cLWG who participated in the revision of the LE. Table 35.1 presents information on the original learning environments and the adapted ones.

Table 35.1 Adapted web-based learning environments (LEs). Only the ones indicated with an * are discussed in this chapter

Title of original LE	Countries involved	Original emphases	Designed by	Target age group	Adapted by
* Nicotine Addiction	Israel, Cyprus	Decision-making and graph interpretation skills	Ben-Gurion University, Israel (BGU LWG)	15–17	Cyprus University of Technology LWG (Cyprus)
* Water Pollution	Germany, Sweden	Conceptual understanding, argumentation	Leibniz Universität Hannover, Germany (LUH LWG)	15–16	Kristianstad University LWG (Sweden)
* Global Warming	Cyprus, Germany	Conceptual understanding, nature of science, reasoning skills	University of Cyprus, Cyprus (UCY LWG)	16	Leibniz Universitat, Hannover LWG (Germany)
* Life in the Universe	Sweden, Cyprus	Conceptual understanding, nature of science, ethics, evidence-based reasoning	Kristianstad University, Sweden (HKr LWG)	15–19	University of Cyprus LWG (Cyprus)
Genetically Modified Plants	Cyprus, Israel	Conceptual understanding, credibility assessment skills	Cyprus University of Technology, Cyprus (CUT LWG)	15–17	Ben-Gurion University LWG (Israel)
Impact of fog on human life	Greece, the Netherlands	Conceptual understanding, self-regulated learning	University of Thessaly, Greece (UTh LWG)	10–12	Twente University (the Netherlands)
My house on the moon	The Netherlands, Greece	Conceptual understanding, development of problem-solving skills	Twente University, the Netherlands (TU LWG)	10–12	University of Thessaly (Greece)

Each LWG documented the adaptation process through a project-wide, agreed-upon adaptation template. Over a period of several months, the cLWGs pairs engaged in a series of synchronous and asynchronous communication which allowed access to the original designers and supported the negotiation of the adaptations. Furthermore, each LWG agreed to document successes, problems, and challenges during the adaptation, planning, and enactment phases. Through the adaptation template, the cLWGs documented, explained, and negotiated changes to the LE so that the revised LE could be perceived as functional and useable in both countries (e.g., Germany and Sweden) or so that the LE were transformed to better fit the new country and school culture (e.g., Greece and the Netherlands).

4 Findings

CoReflect supported seven adaptation efforts in six different countries. We next discuss findings from four of these adaptation efforts in Cyprus, Germany, and Sweden, in the form of four brief case studies. Within each brief case study, we discuss aspects of the adaptation process relating to (a) science content and skills to be taught, (b) pedagogy, and (c) fit with local curricular framework.

4.1 Nicotine Addiction LE Adaptation Process

The Nicotine Addiction LE (Asher et al. 2010) was first designed and enacted in Israel (BGU) and then adapted and enacted in Cyprus (CUT). All changes to the original design were discussed and negotiated among the CUT and BGU LWGs. The two LWGs agreed that they were interested in understanding the adaptation process as it resulted during enactment and due to emerging needs relating to students and the teacher. As such, the negotiation phase focused on identifying, discussing, and modifying only the absolutely necessary elements of the learning environment, leaving the main issue the environment focused on (graph comprehension) as it was originally conceptualized. The latter was important so that the two LWGs could investigate the types of adaptations as a result of the teaching practice. As the following discussion indicates, it soon became obvious that more adaptations were required to address local needs at the teacher and system level.

4.1.1 Science Content and Skills to Be Taught

The driving question of the learning environment designed by the BGU LWG asked students to compare different smoking cessation treatments so that they could help their dad quit smoking. The original design placed emphasis on graph comprehension and synthesis skills embedded in the nicotine decision-making scenario. The topic of neurobiology and drug addiction was initially received with excitement by the CUT Local Working Group and the enacting teacher. However, during the adaptation negotiations between the two LWGs, a main point of debate was that the designers of the original learning environment felt that students did not need much prerequisite knowledge to engage with the socio-scientific problem and that the understanding of the physiology of the human brain could be acquired by exposure to the online materials. On the other hand, the enacting teacher and the CUT LWG saw a need to include sessions incorporating brief experiments, demonstrations, and additional whole-class discussions about the content, in order to ensure that the students had the prerequisite knowledge required to engage in the decision-making activity. The analysis of data drawn from the comparison of the original vs. the adapted LE, teacher interviews, classroom observations, and field notes indicates that the different emphasis of each of the LWGs (Israel, primary emphasis on inquiry skills, secondary emphasis on conceptual understanding; Cyprus, primary emphasis on conceptual understanding, secondary emphasis on inquiry skills) led to the modification of the teaching activity sequence while the students' learning environment was agreed to be held constant. The emphasis on conceptual understanding is a characteristic of the Cypriot educational system, which places high value on concepts and facts, but does not yet emphasize the acquisition of other inquiry skills.

4.1.2 Pedagogy

Initially, the two LWGs agreed to make only minor changes to each LE they were adapting, so that they could focus on investigating enactment-related adaptation. In this context, the CUT LWG identified issues like personalizing the learning environment for the Cypriot students, such as changing the names of the characters to make them more relevant to the students. However, as the discussions proceeded and the teacher begun the enactment, it proved that differentiated emphases required significant adaptations to pedagogy as well, as described in the previous section (Sect. 4.1.1). As part of the emphasis on establishing prerequisite knowledge and promoting conceptual understanding, other modifications included additions (e.g., the definition of the norepinephrine chemical substance) to the STOCHASMOS glossary of the learning environment, so that students could easily retrieve explanations of terms which were identified as important for fostering the understanding of the science behind the decision-making process. Another adaptation of the original activity sequence was the inclusion of whole-class discussions. While the designers of the Nicotine LE advised the teacher to implement small group discussions, the enacting teacher's pedagogical preference was to listen to questions raised at the group level but address them through whole-class discussions. This reflects a feeling of unease with understanding student cognitive processes as students worked on computers in small groups, which is not atypical of other teachers at local or international level.

4.1.3 Fit with Local Curricular Framework

The Cyprus educational system is centralized, with curriculum topics being decided primarily by educational authorities at the level of the local Ministry of Education and Culture. Due to the extensive time required to teach the BGU LE (eight 90-min sessions) and the decision to enact the LE remaining as close to the original design as possible in terms of the learning activities, the adapted LE was taught at 11th grade instead of the 10th grade level initially given by the Israeli LWG. The decision was reached after an analysis of prerequisite skills and knowledge, a discussion with the BGU LWG, and a content analysis of the Cyprus curricular framework for 11th grade biology courses for non majors. At the end, the enactment of the adapted learning environment in Cyprus required seven 90-min sessions. The main adaptations made by the CUT team referred to (a) personalizing the learning environment to make it more relevant to local situations, (b) shifting the emphasis to conceptual understanding of unknown key terms and related background knowledge as a prerequisite to the emphasis to graph comprehension activities, and (c) including more whole-class activities, as requested by the enacting teacher. With these results in mind, a larger scale adoption of this curriculum would require additional discussions with more teachers to understand whether these adaptations were identified as necessary by other local teachers, as well.

4.2 Water Pollution LE Adaptation Process

The Water Pollution LE (Saballus et al. 2010) was first designed and enacted in Germany (LUH) and then adapted before it was enacted in Sweden (HKr). All adaptation measures were discussed and negotiated by the LWGs in an effort to keep the changes to a minimum. The negotiation phase took place before, during, and after the second enactment and used the CoReflect adaptation template. The process adhered to a principle of minimal change and only absolutely necessary alterations were discussed.

4.2.1 Science Content and Skills to Be Taught

The science content in the German LE concurs with curricula and tradition in compulsory school in Sweden, although science content is not traditionally taught using SSI and computers in Sweden or Germany. However, the SSI approach is relevant to Swedish curricula and something that many science teachers in Sweden would be willing to introduce. The Swedish teacher enacting the adopted LE was part of the LWG and participated in the negotiations. She appreciated that chemistry (seen as abstract and irrelevant by many students) was put in a socio-scientific context and felt that the LE would work well in a Swedish context even though it concerned a river in Germany. Interviews of the teacher and selected students indicated that the fact the river was situated in Germany did not make the topic uninteresting or irrelevant to Swedish students.

4.2.2 Pedagogy

The major issue for the adaptation in this case was the prevalent school cultures. The HKr LWG adapted the Water Pollution LE to respond to a different culture of schooling. The German LE reflected a tradition of telling students what to do, whereas many teachers in Sweden follow a Socratic approach, preferring to entice students to work in a predefined direction by asking questions. There were agreed adaptations about this, preceding the second enactment in Germany and the implementation in Sweden. The Swedish teacher provided additional questions at the outset of the teaching sequence. A possible cause for this differentiation is the national curriculum of the Swedish teacher education that supports a problem-based investigative approach, i.e., the use of questions in inquiry-based science teaching. However, the slightly more structured German LE was possible to be adapted into a more open-ended, student-oriented one, which worked well in Sweden. The hands-on lab work was done and reported outside of STOCHASMOS in the first German version, but the connection was made stronger during the adaptation process, and discussion about the lab work became part of STOCHASMOS. Hence, the adaptation phase helped improve the LE in both countries.

4.2.3 Fit with Local Curricular Framework

Results from PISA 2006 (OECD 2006) show that students in Germany perform better, and Swedish students perform better or in level with the OECD average on *identifying scientific issues*, *explaining phenomena scientifically*, and *knowledge about science*. The German students perform better than Swedish students, but the differences are small compared to other participating CoReflect countries. While German students on average perform better than Swedish students, the between-school variance concerning student performance is larger in Germany than in Sweden. The PISA 2009 results (OECD 2009) show the same trend but give no detailed information about science this time. Germany did not participate in the TIMSS 2007 grade 8 evaluation, but there are no significant differences between the Swedish and German 8th grade students' average science achievements. These results indicate a smooth transition from Germany to Sweden.

In Germany the science subjects are not studied integrated, but as the individual subjects Biology, Chemistry, and Physics. In Swedish schools, however, free to choose whether science is studied integrated or in these three subjects. In an integrated situation the teacher has large freedom concerning choice of content. The students who worked with the learning environment were enrolled in a science course which integrated different science subjects. Hence, there were no specific curricular constraints for the adaptation. Also, the students were used to working on their own, in groups, and individually, which was in contrast to the German situation where more traditional teacher-oriented teaching was the norm. The goal-oriented Swedish curriculum was per se no obstacle for the introduction of SSI; on the contrary, it offered large degrees of freedom for the teachers, concerning planning and choice of specific content to cover. An adaptation from a more specified curriculum in Germany to an open goal-oriented one in Sweden came as comparatively straightforward.

4.3 Global Warming LE Adaptation Process

The Global Warming LE (Constantinou et al. 2010) was first designed and enacted in Cyprus (UCY) and then adapted and enacted in Germany (LUH). Following the driving questions "Is global warming man-made or natural? Which position would you support for policy making purposes for your country?" the UCY LE asked 11th grade students to organize a conference in their school to inform schoolmates about the global warming issues. The unit was designed to be taught in about 15 90-minutes sessions.

4.3.1 Science Content and Skills to Be Taught

To achieve curricular coherency (Rosemann et al. 2008), we identified the so-called core concepts, that is, core ideas we associate with the teaching of global warming that are valid and viable throughout the class levels and throughout the different subjects. These concepts could be addressed at lower and higher school levels, varying in knowledge depth or profoundness of knowledge. By referring to core concepts, we assure with the design of our learning unit to meet the given learning objectives of the Cyprus learning unit on a downscaled level. Both LWGs, UCY and LUH, agreed on two concepts as being relevant: (a) the explanations of climate change are based on model assumptions and (b) the awareness of human behavior as one possible main origin. The two concepts were broken down from the LUH group into seven sub-concepts. In a questionnaire given to the UCY LWG by the German LWG, and which functioned as a formative assessment of the adaptation process, the UCY LWG was asked to (a) confirm the importance of the concepts, (b) specify the degree to which the concept is emphasized by the learning environment, and (c) indicate which elements of the learning environment ensure the teaching of the concepts. The UCY LWG strongly agreed with the core concepts that needed to be taught in the global warming unit; however, compared to LUH, they differed in their approach on how to realize some of the sub-concepts, such as "uncertainty about the human impact on the greenhouse effect." While the LUH LWG adapted the web-based learning environment to explicitly address the sub-concepts, the UCY group addressed some of these sub-concepts through whole-class discussions and only included limited information on these issues in the web-based environment they designed.

4.3.2 Pedagogy

There were differences concerning the realization of the core concepts that could partly be explained by the different structure of the LEs. While the UCY LE has a stepwise approach to teach the greenhouse effect, the German students got different roles to prepare for a debate about measures to prevent severe climate changes that may be caused by human behavior. As a result, the German students got all the information about the topic at once so that they could be well prepared for the debate in a self-directed inquiry phase. The role of the German teacher was to monitor and scaffold this part, while according to the original design the Cypriot teachers took active part in classroom discussions and explicitly addressed the core concepts.

4.3.3 Fit with Local Curricular Framework

After a first analysis of the curricular and systemic conditions, including the crucial expert opinions of the cLWG's teachers (Luehmann 2001), it became apparent that it was not achievable to have an adapted LE that was close to the original. In Germany, in particular in the federal state of Lower Saxony, the Global Warming topic is to be taught in the 9th or 10th grade with less complexity than the one that the UCY LWG originally designed. The local German schedule also allows a maximum of eight to ten 45-min lessons for this topic.

To assess the understanding of the core concepts, students' learning outcomes were measured by a multiple-choice questionnaire (15 items, Cronbach Alpha = .71). Data were taken from the enactment of one summer class in Cyprus taught in July 2010 (22 11th grade students) and a German class (23 10th grade students) taught in June 2010. The comparison of mean values (Mann–Whitney-*U*-test, $z = -2.235$, $p = .025$) showed a significant difference between both groups. The German students outperformed the Cypriot students. These results show that the adaptation process was successful and that we reached our ambition to design an adapted learning environment that at least imparts a comparable understanding of the core concepts in a lower school level in less amount of time.

The results are also in line with our assumption that the awareness of key concepts to be taught is one way to adapt a LE in a sustainable and scalable way, even if cultural or systemic limitations necessitate extensive changes. This experience shed a new light on the adaptation of LEs, as it addressed severe obstacles like differences in the scheduled teaching time or the LE's target group. It took the LWGs some effort to get over these obstacles. But this process was very well supported by the design-based research approach (Design-Based Research Collective 2003;

Barab and Squire 2004), as the teachers could bring in their expertise to create a learning scenario that meets the curriculum and the researcher could pay attention to keep the LE close to the original. Based on the experiences made in both countries, the cLWGs saw a need for another revision that could end in a module-based LE, allowing the teacher flexibility to put an emphasis on what might fit the current classroom situation and curricular progressions. The German LE could act like a pre-unit on a downscaled knowledge level, laying out the background for the deeper analysis of the greenhouse effect in the UCY LE at a higher level.

4.4 Life in the Universe LE Adaptation Process

The Astrobiology LE (Hansson et al. 2011) was first designed and enacted in Sweden (HKr) and then adapted and enacted in Cyprus (UCY). The adaptation process was guided by the UCY LWG decision to specifically focus on integrative teaching (NRC 2007). The adapted version provided more explicit scaffolding of argumentation skills through the reflective Work Space template features of the STOCHASMOS platform. As this detailed scaffolding led to an increase in the total time required to enact the adapted unit, the UCY LWG purposefully narrowed the scope of the unit, selected a smaller number of learning objectives, but still focused on fostering conceptual and epistemological understanding and the development of reasoning skills. To reduce complexity, the UCY LWG also modified the pedagogical approach, using the jigsaw method (Aronson 1978) and asking different groups of students to focus on different but complimentary concepts, which at a subsequent phase they discuss with each other.

4.4.1 Science Content and Skills to Be Taught

The adaptation of the original version of the LE focused on enriching students' argumentation skills and intended to explicitly and systematically promote this learning objective. The adaptation of the LE was largely driven by a specific design principle, which was shared across the CoReflect project. This principle derives from the idea that in developing learning environments in science, it is important to take a holistic approach that does not solely focus on content (which is commonplace in the conventional teaching practice) but also integrates additional important components of what is involved in being knowledgeable or proficient in science (NRC 2007). Some of these components, for instance, include reasoning skills (e.g., control of variables and argumentation skills) and understanding of fundamental ideas relevant to how science operates and how scientific knowledge is produced, justified, and organized. Underlying this is the assumption that these components, which are often neglected by conventional science teaching, do not develop spontaneously. Instead, they need to be explicitly addressed through specially designed instruction.

The argumentation activities engage students with the structure of an effective argument and the role of individual components constituting a powerful argument. An additional change that we undertook with respect to the content and the learning objectives of the LE involves the reduction of the breadth of coverage of conceptual ideas and the pursuing of one of the two driving questions of the original learning environment. Specifically, we excluded pieces of information included in the environment that did not bear a connection to the driving question or could not be used in constructing arguments about the corresponding issue (terra formation of Mars).

4.4.2 Pedagogy

The teaching approach that we employed involved the explicit teaching elaboration of argumentation skills in combination with the conceptual elaboration of specific ideas associated with creating and sustaining the function of ecosystems. We focused on a limited set of concepts (significantly narrower compared to the original LE). These ideas were purposefully selected so as to provide a frame for supporting students' attempt to construct arguments about the issue at hand (viability of the terra formation of Mars project). We developed templates that were intended to draw students' attention into these ideas (i.e., energy, temperature, atmosphere, water). Our intention was not to help students gain thorough understanding of the concepts involved. Instead, we aimed at helping them construct a working conceptual framework that could help them contextualize and inform their arguments. An additional characteristic of the instructional approach that we adopted involves the engagement of students with explicit discussion, at various, preselected points which were dispersed throughout the LE. Those were intended to get students to reflect (in the context of their own arguments) on various constituent components of arguments (based on Toulmin's model, Toulmin 1958) and their role in enhancing the quality of an argument.

4.4.3 Fit with Local Curricular Framework

One main constraint that we sought to address through the adaptation process relates to the teaching time needed for the enactment of the learning environment. The Cyprus educational system is centralized; the local authority supplies schools with the national curriculum and the corresponding teaching and learning resources, which are typically followed in a faithful manner. These teaching and learning resources usually consist of a large number of relatively brief units (extending over one to three 80-min sessions). Given the lack of flexibility on the part of the teachers, in terms of substantially deviating from the national curriculum and the corresponding teaching and learning resources, the considerable length of the original learning environment was assumed to confound our attempt to implement it in classroom environments. In addition to the already extensive length of the original version of the LE, this reduction also emerged as a consequence of our decision to supplement the LE with activities associated with the development of argumentation skills. In order to address this, we selected to keep just one of the two driving questions included in the original LE (i.e., does the terra formation of Mars constitute a worthwhile and feasible goal that could be usefully pursued?). Consequently, we removed parts of the LE that solely related to the second driving question (i.e., should we look for and try to contact extraterrestrial life?). Additionally, we sought to decompose the LE into a series of relatively distinct parts that could be enacted independently. This would allow interested teachers to either implement the entire learning environment or only focus on individual sections. This latter possibility would enable teachers to accommodate parts of learning environment in their teaching agenda.

5 Conclusions and Implications

Taken collectively, the outcomes of each of the CoReflect adaptation case studies can contribute to a grounded discussion about the transfer of best practices from one country to another, including what has worked and what did not work as expected during these adaptation processes.

Different adaptation patterns were observed in each of the national cases of adapting and enacting the cLWG learning environment. For example, in Germany the local educational context required the adaptation of the Cypriot learning environment so that it met the expectations set by the State, by making the learning environment less complex and by focusing on one or two core concepts. In Sweden, the adaptations were related to the mainstream school culture, whereas in the other Cyprus case, the adaptation was influenced by predominant teacher epistemology and normative teacher culture. Prioritization of educational objectives also guided the adaptations, as indicated on several occasions. One example is the adaptation of the Swedish environment in Cyprus, which responded to an increased emphasis placed by the cLWG on scaffolding argumentation practices. The findings reported here can be seen as providing an insight in adaptation efforts which may be taken up by small design groups and individual teachers. The reports of the Local Working Groups and of the enacting teachers, in particular, suggest that the adaptation process provided a principled approach that enabled them to identify and modify aspects of the learning environments that needed to be changed to improve teaching at local level. The adoption of adaptation as a way for curricular innovation and as a means to respond to local teacher and student needs would require a discussion at local and national level of what learning environment aspects can be negotiated and what need to be decided at a central, educational authority level.

Stemming out of the interest to understand how the perspectives of the CoReflect project participants differed, a Worldview survey (Cobern 1996) was also developed and administered in all CoReflect participating countries. The aim of this survey, which was not discussed in this chapter, was to provide insights into the adaptation process by identifying differences and similarities of views on nature-of-science-related issues between students from different countries. The results from the Worldview survey indicated the existence of significant differences between the CoReflect countries, as students' predominant answers differed between countries for all categories of statements. The results of the Worldview survey became available after the negotiation phase between the cLWGs. Whenever possible, such surveys can be employed ahead of time to guide the adaptation processes from one context to another.

The investigation of the CoReflect adaptation efforts was made possible by two main driving forces. On one hand, the participatory design approach gave each Local Working Group the ability to bridge theory and practice, by having researchers, practicing teachers, scientists, and educational technologists work together. This allowed each Local Working Group to follow a systematic approach to adaptation while also keeping faithful to learning theory and systemic, practical, and pedagogical constraints. On the other hand, the design-based research approach, with the iterative approach to design and implementation, supported the documentation of the design phases and adaptation processes and allowed the discussion of adaptations using shared language and approaches.

In this work, we set out to understand how best practices can be transferred from one cultural context to another. One of our initial questions related to whether we would be able to meet local teaching and learning demands. The case studies, briefly described in this chapter, suggest that it is indeed possible to adapt a learning environment in a way that it meets local needs, as long as the adaptation process is guided by principled decision-making which makes choices and trade-offs explicit to the adapting team.

The development of validated web-based inquiry learning environments, such as the CoReflect learning environments, is resource – and cost-intensive. For this reason, it is important to use the knowledge encapsulated in the curricular designs and the accompanying materials and acquired in such multinational efforts as a basis for other curricular innovations in Europe. Such actions will support collaboration at European level and will enable collective knowledge building in fostering science teaching and learning. This capacity was significantly supported by the provision of the STOCHASMOS authoring tool, which provided the CoReflect consortium with the technology and a common pedagogical framework to address the adaptation needs, without beginning anew. The ability to reuse and effectively share such learning environments openly across Europe can help promote sophisticated inquiry learning on socio-scientific issues addressing conceptual and epistemological understanding and fostering the development of inquiry skills. We believe that this work can contribute to a more nuanced understanding of adaptation processes and bears implications for both teachers and designers of inquiry and web-based learning environments.

Acknowledgments This work was funded by the "Science in Society" Initiative of the Seventh Framework Research Programme (FP7) of the European Community, under the CoReflect grant (217792) coordinated by the Cyprus University of Technology. Opinions, findings, and conclusions are those of the authors and do not necessarily reflect the views of the funding agency. We would like to thank the remaining lead CoReflect partners (Iris Tabak, Vassilis Kollias, Hans van der Meij, and Marianne Cutler) and all the CoReflect collaborators who worked on these adaptation efforts. We also appreciate the contribution of all members of the Local Working Groups in each country. The names of these LWG members are too many to enumerate here. For more information about the CoReflect project and each LWG, please visit http://www.coreflect.org.

References

Aronson, E. (1978). *The jigsaw classroom*. Beverly Hills: Sage.
Asher, I., Nasser, S., Ganaim, L., &Tabak, I. (2010). Putting the pieces together: The challenge and value of synthesizing disparate graphs in inquiry-based science learning. In K. Gomez, L. Lyons, & J. Radinsky (Eds.), *Learning in the disciplines: Proceedings of the 9th International Conference of the Learning Sciences (ICLS 2010).* (Vol. Volume 2, Short Papers, Symposia, and Selected Abstracts). Chicago: International Society of the Learning Sciences.
Barab, S. A., & Squire, K. (2004). Design-based research: Putting a stake in the ground. *The Journal of the Learning Sciences, 13*(1), 1–14.
Brown, A., & Campione, J. (1996). Psychological theory and the design of innovative learning environments: On procedures, principles, and systems. In L. Schauble & R. Glaser (Eds.), *Innovations in learning: New environments for education* (pp. 289–325). Mahwah: Lawrence Erlbaum.
Carlgren, I. (1999). Professionalism and teachers as designers. *Journal of Curriculum Studies, 31*(1), 43–56.
Cobern, W. W. (1996). Worldview theory and conceptual change in science education. *Science Education, 80*(5), 579–610.
Constantinou, C. P., Papadouris, N., Michael, G., Iordanou, K., Avraam, C., & Siakidou, E. (2010). *Is global warming man-made or natural*? [Web-based learning environment]. Cyprus: University of Cyprus. Retrieved from http://www.coreflect.org
Council, N. R. (2012). *A framework for K-12 science education: Practices, crosscutting concepts, and core ideas*. Washington, DC: The National Academies Press.
Design-Based Research Collective. (2003). Design-based research: An emerging paradigm for educational inquiry. *Educational Researcher, 32*(1), 5–8.
Hansson, L., Redfors, A., & Rosberg, M. (2011). Students' socio-scientific reasoning in an astrobiology context during work with a digital learning environment. *Journal of Science Education and Technology, 20*(4), 388–402.
Hodson, D. (2003). Time for action: Science education for an alternative future. *International Journal of Science Education, 25*(6), 645–670. doi:10.1080/09500690305021.
Konings, K. D., van Zundert, M. J., Brand-Gruwel, S., & van Merrienboer, J. J. G. (2007). Participatory design in secondary education: Is it a good idea? Students' and teachers' opinions on its desirability and feasibility. *Educational Studies, 33*(4), 445–465. doi:10.1080/03055690701423648.
Kyza, E. A., & Constantinou, C. P. (2007). *STOCHASMOS: A web-based platform for reflective, inquiry-based teaching and learning [software].* Cyprus: Learning in Science Group.
Kyza, E. A., Constantinou, C. P., & Spanoudis, G. (2011). Sixth graders' co-construction of explanations of a disturbance in an ecosystem: exploring relationships between grouping, reflective scaffolding, and evidence-based explanations. *International Journal of Science Education, 33*(18), 2489–2525. doi:10.1080/09500693.2010.550951.
Luehmann, A. L. (2001). *Factors affecting secondary science teachers' appraisal and adoption of technology-rich project-based learning environments*. Unpublished Doctoral Dissertation, The University of Michigan, Ann Arbor.
NRC. (2007). *Taking science to school: Learning and teaching science in grades K-8*. Washington, DC: National Academy Press.
OECD (2006). *PISA country profiles*. Available from http://pisacountry.acer.edu.au. Accessed 15 January 2013.
OECD (2009). *PISA 2009 results*. Available from http://www.oecd.org/edu/pisa/2009. Accessed 15 January 2013.
Pintó, R. (2005). Introducing curriculum innovations in science: Identifying teachers' transformations and the design of related teacher education. *Science Education, 89*(1), 1–12. doi:10.1002/sce.20039.
Pinto, R., & Constantinou, C. P. (this book). *On the transfer of teaching-learning materials from one educational setting to another.*

Redfors, A., Hansson, L., Kyza, E. A., Nicolaidou, I., Asher, I., Tabak, I., Papadouris, N., & Avraam, C. (2014). CoReflect: Web-based inquiry learning environments on socio-scientific issues. In C. Bruguière, et al. (Eds.), *Topics and trends in current science education: 9th ESERA conference selected contributions* (Contributions from science education research, Vol. 1, pp. 553–566).

Reiser, B. J., Spillane, J. P., Steinmuller, F., Sorsa, D., Carney, K., & Kyza, E. A. (2000). Investigating the mutual adaptation process in teachers' design of technology-infused curricula. In B. Fishman & S. O'Connor-Divelbiss (Eds.), *Fourth international conference of the learning sciences* (pp. 342–349). Mahwah: Erlbaum.

Rocard, M., Csermely, P., Jorde, D., Lenzen, D., Walberg-Henriksson, H., & Hemmo, V. (2007). *Science education now: A renewed Pedagogy for the future of Europe*: European Commission Directorate-General for Research, Science, Economy and Society.

Roseman, J. E., Linn, M. C., & Koppal, M. (2008). Characterizing curriculum coherence. In Y. Kali, M. C. Linn, & J. E. Roseman (Eds.), *Designing coherent science education* (pp. 13–38). New York: Teachers College Press.

Saballus, U., Sieve, B., Schanze, S., Glumm, J., Goldstein, T., Hentschel, S., Jäger, J., Janssen, O., Manske, M., Misske, R., & Söhlke, M. (2010). *Salinization of the Werra*. [Web-based learning environment]. Germany: Leibniz Universität Hannover. Retrieved from http://www.coreflect.org.

Sadler, T. D. (2009). Situated learning in science education: Socio-scientific issues as contexts for practice. *Studies in Science Education, 45*(1), 1–42. doi:10.1080/03057260802681839.

Singer, J., Marx, R. W., Krajcik, J., & Chambers, J. C. (2000). Constructing extended inquiry projects: Curriculum materials for science education reform. *Educational Psychologist, 35*(3), 165–178.

Squire, K. D., MaKinster, J. G., Barnett, M., Luehmann, A. L., & Barab, S. L. (2003). Designed curriculum and local culture: Acknowledging the primacy of classroom culture. *Science Education, 87*(4), 468–489.

Toulmin, S. E. (1958). *The uses of argument*. Cambridge: Cambridge University Press.

Author Index

C. Bruguière et al. (eds.), *Topics and Trends in Current Science Education: 9th ESERA Conference Selected Contributions*, Contributions from Science Education Research 1,
DOI 10.1007/978-94-007-7281-6, © Springer Science+Business Media Dordrecht 2014

C

D

Subject Index

C. Bruguière et al. (eds.), *Topics and Trends in Current Science Education: 9th ESERA Conference Selected Contributions*, Contributions from Science Education Research 1, DOI 10.1007/978-94-007-7281-6, © Springer Science+Business Media Dordrecht 2014

MIX
Papier aus verantwortungsvollen Quellen
Paper from responsible sources
FSC® C105338

If you have any concerns about our products,
you can contact us on
ProductSafety@springernature.com

In case Publisher is established outside the EU,
the EU authorized representative is:
Springer Nature Customer Service Center GmbH
Europaplatz 3, 69115 Heidelberg, Germany

Printed by Libri Plureos GmbH
in Hamburg, Germany